Liviu C. Florescu · Christiane Godet-Thobie
Young Measures and Compactness in Measure Spaces

Liviu C. Florescu
Christiane Godet-Thobie

Young Measures and Compactness in Measure Spaces

De Gruyter

Mathematics Subject Classification 2010: Primary: 28-01, 28-02, 60-01; Secondary: 28A33, 28C15, 46E30, 46N10, 49J45, 60B05, 60B10.

ISBN 978-3-11-027640-4
e-ISBN 978-3-11-028051-7

Library of Congress Cataloging-in-Publication Data

A CIP catalog record for this book has been applied for at the Library of Congress.

Bibliographic information published by the Deutsche Nationalbibliothek

The Deutsche Nationalbibliothek lists this publication in the Deutsche Nationalbibliografie; detailed bibliographic data are available in the Internet at http://dnb.dnb.de.

Typesetting: Da-TeX Gerd Blumenstein, Leipzig, www.da-tex.de
Printing and binding: Hubert & Co. GmbH & Co. KG, Göttingen
∞ Printed on acid-free paper

Printed in Germany

www.degruyter.com

This book is dedicated
to our wonderful spouses,

Cristina and Roger

Preface

In recent years, technological progress created a great need for complex mathematical models. Many practical problems can be formulated using optimization theory and they hope to obtain an optimal solution. In most cases, such optimal solution can not be found.

So, non-convex optimization problems (arising, e.g., in variational calculus, optimal control, nonlinear evolutions equations) may not possess a classical minimizer because the minimizing sequences have typically rapid oscillations. This behavior requires a relaxation of notion of solution for such problems; often we can obtain a such relaxation by means of Young measures.

The Young measures generalize measurable functions. Thus, a Young measure is herself a measurable application that, to every point t of Ω, associates a probability τ_t on a topological space X; for all Borel set $A \subseteq X$, $\tau_t(A)$ may be interpreted as the probability that the value in t of the "function" $\tau_.$ belongs to A. In the particular case, a measurable application $u : \Omega \to X$ is a Young measure, where, for all $t \in \Omega$, $\tau_t = \delta_{u(t)}$ ($\delta_{u(t)}$ indicates the mass of Dirac in $u(t)$).

Young measures' theory has a long history; it begins with the work of L. C. Young which, in 1937, introduces the so-called "generalized curves" in order to provide extended solutions for some non-convex problems in variational calculus. A milestone in this history is the appearance of the monograph of J. Warga, "Optimal Control of Differential and Functional Equations" (Academic Press, 1972); here is systematically developed a theory of relaxed control in compact metric spaces. The extension of theory on locally compact metric spaces was made by H. Berliocchi and J. M. Lasry in 1973.

The study of Young measures was extended to Polish and Suslin spaces by the works of E. J. Balder (since 1984) and M. Valadier (1990).

Lately, Young measures were the object of an intense research due to their applications in obtaining relaxed solutions; here are some of the areas in which these relaxed solutions find applications: non-convex variational problems and differential inclusions, non-linear homogenization problems, micro-magnetic phenomena in ferro-magnetic materials, Nash equilibrium in games theory, Gamma-convergence, different phenomena in continuum mechanics (as elasticity, microstructures' theory), optimal design and shape optimization problems.

On this subject, recent monographs appeared:

(i) Roubiček, T.—*Relaxation in optimization theory and variational calculus*, Walter de Gruyter, Berlin. New York, 1997.

(ii) Pedregal, P.—*Parametrized Measures and Variational Principles*, Birkhäuser Verlag, Basel. Boston. Berlin, 1997.

(iii) Castaing, Ch., Raynaud de Fitte, P. and Valadier, M.—*Young measures on topological spaces. With applications in control theory and probability theory*, Kluwer Academic Publ. Dordrecht. Boston. London, 2004.

The focus of the first two books is mainly on the applications; therefore, Young measures are used as generalized solutions to non-convex problems of variational calculus, optimization theory, or game theory.

The last monograph considers theoretical aspects of the theory of Young measures as well as the applications in control theory and probability theory. Many of the results presented here make reference to a wide bibliography; thus, the work is difficult to use for beginners.

The literature on the applications of Young measures in various areas (lower semicontinuity, optimal relaxed control, Gamma-convergence and homogenization, differential games, elasticity, hysteresis, etc.) is extremely rich and the existing monographs main focus on applications rather than on theoretical aspects. We found difficult for a young researcher who wants to clarify the theoretical aspects, to go through the extensive bibliography which is usually referred. Thus, our goal was to write a book where to be gathered all the theoretical aspects related to defining of Young measures (measurability, disintegration, stable convergence, compactness), book which to be a useful tool for those interested in theoretical foundations of the theory: the postgraduate students, the students in the doctoral study, but also to all those interested in measure theory and relaxed control.

The developing of Young measures' theory involves some compactness results for measures on abstract spaces and topological spaces. Hence, to achieve our goal, we considered useful to provide a complete set of classical and recent compactness results in measure and function spaces.

The book is organized in three chapters (Weak compactness in measure spaces, Bounded measures on topological spaces, Young measures). For a good comprehension of the subject, we developed in the first two chapters the results used in the third (biting lemma in the abstract measure theory and Prohorov's theorem in the measure theory on topological spaces).

The first chapter covers background material on measure theory in abstract frame. Therefore, we present some results of duality and weakly compactness in $ca(A)$ and $L^1(\lambda)$. However, here we prove some extensions of Dunford–Pettis theorem like biting lemma of Brooks–Chacon or subsequence splitting lemma of H. P. Rosenthal.

In Chapter two, we treat the measure theory on topological spaces. The framework is offered by Suslin spaces; on the one hand, these spaces are Radon and on the other hand, they cover the particular case of a separable Banach space provided with his weak topology. We introduce the narrow topology and then we prove the Prohorov's compactness theorem. In the particular case of Polish spaces, the narrow topology is metrizable; we present the compatible metrics of Dudley and of Lévy–Prohorov. As an application of Prohorov's theorem, we prove in the last paragraph the existence of Wiener's measure on $C[0,1]$.

With some exceptions, in Chapters 1 and 2 are presented classical compactness results for measures on abstract spaces, or on topological spaces. The originality consists in the selection and ordering of these results and the accompanying remarks and examples. However, we note some approaches and new results, such as: the modulus of λ-continuity (1.79) and theorems 1.80 and 1.81, a-convergence of nets in L^1 and the extension of Dunford–Pettis theorem (1.93), a new proof for Rosenthal's Subsequence Splitting Lemma using Biting Lemma, the modulus of narrow compactness (2.65), a-convergence of nets in $ca(\mathcal{B}(T))$, theorem 2.69 and obtaining, as corollary of this theorem, a new proof of Prohorov's compactness theorem. Finally, in the last section of 2 we give a simple and self-contained presentation of Wiener measure (2.6).

Compactness results from the first two chapters are used to study Young measures in Chapter three. We prove the disintegration theorem for product measures and we use it to present Young measures as parametrized measures; the frame is that of a regular Suslin space. We remark that the space of Young measures contains the space of measurable mappings as dense subspace and that the narrow topology is an extension of the topology of convergence in measure. Prohorov's theorem in the case of Young measures highlights the role played by tightness in compactness results. We present a vector version for biting lemma and an extension of this result to some special non-bounded sets of measurable mappings: finite-tight sets. In the seventh paragraph, we will study the two types of products for the Young measures and will give the fiber product lemma.

In the last three sections of the book are presented some applications; thus, Prokhorov's theorem for Young measures was used in the ninth paragraph in the study of strong compactness in $L^p(\mu, E)$. We obtain, as corollaries, the theorems of Visintin–Balder, Rossi–Savaré, Lions–Aubin and Gutman; in the scalar case, the compactness criterion of Riesz–Fréchet–Kolmogorov is obtained.

In the tenth paragraph, we consider some applications of quasiconvexity to the study of gradient Young measures and to the lower semicontinuity. Are studied the Young measures generated by sequences and particularly the gradient Young measures. We pay special attention to quasiconvexity and its various equivalent definitions. The quasiconvexity is essentially used in the Kinderlehrer–Pedregal's characterization of gradient Young measures, but also in the study of lower semicontinuity of energy functional that appears in variational calculus. Finally, in paragraph eleven, we present some results of existence of solutions in relaxed variational calculus.

There are also, in this chapter, some new concepts and results among which: new proofs for theorems 3.30, 3.32 and 3.33, the density result 3.49 and the proof of theorem 3.50, theorems 3.51, 3.66, 3.67 and propositions 3.54 and 3.56, introduction of finite-tight sets (3.75) and use them to obtain extensions of biting lemma (3.84) and Saadoune–Valadier's theorem (3.85), Jordan finite-tight sets (3.91) and their utility in obtain of a compactness result in Sobolev spaces (3.102) and an alternative to Rellich–Kondrachov theorem (3.105).

All results are accompanied by full demonstrations; for many of these results, are given different proofs from those referred in the literature.

The bibliography gives the main references relevant to the content of the book; it is no exhaustive.

Understanding the text requires basic knowledge of general topology, functional analysis and Lebesgue integration that may be found in any textbook on the subject. In rest, all the statements are fully justified and proved.

To conclude, this text is intended as a postgraduate textbook as well as a reference for more experienced researchers.

The book was written over several years of collaboration between authors, with the occasion of stages that the first author has made, as a visiting professor, at the University of Brest.

An important role in setting the ideas and in the organization of book's material was played by discussions with various mathematicians met under these occasions.

First, we mention the authors of monograph "Young measures on topological spaces", C. Castaing, M. Valadier and P. Raynaud de Fitte, that supported and inspired us in writing the last chapter. Also, we have had useful discussions with E. Balder and T. Roubiček at the international conference "*Mesures de Young et Contrôle Stochastique*" (Brest, 2002); in this way, we thank them all.

Iaşi/Brest, December 2011

Liviu C. Florescu, Christiane Godet-Thobie

Contents

Chapter 1

Weak Compactness in Measure Spaces

We will present in this first chapter the main properties of the measure spaces and of the space of integrable functions. We recall the classic results of weak compactness (like Vitali–Hahn–Saks, Radon–Nikodym and Dunford–Pettis theorems) but we will also mention more recent results such as Brooks–Chacon biting lemma or Rosenthal's lemma.

1.1 Measure Spaces

In this introductory section, we recall the definitions and classic properties of the additive and σ-additive measures. We finish this section by the Saks' theorem, the Vitali–Hahn–Saks and Nikodym theorems that we will use in the following sections for a study of weak compactness on $ca(\mathcal{A})$.

For beginning, we will specify the definitions and the notations to be employed henceforward. We consider as known the theory of integration relating to a positive, σ-additive and σ-finite measure.

We designate by X an arbitrary set and by $\mathcal{A}$ a σ–algebra of subsets of X; an $\mathcal{A}$-partition of $A \in \mathcal{A}$ is a partition of A with the elements in $\mathcal{A}$.

According to the usual notations, if μ is a positive, σ-additive and σ-finite measure, we shall denote by $\mathcal{L}^1(\mu) = \mathcal{L}^1(X, \mathcal{A}, \mu)$ the set of all real mappings f defined on X with the property that f is $\mathcal{A}$-measurable and μ-integrable and by $L^1(\mu) = L^1(X, \mathcal{A}, \mu)$ the quotient space $\mathcal{L}^1(\mu)/\doteq$, where $\doteq$ is equality μ – almost everywhere.

In the following, we recall the definition of the signed measures.

Definition 1.1. A set function $\lambda : \mathcal{A} \to \bar{\mathbb{R}} = [-\infty, +\infty]$ is a *finitely additive measure*, or shortly an *additive measure*, if

(i) $\lambda(\emptyset) = 0$,

(ii) $\lambda(A \cup B) = \lambda(A) + \lambda(B)$, for every $A, B \in \mathcal{A}$ with $A \cap B = \emptyset$,

(iii) λ assumes at most one of the values $+\infty$ and $-\infty$.

An additive measure λ on $\mathcal{A}$ is a *σ-additive measure* or a *countably additive measure* if, for every sequence of pairwise disjoint sets $(A_n)_{n\in\mathbb{N}} \subseteq \mathcal{A}$ (i.e.

$A_n \cap A_m = \emptyset$, for every $n \neq m$),

$$\lambda\left(\bigcup_{n=0}^{\infty} A_n\right) = \sum_{n=0}^{+\infty} \lambda(A_n).$$

A σ-additive measure λ is *finite* or *real valued* if its range is contained in $\mathbb{R}$. λ is σ*-finite* if, for every $A \in \mathcal{A}$, there exists a sequence $(A_n)_{n\in\mathbb{N}}$ such that $A = \bigcup_{n\in\mathbb{N}} A_n$ and $\lambda(A_n) \in \mathbb{R}$, for every $n \in \mathbb{N}$.

We will designate by

$ba(\mathcal{A})$—the set of all *real valued bounded additive measures* on $\mathcal{A}$,

$ca(\mathcal{A})$—the set of all *real valued* σ*-additive measures* on $\mathcal{A}$.

$ba(\mathcal{A})$ and $ca(\mathcal{A})$ are vector spaces under the usual addition and scalar multiplication operations.

$ca^+(\mathcal{A})$ ($ba^+(\mathcal{A})$)—the subsets of all *positive measures* of $ca(\mathcal{A})$ ($ba(\mathcal{A})$).

The following properties are easy to demonstrate.

Proposition 1.2. *Let $\lambda : \mathcal{A} \to \bar{\mathbb{R}}$ be an additive measure and let $A, B \in \mathcal{A}$ with $B \subseteq A$.*

(i) *If $|\lambda(A)| < +\infty$, then $|\lambda(B)| < +\infty$.*

(ii) *If $|\lambda(B)| < +\infty$, then $\lambda(A \setminus B) = \lambda(A) - \lambda(B)$.*

Proof. $A = B \cup (A \setminus B)$ and so $\lambda(A) = \lambda(B) + \lambda(A \setminus B)$.

(i) If $\lambda(B) = +\infty(-\infty)$, then $\lambda(A) = +\infty(-\infty)$, what contradicts hypothesis. Therefore, $\lambda(B)$ is finite.

(ii) If $|\lambda(B)| < +\infty$, then $\lambda(A) - \lambda(B) = \lambda(A \setminus B)$. □

Proposition 1.3. *Let λ be a σ-additive measure and let $(A_n)_{n\in\mathbb{N}} \subseteq \mathcal{A}$.*

(i) *If $(A_n)_n$ is an increasing sequence, then $\lambda(\cup_{n=0}^{\infty} A_n) = \lim_{n\to\infty} \lambda(A_n)$.*

(ii) *If $(A_n)_n$ is a decreasing sequence and $|\lambda(A_0)| < +\infty$, then $\lambda(\cap_{n=0}^{\infty} A_n) = \lim_{n\to\infty} \lambda(A_n)$.*

Proof. (i) Let $A = \bigcup_{n\in\mathbb{N}} A_n \in \mathcal{A}$. Firstly, we suppose that there is $n_0 \in \mathbb{N}$ such that $|\lambda(A_{n_0})| = +\infty$. According to (i) of Proposition 1.2, $|\lambda(A)| = +\infty = |\lambda(A_n)|$, for every $n \geq n_0$; since λ assumes at most one of the values $+\infty$ and $-\infty$, $\lambda(A) = \lim_n \lambda(A_n)$.

If, for every $n \in \mathbb{N}, |\lambda(A_n)| < +\infty$, then we define the pairwise disjoint sequence $(B_n)_n \subseteq \mathcal{A}$ letting: $B_0 = A_0, B_n = A_n \setminus A_{n-1}, \quad \forall n \geq 1$; then $A =$

$\cup_{n=0}^{\infty} B_n$ and, using (ii) of Proposition 1.2, we obtain $\lambda(A) = \sum_{n=0}^{\infty} \lambda(B_n) = \lim_n \sum_{k=0}^{n} \lambda(B_k) = \lim_n \lambda(A_n)$.

(ii) If $(A_n)_{n\in\mathbb{N}}$ is decreasing and $|\lambda(A_0)| < +\infty$, then the sequence $(B_n)_n$, where $B_n = A_0 \setminus A_n$, is increasing and the result follows from the first part of the proof. □

Definition 1.4. Let $\lambda : \mathcal{A} \to \bar{\mathbb{R}}$ be a σ-additive measure and let $A \in \mathcal{A}$;

A is called λ-*positive* if, for every $B \in \mathcal{A}$, $\lambda(A \cap B) \geq 0$.

A is called λ-*negative* if, for every $B \in \mathcal{A}, \lambda(A \cap B) \leq 0$.

A is called λ-*null* if it is λ-positive and λ-negative. A is λ-null set if and only if, for every measurable set $B \subseteq A, \lambda(B) = 0$.

For the following two results, see Theorem A, p. 121 in [93].

Proposition 1.5. *Let $\lambda : \mathcal{A} \to (-\infty, +\infty]$ be a σ-additive measure and let $A \in \mathcal{A}$ with $\lambda(A) < 0$. There exists a λ-negative set $B \subseteq A$ such that $\lambda(B) < 0$.*

Proof. If A is λ-negative, then $B = A$.

Otherwise, there exists $C \in \mathcal{A}$ such that $\lambda(C \cap A) > 0$. Let n_1 be the smallest positive integer for which there exists $A_1 \in \mathcal{A}, A_1 \subseteq A$ with $\lambda(A_1) > \frac{1}{n_1}$.

If $A \setminus A_1$ is λ-negative, then $B = A \setminus A_1$; since $\lambda(A) < 0$, $\lambda(A) \in \mathbb{R}$ and then, by Theorem 1.2, $\lambda(A_1) \in \mathbb{R}$ and $\lambda(B) = \lambda(A) - \lambda(A_1) < \lambda(A) < 0$.

If $A \setminus A_1$ is not λ-negative, let n_2 be the smallest positive integer for which there exists $A_2 \in \mathcal{A}, A_2 \subseteq A \setminus A_1$ with $\lambda(A_2) > \frac{1}{n_2}$; obviously, $n_2 > n_1$.

If the above construct does not produce a solution of problem after a finite number of steps, then we obtain a sequence of pairwise disjoint sets $(A_k)_{k\geq 1} \subseteq \mathcal{A}, A_k \subseteq A \setminus \cup_{i=1}^{k-1} A_i$, with $\lambda(A_k) > \frac{1}{n_k}$, for every $k \geq 1$ and $n_k \uparrow +\infty$.

Let $B = A \setminus \bigcup_{k=1}^{\infty} A_k$; then $\lambda(B) = \lambda(A) - \sum_{k=1}^{\infty} \lambda(A_k) < 0$.

For every $C \in \mathcal{A}, C \subseteq B$ and for every $k \in \mathbb{N}^*$, $C \subseteq A \setminus \cup_{i=1}^{k-1} A_i$ so that $\lambda(C) \leq \frac{1}{n_k - 1}$ (n_k is the smallest positive integer for which there is $A_k \subseteq A \setminus \cup_{i=1}^{k-1} A_i$ with $\lambda(A_k) > \frac{1}{n_k}$). Then $\lambda(C) \leq 0$ and therefore B is λ-negative. □

Theorem 1.6 (Hahn decomposition theorem). *Let $\lambda : \mathcal{A} \to \bar{\mathbb{R}}$ be a σ-additive measure; there exists a λ-positive set $H \in \mathcal{A}$ such that $H^c = X \setminus H$ is λ-negative.*

For any other pair $\{H_1, H_1^c\} \subseteq \mathcal{A}$ with H_1 λ-positive and H_1^c λ-negative, $H \Delta H_1$ is a λ-null set.

Proof. First, let us suppose that $\lambda(\mathcal{A}) \subseteq (-\infty, +\infty]$.

Let $a = \inf\{\lambda(A) : A \in \mathcal{A}, A = \lambda\text{-negative}\}$, let $(A_n)_n$ be a sequence of λ-negative sets such that $\lambda(A_n) \to a$ and let $H = X \setminus \bigcup_{n=1}^{\infty} A_n$; then $H^c = \bigcup_{n=1}^{\infty} A_n$.

If we define $(B_n)_{n\in\mathbb{N}^*}$ letting $B_1 = A_1$ and, for every $n \geq 2$, $B_n = A_n \setminus \cup_{i=1}^{n-1} A_i$, then, for every $C \in \mathcal{A}$, $\lambda(C \cap H^c) = \lambda(C \cap \bigcup_{n=1}^{\infty} B_n) = \sum_{n=1}^{\infty} \lambda(C \cap B_n) \leq 0$. Therefore H^c is λ-negative.

Moreover, for every $n \in \mathbb{N}^*$,

$$\lambda(H^c) = \lambda(A_n) + \lambda(H^c \setminus A_n) \leq \lambda(A_n)$$

so that $\lambda(H^c) = a$.

If we suppose that H is not λ-positive, then there exists $C \in \mathcal{A}, C \subseteq H$ such that $\lambda(C) < 0$. The previous proposition assures us on the existence of a λ-negative set $B \subseteq C$ with $\lambda(B) < 0$. Then $H^c \cup B$ is λ-negative and

$$\lambda(H^c \cup B) = \lambda(H^c) + \lambda(B) = a + \lambda(B) < a$$

and this contradicts the definition of a.

In the case where $\lambda(\mathcal{A}) \subseteq [-\infty, +\infty)$, $-\lambda : \mathcal{A} \to (-\infty, +\infty]$ is a σ-additive measure. Let H be a $(-\lambda)$-positive set and H^c be a $(-\lambda)$-negative set; then H^c is λ-positive set and H is λ-negative set.

Let now H_1 λ-positive and H_1^c λ-negative an other pair. For every $B \in \mathcal{A}$ with $B \subseteq H \setminus H_1 = H \cap H_1^c$, $\lambda(B) \leq 0$ and $\lambda(B) \geq 0$, hence $\lambda(B) = 0$. Then $H \setminus H_1$ is a λ-null set. Similarly, $H_1 \setminus H$ is λ-null and then $H \Delta H_1 = (H \setminus H_1) \cup (H_1 \setminus H)$ is a λ-null set. □

Definition 1.7. Every pair of sets $\{H, H^c\} \subseteq \mathcal{A}$, with the property that H is λ-positive and H^c is λ-negative, is called a *Hahn decomposition* of X relatively to the measure λ.

Remark 1.8.

(i) Hahn's decomposition of X relatively to a measure λ is not unique (we can replace H by $H \cup N$ where N is a λ-null set).

(ii) The Hahn decomposition theorem says that, for every σ-additive measure λ, there exists a Hahn decomposition of X relatively to λ.

Proposition 1.9. *Let $\lambda : \mathcal{A} \to \bar{\mathbb{R}}$ be a σ-additive measure, let $\{H, H^c\}$ be a Hahn decomposition of X relatively to λ and let $\lambda^+, \lambda^- : \mathcal{A} \to \bar{\mathbb{R}}_+$ defined by $\lambda^+(A) = \lambda(A \cap H)$, $\lambda^-(A) = -\lambda(A \setminus H)$, for every $A \in \mathcal{A}$.*

Then λ^+, λ^- are two σ-additive positive measures (one of them finite), $\lambda = \lambda^+ - \lambda^-$ and $\lambda^+(H^c) = \lambda^-(H) = 0$.

If λ_1^+, λ_1^- are two other σ-additive positive measures (one of them finite) such that $\lambda = \lambda_1^+ - \lambda_1^-$ and if $\lambda_1^+(H_1^c) = \lambda_1^-(H_1) = 0$ for a set $H_1 \in \mathcal{A}$, then $\lambda^+ = \lambda_1^+$ and $\lambda^- = \lambda_1^-$.

Proof. The first part of the proposition is obvious. We shall prove only the uniqueness of decomposition of λ as difference of positive measures.

The pair $\{H_1, H_1^c\}$ is again a Hahn decomposition of X relatively to λ. Indeed, for every $A \in \mathcal{A}, \lambda(A \cap H_1) = \lambda_1^+(A \cap H_1) - \lambda_1^-(A \cap H_1) = \lambda_1^+(A \cap H_1) \geq 0$ and $\lambda(A \setminus H_1) = \lambda_1^+(A \setminus H_1) - \lambda_1^-(A \setminus H_1) = -\lambda_1^-(A \setminus H_1) \leq 0$. According to Hahn decomposition theorem, $H \Delta H_1$ is a λ-null set. Therefore, for every $A \in \mathcal{A}$, $\lambda^+(A) = \lambda(A \cap H) = \lambda(A \cap H_1) = \lambda_1^+(A)$ so that $\lambda^+ = \lambda_1^+$. Similarly, $\lambda^- = \lambda_1^-$. □

Definition 1.10. We say that the unique pair $\{\lambda^+, \lambda^-\}$ of σ-additive positive measures (one of them finite) with $\lambda = \lambda^+ - \lambda^-$ and $\lambda^+(H^c) = \lambda^-(H) = 0$ for a set $H \in \mathcal{A}$, is the *Jordan decomposition of* λ.

Definition 1.11. A positive measure λ on $\mathcal{A}$ is *concentrated on the set* $D \in \mathcal{A}$ if $\lambda(D) = \lambda(X)$.

Remark 1.12. If λ is a σ-additive measure on $\mathcal{A}$, if $\{H, H^c\}$ is a Hahn decomposition of X relatively to λ and if $\{\lambda^+, \lambda^-\}$ is the Jordan decomposition of λ, then λ^+ is concentrated on the set H and λ^- is concentrated on H^c.

Theorem 1.13. *Let λ be a σ-additive measure on $\mathcal{A}$; there exist $A_m, A_M \in \mathcal{A}$ such that*

$$\lambda(A_m) = \inf\{\lambda(A) : A \in \mathcal{A}\} \leq 0 \leq \sup\{\lambda(A) : A \in \mathcal{A}\} = \lambda(A_M).$$

Every σ-additive measure is bounded either from below or from above.

Proof. Let $\{H, H^c\}$ a Hahn decomposition of X relatively to λ and let $\{\lambda^+, \lambda^-\}$ the Jordan decomposition of λ; for every $A \in \mathcal{A}$, $\lambda^+(A) = \lambda(A \cap H) \leq \lambda(H)$ and $\lambda^-(A) = -\lambda(A \setminus H) \leq \lambda^-(H^c) = \lambda(H^c)$. Then

$$-\lambda(H^c) \leq -\lambda^-(A) \leq \lambda^+(A) - \lambda^-(A) = \lambda(A) \leq \lambda^+(A) \leq \lambda(H).$$

Therefore we can take $A_m = H^c$ and $A_M = H$. □

The following result is a corollary of Theorem 1.13.

Corollary 1.14. *Every measure $\lambda \in ca(\mathcal{A})$ is bounded; therefore*

$$ca(\mathcal{A}) \subseteq ba(\mathcal{A}).$$

Remark 1.15. If $\lambda : \mathcal{A} \to \mathbb{R}$ is only an additive measure, it is not compulsory that λ should be bounded, as it is shown in the following example.

Example 1.16. Let $\sum_{n=0}^{\infty} a_n$ be a conditionally convergent series (a convergent series for which $\sum_{n=0}^{\infty} |a_n| = +\infty$), let $\mathcal{A}$ be the algebra of all sets $A \subseteq \mathbb{N}$ such that A or $\mathbb{N} \setminus A$ is a finite and let $\mu : \mathcal{A} \to \mathbb{R}$ defined by

$$\mu(A) = \begin{cases} \sum_{n\in A} a_n, & A \neq \emptyset, \\ 0, & A = \emptyset. \end{cases}$$

Then μ is an additive measure, but it is not bounded on the algebra $\mathcal{A}$. We notice that $\mathcal{A}$ is not a σ-algebra, but, since all additive function on an algebra can be extended to an additive function on the generated σ-algebra – in our case $\mathcal{P}(\mathbb{N})$ – (see [29], p.185 and [175],1.8.), it is evident that the extension itself is not bounded.

The total variation defined below is introduced in order to define a complete norm on $ba(\mathcal{A})$ or on its subspace $ca(\mathcal{A})$ of σ-additive measures (see Definition III.1.4 and Lemma III.1.6 of [62]).

Theorem 1.17. *For every additive measure λ, let $|\lambda| : \mathcal{A} \to \overline{\mathbb{R}}_+$ defined by*

$$|\lambda|(A) = \sup\left\{\sum_{i=1}^{n} |\lambda(A_i)| : n \in \mathbb{N}^*, \quad \{A_1, \dots, A_n\} = \mathcal{A} - \textit{partition of } A\right\}.$$

Then:

(i) $\sup_{B\in\mathcal{A}, B\subseteq A} |\lambda(B)| \leq |\lambda|(A) \leq 2\sup_{B\in\mathcal{A}, B\subseteq A} |\lambda(B)|$.

(ii) $|\lambda|$ *is additive.*

(iii) *If $\lambda \in ba(\mathcal{A})$, then $|\lambda| \in ba^+(\mathcal{A})$; moreover $|\lambda|$ is the smallest element of the set $\mathfrak{M} = \{\nu \in ba^+(\mathcal{A}) : |\lambda(A)| \leq \nu(A), \quad \forall A \in \mathcal{A}\}$.*

(iv) *If λ is σ-additive, then $|\lambda|$ is σ-additive and $|\lambda|(H) = \lambda(H)$ and $|\lambda|(H^c) = -\lambda(H^c)$, where $\{H, H^c\}$ is a Hahn decomposition of X.*

If $\lambda \in ca(\mathcal{A})$, then $|\lambda| \in ca^+(\mathcal{A})$.

Proof. (i) For every $B \in \mathcal{A}$ with $B \subseteq A$, $\{B, A \setminus B\}$ is an $\mathcal{A}$-partition of A and so $|\lambda|(A) \geq |\lambda(B)| + |\lambda(A \setminus B)| \geq |\lambda(B)|$, from where

$$\sup\{|\lambda(B)| : B \in \mathcal{A}, B \subseteq A\} \leq |\lambda|(A).$$

Now, let $n \in \mathbb{N}^*$ and let $\{A_1, \ldots, A_n\}$ be an $\mathcal{A}$-partition of the set A. We can suppose that $\lambda(A_1), \ldots, \lambda(A_p) \geq 0$ and $\lambda(A_{p+1}), \ldots, \lambda(A_n) < 0$. Then

$$\begin{aligned}\sum_{i=1}^{n} |\lambda(A_i)| &= \sum_{i=1}^{p} \lambda(A_i) - \sum_{j=p+1}^{n} \lambda(A_j) = \lambda\left(\bigcup_{i=1}^{p} A_i\right) - \lambda\left(\bigcup_{j=p+1}^{n} A_j\right) \\ &= \left|\lambda\left(\bigcup_{i=1}^{p} A_i\right)\right| + \left|\lambda\left(\bigcup_{j=p+1}^{n} A_j\right)\right| \\ &\leq 2\sup\{|\lambda(B)| : B \in \mathcal{A}, B \subseteq A\}.\end{aligned}$$

(ii) Let $A, B \in \mathcal{A}$ with $A \cap B = \emptyset$ and let $C = A \cup B$. If $|\lambda|(A) = +\infty$, then $+\infty = \sup\{|\lambda(D)| : D \in \mathcal{A}, D \subseteq A\} \leq \sup\{|\lambda(D)| : D \in \mathcal{A}, D \subseteq C\} \leq |\lambda|(C)$ and so $|\lambda|(C) = +\infty = |\lambda|(A) + |\lambda|(B)$.

In the same way, if $|\lambda|(B) = +\infty$, we have $|\lambda|(C) = +\infty = |\lambda|(A) + |\lambda|(B)$.

Now suppose that $|\lambda|(A) < +\infty$ and $|\lambda|(B) < +\infty$. Then, for every $\varepsilon > 0$, there exists an $\mathcal{A}$-partition $\{A_1, \ldots, A_n\}$ of A and an $\mathcal{A}$-partition of B, $\{B_1, \ldots, B_m\}$, such that

$$|\lambda|(A) - \frac{\varepsilon}{2} < \sum_{i=1}^{n} |\lambda(A_i)| \quad \text{and} \quad |\lambda|(B) - \frac{\varepsilon}{2} < \sum_{j=1}^{m} |\lambda(B_j)|.$$

Then $\{A_1, \ldots, A_n, B_1, \ldots, B_m\}$ is an $\mathcal{A}$-partition of C and therefore

$$|\lambda|(A) + |\lambda|(B) - \varepsilon < \sum_{i=1}^{n} |\lambda(A_i)| + \sum_{j=1}^{m} |\lambda(B_j)| \leq |\lambda|(C),$$

from where, ε being arbitrary,

$$|\lambda|(A) + |\lambda|(B) \leq |\lambda|(C). \tag{1}$$

For every $\mathcal{A}$-partition $\{C_1, \ldots, C_p\}$ of C, let's note $A_i = C_i \cap A$ and $B_i = C_i \cap B$, for all $i = 1, \ldots, p$. Then $A_i, B_i \in \mathcal{A}$ and $C_i = A_i \cup B_i$. Therefore $\{A_1, \ldots, A_p\}$ is an $\mathcal{A}$-partition of A and $\{B_1, \ldots, B_p\}$ is an $\mathcal{A}$-partition of B.

$$\begin{aligned}\sum_{i=1}^{p} |\lambda(C_i)| &= \sum_{i=1}^{p} |\lambda(A_i) + \lambda(B_i)| \\ &\leq \sum_{i=1}^{p} |\lambda(A_i)| + \sum_{i=1}^{p} |\lambda(B_i)| \leq |\lambda|(A) + |\lambda|(B).\end{aligned}$$

As $\{C_1, \ldots, C_p\}$ is an arbitrary partition of C, we have:

$$|\lambda|(C) \leq |\lambda|(A) + |\lambda|(B). \tag{2}$$

From (1) and (2) we have that $|\lambda|$ is additive.

(iii) If $\lambda \in ba(\mathcal{A})$, then λ is bounded; from *(i)*, we obtain that,

$$0 \leq |\lambda|(A) \leq 2\sup\{|\lambda(B)| : B \in \mathcal{A}\} = M < +\infty, \quad \text{for every} \quad A \in \mathcal{A}$$

and so $\sup\{|\lambda|(A) : A \in \mathcal{A}\} \leq M < +\infty$.

Therefore $|\lambda| : \mathcal{A} \to \mathbb{R}_+$ is a positive bounded additive measure on $\mathcal{A}$, which means that $|\lambda| \in ba^+(\mathcal{A})$.

Since $|\lambda(A)| \leq \sup\{|\lambda(B)| : B \in \mathcal{A}, B \subseteq A\} \leq |\lambda|(A)$, for all $A \in \mathcal{A}$, it is clear that $|\lambda| \in \mathfrak{M} = \{\nu \in ba^+(\mathcal{A}) : |\lambda(A)| \leq \nu(A)$, for all $A \in \mathcal{A}\}$.

It remains to show that $|\lambda|$ is the smallest element of $\mathfrak{M}$.

Let $\nu \in \mathfrak{M}$, $A \in \mathcal{A}$ and let $\{A_1, \ldots, A_n\}$ be an $\mathcal{A}$-partition of A; we have

$$\sum_{i=1}^{n} |\lambda(A_i)| \leq \sum_{i=1}^{n} \nu(A_i) = \nu\left(\bigcup_{i=1}^{n} A_i\right) = \nu(A),$$

from where, $|\lambda|(A) \leq \nu(A)$, for every $A \in \mathcal{A}$ and so $|\lambda| \leq \nu$.

(iv) According to (ii), $|\lambda|$ is finite additive. Let $(E_n)_{n\in\mathbb{N}} \subseteq \mathcal{A}$ be a sequence of pairwise disjoint sets and let $E = \bigcup_{n\in\mathbb{N}} E_n$. Then, for every $m \in \mathbb{N}$,

$$|\lambda|(E) \geq |\lambda|(\cup_{n\leq m} E_n) = \sum_{n=0}^{m} |\lambda|(E_n) \quad \text{and so} \quad |\lambda|(E) \geq \sum_{n=0}^{\infty} |\lambda|(E_n).$$

To demonstrate the inverse inequality, let $(F_i)_{i\leq k}$ be an $\mathcal{A}$-partition of E. Then,

$$\begin{aligned} \sum_{i=1}^{k} |\lambda(F_i)| &= \textstyle\sum_{i=1}^{k} |\lambda(F_i \cap (\cup_{n\in\mathbb{N}} E_n)| = \sum_{i=1}^{k} \left|\sum_{n=0}^{\infty} \lambda(F_i \cap E_n)\right| \\ &\leq \textstyle\sum_{n=0}^{\infty} \sum_{i=1}^{k} |\lambda(F_i \cap E_n)| \leq \sum_{n=0}^{\infty} |\lambda|(E_n) \end{aligned}$$

from where $|\lambda|(E) \leq \sum_{n=0}^{\infty} |\lambda|(E_n)$. Therefore $|\lambda|$ is σ-additive.

Let now $\{H, H^c\}$ be a Hahn decomposition of X relatively to λ. Then H is λ-positive and then, for every $\mathcal{A}$ - partition of H, $\{A_1, \ldots, A_n\}$, $\sum_{i=1}^{n} |\lambda(A_i)| = \sum_{i=1}^{n} \lambda(A_i) = \lambda(H)$; so that $|\lambda|(H) = \lambda(H)$. Similarly, $|\lambda|(H^c) = -\lambda(H^c)$.

If $\lambda \in ca(\mathcal{A})$, then λ is σ-additive and bounded (see Corollary 1.14). Therefore $|\lambda|$ is σ-additive and by *(i)*, $|\lambda|$ belongs to $ca(\mathcal{A})$. □

Remark 1.18. Let λ be a σ-additive measure, let $\{H, H^c\}$ be a Hahn decomposition of X relatively to λ and let $\{\lambda^+, \lambda^-\}$ be the Jordan decomposition of λ; then

$$\lambda^+ = \frac{1}{2}(|\lambda| + \lambda), \quad \lambda^- = \frac{1}{2}(|\lambda| - \lambda) \quad \text{and} \quad |\lambda| = \lambda^+ + \lambda^-.$$

Indeed, if we note $\lambda_1^+ = \frac{1}{2}(|\lambda|+\lambda)$ and $\lambda_1^- = \frac{1}{2}(|\lambda|-\lambda)$, then λ_1^+ and λ_1^- are σ-additive positive measures, $\lambda = \lambda_1^+ - \lambda_1^-$ and $\lambda_1^+(H^c) = \frac{1}{2}(-\lambda(H^c)+\lambda(H^c)) = 0 = \lambda^-(H)$. Therefore $\{\lambda_1^+, \lambda_1^-\}$ is the Jordan decomposition of λ and then $\lambda^+ = \lambda_1^+$ and $\lambda^- = \lambda_1^-$.

Definition 1.19. Let $\lambda : \mathcal{A} \to \bar{\mathbb{R}}$ be an additive measure; $|\lambda|$ is called *the total variation* of λ.

Let $\lambda \in ba(\mathcal{A})$; according to previous remark, we say that $\lambda^+(\lambda^-)$ are *the positive variation* (*negative variation*) of λ, where $\lambda^+(\lambda^-) : \mathcal{A} \to \mathbb{R}_+$ is defined by

$$\lambda^+(A) = \frac{1}{2}(\,|\lambda|(A) + \lambda(A)) \quad \text{for every} \quad A \in \mathcal{A}$$
$$(\lambda^-(A) = \frac{1}{2}(\,|\lambda|(A) - \lambda(A)) \quad \text{for every} \quad A \in \mathcal{A}).$$

Obviously, $\lambda^+, \lambda^- \in ba^+(\mathcal{A})$, $\lambda = \lambda^+ - \lambda^-$ and $|\lambda| = \lambda^+ + \lambda^-$.

If $\lambda \in ca(\mathcal{A})$, then $\lambda^+, \lambda^- \in ca^+(\mathcal{A})$.

Remark 1.20. It results that every bounded additive measure is a difference between two bounded positive additive measures. This decomposition is not unique. Indeed, if $\lambda \in ba(\mathcal{A})$, for all $\mu \in ba^+(\mathcal{A})$, $\lambda = (\lambda^+ + \mu) - (\lambda^- + \mu)$ is another decomposition of λ.

In the following results, we will mention some direct consequences of Theorem 1.13.

Corollary 1.21. *If* $\lambda \in ca(\mathcal{A})$, *then the sets* A_m *and* A_M, *introduced in Theorem* 1.13, *have the following properties:*

(i) $\lambda(A) \geq 0, \quad \forall A \in \mathcal{A}, A \subseteq A_M,$

$\lambda(A) \leq 0, \quad \forall A \in \mathcal{A}, A \subseteq A_m.$

(ii) $\lambda(A) = 0, \quad \forall A \in \mathcal{A}, A \subseteq A_m \cap A_M.$

(iii) $\lambda(A \setminus (A_m \cup A_M)) = 0, \quad \forall A \in \mathcal{A}.$

(iv) $\lambda(A) = \lambda(A \cap A_M) + \lambda(A \cap A_m), \quad \forall A \in \mathcal{A}.$

Proof. (i) This point is demonstrated in the previous proposition.

(ii) is a consequence of (i).

(iii) Suppose that there exists a set $A_0 \in \mathcal{A}$ such that $\lambda(A_0 \setminus (A_m \cup A_M)) \neq 0$. Let $B_0 = A_0 \setminus (A_m \cup A_M)$.

If $\lambda(B_0) > 0$, let $B_1 = B_0 \cup A_M$; then $\lambda(B_1) = \lambda(B_0) + \lambda(A_M) > \lambda(A_M)$ which contradicts the maximality of A_M.

If $\lambda(B_0) < 0$, let $B_1 = B_0 \cup A_m$; then $\lambda(B_1) = \lambda(B_0) + \lambda(A_m) < \lambda(A_m)$ which contradicts the minimality of A_m. Therefore (iii) is satisfied.

(iv) For every $A \in \mathcal{A}, \lambda(A) = \lambda(A \cap A_M) + \lambda(A \setminus A_M) = \lambda(A \cap A_M) + \lambda((A \setminus A_M) \cap A_m) + \lambda((A \setminus A_M) \setminus A_m) = \lambda(A \cap A_M) + \lambda(A \cap A_m) - \lambda(A_m \cap A_M) + \lambda[A \setminus (A_m \cup A_M)]$.

According to (ii) and (iii), the last two terms are null. □

Corollary 1.22. *Let $\lambda \in ca(\mathcal{A})$ and A_m, A_M the already defined sets. Then, for every $A \in \mathcal{A}$,*

(i) $\lambda(A \cap A_M) = \sup\{\lambda(E) : E \in \mathcal{A}, E \subseteq A\},$

$\lambda(A \cap A_m) = \inf\{\lambda(E) : E \in \mathcal{A}, E \subseteq A\}.$

(ii) $|\lambda|(A) = \lambda(A \cap A_M) - \lambda(A \cap A_m).$

(iii) $\lambda^+(A) = \lambda(A \cap A_M), \lambda^-(A) = -\lambda(A \cap A_m).$

Proof. (i) Obviously, $\lambda(A \cap A_M) \leq \sup\{\lambda(E) : E \in \mathcal{A}, E \subseteq A\}$.

We suppose that $\lambda(A \cap A_M) < \sup\{\lambda(E) : E \in \mathcal{A}, E \subseteq A\}$. Then there exists $E_0 \in \mathcal{A}, E_0 \subseteq A$ such that $\lambda(A \cap A_M) < \lambda(E_0)$. $\lambda(A_M) = \lambda(A \cap A_M) + \lambda(A_M \setminus A) < \lambda(E_0) + \lambda(A_M \setminus A) = \lambda(E_0 \cup (A_M \setminus A))$, which contradicts the maximality of the set A_M. The second equality is proved in a similar manner.

(ii) For every $A \in \mathcal{A}, \{A \cap A_M, A \setminus A_M\}$ is an $\mathcal{A}$-partition of A; therefore $|\lambda|(A) \geq |\lambda(A \cap A_M)| + |\lambda(A \setminus A_M)|$. According to (iv) of Corollary 1.21 $\lambda(A \setminus A_M) = \lambda((A \setminus A_M) \cap A_M) + \lambda((A \setminus A_M) \cap A_m) = \lambda(A \cap A_m \setminus A_M)$; by (ii) of Theorem 1.21, $\lambda(A \cap A_m \setminus A_M) = \lambda(A \cap A_m \cap A_M) + \lambda(A \cap A_m \setminus A_M) = \lambda(A \cap A_m)$. Finally, using (i) of Theorem 1.21,

$$|\lambda|(A) \geq \lambda(A \cap A_M) - \lambda(A \cap A_m). \tag{1}$$

If we note $\lambda_1(A) = \lambda(A \cap A_M) - \lambda(A \cap A_m)$, then λ_1 is a positive measure and, according to (iv) and (i) of Theorem 1.21, for every $A \in \mathcal{A}$,

$$|\lambda(A)| = |\lambda(A \cap A_M) + \lambda(A \cap A_m)| \leq \lambda(A \cap A_M) - \lambda(A \cap A_m) = \lambda_1(A).$$

According to (iii) of Theorem 1.17,

$$|\lambda|(A) \leq \lambda_1(A) = \lambda(A \cap A_M) - \lambda(A \cap A_m). \tag{2}$$

By (1) and (2), $|\lambda|(A) = \lambda(A \cap A_M) - \lambda(A \cap A_m)$.

(iii) According to (ii) and to (iv) of Corollary 1.21, $\lambda^+(A) = \frac{1}{2}(|\lambda|(A) + \lambda(A)) = \frac{1}{2}(\lambda(A \cap A_M) - \lambda(A \cap A_m) + \lambda(A \cap A_M) + \lambda(A \cap A_m)) = \lambda(A \cap A_M)$ and $\lambda^-(A) = \frac{1}{2}(|\lambda|(A) - \lambda(A)) = \frac{1}{2}(\lambda(A \cap A_M) - \lambda(A \cap A_m) - \lambda(A \cap A_M) - \lambda(A \cap A_m)) = -\lambda(A \cap A_m)$. □

Theorem 1.17 allows us to introduce a norm on $ba(\mathcal{A})$ equivalent to the norm $\|\cdot\|_\infty$ of the uniform convergence.

Theorem 1.23. *The applications $\|\cdot\|, \|\cdot\|_\infty : ba(\mathcal{A}) \to \mathbb{R}_+$ defined as $\|\lambda\| = |\lambda|(X) = \lambda^+(X) + \lambda^-(X)$ and $\|\lambda\|_\infty = \sup_{A\in\mathcal{A}}(|\lambda(A)|)$ are two equivalent norms on $ba(\mathcal{A})$.*

The spaces $(ba(\mathcal{A}), \|\cdot\|)$ and $(ba(\mathcal{A}), \|\cdot\|_\infty)$ are Banach spaces; $ca(\mathcal{A})$ is a Banach subspace of $ba(\mathcal{A})$.

Proof. Firstly, we show that $\|\cdot\|, \|\cdot\|_\infty$ are norms. According to Theorem 1.17(i), for every $\lambda \in ba(\mathcal{A})$

$$\|\lambda\|_\infty \leq \|\lambda\| \leq 2\|\lambda\|_\infty.$$

Therefore we have $\|\lambda\| = 0 \Leftrightarrow \|\lambda\|_\infty = 0 \Leftrightarrow \sup_{A\in\mathcal{A}} |\lambda(A)| = 0$ what comes back to $\lambda = 0$. For every $a \in \mathbb{R}$, $\|a\lambda\| = |a\lambda|(X) = \sup\{\sum_{i=1}^n |a\lambda(A_i)| : \{A_1, \ldots, A_n\} = \mathcal{A}\text{-partition of } X\} = |a| \cdot \|\lambda\|$ and $\|a\lambda\|_\infty = \sup_{A\in\mathcal{A}} |a\lambda(A)| = |a| \cdot \|\lambda\|_\infty$.

Now we prove the triangular inequality; according to Theorem 1.17, for every $\varepsilon > 0$, there exists an $\mathcal{A}$-partition of X, $\{A_1, \ldots, A_n\}$, such that

$$\begin{aligned}\|\lambda + \mu\| - \varepsilon = |\lambda + \mu|(X) - \varepsilon &< \sum_{i=1}^n |\lambda(A_i) + \mu(A_i)| \\ &\leq \sum_{i=1}^n |\lambda(A_i)| + \sum_{i=1}^n |\mu(A_i)| \leq \|\lambda\| + \|\mu\|,\end{aligned}$$

from where $\|\lambda + \mu\| \leq \|\lambda\| + \|\mu\|$.

In the same way for $\|\cdot\|_\infty$, for every $\varepsilon > 0$, there exists $A_\varepsilon \in \mathcal{A}$ such that

$$\|\lambda + \mu\|_\infty - \varepsilon < |(\lambda + \mu)(A_\varepsilon)| \leq |\lambda(A_\varepsilon)| + |\mu(A_\varepsilon)| \leq \|\lambda\|_\infty + \|\mu\|_\infty$$

from where $\|\lambda + \mu\|_\infty \leq \|\lambda\|_\infty + \|\mu\|_\infty$.

Therefore, in the light of the inequalities mentioned at the beginning of the demonstration, $\|\cdot\|$ and $\|\cdot\|_\infty$ are equivalent norms on $ba(\mathcal{A})$.

It is evident that

$$\lambda_n \xrightarrow{\|\cdot\|} \lambda \Leftrightarrow \lambda_n \xrightarrow{\|\cdot\|_\infty} \lambda \Leftrightarrow \lambda_n \underset{\mathcal{A}}{\overset{u}{\longrightarrow}} \lambda.$$

Similarly, if (λ_n) is a sequence $\|\cdot\|$-Cauchy (and so $\|\cdot\|_\infty$-Cauchy), then $(\lambda_n(A))_{n\in\mathbb{N}}$ is a Cauchy sequence, uniformly in $A \in \mathcal{A}$; then there exists $\lambda : \mathcal{A} \to \mathbb{R}$ such that $\lambda_n \underset{\mathcal{A}}{\overset{u}{\longrightarrow}} \lambda$. Therefore λ is additive and bounded on $\mathcal{A}$; that is to say $\lambda \in ba(\mathcal{A})$. (λ_n) converges to λ in $(ba(\mathcal{A}), \|\cdot\|)$ and also in $(ba(\mathcal{A}), \|\cdot\|_\infty)$ so that $(ba(\mathcal{A}), \|\cdot\|)$ and $(ba(\mathcal{A}), \|\cdot\|_\infty)$ are Banach spaces.

Finally, in order to establish that $ca(\mathcal{A})$ is a closed subspace of $(ba(\mathcal{A}), \|\cdot\|)$, let $\lambda \in ba(\mathcal{A})$ and $(\lambda_n) \subseteq ca(\mathcal{A})$ such that $\lambda_n \xrightarrow{\|\cdot\|} \lambda$; we show that $\lambda \in ca(\mathcal{A})$.

Let $\{A_p : p \in \mathbb{N}^*\} \subseteq \mathcal{A}$ be a pairwise disjoint family of sets and let $A = \bigcup_{p=1}^\infty A_p \in \mathcal{A}$. Since $\lambda_n \underset{\mathcal{A}}{\overset{u}{\longrightarrow}} \lambda$, for every $\varepsilon > 0$, there exists $n_0 \in \mathbb{N}$ such that

$$|\lambda_n(B) - \lambda(B)| < \varepsilon, \quad \forall n \geq n_0, \quad \forall B \in \mathcal{A}. \tag{1}$$

Because $\lambda_{n_0} \in ca(\mathcal{A})$, there exists $n_1 > n_0$ such that, for every $n \geq n_1$

$$\left|\lambda_{n_0}(A) - \sum_{k=1}^n \lambda_{n_0}(A_k)\right| < \varepsilon. \tag{2}$$

Then, for all $n \geq n_1$,

$$\begin{aligned}\left|\lambda(A) - \sum_{k=1}^n \lambda(A_k)\right| &= \left|\lambda\left(A \setminus \bigcup_{k=1}^n A_k\right)\right| \leq \left|\lambda\left(A \setminus \bigcup_{k=1}^n A_k\right) - \lambda_{n_0}\left(A \setminus \bigcup_{k=1}^n A_k\right)\right| \\ &\quad + \left|\lambda_{n_0}\left(A \setminus \bigcup_{k=1}^n A_k\right)\right| = \left|\lambda\left(A \setminus \bigcup_{k=1}^n A_k\right) - \right. \\ &\quad \left. \lambda_{n_0}\left(A \setminus \bigcup_{k=1}^n A_k\right)\right| + \left|\lambda_{n_0}(A) - \sum_{k=1}^n \lambda_{n_0}(A_k)\right|.\end{aligned}$$

From (1) and (2), we obtain

$$\left|\lambda(A) - \sum_{k=1}^n \lambda(A_k)\right| < 2\varepsilon, \quad \forall n \geq n_1, \tag{3}$$

so that $\lambda \in ca(\mathcal{A})$. □

Before mentioning the definition of the integral in relation to a signed measure, we need to clarify a number of notations and properties of the integral relatively to a positive measure.

Let λ be a positive σ-additive measure on $\mathcal{A}$ and let $f : X \to \mathbb{R}$ be an $\mathcal{A}$-measurable mapping. We recall that f is λ-integrable if $f^+ = \sup\{f, 0\}$ and $f^- = \sup\{-f, 0\}$ are λ-integrable. Then, $\int_X f d\lambda = \int_X f^+ d\lambda - \int_X f^- d\lambda$; let $\mathcal{L}^1(\lambda)$ be the set of all λ-integrable mappings and let $L^1(\lambda) = L^1(X, \mathcal{A}, \lambda)$ be the quotient space $\mathcal{L}^1(\lambda)/\doteq$, where $\doteq$ is equality λ-almost everywhere.

If at least one of the two functions f^+ and f^- is λ-integrable, the difference of the integrals is always defined and will be marked by $\int_X f d\lambda$.

Definition 1.24. Let $\lambda \in ca(\mathcal{A})$ and let $f : X \to \mathbb{R}$ be an $\mathcal{A}$-measurable mapping; we say that f *is λ-integrable* if $f \in \mathcal{L}^1(\lambda^+) \cap \mathcal{L}^1(\lambda^-)$. Let us denote

$$\mathcal{L}^1(\lambda) = \mathcal{L}^1(\lambda^+) \cap \mathcal{L}^1(\lambda^-), L^1(\lambda) = L^1(\lambda^+) \cap L^1(\lambda^-) \quad \text{and}$$
$$\int_A f d\lambda = \int_A f d\lambda^+ - \int_A f d\lambda^-, \quad \text{for every} \quad A \in \mathcal{A},$$

where f marks, according to the context, the function f or the equivalence class of a function f. It is clear that $L^1(\lambda)$ is a vector space and that $\int_A$ is a linear operator on $L^1(\lambda)$.

Proposition 1.25. *Let $\lambda \in ca^+(\mathcal{A})$ and $f : X \to \mathbb{R}$ be an $\mathcal{A}$-measurable mapping such that at least one of mappings f^+ and f^- is λ-integrable. Let $\mu(A) = \int_A f^+ d\lambda - \int_A f^- d\lambda$. Then, μ is a σ-additive measure on $\mathcal{A}$ and $\mu^+(A) = \int_A f^+ d\lambda$, $\mu^-(A) = \int_A f^- d\lambda$.*

If $f \in \mathcal{L}^1(\lambda)$, then $\mu \in ca(\mathcal{A})$ and $\|\mu\| = |\mu|(X) = \int_X |f| d\lambda$.

Proof. Let $H = \{x \in X : f(x) \geq 0\}$; $H \in \mathcal{A}$, $f^+ \chi_H = f^+$ and $f^- \chi_H = 0$. $\int_A f^+ d\lambda = \int_A f^+ \chi_H d\lambda = \int_{A \cap H} f^+ d\lambda$ and $\int_{A \cap H} f^- d\lambda = \int_A f^- \chi_H d\lambda = 0$.

Therefore $\mu(A \cap H) = \int_A f^+ d\lambda$. If $B \subset A$, then $\mu(B) \leq \int_B f^+ d\lambda \leq \int_A f^+ d\lambda$. Then, according to Corollary 1.22, $\int_A f^+ d\lambda = \sup\{\mu(B) : B \in \mathcal{A}, B \subset A\} = \mu^+(A) = \mu(A \cap H)$, which leads to $\int_A f^- d\lambda = \mu^-(A) = -\mu(A \cap H^c)$.

If $f \in \mathcal{L}^1(\lambda)$, then $\mu(A) = \int_A f d\lambda$, for every $A \in \mathcal{A}$, hence $\mu \in ca(\mathcal{A})$. According to Definition 1.19, $\|\mu\| = |\mu|(X) = \mu^+(X) + \mu^-(X) = \int_X f^+ d\lambda + \int_X f^- d\lambda = \int_X |f| d\lambda$. □

Proposition 1.26.

(i) $L^1(\lambda) = L^1(|\lambda|)$ *and* $|\int_A f d\lambda| \le \int_A |f| d|\lambda|, \quad \forall A \in \mathcal{A}, \quad \forall f \in L^1(\lambda)$.

(ii) $|\lambda|(A) = \sup\{|\int_A f d\lambda| : f \in L^1(\lambda), |f| \le 1\}$.

(iii) *The mapping* $\|\cdot\|_1 : L^1(\lambda) \to \mathbb{R}_+, \|f\|_1 = \int_X |f| d|\lambda|$ *is a norm on* $L^1(\lambda)$ *and* $(L^1(\lambda), \|\cdot\|_1)$ *is a Banach space.*

Proof. (i) Let $f \in L^1(\lambda)$; then f is $\mathcal{A}$-measurable and $f \in L^1(\lambda^+) \cap L^1(\lambda^-)$. $\int_X |f| d|\lambda| = \int_X |f| d\lambda^+ + \int_X |f| d\lambda^- < +\infty$ and then $f \in L^1(|\lambda|)$.

Reciprocally, if $f \in L^1(|\lambda|), \int_X |f| d\lambda^+ \le \int_X |f| d|\lambda| < +\infty$ and so $f \in L^1(\lambda^+)$; similarly, $f \in L^1(\lambda^-)$ and, therefore, $f \in L^1(\lambda)$.

The inequality follows immediately:

$$\begin{aligned}\left|\int_A f d\lambda\right| &= \left|\int_A f d\lambda^+ - \int_A f d\lambda^-\right| \\ &\le \left|\int_A f d\lambda^+\right| + \left|\int_A f d\lambda^-\right| \le \int_A |f| d\lambda^+ + \int_A |f| d\lambda^- \\ &= \int_A |f| d|\lambda|.\end{aligned}$$

(ii)

$$|\lambda|(A) = \sup\left\{\sum_{i=1}^n |\lambda(A_i)| : n \in \mathbb{N}^* : \{A_1, \dots, A_n\} = \mathcal{A} - \text{partition of } A\right\}.$$

Let $\{A_1, \dots, A_n\}$ be an arbitrary partition of A. We can assume that $\lambda(A_i) \ge 0$, for every $i = 1, \dots, p$ and $\lambda(A_j) < 0$, for every $j = p+1, \dots, n$. Let $f = \chi_{\cup_1^p A_i} - \chi_{\cup_{p+1}^n A_j}$; then $|f| = 1$, f is $\mathcal{A}$-measurable and

$$\begin{aligned}\sum_1^n |\lambda(A_i)| &= \lambda(\cup_1^p A_i) - \lambda(\cup_{p+1}^n A_j) = \int_A f d\lambda \\ &\le \sup\left\{\left|\int_A f d\lambda\right| : f \in L^1(\lambda), |f| \le 1\right\}.\end{aligned}$$

We have therefore $|\lambda|(A) \le \sup\left\{|\int_A f d\lambda| : f \in L^1(\lambda), |f| \le 1\right\}$.

On the other hand, for every $f \in L^1(\lambda)$ with $|f| \le 1$, by (i), we have

$$\left|\int_A f d\lambda\right| \le \int_A |f| d|\lambda| \le \int_A 1 d|\lambda| = |\lambda|(A),$$

from where $\sup\{|\int_A f d\lambda| : f \in L^1(\lambda), |f| \le 1\} \le |\lambda|(A)$.

(iii) Obviously, $\|\cdot\|_1$ is a norm on $L^1(\lambda) = L^1(|\lambda|)$. It is easy to see that $(L^1(|\lambda|), \|\cdot\|_1)$ is a Banach space. □

In the following, we will present Saks' theorem which will be useful in the study of the weak compactness on $(ca(\mathcal{A}), \|\cdot\|)$ (see [62], Lemma III.7.1 and [57], Theorem 8, p. 86).

Theorem 1.27 (Saks). *Let $\lambda : \mathcal{A} \to \overline{\mathbb{R}}_+$ be a positive and σ-additive measure. The mapping $d_\lambda : \mathcal{A} \times \mathcal{A} \to \mathbb{R}_+$, defined by $d_\lambda(A, B) = \arctan(\lambda(A \triangle B))$, for every $(A, B) \in \mathcal{A} \times \mathcal{A}$, is a pseudo-metric on $\mathcal{A}$. The pseudo-metric space $(\mathcal{A}, d_\lambda)$ is complete and the binary operations $(A, B) \mapsto A \cup B$, $(A, B) \mapsto A \cap B$ and $(A, B) \mapsto A \setminus B$ are continuous maps on $(\mathcal{A} \times \mathcal{A}, d_\lambda \times d_\lambda)$.*

Proof. Let's recall that if $\lambda(A \triangle B)$ is finite, $d_\lambda(A, B) = \arctan(\lambda(A \triangle B)) \in [0, \frac{\pi}{2}[$ and if $\lambda(A \triangle B) = +\infty$, then $d_\lambda(A, B) = \frac{\pi}{2}$.

We can put $D_\lambda(A, B) = \lambda(A \triangle B)$; obviously $d_\lambda = \arctan D_\lambda$ is a pseudo-metric on $\mathcal{A}$. In the following we say that d_λ is *the pseudo-metric associated* to λ.

It remains to demonstrate that $(\mathcal{A}, d_\lambda)$ is complete. Let $(A_n)_{n\in\mathbb{N}}$ be a Cauchy sequence in $(\mathcal{A}, d_\lambda)$; for every $\varepsilon > 0$, there exists $n_0 \in \mathbb{N}$ such that, for all $m, n \geq n_0$, $D_\lambda(A_m, A_n) = \lambda(A_m \triangle A_n) < \varepsilon$.

Step by step, we define a strictly increasing sequence of integers $(k_n)_{n\in\mathbb{N}}$ such that, for every $n \in \mathbb{N}$, $\lambda(A_{k_n} \triangle A_{k_{n+1}}) < \frac{1}{2^n}$.

Let $N = \limsup_n (A_{k_n} \triangle A_{k_{n+1}}) = \bigcap_{n=0}^\infty \bigcup_{i=n}^\infty (A_{k_i} \triangle A_{k_{i+1}}) \in \mathcal{A}$; for every $n \in \mathbb{N}$, $N \subseteq \bigcup_{i=n}^\infty (A_{k_i} \triangle A_{k_{i+1}})$ and then $\lambda(N) \leq \sum_{i=n}^\infty \frac{1}{2^i} = \frac{1}{2^{n-1}}$. Therefore $\lambda(N) = 0$.

Let $A = \liminf_n A_{k_n} = \bigcup_{n=0}^\infty \bigcap_{i=n}^\infty A_{k_i} \in \mathcal{A}$; then

$$X \setminus N \subseteq A \cup \left[\liminf_n (X \setminus A_{k_n})\right]. \tag{1}$$

Indeed, for every $x \in X \setminus N = \bigcup_{n=0}^\infty \bigcap_{i=n}^\infty [X \setminus (A_{k_i} \triangle A_{k_{i+1}})]$, there exists $n_0 \in \mathbb{N}$ such that, for all $i \geq n_0$, $x \notin A_{k_i} \triangle A_{k_{i+1}}$, or $\chi_{A_{k_i}}(x) = \chi_{A_{k_{i+1}}}(x)$; therefore $\chi_{A_{k_i}}(x) = 1$, for every $i \geq n_0$, or $\chi_{A_{k_i}}(x) = 0$, for every $i \geq n_0$, from where $x \in (\liminf_n A_{k_n}) \cup [\liminf_n (X \setminus A_{k_n})]$.

From (1), we have

$$A_{k_p} \triangle A \subseteq N \cup \left[\liminf_n (A_{k_p} \triangle A_{k_n})\right], \quad \text{for every} \quad p \in \mathbb{N}. \tag{2}$$

Indeed, from (1), one gets that, for every $p \in \mathbb{N}$,

$$(A_{k_p} \setminus A) \setminus N \subseteq A_{k_p} \cap \left[\liminf_n (X \setminus A_{k_n})\right] = \liminf_n (A_{k_p} \setminus A_{k_n}) \quad \text{and}$$
$$(A \setminus A_{k_p}) \setminus N \subseteq \left(\liminf_n A_{k_n}\right) \setminus A_{k_p} = \liminf_n (A_{k_n} \setminus A_{k_p}).$$

From (2), we have,

$$
\begin{aligned}
\lambda(A_{k_p} \triangle A) &\le \lambda(\liminf_n (A_{k_p} \triangle A_{k_n})) \le \liminf_n \lambda(A_{k_p} \triangle A_{k_n}) \\
&\le \liminf_n \left[\lambda(A_{k_p} \triangle A_{k_{p+1}}) + \cdots + \lambda(A_{k_{n-1}} \triangle A_{k_n})\right] \\
&\le \liminf_n \left(\frac{1}{2^p} + \cdots + \frac{1}{2^{n-1}}\right) = \frac{1}{2^{p-1}}, \quad \text{for every} \quad p \in \mathbb{N}
\end{aligned}
$$

hence we obtain

$$
d_\lambda(A_{k_p}, A) \le \arctan\left(\frac{1}{2^{p-1}}\right), \quad \text{for every} \quad p \in \mathbb{N}. \tag{3}
$$

From (3), it results that $A_{k_p} \xrightarrow{d_\lambda} A$ and consequently $A_n \xrightarrow{d_\lambda} A$.

Therefore $(\mathcal{A}, d_\lambda)$ is complete.

Suppose now that $A_n \xrightarrow{d_\lambda} A$ and $B_n \xrightarrow{d_\lambda} B$.

From the following inclusions:

$$
\begin{aligned}
(A_n \cup B_n) \triangle (A \cup B) &\subseteq (A_n \triangle A) \cup (B_n \triangle B), \\
(A_n \cap B_n) \triangle (A \cap B) &\subseteq (A_n \triangle A) \cup (B_n \triangle B), \\
(A_n \setminus B_n) \triangle (A \setminus B) &\subseteq (A_n \triangle A) \cup (B_n \triangle B),
\end{aligned}
$$

we obtain

$$
\begin{aligned}
D_\lambda(A_n \cup B_n, A \cup B) &\le D_\lambda(A_n, A) + D_\lambda(B_n, B), \\
D_\lambda(A_n \cap B_n, A \cap B) &\le D_\lambda(A_n, A) + D_\lambda(B_n, B), \\
D_\lambda(A_n \setminus B_n, A \setminus B) &\le D_\lambda(A_n, A) + D_\lambda(B_n, B).
\end{aligned}
$$

The inequality $\arctan(x + y) \le \arctan x + \arctan y$ implies that

$$
\begin{aligned}
d_\lambda(A_n \cup B_n, A \cup B) &\le d_\lambda(A_n, A) + d_\lambda(B_n, B), \\
d_\lambda(A_n \cap B_n, A \cap B) &\le d_\lambda(A_n, A) + d_\lambda(B_n, B), \\
d_\lambda(A_n \setminus B_n, A \setminus B) &\le d_\lambda(A_n, A) + d_\lambda(B_n, B),
\end{aligned}
$$

from where it results that

$$
\begin{aligned}
(A_n \cup B_n) &\xrightarrow{d_\lambda} (A \cup B), \\
(A_n \cap B_n) &\xrightarrow{d_\lambda} (A \cap B), \\
(A_n \setminus B_n) &\xrightarrow{d_\lambda} (A \setminus B).
\end{aligned}
$$

which demonstrates the continuity of the applications $\cup$, $\cap$ and $\setminus$. □

Definition 1.28. Let $\lambda : \mathcal{A} \to \overline{\mathbb{R}}_+$ be an additive measure and let $\mu \in ba(\mathcal{A})$; we say that μ est *absolutely continuous* with respect to λ if, for every $A \in \mathcal{A}$,

$$\lambda(A) = 0 \Longrightarrow \mu(A) = 0.$$

We note this by $\mu \ll \lambda$.

Remark 1.29. Let $\lambda \in ba^+(\mathcal{A})$ and $\mu \in ba(\mathcal{A})$; then

$$\mu \ll \lambda \Leftrightarrow |\mu| \ll \lambda \Leftrightarrow \mu^+ \ll \lambda \quad \text{and} \quad \mu^- \ll \lambda.$$

Indeed, if $\mu \ll \lambda$, then, for every $A \in \mathcal{A}$ with $\lambda(A) = 0$ and for every $B \in \mathcal{A}$, $B \subseteq A$, we have $\lambda(B) = 0$ and so $\mu(B) = 0$. Then, by (i) of Theorem 1.17, $|\mu|(A) = 0$, from where $|\mu| \ll \lambda$. The implication $|\mu| \ll \lambda \Rightarrow \mu^+ \ll \lambda$ and $\mu^- \ll \lambda$ is obvious and, from $\mu^+ \ll \lambda, \mu^- \ll \lambda$ and $\mu = \mu^+ - \mu^-$, it results immediately that $\mu \ll \lambda$.

The following proposition shows that, for the real σ-additive measures, the property of a measure to be absolutely continuous with respect to another one is a property of continuity (see Definition III.4.12 and Lemma III.4.13 of [62]).

Proposition 1.30. *Let $\lambda : \mathcal{A} \to \overline{\mathbb{R}}_+$ be a σ-additive measure and let $\mu \in ca(\mathcal{A})$; then the following properties are equivalent:*

(i) *μ is absolutely continuous with respect to λ ($\mu \ll \lambda$),*

(ii) *for every $\varepsilon > 0$, there exists $\delta > 0$ such that, for every $A \in \mathcal{A}$ satisfying $\lambda(A) < \delta$, we have $|\mu|(A) < \varepsilon$,*

(iii) *$\mu : (\mathcal{A}, d_\lambda) \to \mathbb{R}$ is a d_λ-continuous function.*

Proof. (i) $\Longrightarrow$ (ii). Let us suppose that (ii) is not satisfied. There exist $\varepsilon > 0$ and $(A_n)_n \subseteq \mathcal{A}$ such that $\lambda(A_n) < \frac{1}{2^n}$ and $|\mu|(A_n) \geq \varepsilon$, for every $n \in \mathbb{N}$. Let $A = \limsup_n A_n = \bigcap_{n=1}^{\infty} \bigcup_{k=n}^{\infty} A_k \in \mathcal{A}$. Then, for every $n \in \mathbb{N}$, $|\mu|(\bigcup_{k=n}^{\infty} A_k) \geq \varepsilon$ and $\lambda(\bigcup_{k=n}^{\infty} A_k) \leq \sum_{k=n}^{\infty} \lambda(A_k) \leq \sum_{k=n}^{\infty} \frac{1}{2^n} = \frac{1}{2^{n-1}}$. We have then $\lambda(A) = \lim_n \lambda(\bigcup_{k=n}^{\infty} A_k) = 0$ and $|\mu|(A) = \lim_{n\to\infty} |\mu|(\bigcup_{k=n}^{\infty} A_k) \geq \varepsilon$. Therefore, (i) is not satisfied.

(ii) $\Longrightarrow$ (iii). According to (ii), for every $\varepsilon > 0$, there exists $\delta > 0$ such that, for every $B \in \mathcal{A}$ such that $\lambda(B) < \delta, |\mu|(B) < \varepsilon$ and, by continuity of the mapping $\tan$ in 0, there exists $\eta > 0$ such that $\arctan \lambda(B) < \eta$ implies $\lambda(B) < \delta$ and so $|\mu|(B) < \varepsilon$.

Let now $A \in \mathcal{A}$ and let $B \in \mathcal{A}$ with $d_\lambda(A, B) < \eta$; then $\lambda(A \triangle B) < \delta$ and so $|\mu|(A \triangle B) < \varepsilon$. Therefore

$$\begin{aligned}|\mu(A) - \mu(B)| &= |\mu(A) - \mu(A \cap B) + \mu(A \cap B) - \mu(B)| \\ &= |\mu(A \setminus B) - \mu(B \setminus A)| \leq |\mu(A \setminus B)| + |\mu(B \setminus A)| \\ &\leq |\mu|(A \triangle B) < \varepsilon\end{aligned}$$

and so μ is d_λ-continuous in A.

(iii) $\Longrightarrow$ (i). Since μ is continuous on $\mathcal{A}$, it is continuous at $\emptyset \in \mathcal{A}$. Then, for every $\varepsilon > 0$, there exists $\delta \in]0, 1[$ such that, for every $A \in \mathcal{A}$ satisfying $d_\lambda(A, \emptyset) = \lambda(A) < \delta, |\mu(A)| < \varepsilon$. Let $A \in \mathcal{A}$ with $\lambda(A) = 0$; then $d_\lambda(A, \emptyset) < \delta$ and hence $|\mu(A)| < \varepsilon$ and, as ε is arbitrary, $\mu(A) = 0$. Therefore $\mu \ll \lambda$. □

Proposition 1.31. *Let $\lambda : \mathcal{A} \to \overline{\mathbb{R}}_+$ be a σ-additive measure and let $\mu : \mathcal{A} \to \mathbb{R}$ be an additive measure. If μ is d_λ-continuous, then $\mu \in ca(\mathcal{A})$.*

Proof. Let $(A_n)_{n\in\mathbb{N}} \subseteq \mathcal{A}$ be a sequence of pairwise disjoint sets and let $A = \bigcup_{n=1}^\infty A_n \in \mathcal{A}$.

Since μ is d_λ-continuous, for every $\varepsilon > 0$, there exists $\delta > 0$ such that, for every B and C of $\mathcal{A}$ satisfying $\lambda(B \triangle C) < \delta, |\mu(B) - \mu(C)| < \varepsilon$. Since λ is σ-additive, there exists $n_0 \in \mathbb{N}$ such that, for every $n \geq n_0$,

$$\left|\lambda(A) - \sum_{k=1}^n \lambda(A_k)\right| = \lambda\left(A \setminus \bigcup_{k=1}^n A_k\right) = \lambda\left(A \triangle \bigcup_{k=1}^n A_k\right) < \delta.$$

Then $|\mu(A) - \sum_1^n \mu(A_k)| = |\mu(A) - \mu(\cup_1^n A_k)| < \varepsilon$ and so $\mu \in ca(\mathcal{A})$. □

Remark 1.32. The result of Proposition 1.30 asserts that, if μ is σ- additive, then the absolute continuity of μ with respect to λ is equivalent to the d_λ-continuity of μ; this result is no longer valid if μ is only additive.

In fact, let μ be as in the example of Remark 1.16; μ is additive, it is not bounded and, according to Corollary 1.14, its extension to $\mathcal{P}(\mathbb{N})$, still noted μ, is not σ - additive.

Let $\lambda : \mathcal{P}(\mathbb{N}) \to \mathbb{R}_+, \lambda(A) = \sum_{n\in A} \delta_n(A)$, where δ_n is the Dirac measure that gives to singleton set $\{n\}$ the measure 1. Then λ is a σ-additive measure and, as μ is not σ-additive, according to Proposition 1.31, μ is not d_λ- continuous. However μ is absolutely continuous with respect to λ. Indeed, let $A \in \mathcal{P}(\mathbb{N})$ with $\lambda(A) = 0$; then $A = \emptyset$ and therefore $\mu(A) = 0$.

In the case where $\mu \in ba(\mathcal{A}) \setminus ca(\mathcal{A})$, we have the following implications among the conditions of Proposition 1.30: (ii)$\Longleftrightarrow$(iii) $\Longrightarrow$(i).

In relation to Proposition 1.30, if λ is σ-additive, then in order to avoid the use of the difficult formulation of "d_λ-continuity" or the longer one "absolutely continuous with respect to λ", we will give the following definition:

Definition 1.33. A measure $\mu \in ca(\mathcal{A})$, continuous on the space $(\mathcal{A}, d_\lambda)$, is called *λ-continuous.*

We will note by $ca_\lambda(\mathcal{A})$ the subset of all λ-continuous measure of $ca(\mathcal{A})$.

We can find the following theorem in [57] (see Theorem 9, p. 87).

Theorem 1.34. *Let $\lambda : \mathcal{A} \to \overline{\mathbb{R}}_+$ be a σ-additive measure, let d_λ be the pseudo-metric associated and let $\mathcal{K}$ be a family of measures of $ca(\mathcal{A})$; then the following properties are equivalent:*

(i) *the family $\mathcal{K}$ is d_λ-equicontinuous at some $E \in \mathcal{A}$.*

(ii) *the family $\mathcal{K}$ is d_λ-equicontinuous at the point $\emptyset \in \mathcal{A}$.*

(iii) *the family $\mathcal{K}$ is uniformly d_λ-equicontinuous on $\mathcal{A}$.*

Each of these conditions entails the following:

(iv) *the family $\mathcal{K}$ is uniformly σ-additive.*

Proof. (i)$\Longrightarrow$(ii). Let $\mathcal{K}$ be d_λ-equicontinuous at $E \in \mathcal{A}$; then, for every $\varepsilon > 0$, there exists $\delta > 0$ such that, for every $A \in \mathcal{A}$ with $d_\lambda(A, E) < \delta$ and for every $\mu \in \mathcal{K}$, $|\mu(A) - \mu(E)| < \varepsilon$. Let $A \in \mathcal{A}$ such that $d_\lambda(A, \emptyset) < \delta$, that is $\lambda(A) < \eta = \tan(\delta)$; then

$$D_\lambda(A \cup E, E) = \lambda\left[(A \cup E) \triangle E\right] = \lambda(A \setminus E) \le \lambda(A) < \eta \quad \text{and}$$
$$D_\lambda(E \setminus A, E) = \lambda\left[(E \setminus A) \triangle E\right] = \lambda(A \cap E) \le \lambda(A) < \eta.$$

Therefore, for every $\mu \in \mathcal{K}$, $|\mu(A \cup E) - \mu(E)| < \varepsilon$ and $|\mu(E) - \mu(E \setminus A)| < \varepsilon$. We have then:

$$\begin{aligned} |\mu(A)| &= |\mu(A \cup E) - \mu(E \setminus A)| \\ &\le |\mu(A \cup E) - \mu(E)| + |\mu(E) - \mu(E \setminus A)| < 2\varepsilon. \end{aligned}$$

$\mathcal{K}$ is therefore d_λ-equicontinuous at $\emptyset$.

(ii) $\Longrightarrow$(iii). $\mathcal{K}$ being d_λ-equicontinuous at $\emptyset$, for every $\varepsilon > 0$, there exists $\delta > 0$ such that, for every $E \in \mathcal{A}$ satisfying $d_\lambda(E, \emptyset) = \arctan \lambda(E) < \delta$, we have $|\mu(E)| < \varepsilon$, for every $\mu \in \mathcal{K}$.

If $d_\lambda(C, D) = \arctan \lambda(C \triangle D) < \delta$ then $d_\lambda(C \setminus D, \emptyset) < \delta$ and $d_\lambda(D \setminus C, \emptyset) < \delta$.

Then, for every $\mu \in \mathcal{K}$,

$$|\mu(C) - \mu(D)| = |\mu(C \setminus D) + \mu(C \cap D) - \mu(C \cap D) - \mu(D \setminus C)|$$
$$\leq |\mu(C \setminus D)| + |\mu(D \setminus C)| < 2\varepsilon.$$

Therefore $\mathcal{K}$ is uniformly d_λ-equicontinuous on $\mathcal{A}$.

Obviously, (iii) $\Longrightarrow$(i).

(ii) $\Longrightarrow$(iv). Let $\mathcal{K} \subseteq ca(\mathcal{A})$ be a family satisfying (ii), let $(A_n)_{n\in\mathbb{N}} \subseteq \mathcal{A}$ be a sequence of pairwise disjoint sets and let $A = \cup_1^\infty A_n$. Then for every $\varepsilon > 0$, there exists $\delta > 0$ such that, for all $E \in \mathcal{A}$ with $\lambda(E) < \delta$, we have $|\mu(E)| < \varepsilon$, for every $\mu \in \mathcal{K}$. λ being σ-additive and positive, there exists $n_0 \in \mathbb{N}$ such that $|\lambda(A) - \sum_{k=1}^n \lambda(A_k)| = \lambda(A \setminus \cup_{k=1}^n A_k) < \delta$, for every $n \geq n_0$.

Then, for every $\mu \in \mathcal{K}$, $|\mu(A) - \sum_{k=1}^n \mu(A_k)| = |\mu(A \setminus \cup_{k=1}^n A_k)| < \varepsilon$ from where it results that $\mathcal{K}$ is uniformly σ-additive. □

In a consistent manner with Theorem 1.33, we give the following definition:

Definition 1.35. A family of measures $\mathcal{K} \subseteq ca(\mathcal{A})$, d_λ-equicontinuous at $\emptyset$ (and therefore on $\mathcal{A}$) is called λ-*equicontinuous.*

We need to emphasize that the definitions of λ-continuity and λ-equicontinuity refer only to the real σ-additive measures, meaning that they do not refer to the σ-additive measures taking at most one of the values $+\infty$ or $-\infty$. However, the previous results can be extended by replacing the σ-additive and positive measure λ by a σ-additive measure λ of finite total variation $|\lambda|$ and the measure $\mu \in ca(A)$ by a σ-additive measure with values in a Banach space.

We will now give a very important result of equicontinuity (Vitali–Hahn–Saks theorem) which allows us to establish the analogue of the uniform boundedness principle from Functional Analysis for the Measure Theory (see Theorem III.7.2 and Corollary III.7.3 in [62] or Theorem 2.53 of [85]).

Theorem 1.36 (Vitali–Hahn–Saks). *Let $\lambda : \mathcal{A} \to \overline{\mathbb{R}}_+$ be a σ-additive measure and let $(\mu_n)_{n\in\mathbb{N}} \subseteq ca(\mathcal{A})$ be a sequence of λ-continuous measures.*

Assume that $\lim_n \mu_n(A) = \mu(A) \in \mathbb{R}$ *exists, for every $A \in \mathcal{A}$; then:*

(i) $\{\mu_n : n \in \mathbb{N}\}$ *is λ-equicontinuous,*

(ii) $\mu \in ca(\mathcal{A})$,

(iii) μ *is λ-continuous.*

Proof. According to Proposition 1.30, $\mu_n : (\mathcal{A}, d_\lambda) \to \mathbb{R}$ is a continuous function, for every $n \in \mathbb{N}$.

For every $\varepsilon > 0$ and for all couple $(n, m) \in \mathbb{N} \times \mathbb{N}$, let's note

$$\mathcal{A}_{n,m}(\varepsilon) = \{A \in \mathcal{A} : |\mu_n(A) - \mu_m(A)| \leq \varepsilon\}.$$

$\mathcal{A}_{n,m}(\varepsilon)$ are closed sets in the complete space $(\mathcal{A}, d_\lambda)$; then, for every $p \in \mathbb{N}$,

$$\mathcal{A}_p(\varepsilon) = \bigcap_{m,n \geq p} \mathcal{A}_{n,m}(\varepsilon)$$

is a closed set in $(\mathcal{A}, d_\lambda)$.

Since $\lim_n \mu_n(A) \in \mathbb{R}$, for every $A \in \mathcal{A}$,

$$\mathcal{A} = \bigcup_{p=1}^{\infty} \mathcal{A}_p(\varepsilon).$$

According to Baire theorem, there exists $p_0 \in \mathbb{N}$ such that $\mathcal{A}_{p_0}(\varepsilon)$ has nonempty interior in $(\mathcal{A}, d_\lambda)$. Therefore, there exists $A_0 \in \mathcal{A}$, there exists $r > 0$ such that the ball $S(A_0, \arctan r) \subseteq \mathcal{A}_{p_0}(\varepsilon)$, i.e.,

$$|\mu_n(A) - \mu_m(A)| < \varepsilon, \quad \forall A \in \mathcal{A} \quad \text{with} \quad \lambda(A \triangle A_0) < r, \quad \forall m, n \geq p_0. \tag{1}$$

Since the set $\{\mu_1, \ldots, \mu_{p_0}\}$ is λ-equicontinuous, there exists $\delta \in]0, r[$ such that

$$|\mu_n(B)| < \varepsilon, \quad \text{for all} \quad B \in \mathcal{A} \quad \text{with} \quad \lambda(B) < \delta, \quad \forall n = 1, \ldots, p_0. \tag{2}$$

Let $A \in \mathcal{A}$ with $\lambda(A) < \delta$; then

$$\lambda((A \cup A_0) \triangle A_0) = \lambda(A \setminus A_0) \leq \lambda(A) < \delta < r \quad \text{and}$$
$$\lambda((A_0 \setminus A) \triangle A_0) = \lambda(A_0 \cap A) \leq \lambda(A) < \delta < r.$$

By (1), we have

$$|\mu_n(A \cup A_0) - \mu_{p_0}(A \cup A_0)| < \varepsilon, \quad \forall n \geq p_0 \tag{3}$$

and

$$|\mu_n(A_0 \setminus A) - \mu_{p_0}(A_0 \setminus A)| < \varepsilon, \quad \forall n \geq p_0. \tag{4}$$

By (2), (3) and (4), we deduct that, for every $n \geq p_0$,

$$\begin{aligned}|\mu_n(A)| &= |\mu_{p_0}(A) + [\mu_n(A) - \mu_{p_0}(A)]| \\ &\leq |\mu_{p_0}(A)| + |\mu_n(A \cup A_0) - \mu_{p_0}(A \cup A_0) + \mu_{p_0}(A_0 \setminus A) - \mu_n(A_0 \setminus A)| \\ &\leq |\mu_{p_0}(A)| + |\mu_n(A \cup A_0) - \mu_{p_0}(A \cup A_0)| + |\mu_{p_0}(A_0 \setminus A) - \mu_n(A_0 \setminus A)| \\ &< 3\varepsilon.\end{aligned}$$

Therefore, according to (2), $|\mu_n(A)| < 3\varepsilon$, for every $n \in \mathbb{N}$. $\{\mu_n : n \in \mathbb{N}\}$ is therefore d_λ-equicontinuous at $\emptyset$ and so it is λ-equicontinuous.

Since (μ_n) converges punctually to μ, μ is additive on $\mathcal{A}$.

We still need to show that μ is λ-continuous.

According to Theorem 1.34, since $\{\mu_n\}$ is λ-equicontinuous, it is uniformly d_λ-equicontinuous on $\mathcal{A}$. Therefore, for every $\varepsilon > 0$, there exists $\delta > 0$ such that, for all A and B of $\mathcal{A}$ with $\lambda(A \triangle B) < \delta$, for any $n \in \mathbb{N}$, $|\mu_n(A) - \mu_n(B)| < \varepsilon$. Now let n tend to ∞; so we obtain $|\mu(A) - \mu(B)| \leq \varepsilon$, from where it results that μ is uniformly - d_λ-continuous and so it is d_λ-continuous.

From Proposition 1.31, it results that $\mu \in ca(\mathcal{A})$ and, according to Proposition 1.30, μ is λ-continuous. □

The previous theorem accepts as corollary the following result (see Corollary III.7.4 and Theorem IV.9.8 of [62] and [57], p. 90):

Theorem 1.37 (Nikodym). *Let $(\mu_n) \subseteq ca(\mathcal{A})$ be a sequence of measures such that, for every $E \in \mathcal{A}$, there exists* $\lim_n \mu_n(E) = \mu(E) \in \mathbb{R}$. *Then:*

(i) $\mu \in ca(\mathcal{A})$,

(ii) $\{\mu_n : n \in \mathbb{N}\}$ *is uniformly σ-additive and*

(iii) $\{\mu_n : n \in \mathbb{N}\}$ *is bounded in the space $(ca(\mathcal{A}), \|\cdot\|)$.*

Proof. (i) + (ii) Let $\lambda : \mathcal{A} \to \mathbb{R}_+$ be defined by

$$\lambda(A) = \sum_{n=1}^{\infty} \frac{1}{2^n} \cdot \frac{|\mu_n|(A)}{1 + \|\mu_n\|}, \quad \forall A \in \mathcal{A}.$$

It is clear that, for all $A \in \mathcal{A}$, $\lambda(A) < \sum_{n=1}^{\infty} \frac{1}{2^n} = 1$.

Let $\{E_p : p \in \mathbb{N}\} \subseteq \mathcal{A}$ be a family of disjoint sets and let $E = \bigcup_{p=1}^{\infty} E_p$. For all $\varepsilon > 0$ let $n_0 \in \mathbb{N}$ such that $\frac{1}{2^{n_0-2}} < \varepsilon$. Since $\mu_n \in ca(\mathcal{A})$, $|\mu_n| \in ca(\mathcal{A})$ and then there exists $k_0 \in \mathbb{N}$ such that, for every $k \geq k_0$ and every $n = 1, \dots, n_0$,

$$\left| |\mu_n|(E) - \sum_{p=1}^{k} |\mu_n|(E_p) \right| < \frac{\varepsilon}{2}(1 + \|\mu_n\|).$$

Then

$$\left|\lambda(E)-\sum_{p=1}^{k}\lambda(E_p)\right| = \left|\sum_{n=1}^{\infty}\left(\frac{1}{2^n}\cdot\frac{|\mu_n|(E)}{1+\|\mu_n\|}-\sum_{p=1}^{k}\frac{1}{2^n}\cdot\frac{|\mu_n|(E_p)}{1+\|\mu_n\|}\right)\right|$$
$$\leq \sum_{n=1}^{n_0}\frac{1}{2^n}\left|\frac{|\mu_n|(E)}{1+\|\mu_n\|}-\sum_{p=1}^{k}\frac{|\mu_n|(E_p)}{1+\|\mu_n\|}\right| + \sum_{n=n_0+1}^{\infty}\frac{1}{2^n}\cdot 2$$
$$< \sum_{n=1}^{n_0}\frac{1}{2^n}\cdot\frac{\varepsilon}{2}+\frac{1}{2^{n_0-1}} < \frac{\varepsilon}{2}+\frac{\varepsilon}{2}=\varepsilon.$$

Therefore $\lambda \in ca^+(\mathcal{A})$.

Moreover, it is evident that, for every $n \in \mathbb{N}$, $\mu_n \ll \lambda$. We are, therefore, in the conditions to apply the Vitali–Hahn–Saks theorem. Therefore $\mu \in ca(\mathcal{A})$, $\{\mu_n : n \in \mathbb{N}\}$ is λ-equicontinuous and, according to (iv) of Theorem 1.34, $\{\mu_n : n \in \mathbb{N}\}$ is uniformly σ-additive.

(iii) As, for every $A \in \mathcal{A}$, $\mu_n(A) \to \mu(A) \in \mathbb{R}$, we have:

$$\sup_n |\mu_n(A)| < +\infty \quad \text{for every} \quad A \in \mathcal{A}. \tag{1}$$

Suppose that the family $\{\mu_n : n \in \mathbb{N}\}$ is not bounded in the space $(ca(\mathcal{A}), \|\cdot\|)$. Then $\sup_{n\in\mathbb{N}} \|\mu_n\| = +\infty$.

According to Theorem 1.23, for every $n \in \mathbb{N}$, $\|\mu_n\| \leq 2\|\mu_n\|_\infty$.

We have therefore

$$\sup_n \|\mu_n\|_\infty = \sup_{n\in\mathbb{N}}(\sup_{A\in\mathcal{A}} |\mu_n(A)|) = +\infty. \tag{2}$$

By (1), $\sup_k |\mu_k(X)| < +\infty$. Let $\Sigma_1 = \sup_k |\mu_k(X)| + 1$; according to (2), there exists $n_1 \in \mathbb{N}$ such that

$$\sup_{A\in\mathcal{A}} |\mu_{n_1}(A)| > \Sigma_1$$

and therefore there exists $A_1 \in \mathcal{A}$ such that $|\mu_{n_1}(A_1)| > \Sigma_1$.

$$|\mu_{n_1}(X \setminus A_1)| = |\mu_{n_1}(X) - \mu_{n_1}(A_1)|$$
$$\geq |\mu_{n_1}(A_1)| - |\mu_{n_1}(X)| \geq |\mu_{n_1}(A_1)| - \sup_k |\mu_k(X)| > 1.$$

Let's note $B_1 = X \setminus A_1$. We have obtained an $\mathcal{A}$- partition (A_1, B_1) of X such that

$$|\mu_{n_1}(A_1)| \geq 1, \quad |\mu_{n_1}(B_1)| \geq 1. \tag{3}$$

By (2), we have that

$$\sup_{n\in\mathbb{N}}(\sup_{A\in\mathcal{A}}|\mu_n(A\cap A_1)|) = +\infty. \tag{4}$$

or

$$\sup_{n\in\mathbb{N}}(\sup_{A\in\mathcal{A}}|\mu_n(A\cap B_1)|) = +\infty. \tag{5}$$

If (4) is satisfied, then we note $C_1 = B_1$ (otherwise $C_1 = A_1$). Because all finite subset of $ca(\mathcal{A})$ is uniformly bounded on $\mathcal{A}$ (this is an immediate consequence of Corollary 1.14), we have

$$\sup_{n>n_1}(\sup_{A\in\mathcal{A}}|\mu_n(A\cap A_1)|) = +\infty.$$

Can one restart this procedure by applying (1) to A_1.

Let $\Sigma_2 = \sup_k|\mu_k(A_1)|+2$; there exist $n_2 > n_1$ and an $\mathcal{A}$ - partition (A_2, B_2) of A_1 such that

$$|\mu_{n_2}(A_2)| \geq 2, \quad |\mu_{n_2}(B_2)| \geq 2$$

and

$$\sup_{n>n_2}(\sup_{A\in\mathcal{A}}|\mu_n(A\cap A_2)|) = +\infty. \tag{6}$$

or

$$\sup_{n>n_2}(\sup_{A\in\mathcal{A}}|\mu_n(A\cap B_2)|) = +\infty. \tag{7}$$

If (6) is satisfied, we note $C_2 = B_2$ (otherwise $C_2 = A_2$). $C_2 = B_2 \subseteq X \setminus C_1$.

Continuing in this fashion, we define a strictly increasing sequence of integers $(n_p)_{p\in\mathbb{N}}$ tending to infinity and a sequence of pairwise disjoint sets $(C_p)_{p\in\mathbb{N}} \subseteq \mathcal{A}$ such that,

$$|\mu_{n_p}(C_p)| \geq p, \quad \text{for every} \quad p \in \mathbb{N}. \tag{8}$$

Let $C = \bigcup_1^\infty C_p \in \mathcal{A}$; by (ii), (μ_n) are uniformly σ-additive. Therefore, for $\varepsilon = 1$, there exists $k_0 \in \mathbb{N}$ such that

$$|\mu_n(C) - \sum_{i=1}^{k}\mu_n(C_i)| < 1, \quad \text{for every} \quad k \geq k_0 \quad \text{and for all} \quad n \in \mathbb{N}.$$

Then, for $p > k_0$ and for all $n \in \mathbb{N}$,

$$\begin{aligned}|\mu_n(C_p)| &= \left|\sum_{i=1}^{p} \mu_n(C_i) - \sum_{i=1}^{p-1} \mu_n(C_i)\right| \\ &\le \left|\sum_{i=1}^{p} \mu_n(C_i) - \mu_n(C)\right| + \left|\mu_n(C) - \sum_{i=1}^{p-1} \mu_n(C_i)\right| < 2,\end{aligned}$$

which is incompatible with (8).

Therefore, $\sup_{n\in\mathbb{N}} \|\mu_n\| < +\infty$ and (iii) is proved. □

We will now give Nikodym's theorem on uniform boundedness; the proof follows that of [57], p. 80.

Theorem 1.38 (Uniform boundedness Nikodym's theorem). *Suppose that $(\mu_n) \subseteq ca(\mathcal{A})$ is a sequence of measures such that*

$$sup_n|\mu_n(A)| < +\infty, \quad \textit{for every} \quad A \in \mathcal{A}.$$

Then $\{\mu_n : n \in \mathbb{N}\}$ is bounded in the space $(ca(\mathcal{A}), \|\cdot\|)$.

Proof. We suppose that $\{\mu_n : n \in \mathbb{N}\}$ is unbounded in $(ca(\mathcal{A}), \|\cdot\|)$. Since the norms $\|\cdot\|$ and $\|\cdot\|_\infty$ are equivalent on $ca(\mathcal{A})$ (cf. Theorem 1.23), this means that

$$\sup_{n\in\mathbb{N}}(\sup_{A\in\mathcal{A}} |\mu_n(A)|) = +\infty. \tag{1}$$

By hypothesis

$$\sup_n |\mu_n(X)| < +\infty. \tag{2}$$

Then, for all $\rho > 0$, there exist $n_\rho \in \mathbb{N}$ and $A_\rho \in \mathcal{A}$ such that $|\mu_{n_\rho}(A_\rho)| > \rho$. We also have

$$\begin{aligned}|\mu_{n_\rho}(X \setminus A_\rho)| &= |\mu_{n_\rho}(X) - \mu_{n_\rho}(A_\rho)| \\ &\ge |\mu_{n_\rho}(A_\rho)| - |\mu_{n_\rho}(X)| \ge |\mu_{n_\rho}(A_\rho)| - \sup_k |\mu_k(X)|.\end{aligned}$$

Therefore, if $\rho = \sup_n |\mu_n(X)| + \alpha$, then we have an $\mathcal{A}$-partition of X: $(A_\rho, B_\rho = X \setminus A_\rho)$ such that

$$|\mu_{n_\rho}(A_\rho)| \ge \alpha, \quad |\mu_{n_\rho}(B_\rho)| \ge \alpha. \tag{3}$$

Let us take $\alpha = 2$. According to (1), we have either

$$\sup_{n\in\mathbb{N}}(\sup_{A\in\mathcal{A}} |\mu_n(A \cap A_\rho)|) = +\infty, \quad \text{or} \tag{4}$$

$$\sup_{n\in\mathbb{N}}(\sup_{A\in\mathcal{A}} |\mu_n(A \cap B_\rho)|) = +\infty. \tag{5}$$

Suppose that (4) were achieved and let us note $n_1 = n_\rho$, $A_1 = A_\rho$ and $B_1 = B_\rho$ (otherwise, we note $A_1 = B_\rho$ and $B_1 = A_\rho$).

We restart the operation by replacing (1) with (4), X by A_1, ρ and α by $\rho_2 = \sup |\mu_n(A_1)| + \alpha_2$ where $\alpha_2 = 3 + \sup_n |\mu_n(B_1)|$.

Then, there exist $n_2 > n_1$ and $A_{\rho_2} \in \mathcal{A}$, $A_{\rho_2} \subseteq A_1$ such that $|\mu_{n_2}(A_{\rho_2})| \geq \rho_2$. We have then $|\mu_{n_2}(A_1 \setminus A_{\rho_2})| \geq |\mu_{n_2}(A_{\rho_2})| - |\mu_{n_2}(A_1)| \geq \rho_2 - \sup|\mu_n(A_1)| = \alpha_2$. Since $\alpha_2 \geq 3 + |\mu_{n_2}(B_1)|$, we obtain an $\mathcal{A}$-partition $(A_{\rho_2}, A_1 \setminus A_{\rho_2})$ of A_1 such that

$$|\mu_{n_2}(A_{\rho_2})| \geq 3 + |\mu_{n_2}(B_1)| \quad \text{and} \quad |\mu_{n_2}(A_1 \setminus A_{\rho_2})| \geq 3 + |\mu_{n_2}(B_1)|. \tag{6}$$

As previously, we note $A_2 = A_{\rho_2}$ if $\sup_{n\in\mathbb{N}}(\sup_{A\in\mathcal{A}} |\mu_n(A \cap A_{\rho_2})|) = +\infty$ and $B_2 = A_1 \setminus A_2$ (otherwise, we note $A_2 = A_1 \setminus A_{\rho_2}$).

Repeating, we thus define a strictly increasing sequence $(n_p)_{p\in N^*} \subseteq \mathbb{N}$ and a sequence of pairwise disjoint sets $(B_p)_{p\in N^*} \subseteq \mathcal{A}$ such that

$$|\mu_{n_p}(B_p)| > \sum_{k=1}^{p-1} |\mu_{n_p}(B_k)| + p + 1, \quad \text{for every} \quad p \geq 2, \text{ and} \tag{7}$$

$$|\mu_{n_1}(B_1)| > 2 = 1 + 1.$$

Let now $\{N_k, k \in N^*\}$ be a countable partition of infinite subsets of N^*. Because

$$\sum_{k=1}^{\infty} |\mu_{n_1}| \left(\bigcup_{p\in N_k} B_p \right) \leq |\mu_{n_1}| \left(\bigcup_{p\in N^*} B_p \right) \leq |\mu_{n_1}|(X) < +\infty,$$

there exists k_1 such that $|\mu_{n_1}|(\bigcup_{p\in N_{k_1}} B_p) < 1$; if p_1 is the smallest integer of N_{k_1} then $n_{p_1} > n_1$. We note C_1 the first set of the sequence $(B_p)_{p\in N_{k_1}}$; we have $C_1 = B_{p_1}$. By replacing N^* with N_{k_1}, $\bigcup_{p\in N^*} B_p$ by $\bigcup_{p\in N_{k_1}} B_p$ and μ_{n_1} by $\mu_{n_{p_1}}$, we define a second subsequence $(B_p)_{p\in N_{k_2}}$ extracted from the sequence $(B_p)_{p\in N_{k_1}}$ such that $|\mu_{n_{p_1}}|(\bigcup_{p\in N_{k_2}} B_p) < 1$ and, if we note by C_2

the first set of the subsequence $(B_p)_{p\in N_{k_2}}$, by p_2 the smallest integer of N_{k_2} then $n_{p_2} > n_{p_1} > n_1$ and $C_2 = B_{p_2}$. By reiterating the process, we define a decreasing sequence of infinite subsets of N^*, $(N_{k_j})_{j\in N^*}$, a strictly increasing sequence of integers $(p_j)_{j\in\mathbb{N}^*}$ and a sequence of pairwise disjoint sets of $\mathcal{A}$, $(C_j)_{j\in\mathbb{N}^*}$, such that, for all $j > 2$,

$$|\mu_{n_{p_{j-1}}}|\left(\bigcup_{p\in\mathbb{N}_{k_j}} B_p\right) < 1. \tag{8}$$

If we note $C = \bigcup_{j\in N^*} C_j$, then we have from (8),

$$|\mu_{n_{p_j}}|\left(\bigcup_{h=j+1}^{\infty} C_h\right) \le |\mu_{n_{p_j}}|\left(\bigcup_{p\in N_{k_{j+1}}} B_p\right) < 1, \quad \forall j \in \mathbb{N}$$

and by (7) and by inequality $p_{j-1} \le p_j - 1$,

$$|\mu_{n_{p_j}}(C_j)| = |\mu_{n_{p_j}}(B_{p_j})| > \sum_{k=1}^{p_j-1} |\mu_{n_{p_j}}(B_k)| + p_j + 1 > \sum_{h=0}^{j-1} |\mu_{n_{p_j}}(C_h)| + p_j + 1$$

from where

$$|\mu_{n_{p_j}}(C_j)| - \sum_{h=0}^{j-1} |\mu_{n_{p_j}}(C_h)| > p_j + 1 > j + 1.$$

Therefore, for every $j \in \mathbb{N}$,

$$|\mu_{n_{p_j}}(C)| \ge |\mu_{n_{p_j}}(C_j)| - \left|\mu_{n_{p_j}}\left(\bigcup_{h=0}^{j-1} C_h\right)\right| - \left|\mu_{n_{p_j}}\left(\bigcup_{h=j+1}^{\infty} C_h\right)\right| > j,$$

what implies that $\sup_{j\in N^*} |\mu_{n_{p_j}}(C)| = +\infty$ and this is incompatible with the hypothesis $\sup_{n\in N^*} |\mu_n(C)| < +\infty$. □

Remark 1.39.

(i) In [62], we can find a different proof of this theorem based on the use of Baire's theorem.

(ii) Nikodym's Theorem 1.38 is not true if $\mathcal{A}$ is only an algebra and not a σ-algebra of subsets of X, as shown by the example given in Remark 1.16: let $\mathcal{A} = \{A \subseteq \mathbb{N} : A \text{ finite or } \mathbb{N} \setminus A \text{ finite }\}$, let δ_n be the Dirac measure supported on the singleton $\{n\}$ and let $\mu_n(A) = n[\delta_{n+1}(A) - \delta_n(A)]$, for every $A \in \mathcal{A}$. Then, for every $A \in \mathcal{A}$, $\sup_n |\mu_n(A)| < \infty$ and $\sup_n \sup_{A \in \mathcal{A}} |\mu_n(A)| = +\infty$.

1.2 Radon–Nikodym Theorem. The Dual of L^1

Firstly, we will present the Radon–Nikodym's theorem to identify the absolutely continuous measures with respect to a positive measure and then we will give the Lebesgue's decomposition theorem. We will conclude this part with the theorem of isomorphism between $L^\infty(\lambda)$ and the dual space of $L^1(\lambda)$.

Let $\mathcal{A}$ be a σ-algebra on X, let $\lambda : \mathcal{A} \to \overline{\mathbb{R}}_+$ be a σ-additive and σ-finite positive measure and let $(L^1(\lambda), \|\cdot\|_1)$ be the Banach space of all integrable mappings with respect to λ provided with norm $\|\cdot\|_1 : \|f\|_1 = \int_X |f| d\lambda$.

Let $\psi : \mathcal{L}^1(\lambda) \to ca(\mathcal{A})$ be the function defined by $\psi(f) = \mu_f$, where, for all $E \in \mathcal{A}$, $\mu_f(E) = \int_E f d\lambda$. Then ψ is a linear application.

For every $f \in \mathcal{L}^1(\lambda)$, let

$$A_- = \{x \in X : f(x) < 0\} \quad \text{and}$$
$$A_+ = \{x \in X : f(x) \geq 0\}.$$

Then $A_-, A_+ \in \mathcal{A}$ and, for every $E \in \mathcal{A}$,

$$\mu_f(E) = \int_{E \cap A_-} f d\lambda + \int_{E \cap A_+} f d\lambda \leq \int_{A_+} f d\lambda = \mu_f(A_+).$$

Therefore $\mu_f(A_+) = \sup\{\mu_f(E) : E \in \mathcal{A}\}$.

Similarly, $\mu_f(A_-) = \inf\{\mu_f(E) : E \in \mathcal{A}\}$.

According to Theorem 1.13, Corollary 1.22 or Proposition 1.25,

$$\mu_f^+(E) = \mu_f(E \cap A_+) = \int_{E \cap A_+} f d\lambda = \int_E f^+ d\lambda \quad \text{and}$$

$$\mu_f^-(E) = -\mu_f(E \cap A_-) = -\int_{E \cap A_-} f d\lambda = \int_E f^- d\lambda,$$

where $f^+ = \sup\{f, 0\}$ and $f^- = \sup\{-f, 0\}$.

It results that

$$|\mu_f|(E) = \mu_f^+(E) + \mu_f^-(E) = \int_E (f^+ + f^-)d\lambda = \int_E |f|d\lambda$$

from where

$$\|\psi(f)\| = \|\mu_f\| = |\mu_f|(X) = \int_X |f|d\lambda = \|f\|_1, \quad \text{for all} \quad f \in \mathcal{L}^1(\lambda).$$

Due to the absolute continuity of integral, for all $f \in \mathcal{L}^1(\lambda)$, for every $\varepsilon > 0$, there exists $\delta > 0$ such that, for all $E \in \mathcal{A}$ with $\lambda(E) < \delta$, we have $\int_E |f|d\lambda < \varepsilon$ or $|\mu_f|(E) < \varepsilon$. According to Proposition 1.30 (ii), it results that $\mu_f \ll \lambda$.

Let $f_1, f_2 \in [f]$, where $[f]$ is the equivalence class of function $f \in \mathcal{L}^1(\lambda)$; it is easy to see that, for all $E \in \mathcal{A}$, $\mu_{f_1}(E) = \mu_{f_2}(E)$, $\mu_{f_1}^+(E) = \mu_{f_2}^+(E) = \int_E f_1{}^+d\lambda = \int_E f_2^+d\lambda$ and $|\mu_{f_1}|(E) = \mu_{f_1}^+(E) + \mu_{f_1}^-(E) = \mu_{f_2}^+(E) + \mu_{f_2}^-(E) = |\mu_{f_2}|(E) = \int_E |f|d\lambda$.

Therefore, we can define $\Psi : L^1(\lambda) \to ca(\mathcal{A})$ by $\Psi([f]) = \psi(f) = \mu_f$. Ψ embeds the space $L^1(\lambda)$ into the subspace $ca_\lambda(\mathcal{A}) \subseteq ca(\mathcal{A})$ of all σ-additive measures, absolutely continuous with respect to λ.

The following theorem, allows us to show that $\Psi(L^1(\lambda)) = ca_\lambda(\mathcal{A})$; so that Ψ is an isomorphism between $(L^1(\lambda), \|\cdot\|_1)$ and $(ca_\lambda(\mathcal{A}), \|\cdot\|)$, where $\|\cdot\|$ is the norm defined in Theorem 1.23 (for Radon–Nikodym theorem, may consult references [62], Theorem III.10.2 and [85], Theorems 1.101 and 1.111).

Theorem 1.40 (Radon–Nikodym's theorem). *Let $\lambda : \mathcal{A} \to \overline{\mathbb{R}}_+$ be a σ-additive and σ-finite positive measure and let μ be a σ-additive, σ-finite and λ-continuous measure on $\mathcal{A}$.*

There exists an $\mathcal{A}$-measurable application $f : X \to \mathbb{R}$ such that

$$\mu(A) = \int_A f d\lambda, \quad \textit{for every} \quad A \in \mathcal{A}.$$

The function f is unique modulo the equality λ-almost everywhere.

In this statement, it is convenient to remark that, according to notations in Proposition 1.25, $\int_A f d\lambda = \int_A f^+ d\lambda - \int_A f^- d\lambda$ shows that at least one of the functions f^+ or f^- is integrable on $\mathcal{A}$. The proof from below follows the ideas of [93] (see Theorems A and B, sect. 31, p. 128); it will be achieved in several stages and require the following lemma:

Lemma 1.41. *Let λ and μ be two σ-additive positive measures defined on $\mathcal{A}$; let us suppose that λ is finite, that μ is λ-continuous and $\mu(X) > 0$.*

There exist $\varepsilon > 0$ and $A \in \mathcal{A}$ such that $\lambda(A) > 0$ and $(\mu - \varepsilon\lambda)^+$ is concentrated on the set A.

Proof. Suppose that, for every $n \in \mathbb{N}^*$, $(\mu - \frac{1}{n} \cdot \lambda)^+$ is concentrated on the set $A_n \in \mathcal{A}$ and let $D = \bigcup_{n \in N} A_n$. For every $n \in \mathbb{N}^*$, $D^c \subset A_n^c$; therefore, $(\mu - \frac{1}{n} \cdot \lambda)^+(D^c) = 0$ and then $(\mu - \frac{1}{n} \cdot \lambda)(D^c) \leq 0$. Therefore

$$0 \leq \mu(D^c) \leq \frac{1}{n} \cdot \lambda(D^c), \quad \text{for every} \quad n \in \mathbb{N}^*.$$

Because $\lambda(D^c) < +\infty$, $\mu(D^c) = 0$. Since $\mu(X) > 0$, $\mu(D) > 0$ and, as $\mu \ll \lambda$, $\lambda(D) > 0$. So, there is at least one $n \in \mathbb{N}^*$ such that $\lambda(A_n) > 0$. $\varepsilon = \frac{1}{n}$ and A_n satisfy then the conditions of lemma.

Proof of theorem. (a) Firstly, we suppose that λ and μ are positive and finite. Let

$$\mathcal{F} = \left\{ f : X \to \mathbb{R}_+ : f \in \mathcal{L}^1(\lambda) \text{ and } \int_A f d\lambda \leq \mu(A), \text{ for every } A \in \mathcal{A} \right\}.$$

Let $\alpha = \sup \left\{ \int_X f d\lambda : f \in \mathcal{F} \right\}$. Because μ is finite, $\alpha \leq \mu(X) < +\infty$.
Let $(f_n)_{n \in \mathbb{N}^*} \subseteq \mathcal{F}$ such that $\sup_n \int_X f_n d\lambda = \alpha$ and let $\bar{f} = \sup_n f_n$.
For every $n \in \mathbb{N}^*$, let $h_n = \sup\{f_1, f_2, \cdots, f_n\} \in \mathcal{L}^1(\lambda)$; then $h_n \uparrow \bar{f}$.
According to Lebesgue monotone convergence theorem,

$$\int_A \bar{f} d\lambda = \lim_n \int_A h_n d\lambda, \quad \text{for every} \quad A \in \mathcal{A} \quad \text{and therefore} \tag{1}$$

$$\int_X \bar{f} d\lambda \geq \sup_n \int_X f_n d\lambda = \alpha < +\infty. \tag{2}$$

Now, we show that $\bar{f} \in \mathcal{F}$.
Let A be arbitrary in $\mathcal{A}$ and let $n \in \mathbb{N}^*$; we note

$A_1 = \{x \in A : h_n(x) = f_1(x)\}$
$A_2 = \{x \in A : h_n(x) = f_2(x) > f_1(x)\}$
$\vdots$
$A_n = \{x \in A : h_n(x) = f_n(x) > f_i(x) \quad \text{for every} \quad i < n\}$

The sets $(A_i)_{i \leq n} \subseteq \mathcal{A}$ are disjoint and $\cup_{i \leq n} A_i = A$. So we have:

$$\begin{aligned} \int_A h_n d\lambda &= \int_{A_1} f_1 d\lambda + \int_{A_2} f_2 d\lambda + \cdots + \int_{A_n} f_n d\lambda \\ &\leq \mu(A_1) + \mu(A_2) + \cdots + \mu(A_n) \leq \mu(A). \end{aligned}$$

Then $h_n \in \mathcal{F}$, for every $n \in \mathbb{N}^*$. According to (1)

$$\int_A \bar{f} d\lambda = \lim_{n \to \infty} \int_A h_n d\lambda \leq \mu(A). \tag{3}$$

By (3), $\int_X \bar{f}\, d\lambda \le \mu(X) < +\infty$. Therefore $\bar{f}$ is λ-integrable and, using again (3), $\bar{f} \in \mathcal{F}$. Then

$$\int_X \bar{f}\, d\lambda \le \alpha. \tag{4}$$

By (2) and (4),

$$\int_X \bar{f}\, d\lambda = \alpha. \tag{5}$$

Now, we will show that $\bar{f}$ is the solution to the stated problem.

Let $\nu = \mu - \bar{f} \cdot \lambda$, where $\bar{f} \cdot \lambda$ indicates the positive measure which, to any $E \in \mathcal{A}$, associates $\int_E \bar{f}\, d\lambda$. Since $\bar{f} \in \mathcal{F}$, ν is a measure positive, finite and λ-continuous. Therefore, according to the previous lemma, if $\nu(X) > 0$, then there exist $\varepsilon > 0$ and $A \in \mathcal{A}$ such that $\lambda(A) > 0$ and $(\nu - \varepsilon\lambda)^+$ is concentrated on A.

Since $(\nu - \varepsilon\lambda)^-$ is concentrated on A^c, for every $B \in \mathcal{A}$,

$$0 \le (\nu - \varepsilon\lambda)^+(B) = (\nu - \varepsilon\lambda)^+(B \cap A) = (\nu - \varepsilon\lambda)(A \cap B).$$

Therefore we have

$$\varepsilon\lambda(A \cap B) \le \nu(A \cap B) = \mu(A \cap B) - \int_{A\cap B} \bar{f}\, d\lambda, \quad \text{for every} \quad B \in \mathcal{A}. \tag{6}$$

The function $g = \bar{f} + \varepsilon\chi_A$ is λ-integrable and, by (6),

$$\begin{aligned}\int_B g\, d\lambda &= \int_B \bar{f}\, d\lambda + \varepsilon\lambda(A \cap B) = \int_{A\cap B} \bar{f}\, d\lambda + \varepsilon\lambda(A \cap B) + \int_{B\setminus A} \bar{f}\, d\lambda \\ &\le \mu(A \cap B) + \mu(B \setminus A) = \mu(B), \quad \text{for every} \quad B \in \mathcal{A}.\end{aligned}$$

Consequently, $g \in \mathcal{F}$ and then $\int_X g\, d\lambda \le \alpha$. On the other hand, according to (5),

$$\int_X g\, d\lambda = \int_X \bar{f}\, d\lambda + \varepsilon\lambda(A) = \alpha + \varepsilon\lambda(A) > \alpha$$

which is impossible with the choice of $\bar{f}$.

Therefore $\nu(X) = 0$ and so $\mu = \bar{f} \cdot \lambda$.

The uniqueness of f is easy to demonstrate. Let us suppose that, for every $A \in \mathcal{A}$, $\mu(A) = \int_A f_1\, d\lambda = \int_A f_2\, d\lambda$ and let $B = \{x \in X : f_1(x) > f_2(x)\}$. Then $\int_B (f_1 - f_2)\, d\lambda = 0$ and so $\lambda(B) = 0$. It means that $f_1 \le f_2$ λ-a.e. Similarly, $f_2 \le f_1$ a.e. and then $f_1 = f_2$, λ-almost everywhere.

The theorem is therefore demonstrated in the case (a).

(b) Let now λ and μ be positive and σ-finite. There exists an $\mathcal{A}$-partition $(X_n)_{n\in\mathbb{N}}$ of X such that $\lambda(X_n) < +\infty$ and $\mu(X_n) < +\infty$, for every $n \in \mathbb{N}$. For every n, let $\mathcal{A}_n = \{A \cap X_n : A \in \mathcal{A}\}$, $\lambda_n(A \cap X_n) = \lambda(A \cap X_n)$ and $\mu_n(A \cap X_n) = \mu(A \cap X_n)$. Then $\mu_n \ll \lambda_n$, for every $n \in \mathbb{N}$. Therefore, by (a), there exists $f_n \in \mathcal{L}^1_+(X_n, \mathcal{A}_n, \lambda_n)$ such that $\mu_n = f_n \cdot \lambda_n$.

We define $f : X \to \mathbb{R}_+$ letting $f(x) = f_n(x)$, if $x \in X_n$. Then, f is $\mathcal{A}$-measurable and positive and, for every $A \in \mathcal{A}$,

$$\int_A f d\lambda = \sum_{n\in\mathbb{N}} \int_{A\cap X_n} f d\lambda = \sum_{n\in\mathbb{N}} \int_{A\cap X_n} f_n d\lambda_n = \sum_{n\in\mathbb{N}} \mu_n(A_n) = \mu(A).$$

The proof of the uniqueness of f is identical to that of (a).

(c) Now let λ be a positive σ-finite measure and let μ be σ-finite signed measure with $\mu \ll \lambda$. Then $\mu = \mu^+ - \mu^-$, where μ^+ and μ^- are σ-finite. According to Remark 1.29, μ^+ and μ^- satisfy the hypothesis of (b). Therefore, there exist two $\mathcal{A}$-measurable mappings g and h such that

$$\mu^+(A) = \int_A g d\lambda \quad \text{and} \quad \mu^-(A) = \int_A h d\lambda, \quad \text{for every} \quad A \in \mathcal{A}.$$

Since μ can take only one of the values $+\infty$ or $-\infty$, at least one of the mappings g or h is λ-integrable. If $f = g - h$, then we have

$$\mu(A) = \int_A f d\lambda, \text{ for every } A \in \mathcal{A}. \qquad \square$$

Definition 1.42. The unique function f for which $\mu(E) = \int_E f d\lambda$, for every $E \in \mathcal{A}$ is called *Radon–Nikodym derivative* of μ with respect to λ and it is denoted by $f = \frac{d\mu}{d\lambda}$. One will also write $\mu = f \cdot \lambda$.

In the sequel, in this section, we will give some important consequences of Radon–Nikodym theorem. If we restrict ourselves to integrable functions, we have the already announced isomorphism theorem:

Theorem 1.43. *Let λ be a σ-additive, σ-finite and positive measure. Then, the mapping $\Psi : L^1(\lambda) \to ca(\mathcal{A})$, defined by $\Psi(f) = \mu = f \cdot \lambda$, for every $f \in L^1(\lambda)$, is an isometric isomorphism between the normed spaces $L^1(\lambda)$ and $ca_\lambda(\mathcal{A})$.*

Proof. Obviously, Ψ is a bijective linear map. According to Radon–Nikodym theorem, $\Psi(L^1(\lambda)) = ca_\lambda(\mathcal{A})$. Moreover, according to Proposition 1.25,

$$\|f\|_1 = \int_X |f| d\lambda = |\mu|(X) = \|\mu\| = \|\Psi(f)\|, \quad \text{for every } f \in L^1(\lambda). \qquad \square$$

Example 1.44. Let $\lambda : \mathcal{P}(\mathbb{N}) \to \overline{\mathbb{R}}_+, \lambda(A) = \sum_{n \in A} \delta_n(A)$, be the measure given in Remark 1.32. Then λ is σ-additive and positive. We can easily see that $\mu \ll \lambda$, for every $\mu \in ba(\mathcal{P}(\mathbb{N}))$. The space of λ-integrable functions, $L^1(\lambda)$, is the space of all absolutely summable sequences, i.e.

$$L^1(\lambda) = \ell^1 = \left\{ (x_n)_{n \in \mathbb{N}} \subseteq \mathbb{R} : \sum_{n \in \mathbb{N}} |x_n| < +\infty \right\}.$$

If $x = (x_n)_{n \in \mathbb{N}} \in \ell^1$, then $\Psi(x) = \mu_x$ where

$$\mu_x(E) = \int_E x d\lambda = \sum_{n \in E} x_n.$$

For every $\mu \in ca(\mathcal{P}(\mathbb{N})) = ca_\lambda(\mathcal{P}(\mathbb{N}))$ and every $n \in \mathbb{N}$, let's define x_n by $x_n = \mu(\{n\})$.

Then $x = (x_n)_{n \in \mathbb{N}} \in \ell^1$ and $\Psi(x) = \mu_x$.

Definition 1.45. Two signed measures $\lambda, \mu \in ca(\mathcal{A})$ are called *singular* or *orthogonal* if there exists a set $D \in \mathcal{A}$ such that $|\lambda|$ is concentrated on D and $|\mu|$ is concentrated on $D^c = X \setminus D$. This is denoted by $\lambda \perp \mu$.

We remark that, if $\{\lambda^+, \lambda^-\}$ is the Jordan decomposition of λ, then $\lambda^+ \perp \lambda^-$.

Proposition 1.46. *The measures $\lambda, \mu \in ca(\mathcal{A})$ are singular if and only if, for every $\varepsilon > 0$, there exists $A_\varepsilon \in \mathcal{A}$ such that $|\lambda|(A_\varepsilon) < \varepsilon$ and $|\mu|(X \setminus A_\varepsilon) < \varepsilon$.*

Proof. The implication "$\Longrightarrow$" is obvious.

$\Longleftarrow$: For every $n \in \mathbb{N}$, let $\varepsilon = \frac{1}{2^n}$; there exists $A_n \in \mathcal{A}$ such that $|\lambda|(A_n) < \frac{1}{2^n}$ and $|\mu|(X \setminus A_n) < \frac{1}{2^n}$. Let

$$A = \limsup_n A_n = \bigcap_{n=1}^{\infty} \bigcup_{k=n}^{\infty} A_k.$$

$$|\lambda|(A) = \lim_n |\lambda| \left(\bigcup_{k=n}^{\infty} A_k \right) \leq \lim_n \sum_{k=n}^{\infty} |\lambda|(A_k) \leq \lim_n \frac{1}{2^{n-1}} = 0.$$

$$|\mu|(X \setminus A) = |\mu| \left(\liminf_n (X \setminus A_n) \right) \leq \liminf_n |\mu|(X \setminus A_n) = 0. \qquad \square$$

Theorem 1.47 (Lebesgue's decomposition theorem; see Theorem III.4.14 in [62] and 1.115 in [85]). *Let λ be a σ-additive, σ-finite and positive measure on $\mathcal{A}$ and let μ be a σ-additive and σ-finite measure on $\mathcal{A}$. Then there exist two σ-additive and σ-finite signed measures ρ, ν on $\mathcal{A}$, such that*

(i) $\mu = \rho + \nu$.

(ii) ρ *is* λ*-continuous.*

(iii) ν *and* λ *are singular measures.*

These two measures are uniquely determined.

Proof. (a) Firstly, let $\lambda, \mu \in ca^+(\mathcal{A})$. If $\tau = \lambda + \mu$, then it is clear that $\mu \ll \tau$. Therefore, from Theorem 1.40, there exists $g \in \mathcal{L}^1(\tau)$ such that

$$\mu(A) = \int_A g d\tau = \int_A g d\lambda + \int_A g d\mu.$$

For all $A \in \mathcal{A}$, $0 \leq \mu(A) \leq \tau(A)$; so that, τ-a.e., $0 \leq g \leq 1$ which means that $\{x \in X : g(x) < 0\} \cup \{x \in X : g(x) > 1\}$ is a set of τ-measure zero and consequently of μ-measure zero.

Let now $B = \{x \in X : g(x) = 1\}$ and $B' = \{x \in X : 0 < g(x) < 1\}$. Then $\mu(B) = \int_B g d\lambda + \int_B g d\mu = \lambda(B) + \mu(B)$; since μ is finite, $\lambda(B) = 0$. Let's put, for every $A \in \mathcal{A}$, $\nu(A) = \mu(A \cap B)$ and $\rho(A) = \mu(A \cap B')$.

Then ν is concentrated on B and $\lambda(B) = 0$; therefore ν and λ are singular measures. Since $(B \cup B')^c$ is a set of μ-measure zero,

$$\mu(A) = \mu(A \cap B') + \mu(A \cap B) = \rho(A) + \nu(A).$$

Now let us show that $\rho \ll \lambda$.

For any $A \in \mathcal{A}$ with $\lambda(A) = 0$, $\rho(A) = \mu(A \cap B') = \int_{A \cap B'} g d\mu$. Therefore $\int_{A \cap B'} (1 - g) d\mu = 0$ and, since $g < 1$ on B', $\mu(A \cap B') = \rho(A) = 0$.

In order to demonstrate the uniqueness, let's put $\mu = \rho + \nu = \rho_1 + \nu_1$. Then $\rho - \rho_1 = \nu_1 - \nu$. The measures $\nu_1 - \nu$ and λ are singular; therefore there exists $D \in \mathcal{A}$ such that $|\nu_1 - \nu|$ is concentrated on the set D and $\lambda(D) = 0$. On the other hand $\rho - \rho_1 \ll \lambda$. Consequently, since $\lambda(D) = 0$, $|\rho - \rho_1|(D) = |\nu_1 - \nu|(D) = 0$; because $|\nu_1 - \nu|$ is concentrated on the set D, it follows that $\nu_1 = \nu$ and then $\rho_1 = \rho$.

(b) Now we suppose that λ and μ are two σ-additive, σ-finite and positive measures. Then, we will appeal to the previous case, considering an $\mathcal{A}$-partition of $X = \bigcup_{n=1}^{\infty} X_n$ with λ and μ be finite on every X_n. If λ_n and μ_n are the restrictions of λ and μ respectively to $(X_n, \mathcal{A}_n)(\mathcal{A}_n = \mathcal{A} \cap X_n)$. Then λ_n and μ_n belong to $ca^+(\mathcal{A}_n)$ and, from (a), there exist ρ_n and ν_n such that $\mu_n = \rho_n + \nu_n$ where ρ_n and ν_n satisfy the properties (ii) and (iii); i.e. $\rho_n \ll \lambda_n$ and $\nu_n \perp \lambda_n$.

Let $\rho, \nu : \mathcal{A} \to \mathbb{R}_+$, defined by

$$\rho(A) = \sum_1^{\infty} \rho_n(A \cap X_n) \quad \text{and} \quad \nu(A) = \sum_1^{\infty} \nu_n(A \cap X_n).$$

Then, for all $A \in \mathcal{A}$, $\mu(A) = \sum_1^\infty \mu(A \cap X_n) = \sum_1^\infty \mu_n(A \cap X_n) = \sum_1^\infty \rho_n(A \cap X_n) + \sum_1^\infty \nu_n(A \cap X_n) = \rho(A) + \nu(A)$ therefore

$$\mu = \rho + \nu. \tag{4}$$

It is easy to see that ρ and ν are σ-additive. Indeed, if $(B_i)_{i\in\mathbb{N}} \subseteq \mathcal{A}$ is a sequence of pairwise disjoint sets and if $B = \bigcup_1^\infty B_i$, then

$$\begin{aligned}\rho(B) &= \sum_{n=1}^{\infty} \rho_n(B \cap X_n) = \sum_{n=1}^{\infty}\sum_{i=1}^{\infty} \rho_n(B_i \cap X_n) \\ &= \sum_{i=1}^{\infty}\sum_{n=1}^{\infty} \rho_n(B_i \cap X_n) = \sum_{i=1}^{\infty} \rho(B_i).\end{aligned}$$

Similarly, we demonstrate that ν is σ-additive.

We still need to demonstrate that $\rho \ll \lambda$ and $\nu \perp \lambda$.

Let $A \in \mathcal{A}$ with $\lambda(A) = 0$; for every $n \in \mathbb{N}^*$, $\lambda(A \cap X_n) = \lambda_n(A \cap X_n) = 0$ and therefore, since $\rho_n \ll \lambda_n$, $\rho_n(A \cap X_n) = 0$.

Hence $\rho(A) = \sum_{n=1}^\infty \rho_n(A \cap X_n) = 0$; which means that $\rho \ll \lambda$.

For every $n \in \mathbb{N}^*$, ν_n and λ_n are singular measures; so there exists $D_n \in \mathcal{A}_n$ such that ν_n is concentrated on D_n and $\lambda_n(D_n) = 0$. If $D = \bigcup_{n\in N} D_n$ then it is clear that $\lambda(D) = 0$ and ν is concentrated on D. Therefore we obtain $\nu \perp \lambda$.

(c) Let now λ be a σ-additive, σ-finite and positive measure and let μ be a σ-additive and σ-finite signed measure on $\mathcal{A}$. With the help of the relation $\mu = \mu^+ - \mu^-$, the proof ends in an obvious way. □

Definition 1.48. The decomposition presented in the previous theorem is named *Lebesgue decomposition* of μ relatively to λ.

We will end this section with a final important application of the Radon–Nikodym theorem: the identification of the dual of $L^1(\lambda)$.

Theorem 1.49 (Theorem IV.8.5 in [62]). *Let $\lambda : \mathcal{A} \to \overline{\mathbb{R}}_+$ be a σ-additive, σ-finite and positive measure on $\mathcal{A}$.*

There is an isometric isomorphism between the topological dual $[L^1(\lambda)]^*$ of $L^1(\lambda)$ and $L^\infty(\lambda)$.

Proof. (a) Firstly, we suppose that $\lambda \in ca^+(\mathcal{A})$, which means that $\lambda(X) < +\infty$.

For every $f \in L^\infty(\lambda)$, let us define a map $F_f : L^1(\lambda) \to \mathbb{R}$ by $F_f(g) = \int_X fg d\lambda$, for every $g \in L^1(\lambda)$. Then F_f is linear and $|F_f(g)| = |\int_X fg d\lambda| \le \|f\|_\infty \|g\|_1$. Therefore F_f is a linear continuous map and $\|F_f\| \le \|f\|_\infty$. We will show that $\|F_f\| = \|f\|_\infty$.

For every $\varepsilon > 0$, let $g(x) = \begin{cases} +1, & f(x) \geq \|f\|_\infty - \varepsilon, \\ -1, & f(x) \leq -\|f\|_\infty + \varepsilon, \\ 0, & \text{elsewhere.} \end{cases}$

Let $A = \{x \in X : |f(x)| \geq \|f\|_\infty - \varepsilon\}$; then $|g| = \chi_A$. Therefore $g \in L^1(\lambda)$ and $\|g\|_1 = \int |g| d\lambda = \lambda(A)$.

Hence $|F_f(g)| = |\int_X fg d\lambda| = \int_A |f| d\lambda \geq \lambda(A)(\|f\|_\infty - \varepsilon) = \|g\|_1(\|f\|_\infty - \varepsilon)$, from where $\|F_f\| \geq \|f\|_\infty - \varepsilon$. Because $\varepsilon > 0$ is arbitrary, it follows that $\|F_f\| = \|f\|_\infty$.

Let now $T : L^\infty(\lambda) \to [L^1(\lambda)]^*$ be the mapping defined by $T(f) = F_f$, for every $f \in L^\infty(\lambda)$. Obviously, T is a linear continuous mapping of norm 1 and therefore it is an injective mapping.

We still have to demonstrate that T is surjective.

For every $x^* \in [L^1(\lambda)]^*$, let $\mu : \mathcal{A} \to \mathbb{R}$ be defined by $\mu(A) = x^*(\chi_A)$, for all $A \in \mathcal{A}$. It is obvious that μ is additive and because

$$|\mu(A)| \leq \|x^*\| \cdot \|\chi_A\|_1 = \|x^*\| \cdot \lambda(A) \leq \|x^*\| \cdot \lambda(X), \quad \text{for every} \quad A \in \mathcal{A}, \tag{1}$$

μ is bounded; therefore $\mu \in ba(\mathcal{A})$. It is clear that $\mu \ll \lambda$; according to Proposition 1.31, μ is a σ-additive measure.

Therefore, according to the Radon–Nikodym theorem (Theorem 1.40), μ admits a Radon–Nikodym derivative f with respect to λ; it means that $f \in L^1(\lambda)$ and $x^*(\chi_A) = \mu(A) = \int_A f d\lambda$, for every $A \in \mathcal{A}$.

Let $\mathcal{E}(\mathcal{A})$ be the space of all $\mathcal{A}$-simple functions; it is clear that we have

$$x^*(g) = \int_X fg d\lambda, \quad \text{for every} \quad g \in \mathcal{E}(\mathcal{A}). \tag{2}$$

If we prove that $f \in L^\infty(\lambda)$, then T is surjective. Indeed, in this case $F_f : L^1(\lambda) \to \mathbb{R}$, defined by $F_f(g) = \int_X fg d\lambda$, is continuous on $L^1(\lambda)$; because it coincides with x^* on $\mathcal{E}(\mathcal{A})$ which is dense in $L^1(\lambda)$, $x^* = F_f = T(f)$.

Let's prove that $f \in L^\infty(\lambda)$. For every $n \in \mathbb{N}^*$, let $A_n = \{x \in X : |f(x)| \geq n\}$. Then $n \cdot \lambda(A_n) \leq \int_{A_n} f d\lambda = x^*(\chi_{A_n}) \leq \|x^*\| \cdot \lambda(A_n)$. Therefore $\|x^*\| \geq n$, for every $n \in \mathbb{N}$ with $\lambda(A_n) \neq 0$. Because $\|x^*\|$ is finite, there exists $n_0 \in \mathbb{N}$ such that $\lambda(A_{n_0}) = 0$. Therefore $|f| \leq n_0$, λ-a.e. which means that $f \in L^\infty(\lambda)$.

We have demonstrated the surjectivity of T which ends the proof in the case where λ is a finite measure on $\mathcal{A}$.

(b) Let now λ be a σ-additive, σ-finite and positive measure on $\mathcal{A}$. Just as in the case (a), $T : L^\infty(\lambda) \to [L^1(\lambda)]^*$ is defined by $T(f) = F_f$, where $F_f : L^1(\lambda) \to \mathbb{R}$, $F_f(g) = \int_X fg d\lambda$, for every $g \in L^1(\lambda)$. Because $\|F_f\| \leq \|f\|_\infty$, T is a linear and continuous mapping.

Let $N(T) = \{f \in L^\infty(\lambda) : T(f) = 0\}$; then $f \in N(T) \Leftrightarrow F_f = 0 \Leftrightarrow \int_X fg d\lambda = 0$, for every $g \in L^1(\lambda) \Leftrightarrow f = 0, \lambda$-a.e. Therefore $N(T) = \{0\}$ and so T is injective.

We still need to proof the surjectivity of T. Because λ is σ-finite, there exists an increasing sequence $(E_n)_{n\in\mathbb{N}} \subseteq \mathcal{A}$ whose union is X such that $\lambda(E_n) < +\infty$, for every $n \in \mathbb{N}$.

For every $n \in \mathbb{N}$, let $L^1_{E_n}(\lambda)$ be the Banach space of all integrable functions on E_n and, for every $g \in L^1_{E_n}(\lambda)$, let $\bar{g} \in L^1(\lambda)$ be the extension of g to X that takes the value 0 on $X \setminus E_n$.

For every $x^* \in [L^1(\lambda)]^*$ let $x^*_n : L^1_{E_n}(\lambda) \to \mathbb{R}$ defined by $x^*_n(g) = x^*(\bar{g})$. Then x^*_n is linear and continuous on $L^1_{E_n}(\lambda)$ and so, by (a), there exists $f_n \in L^\infty_{E_n}(\lambda)$ such that

$$x^*_n(g) = \int_{E_n} f_n g d\lambda, \quad \text{for every} \quad g \in L^1_{E_n}(\lambda). \tag{3}$$

Obviously, $\|f_n\|_\infty = \|F_{f_n}\| = \|x^*_n\| \le \|x^*\|$, for every $n \in \mathbb{N}$.

For every $n \in \mathbb{N}$ and every $g \in L^1_{E_n}(\lambda)$, $x^*_n(g) = x^*_{n+1}(g)$; hence

$$\int_{E_n} f_n g d\lambda = \int_{E_n} f_{n+1} g d\lambda, \quad \text{for every} \quad g \in L^1_{E_n}(\lambda).$$

Then $f_n = f_{n+1}$, λ-a.e. on E_n and therefore there exists $f : X \to \mathbb{R}$, $f = \lim_n \bar{f}_n$, λ-a.e. on X. For every $n \in \mathbb{N}$, $f \restriction E_n = f_n$ and so

$$\lambda(|f| > \|x^*\|) = \lim_n \lambda(|f_n| > \|x^*\|) = 0.$$

Therefore $|f| \le \|x^*\|$, λ-a.e. and then $\|f\|_\infty \le \|x^*\| < +\infty$ which means that $f \in L^\infty(\lambda)$.

According to (3), for every $n \in \mathbb{N}$ and every $g \in L^1(\lambda)$

$$x^*(g \cdot \chi_{E_n}) = \int_X g \cdot f \cdot \chi_{E_n} d\lambda.$$

On the other hand $g \cdot \chi_{E_n} \xrightarrow{\|\cdot\|_1} g$; therefore

$$x^*(g) = \lim_n x^*(g \cdot \chi_{E_n}) = \lim_n \int_X f \cdot g \cdot \chi_{E_n} d\lambda = \int_X fg d\lambda.$$

Hence $x^* = F_f = T(f)$, where $f \in L^\infty(\lambda)$. □

Remark 1.50.

(i) Using an analogous method, we can demonstrate the following statement: Let $1 < p < +\infty$ and q such that $\frac{1}{p} + \frac{1}{q} = 1$. For every σ-additive, σ-finite and positive measure $\lambda : \mathcal{A} \to \overline{\mathbb{R}}_+$ and for every $f \in L_q(\lambda)$, the mapping $F_f : L_p \to \mathbb{R}$, defined by: $F_f(g) = \int_X fg d\lambda$, is a linear continuous mapping on L_p. The mapping $f \mapsto F_f$ is an isometric isomorphism between the Banach spaces L_q and the dual $(L_p)^*$ of L_p.

(ii) If $\lambda : \mathcal{A} \to \overline{\mathbb{R}}_+$ is a σ-additive and positive measure, but not a σ-finite one then J. Schwartz ([151]) showed, using a totally different method, that this result is still true for $1 \leq p < +\infty$.

Example 1.51. In Example 1.44, we have introduced the measure λ on $\mathbb{N}$ and we showed that $L^1(\lambda)$ is the space ℓ^1 of all absolutely summable sequences. From the previous theorem, it follows that the dual space of ℓ^1 is $L^\infty(\lambda)$, the space of all bounded sequences; this space is noted with ℓ^∞.

1.3 Convergences in $L^1(\lambda)$ and $ca(\mathcal{A})$

In this section, we will characterize the strong convergence in $L^1(\lambda)$ through of convergence in measure and weak convergence. In the last part, we will present the weak convergence in $L^1(\lambda)$ and in $ca(\mathcal{A})$.

We first recall the definition of the strong convergence and that of the convergence in measure, as well as the notion of λ-equicontinuity of a family of functions.

Let $\lambda : \mathcal{A} \to \overline{\mathbb{R}}_+$ be a σ-additive and positive measure and let $(f_n)_{n\in\mathbb{N}} \subseteq L^1(\lambda)$ and $f \in L^1(\lambda)$; we say that:

(i) (f_n) *is strongly convergent* to f $\Longleftrightarrow \|f_n - f\|_1 = \int_X |f_n - f| d\lambda \xrightarrow[n\to\infty]{} 0.$ The strong convergence (or norm convergence) of the sequence (f_n) to f is noted: $f_n \xrightarrow{\|\cdot\|_1} f$

(ii) (f_n) *converges in measure* to $f \Leftrightarrow \lambda(\{x : |f_n(x) - f(x)| \geq \varepsilon\}) \xrightarrow[n\to\infty]{} 0,$ for every $\varepsilon > 0$. The convergence in measure of the sequence (f_n) to f is denoted by $f_n \xrightarrow{\lambda} f$.

Definition 1.52. A family $\mathcal{F} \subseteq L^1(\lambda)$ of integrable functions is λ-*equicontinuous* if, for every $\varepsilon > 0$, there exists $\delta > 0$ such that, for any $E \in \mathcal{A}$ with $\lambda(E) < \delta$ and for all $f \in \mathcal{F}$, $|\int_E f d\lambda| < \varepsilon$.

Remark 1.53.

(i) This definition is coherent with Definition 1.35. Indeed, let us denote, for every $f \in \mathcal{F}$, $\mu_f = \Psi(f)$, where $\mu_f : \mathcal{A} \to \mathbb{R}$, $\mu_f(E) = \int_E f d\lambda$; then $\mathcal{F}$ is λ-equicontinuous if and only if $\{\mu_f : f \in \mathcal{F}\}$ is λ-equicontinuous.

(ii) According to Remark 1.29 and to (i), $\mathcal{F} \subseteq L^1(\lambda)$ is λ-equicontinuous if, for every $\varepsilon > 0$, there exists $\delta > 0$ such that $\int_E |f| d\lambda < \varepsilon$, for every $E \in \mathcal{A}$ verifying $\lambda(E) < \delta$ and for all $f \in \mathcal{F}$.

(iii) It is obvious that all finite family of functions $\mathcal{F} \subseteq L^1(\lambda)$ is λ-equicontinuous.

The following theorem gives us the conditions that will be added to the convergence in measure in order to obtain the strong convergence in $L^1(\lambda)$ (see Theorem III.6.15 of [62]).

Theorem 1.54 (Vitali). *Let $\lambda : \mathcal{A} \to \overline{\mathbb{R}}_+$ be a σ-additive and positive measure, let $(f_n)_{n\in\mathbb{N}} \subseteq L^1(\lambda)$ and let $f : X \to \mathbb{R}$ be an $\mathcal{A}$-measurable function. Then $f \in L^1(\lambda)$ and $f_n \xrightarrow{\|\cdot\|_1} f$ if and only if the following conditions are accomplished:*

(i) $f_n \xrightarrow{\lambda} f$,

(ii) *$\{f_n : n \in \mathbb{N}\}$ is λ-equicontinuous,*

(iii) *for each $\varepsilon > 0$, there exists $A \in \mathcal{A}$, with $\lambda(A) < +\infty$ such that, $\int_{X\setminus A} |f_n| d\lambda < \varepsilon$, for every $n \in \mathbb{N}$.*

Proof. Firstly, we suppose that $f \in L^1(\lambda)$ and $f_n \xrightarrow{\|\cdot\|_1} f$.

The convergence in measure results immediately from the following inequality: $\lambda(|f_n - f| \geq \varepsilon) \leq \frac{1}{\varepsilon}\|f_n - f\|_1$, for all n and for all $\varepsilon > 0$.

Let's show that $\{f_n : n \in \mathbb{N}\}$ is λ-equicontinuous. As $f_n \xrightarrow{\|\cdot\|_1} f$, for every $\varepsilon > 0$, there exists $n_0 \in \mathbb{N}$ such that, for all $n \geq n_0$,

$$\|f_n - f\|_1 < \varepsilon. \tag{1}$$

According to Remark 1.53(iii), there exists $\delta > 0$ such that, for all $E \in \mathcal{A}$ such that $\lambda(E) < \delta$,

$$\int_E |f| d\lambda < \varepsilon \tag{2}$$

and

$$\int_E |f_n| d\lambda < \varepsilon, \quad \text{for every} \quad n \leq n_0 - 1. \tag{3}$$

From (1) and (2), for every $n \geq n_0$,

$$\int_E |f_n| d\lambda \leq \int_E |f_n - f| d\lambda + \int_E |f| d\lambda \leq \|f_n - f\|_1 + \int_E |f| d\lambda < 2\varepsilon. \tag{4}$$

(3) and (4) show that $\{f_n : n \in \mathbb{N}\}$ is λ-equicontinuous.

Similarly, there is $A \in \mathcal{A}$ with $\lambda(A) < +\infty$ such that

$$\int_{X \setminus A} |f| d\lambda < \varepsilon \quad \text{and} \tag{5}$$

$$\int_{X \setminus A} |f_n| d\lambda < \varepsilon, \quad \text{for every} \quad n \leq n_0 - 1. \tag{6}$$

From (1) and (5), we obtain, for every $n \geq n_0$,

$$\int_{X \setminus A} |f_n| d\lambda \leq \int_{X \setminus A} |f_n - f| d\lambda + \int_{X \setminus A} |f| d\lambda < \|f_n - f\|_1 + \varepsilon < 2\varepsilon. \tag{7}$$

(6) and (7) lead to Condition (iii).

Reciprocally, we assume (i), (ii) and (iii).

According to (iii), for every $\varepsilon > 0$, there exists $A_0 \in \mathcal{A}$ with $0 < \lambda(A_0) < +\infty$ satisfying

$$\int_{X \setminus A_0} |f_n| d\lambda < \varepsilon, \quad \text{for every} \quad n \in \mathbb{N}. \tag{8}$$

Since $f_n \xrightarrow{\lambda} f$, there exists a subsequence $(f_{k_n})_n$ of $(f_n)_n$ such that $f_{k_n} \to f$ λ-a.e. According to Fatou's lemma, we obtain

$$\int_{X \setminus A_0} |f| d\lambda \leq \liminf_n \textstyle\int_{X \setminus A_0} |f_{k_n}| d\lambda \leq \varepsilon \quad \text{and therefore}$$

$$\int_{X \setminus A_0} |f| d\lambda \leq \varepsilon. \tag{9}$$

The family $\{f_n : n \in \mathbb{N}\}$ is λ-equicontinuous; then there exists $\delta \in]0, \varepsilon[$ such that, for every $E \in \mathcal{A}$ satisfying $\lambda(E) < \delta$,

$$\int_E |f_n| d\lambda < \varepsilon, \quad \text{for every} \quad n \in \mathbb{N}. \tag{10}$$

Since the sequence (f_p) is convergent in measure to f,

$$\lim_p \lambda\left(|f_p - f| > \frac{\varepsilon}{\lambda(A_0)}\right) = 0,$$

Consequently, we can find $p_0 \in \mathbb{N}$ such that $\lambda(|f_p - f| > \frac{\varepsilon}{\lambda(A_0)}) < \delta < \varepsilon$, for every $p \geq p_0$; let $\varepsilon_1 = \frac{\varepsilon}{\lambda(A_0)}$.

According to (10), we have

$$\int_{(|f_p - f| > \varepsilon_1)} |f_n| d\lambda < \varepsilon, \quad \text{for every} \quad p \geq p_0 \quad \text{and every} \quad n \in \mathbb{N}. \tag{11}$$

Applying again the Fatou's lemma to the sequence (f_{k_n}) on the sets $(|f_p - f| > \varepsilon_1)$ with $p \geq p_0$, we obtain

$$\int_{(|f_p - f| > \varepsilon_1)} |f| d\lambda \leq \liminf_n \int_{(|f_p - f| > \varepsilon_1)} |f_{k_n}| d\lambda \leq \varepsilon$$

and therefore

$$\int_{(|f_p - f| > \varepsilon_1)} |f| \leq \varepsilon, \quad \text{for every} \quad p \geq p_0. \tag{12}$$

Then, by (8), (9), (11) and (12), for every $p \geq p_0$,

$$\begin{aligned}
\int_X |f_p - f| d\lambda &= \int_{X \setminus A_0} |f_p - f| d\lambda + \int_{A_0} |f_p - f| d\lambda \\
&\leq \int_{X \setminus A_0} |f_p| d\lambda + \int_{X \setminus A_0} |f| d\lambda + \int_{A_0 \cap (|f_p - f| > \varepsilon_1)} |f_p - f| d\lambda \\
&\quad + \int_{A_0 \cap (|f_p - f| \leq \varepsilon_1)} |f_p - f| d\lambda \\
&\leq 2\varepsilon + \int_{(|f_p - f| > \varepsilon_1)} |f_p| d\lambda + \int_{(|f_p - f| > \varepsilon_1)} |f| d\lambda + \varepsilon_1 \lambda(A_0) \\
&< 2\varepsilon + 2\varepsilon + \frac{\varepsilon}{\lambda(A_0)} \cdot \lambda(A_0) = 5\varepsilon.
\end{aligned}$$

Therefore $f_p - f \in L^1(\lambda)$ from where $f \in L^1(\lambda)$ and

$$\|f_p - f\|_1 < 5\varepsilon, \quad \text{for every} \quad p \geq p_0. \tag{13}$$

Hence $f_p \xrightarrow{\|\cdot\|_1} f$. □

Remark 1.55. According to Corollary 1.14, if $\lambda \in ca^+(\mathcal{A})$, then $\lambda(X) < +\infty$; therefore the Condition (iii) of previous theorem is satisfied. We have, therefore, the following theorem:

Theorem. *Let $\lambda \in ca^+(\mathcal{A})$; the sequence $(f_n) \subseteq L^1(\lambda)$ is strongly convergent to $f \in L^1(\lambda)$ if and only if it is convergent in measure to f and it is λ-equicontinuous.*

As we show in the following examples, the convergence in measure or the λ-equicontinuity, separately taken, do not lead to the strong convergence.

Example 1.56.

(a) Let λ be the Lebesgue measure on $[0,1]$ and let us define the mapping $f_n : [0,1] \to \mathbb{R}$, letting $f_n = n\chi_{[0,\frac{1}{n}]}$. Then $f_n \xrightarrow{\lambda} 0$ but $\|f_n - 0\|_1 = 1 \nrightarrow 0$ and therefore $\{f_n : n \in \mathbb{N}\}$ it is not λ-equicontinuous.

(b) Let $(r_n)_{n\in\mathbb{N}}$ be the Rademacher's sequence: for every $n \in \mathbb{N}$, $r_n : [0,1] \to \mathbb{R}$ is defined by

$$r_n(t) = \begin{cases} +1, & t \in \bigcup_{k=0}^{2^{n-1}-1} \left[\frac{2k}{2^n}, \frac{2k+1}{2^n}\right[, \\ -1, & t \in \bigcup_{k=0}^{2^{n-1}-1} \left[\frac{2k+1}{2^n}, \frac{2k+2}{2^n}\right[. \end{cases}$$

For any $\varepsilon > 0$, there exists $\delta = \varepsilon$ such that, for every $A \in \mathcal{A}$ with $\lambda(A) < \delta$, $\int_A |r_n| d\lambda = \lambda(A) < \delta = \varepsilon$. Therefore, the sequence (r_n) is λ-equicontinuous but it does not converge strongly in $L^1(\lambda)$.

In fact, if we assume that (r_n) is strongly convergent in $L^1(\lambda)$, then it is a Cauchy sequence in $L^1(\lambda)$; but it is not, because, for every $n \in \mathbb{N}$,

$$r_n(t) - r_{n+1}(t) = \begin{cases} 0 & , t \in \bigcup_{k=0}^{2^{n-1}-1} \left[\frac{4k}{2^{n+1}}, \frac{4k+1}{2^{n+1}}\right[\cup \left[\frac{4k+3}{2^{n+1}}, \frac{4k+4}{2^{n+1}}\right[, \\ 2 & , t \in \bigcup_{k=0}^{2^{n-1}-1} \left[\frac{4k+1}{2^{n+1}}, \frac{4k+2}{2^{n+1}}\right[, \\ -2, & t \in \bigcup_{k=0}^{2^{n-1}-1} \left[\frac{4k+2}{2^{n+1}}, \frac{4k+3}{2^{n+1}}\right[\end{cases}$$

and then

$$\int_{[0,1]} |r_n - r_{n+1}| d\lambda = 4 \cdot 2^{n-1} \cdot \frac{1}{2^{n+1}} = 1 \nrightarrow 0.$$

According to the previous theorem, one can also deduce that the Rademacher's sequence (r_n) is not convergent in measure on $[0,1]$.

In the following theorem, we will characterize the weak convergence in $L^1(\lambda)$ (as references, see Theorem IV.8.7 of [62] or corollary on page 91 of [57]).

Theorem 1.57. *Let $\lambda : \mathcal{A} \to \overline{\mathbb{R}}_+$ be a σ-additive and σ-finite measure; a sequence $(f_n) \subseteq L^1(\lambda)$ is weakly convergent to $f \in L^1(\lambda)$ if and only if*

$$\int_A f_n d\lambda \to \int_A f d\lambda, \quad \textit{for every} \quad A \in \mathcal{A}. \tag{*}$$

Proof. According to Theorem 1.49, (f_n) is weakly convergent to f if and only if, for every $g \in L^\infty(\lambda)$,

$$\int_X f_n g d\lambda \to \int_X f g d\lambda.$$

Since, for all $A \in \mathcal{A}$, $\chi_A \in L^\infty(\lambda)$, the condition $(*)$ is verified.

To proof the sufficiency of $(*)$, we will first show that $(f_n)_{n\in\mathbb{N}}$ is bounded in $L^1(\lambda)$.

For every $h \in L^1(\lambda)$, let $\mu_h : \mathcal{A} \to \mathbb{R}$ be defined by $\mu_h(A) = \int_A h \, d\lambda$, for every $A \in \mathcal{A}$ and let $\Psi : L^1(\lambda) \to ca(\mathcal{A})$, defined by $\Psi(h) = \mu_h$. According to Theorem 1.43, Ψ is an isometric isomorphism between $L^1(\lambda)$ and $ca_\lambda(\mathcal{A}) \subseteq ca(\mathcal{A})$. Therefore, for every $h \in L^1(\lambda)$, $\|\Psi(h)\| = \|\mu_h\| = |\mu_h|(X) = \|h\|_1$.

Let's note $\mu_n = \Psi(f_n)$ and $\mu = \Psi(f)$. By $(*)$, for every $A \in \mathcal{A}$, $\mu_n(A) \to \mu(A)$. Therefore, according to Nikodym's Theorem 1.37, we obtain that (μ_n) is bounded in $ca(\mathcal{A})$ and, because Ψ is an isometric isomorphism, (f_n) is bounded in $L^1(\lambda)$. Let $M > 0$ such that

$$\|f_n\|_1 \leq M, \quad \text{for every} \quad n \in \mathbb{N}. \tag{1}$$

From $(*)$, it follows immediately that, for every $h \in \mathcal{E}(\mathcal{A})$,

$$\int_X f_n h d\lambda \to \int_X f h d\lambda. \tag{2}$$

Let g be arbitrary in $L^\infty(\lambda)$. We must show that

$$\int_X f_n g d\lambda \to \int_X f g d\lambda.$$

Since $\mathcal{E}(\mathcal{A})$ is dense in $L^\infty(\lambda)$, for every $\varepsilon > 0$, there exists $h \in \mathcal{E}(\mathcal{A})$ such that

$$\|g - h\|_\infty < \varepsilon. \tag{3}$$

By (2), there exists $n_0 \in \mathbb{N}$ such that, for every $n \geq n_0$,

$$\left| \int_X f_n h d\lambda - \int_X f h d\lambda \right| < \varepsilon. \tag{4}$$

Then, by (4), (1) and (3), for every $n \geq n_0$, we have

$$\begin{aligned} \left| \int_X f_n g d\lambda - \int_X f g d\lambda \right| &\leq \left| \int_X f_n g d\lambda - \int_X f_n h d\lambda \right| + \left| \int_X f_n h d\lambda - \int_X f h d\lambda \right| \\ &\quad + \left| \int_X f h d\lambda - \int_X f g d\lambda \right| \\ &\leq \|f_n\|_1 \cdot \|g - h\|_\infty + \varepsilon + \|f\|_1 \cdot \|g - h\|_\infty \\ &\leq M \cdot \varepsilon + \varepsilon + \|f\|_1 \cdot \varepsilon = \varepsilon(M + \|f\|_1 + 1) \end{aligned}$$

who shows that (f_n) converges weakly to f. □

Corollary 1.58 (Theorem IV.9.5 of [62] or Theorem 11, p. 90, of [57]). *Let $(\mu_n) \subseteq ca(\mathcal{A})$ and let $\mu \in ca(\mathcal{A})$. (μ_n) is weakly convergent to μ in $ca(\mathcal{A})$ if and only if $\mu_n(A) \to \mu(A)$, for every $A \in \mathcal{A}$.*

Proof. If we suppose that (μ_n) is weakly convergent to μ, then, for any continuous linear functional x^* on $ca(\mathcal{A})$, $x^*(\mu_n)$ converges to $x^*(\mu)$ for n tending to infinity. If $x_A^* : ca(\mathcal{A}) \to \mathbb{R}$ is defined by: $x_A^*(\lambda) = \lambda(A)$, for every $\lambda \in ca(\mathcal{A})$, then x_A^* is a linear functional on $ca(\mathcal{A})$. For every $\lambda \in ca(\mathcal{A})$, $|x_A^*(\lambda)| = |\lambda(A)| \leq |\lambda|(A) \leq \|\lambda\|$. Therefore $x_A^* \in [ca(\mathcal{A})]^*$ and then $x_A^*(\mu_n) \to x_A^*(\mu)$ which means that $\mu_n(A) \to \mu(A)$, for every $A \in \mathcal{A}$.

Conversely, let $(\mu_n) \subseteq ca(\mathcal{A})$ and let $\mu \in ca(\mathcal{A})$ with, $\mu_n(A) \to \mu(A)$, for every $A \in \mathcal{A}$. The Nikodym's Theorem 1.37 says that (μ_n) is bounded in $(ca(\mathcal{A}), \|.\|)$. Let now $\lambda : \mathcal{A} \to \mathbb{R}_+$, defined by:

$$\lambda(A) = \sum_{n=1}^{\infty} \frac{1}{2^n} |\mu_n|(A), \quad \text{for every} \quad A \in \mathcal{A}.$$

Therefore, through a similar proof to that of Nikodym's theorem, we show that $\lambda \in ca^+(\mathcal{A})$. Because, for all $n \in \mathbb{N}$ and for all $A \in \mathcal{A}$, $|\mu_n(A)| \leq 2^n \cdot \lambda(A)$, $\mu_n \ll \lambda$, for every $n \in \mathbb{N}$. Therefore, according to Radon–Nikodym's theorem (see Theorem 1.40), for every $n \in \mathbb{N}$, there exists $f_n \in L^1(\lambda)$ such that, $\mu_n(A) = \int_A f_n d\lambda$, for every $A \in \mathcal{A}$.

As stated by Vitali–Hahn–Saks' theorem (see Theorem 1.36 (iii)), $\mu \ll \lambda$ and so, from Radon–Nikodym's theorem, there exists $f \in L^1(\lambda)$ such that, $\mu(A) = \int_A f d\lambda$, for every $A \in \mathcal{A}$.

Then, for all $A \in \mathcal{A}$, $\int_A f_n d\lambda = \mu_n(A) \to \mu(A) = \int_A f d\lambda$ and so, according to Theorem 1.57, the sequence (f_n) is weakly convergent to f in the space $L^1(\lambda)$. Since $L^1(\lambda)$ is a normed space isometrically isomorphic with $ca_\lambda(\mathcal{A})$ (see Theorem 1.43), (μ_n) converges weakly to μ in $ca_\lambda(\mathcal{A})$ and so in $ca(\mathcal{A})$. □

Corollary 1.59 (Theorem IV.9.4 of [62] or Theorem 12, p. 91, [57]). *$ca(\mathcal{A})$ is sequentially weakly complete.*

Proof. Let $(\mu_n) \subseteq ca(\mathcal{A})$ be a weakly Cauchy sequence; this means that, for every $x^* \in [ca(\mathcal{A})]^*$, $(x^*(\mu_n))_{n\in\mathbb{N}}$ is a Cauchy sequence in $\mathbb{R}$.

Let now, for every $A \in \mathcal{A}$, $x_A^* \in [ca(\mathcal{A})]^*$, defined by $x_A^*(\lambda) = \lambda(A)$, for every $\lambda \in ca(\mathcal{A})$. Then $x_A^*(\mu_n) = \mu_n(A)$ and $(\mu_n(A))_{n\in\mathbb{N}}$ is a Cauchy sequence in $\mathbb{R}$, for every $A \in \mathcal{A}$. We note, for all $A \in \mathcal{A}$, $\mu(A) = \lim_n \mu_n(A)$. According to Nikodym's theorem (see Theorem 1.37), $\mu \in ca(\mathcal{A})$ and since $\mu_n(A) \to \mu(A)$, for all $A \in \mathcal{A}$, the previous corollary assures that (μ_n) converges weakly to μ. □

Remark 1.60. Let $\lambda : \mathcal{A} \to \overline{\mathbb{R}}_+$ be a σ-additive and σ-finite measure. $ca_\lambda(\mathcal{A})$ is a strongly closed subspace of $ca(\mathcal{A})$; therefore it is a weakly closed one. It follows that $ca_\lambda(\mathcal{A})$ is weakly sequentially complete and therefore $L^1(\lambda)$ is weakly sequentially complete, too.

It is known that the weak topology coincides with the strong topology on a normed space if and only if it is of finite dimension. However one may happen in an infinite dimensional space that every weakly convergent sequence to be strongly convergent. This phenomenon occurs in the space ℓ^1.

Theorem 1.61 (Schur; see Corollary IV.8.14 of [62]). *Let $\lambda : \mathcal{P}(\mathbb{N}) \to \overline{\mathbb{R}}_+$ be the measure of Example 1.44 and let $L^1(\lambda) = \ell^1$ be the space of all absolutely convergent sequences $x = (x_n) \subseteq \mathbb{R}$, with the norm $\|x\| = \sum_{n\in\mathbb{N}} |x_n|$.*

A sequence $(x^p)_{p\in\mathbb{N}} \subseteq (\ell^1, \|\cdot\|_1)$ is strongly convergent to $x \in \ell^1$ if and only if (x^p) is weakly convergent to x.

Proof. For all normed space, the strong convergence implies the weak convergence.

Conversely, let $(x^p)_{p\in\mathbb{N}}$ be a sequence weakly convergent to $x = (x_n)_{n\in\mathbb{N}}$ in ℓ^1; we note that, for every $p \in \mathbb{N}$, $x^p = (x_n^p)_{n\in\mathbb{N}}$. We prove that the sequence $(x^p)_{p\in\mathbb{N}}$ is strongly convergent to $x = (x_n)_{n\in\mathbb{N}}$ by using the Vitali's Theorem 1.54.

For every $n \in \mathbb{N}$, $\{n\} \in \mathcal{P}(\mathbb{N})$ and so, according to Theorem 1.57,

$$\int_{\{n\}} x^p d\lambda \to \int_{\{n\}} x d\lambda,$$

which means that $(x_n^p)_p$ converges to x_n, for every $n \in \mathbb{N}$.

We show that $x_n^p \xrightarrow[p\to\infty]{} x_n$ uniformly in $n \in \mathbb{N}$.

For every $p \in \mathbb{N}$, let $\mu_p, \mu : \mathcal{P}(\mathbb{N}) \to \mathbb{R}$ defined by $\mu_p(A) = \int_A x^p d\lambda = \sum_{n\in A} x_n^p$ and $\mu(A) = \int_A x d\lambda = \sum_{n\in A} x_n$, for every $A \subseteq \mathbb{N}$.

Then $(\mu_p) \subseteq ca(\mathcal{P}(\mathbb{N}))$ and, according to Theorem 1.57, $\mu_p(A) = \int_A x^p d\lambda$ converges to $\int_A x d\lambda = \mu(A)$, for every $A \in \mathcal{P}(\mathbb{N})$.

According to Nikodym's theorem (1.37), the family $\{\mu_p : p \in \mathbb{N}\}$ is uniformly σ-additive, and from Theorems 1.36 and 1.34, $\{\mu_p : p \in \mathbb{N}\}$ is λ-equicontinuous. Then the condition (ii) of Vitali's theorem is satisfied. Now, since $\{\mu_p : p \in \mathbb{N}\}$ is uniformly σ-additive, for every $\varepsilon > 0$, there exists $n_0 \in \mathbb{N}$ such that, for all $n \geq n_0$,

$$\left|\sum_{k=1}^{n} \mu_p(\{k\}) - \mu_p(\mathbb{N})\right| < \varepsilon, \quad \text{for all} \quad p \in \mathbb{N} \tag{1}$$

what is equivalent with

$$\begin{aligned} |\mu_p(\{n+1, n+2, \dots\})| &= \left|\int_{\mathbb{N}\setminus\{0,1,\dots,n\}} x^p d\lambda\right| \\ &= \left|\sum_{k>n} x_k^p\right| < \varepsilon, \qquad \forall p \in \mathbb{N}, \quad \forall n \geq n_0. \end{aligned} \tag{2}$$

Therefore

$$|x_n^p| = \left|\sum_{k>n-1} x_k^p - \sum_{k>n} x_k^p\right| < 2\varepsilon, \text{ for every } p \in \mathbb{N}, \text{ and every } n \geq n_0 + 1. \tag{3}$$

We recall that $x \in \ell^1$; so we can choose n_0 such as

$$|x_n| < \varepsilon, \quad \text{for every} \quad n \geq n_0 + 1. \tag{4}$$

By (3) and (4), we have,

$$|x_n^p - x_n| < 3\varepsilon, \quad \text{for every} \quad p \in \mathbb{N}, \quad \text{and every} \quad n \geq n_0 + 1. \tag{5}$$

Since (x_n^p) is punctually convergent to x_n, there exists $p_0 \in \mathbb{N}$ such that

$$|x_n^p - x_n| < \varepsilon, \quad \text{for every} \quad p \geq p_0, \quad \text{and every} \quad n \leq n_0. \tag{6}$$

By (5) and (6), we obtain that, for all $p \geq p_0$ and all $n \in \mathbb{N}$, $|x_n^p - x_n| < 3\varepsilon$ which means that $(x^p) = (x_n^p)_p$ converges uniformly in n to $x = (x_n)$ in $(\mathbb{N}, \mathcal{P}(\mathbb{N}), \lambda)$.

Therefore (x^p) is convergent in measure to x.

Moreover, by (2), the condition (iii) of Vitali's theorem is satisfied and so, the sequence (x^p) is strongly convergent to x. □

1.4 Weak Compactness in $ca(\mathcal{A})$ and $L^1(\lambda)$

$L^p(\lambda)$ is a reflexive space for all $p \in]1, +\infty[$. This is an easy consequence of the theorem given in Remark 1.50. In this situation, a set $K \subseteq L^p(\lambda)$ is relatively weakly compact if and only if K is a bounded set. For the space $L^1(\lambda)$, which is not reflexive, the issue of weak compactness is more complicated.

We begin with the following lemma (see lemma at page 91 in [57]):

Lemma 1.62. *Let $\mathcal{G}$ be an algebra which generates the σ-algebra $\mathcal{A}$ and let $(\mu_n) \subseteq ca(\mathcal{A})$ be a sequence of measures. The limit $\lim_n \mu_n(A) \in \mathbb{R}$ exists, for all $A \in \mathcal{A}$, if and only if the following two conditions are verified:*

(1) $\lim_n \mu_n(E) \in \mathbb{R}$, *for every* $E \in \mathcal{G}$,

(2) $\{\mu_n : n \in \mathbb{N}\}$ *is uniformly* σ*-additive.*

Proof. The necessity of (1) is obvious and that of condition (2) is a consequence of the Nikodym's theorem (see Theorem 1.37).

To show that the conditions (1) and (2) are also sufficient, we use the fact that σ-algebra generated by an algebra is just the monotone class generated by this algebra [93, Th.B, p. 27].

Let $\mathcal{L} = \{A \in \mathcal{A} :$ there exists $\lim_n \mu_n(A) \in \mathbb{R}\}$. Then,

(a) $\mathcal{G} \subseteq \mathcal{L} \subseteq \mathcal{A}$ and

(b) $\mathcal{L}$ is a monotone class of sets. Indeed, let $(E_m) \subseteq \mathcal{L}$ be a monotone sequence such that $E = \lim_m E_m$. If we suppose that the sequence is increasing, then $E = \bigcup_0^\infty E_m$ and, since $\{\mu_n : n \in \mathbb{N}\}$ is uniformly $\sigma -$ additive, for every $\varepsilon > 0$, there exists $m_0 \in \mathbb{N}$ such that

$$\left|\mu_n(E) - \sum_{i=0}^{m-1} \mu_n(E_{i+1} \setminus E_i)\right| < \varepsilon, \quad \text{for every } m \geq m_0 \text{ and every } n \in \mathbb{N}$$

which can be written:

$$|\mu_n(E) - \mu_n(E_m)| < \varepsilon, \quad \text{for every} \quad m \geq m_0 \quad \text{and every} \quad n \in \mathbb{N}. \quad (1)$$

It means that $(\mu_n(E_m))_m$ converges to $\mu_n(E)$, uniformly in $n \in \mathbb{N}$.

We have the same result if we assume that the sequence $(E_m)_m$ is decreasing to E.

If $E_{m_0} \in \mathcal{L}$, then $(\mu_n(E_{m_0}))_{n\in\mathbb{N}}$ is a Cauchy sequence; then there exists $n_0 \in \mathbb{N}$ such that,

$$|\mu_p(E_{m_0}) - \mu_q(E_{m_0})| < \varepsilon, \quad \text{for every} \quad p, q \geq n_0. \tag{2}$$

Therefore, by (1) and (2), for every $p, q \geq n_0$,

$$\begin{aligned}|\mu_p(E) - \mu_q(E)| \leq |\mu_p(E) - \mu_p(E_{m_0})| + |\mu_p(E_{m_0}) - \mu_q(E_{m_0})| \\ + |\mu_q(E_{m_0}) - \mu_q(E)| < 3\varepsilon.\end{aligned}$$

Then $(\mu_p(E))_{p\in\mathbb{N}}$ is convergent in $\mathbb{R}$ and therefore $E \in \mathcal{L}$; this proves (b).

(c) $\mathcal{L}$ is a monotone class containing $\mathcal{G}$. Then the monotone class generated by $\mathcal{G}$, $\mathcal{M}(\mathcal{G})$, is contained in $\mathcal{L}$. But, the σ-algebra $\sigma(\mathcal{G}) = \mathcal{A}$ is just the monotone class $\mathcal{M}(\mathcal{G}) \subseteq \mathcal{L} \subseteq \mathcal{A}$. Which proves the result. □

Remark 1.63. As we show in the following example, the condition of "uniformly σ-additivity" in the previous lemma is indispensable.

Let $\mathcal{G} = \{A \subseteq \mathbb{N} : A \text{ finite or } \mathbb{N} \setminus A \text{ finite}\}$; $\mathcal{G}$ is an algebra that generates the σ-algebra $\mathcal{P}(\mathbb{N})$. For every $n \in \mathbb{N}$, let $\mu_n = \delta_n$ the unit mass concentrated at n. Then $(\mu_n) \subseteq ca(\mathcal{P}(\mathbb{N}))$ and

$$\lim_{n\to+\infty} \mu_n(E) = \begin{cases} 0, & E \quad \text{is finite}, \\ 1, & X \setminus E \quad \text{is finite}, \end{cases} \qquad \text{for all} \quad E \in \mathcal{G}.$$

If P is the set of all even numbers of $\mathbb{N}$, then the limit $\lim_{n\to+\infty} \mu_n(P)$ doesn't exist. It is clear that $\{\mu_n : n \in \mathbb{N}\}$ is not uniformly σ-additive on $\mathbb{N}$.

Using the previous lemma, we can give the following weak compactness result on $ca(\mathcal{A})$ (see Theorems IV.9. 1 and 2 of [62] and Theorem 13, p. 92, of [57]).

Theorem 1.64. *Let $\mathcal{K} \subseteq ca(\mathcal{A})$; the following conditions are equivalent:*

(i) *$\mathcal{K}$ is relatively weakly compact in $(ca(\mathcal{A}), \|.\|)$,*

(ii) *$\mathcal{K}$ is bounded and uniformly σ-additive,*

(iii) *$\mathcal{K}$ is bounded and there exists a measure $\lambda \in ca^+(\mathcal{A})$ such that $\mathcal{K}$ is λ-equicontinuous.*

Proof. (i)$\Longrightarrow$(iii) We start by noticing that, in every normed space $(X, \|.\|)$, the relatively weakly compact sets are bounded. Let then $M > 0$ such that, for every $\mu \in \mathcal{K}$, $\|\mu\| \leq M$. Let us show that

$$\begin{array}{l}\text{for every } \varepsilon > 0, \text{ there exist } \{\mu_1, \ldots, \mu_n\} \subseteq \mathcal{K} \text{ and } \delta > 0 \text{ such that} \\ |\mu_k|(E) < \delta, \text{ for every } k = 1, \ldots, n, \Longrightarrow |\mu(E)| < \varepsilon, \text{ for every } \mu \in \mathcal{K}.\end{array} \tag{*}$$

By reducing it to absurd, let us suppose that the condition $(*)$ is not satisfied: there is $\varepsilon > 0$ such that, if we fix $\mu_1 \in \mathcal{K}$ and $\delta_1 = \frac{1}{2}$, there exist $E_1 \in \mathcal{A}$ and $\mu_2 \in \mathcal{K}$ with $|\mu_1|(E_1) < \frac{1}{2}$ and $|\mu_2(E_1)| \geq \varepsilon$. Similarly, for $\{\mu_1, \mu_2\} \subseteq \mathcal{K}$ and $\delta_2 = \frac{1}{2^2}$, there exist $E_2 \in \mathcal{A}$ and $\mu_3 \in \mathcal{K}$ with $|\mu_1|(E_2) < \frac{1}{2^2}$, $|\mu_2|(E_2) < \frac{1}{2^2}$ and $|\mu_3(E_2)| \geq \varepsilon$.

Recursively, for $\{\mu_1, \ldots, \mu_{n-1}\} \subseteq \mathcal{K}$ and $\delta_{n-1} = \frac{1}{2^{n-1}}$, there exist $E_{n-1} \in \mathcal{A}$ and $\mu_n \in \mathcal{K}$ with $|\mu_i|(E_{n-1}) < \frac{1}{2^{n-1}}, \quad \forall i = 1, \ldots, n-1$, and $|\mu_n(E_{n-1})| \geq \varepsilon$.

Therefore, we have found two sequences: $(\mu_n) \subseteq \mathcal{K}$ and $(E_n) \subseteq \mathcal{A}$ such that, for every $n \in \mathbb{N}$,

$$|\mu_i|(E_n) < \frac{1}{2^n}, \quad \forall i = 1, \ldots, n \quad \text{and} \quad |\mu_{n+1}(E_n)| \geq \varepsilon. \tag{1}$$

Since the set $\mathcal{K}$ is weakly sequentially compact, there exists a subsequence (μ_{k_n}) of (μ_n) weakly convergent to a measure $\mu \in ca(\mathcal{A})$. According to Corollary 1.58, for every $A \in \mathcal{A}$, $(\mu_{k_n}(A))_n$ converges to $\mu(A)$.

By (1), for any $n \in \mathbb{N}$,

$$|\mu_{k_i}|\left(E_{k_{n+1}-1}\right) < \frac{1}{2^{k_{n+1}-1}}, \ \forall i = 1, \ldots, n \text{ and } \left|\mu_{k_{n+1}}\left(E_{k_{n+1}-1}\right)\right| \geq \varepsilon. \tag{2}$$

Let

$$\lambda = \sum_{n=1}^{\infty} \frac{1}{2^n} |\mu_{k_n}| \in ca^+(\mathcal{A}).$$

By Vitali–Hahn–Saks theorem (see Theorem 1.36), the family $\{\mu_{k_n} : n \in \mathbb{N}\}$ is λ–equicontinuous and therefore there exists $\delta > 0$ such that

$$\left|\mu_{k_n}(A)\right| < \varepsilon, \text{ for every } A \in \mathcal{A} \text{ with } \lambda(A) < \delta, \text{ and for every } n \in \mathbb{N}. \tag{3}$$

For all $p \in \mathbb{N}$,

$$\begin{aligned}\lambda\left(E_{k_{p+1}-1}\right) &= \sum_{n=1}^{\infty} \frac{1}{2^n} |\mu_{k_n}|\left(E_{k_{p+1}-1}\right) \leq \sum_{n=1}^{p} \frac{1}{2^n} |\mu_{k_n}|\left(E_{k_{p+1}-1}\right) + \sum_{n=p+1}^{\infty} \frac{M}{2^n} \\ &\leq p \cdot \frac{1}{2} \cdot \frac{1}{2^{k_{p+1}-1}} + \frac{M}{2^p} = \frac{p}{2^{k_{p+1}}} + \frac{M}{2^p}.\end{aligned}$$

Therefore $\lambda(E_{k_{p+1}-1}) \to 0$ and so there exists $p_0 \in \mathbb{N}$ such that, for all $p \geq p_0$, $\lambda(E_{k_{p+1}-1}) < \delta$. According to (3),

$$\left|\mu_{k_n}\left(E_{k_{p+1}-1}\right)\right| < \varepsilon, \quad \text{for all} \quad n \in \mathbb{N} \quad \text{and for all} \quad p \geq p_0.$$

For $n = p + 1$, we obtain,

$$\left|\mu_{k_{p+1}}\left(E_{k_{p+1}-1}\right)\right| < \varepsilon, \quad \text{for every} \quad p \geq p_0 \tag{4}$$

which contradicts (2). Then, we have demonstrated $(*)$.

Now we use inductively the condition $(*)$:

$$\begin{cases} \varepsilon = 1, \text{ there exist } \{\mu_1, \ldots, \mu_{n_1}\} \subseteq \mathcal{K} \text{ such that, if } A \in \mathcal{A} \text{ satisfies} \\ |\mu_1|(A) = \ldots = |\mu_{n_1}|(A) = 0, \text{ then } |\mu(A)| < 1, \text{ for any } \mu \in \mathcal{K}. \end{cases}$$

$$\begin{cases} \varepsilon = \frac{1}{2}, \text{ there exist } \{\mu_{n_1+1}, \ldots, \mu_{n_2}\} \subseteq \mathcal{K} \text{ such that, if } A \in \mathcal{A} \text{ satisfies} \\ |\mu_{n_1+1}|(A) = \ldots = |\mu_{n_2}|(A) = 0, \text{ then } |\mu(A)| < \frac{1}{2}, \text{ for any } \mu \in \mathcal{K}. \end{cases}$$

....

$$\begin{cases} \varepsilon = \frac{1}{p}, \text{ there exist } \{\mu_{n_{p-1}+1}, \ldots, \mu_{n_p}\} \subseteq \mathcal{K} \text{ s. t., if } A \in \mathcal{A} \text{ satisfies} \\ |\mu_{n_{p-1}+1}|(E) = \ldots = |\mu_{n_p}|(E) = 0, \text{ then } |\mu(E)| < \frac{1}{p}, \text{ for any } \mu \in \mathcal{K}. \end{cases}$$

....

For every $p \in \mathbb{N}$, let $\nu_p = \mu_{n_{p-1}+1}$. Then the sequence $(\nu_p)_p \subseteq \mathcal{K}$ satisfies:

$$|\nu_p|(A) = 0, \text{ for any } p \in \mathbb{N} \Longrightarrow \mu(A) = 0, \text{ for every } \mu \in \mathcal{K}. \tag{5}$$

Let now $\lambda = \sum_{p=1}^{\infty} \frac{1}{2^p} |\nu_p|$; $\lambda \in ca^+(\mathcal{A})$. By (5),

$$\mu \ll \lambda, \quad \text{for every} \quad \mu \in \mathcal{K}. \tag{6}$$

Let us show, by reducing it to absurd, that $\mathcal{K}$ is λ-equicontinuous.

If $\mathcal{K}$ is not λ-equicontinuous, according to Theorem 1.34, it results that $\mathcal{K}$ is not λ-equicontinuous at $\emptyset$ and therefore there exists $\varepsilon > 0$ such that, for every $\delta > 0$, there exist $F_\delta \in \mathcal{A}$ and $\theta_\delta \in \mathcal{K}$ satisfying:

$$\lambda(F_\delta) < \delta \quad \text{and} \quad |\theta_\delta(F_\delta)| \geq \varepsilon.$$

For $\delta = \frac{1}{n}$, $n \in \mathbb{N}$, we obtain two sequences $(F_n)_n \subseteq \mathcal{A}$, $(\theta_n)_n \subseteq \mathcal{K}$ such that

$$\lambda(F_n) \to 0 \quad \text{and} \quad |\theta_n(F_n)| \geq \varepsilon, \quad \text{for all} \quad n \in \mathbb{N}. \tag{7}$$

Because $\mathcal{K}$ is weakly sequentially compact (see Eberlein–Šmulian theo rem [62], V.6.1), perhaps extracting a subsequence, we may assume that $(\theta_n)_n$ is weakly convergent to a measure $\theta \in ca(\mathcal{A})$. By (6), $\theta_n \ll \lambda$, for all $n \in \mathbb{N}$; then $(\theta_n)_n$ satisfies the hypotheses of the Vitali–Hahn–Saks theorem (see Theorem 1.36). Therefore (θ_n) is λ-equicontinuous and this contradicts (7).

(iii)$\Longrightarrow$(ii) The implication is a consequence of Theorem 1.34.

(ii)$\Longrightarrow$(i) Let $\mathcal{K} \subseteq ca(\mathcal{A})$ be a bounded and uniformly σ-additive family of measures; let $(\mu_n)_{n\in\mathbb{N}}$ be any sequence in $\mathcal{K}$ and let

$$\lambda = \sum_{n=1}^{\infty} \frac{1}{2^n}|\mu_n|; \quad \lambda \in ca^+(\mathcal{A}).$$

For every $n \in \mathbb{N}, \mu_n \ll \lambda$ and, according to Radon–Nikodym (Theorem 1.40), for any $n \in \mathbb{N}$, there exists a function $f_n \in L^1(\lambda)$ such that

$$\mu_n(E) = \int_E f_n d\lambda, \quad \text{for every} \quad E \in \mathcal{A}.$$

For all $n \in \mathbb{N}$, there exists a sequence of simple functions $(f_p^n)_{p\in\mathbb{N}} \subseteq \mathcal{E}(\mathcal{A})$ such that $f_p^n \xrightarrow[p\to\infty]{} f_n$, λ-a.e. If

$$f_n^p = \sum_{k=1}^{m(n,p)} a_k^{n,p} \chi_{A_k^{n,p}},$$

then, for every $n \in \mathbb{N}$, let us denote with $\mathcal{F}_n = \{A_k^{n,p} : p \in \mathbb{N}, k = 1, \ldots, m(n,p)\}$ and let $\mathcal{F} = \bigcup_{n=1}^{\infty} \mathcal{F}_n \subseteq \mathcal{A}$. $\mathcal{F}$ is a countably family of sets and therefore the algebra $\mathcal{B}$ generated by $\mathcal{F}$, is also countable.

So, let $\mathcal{B} = \{E_1, \ldots, E_p, \ldots\}$.

By using a diagonal method, we obtain a subsequence $(\mu_{k_n})_{n\in\mathbb{N}}$ such that $(\mu_{k_n}(E_p))_{n\in\mathbb{N}}$ is convergent in $\mathbb{R}$, for every $p \in \mathbb{N}$.

$(\mu_{k_n})_n$ being still uniformly σ-additive, $(\mu_{k_n}(E))_{n\in\mathbb{N}}$ is convergent, for every $E \in \sigma(\mathcal{B})$ - the σ-algebra generated by $\mathcal{B}$ (see Lemma 1.62).

We note by $\sigma_1(\mathcal{B})$ the λ-completion of $\sigma(\mathcal{B})$ and by $L^1(\lambda, \sigma_1(\mathcal{B}))$ the space of all $\sigma_1(\mathcal{B})$-measurable functions f, with $\int_X |f| d\lambda < +\infty$.

Then f_n^p is $\sigma(\mathcal{B})$-measurable, for all $p, n \in \mathbb{N}$; therefore, for every n, f_n is $\sigma_1(\mathcal{B})$-measurable. Hence we have: $(f_n) \subseteq L^1(\lambda, \sigma_1(\mathcal{B})) \subseteq L^1(\lambda)$ and, according to Theorem 1.57, (f_{k_n}) is weakly convergent in $L^1(\lambda, \sigma_1(\mathcal{B}))$ (we need to remark that $\int_E f_{k_n} d\lambda \to 0$, for every $E \in \mathcal{A}$ with $\lambda(E) = 0$, and therefore $(\int_E f_{k_n} d\lambda)_{n\in\mathbb{N}}$ is convergent, for every $E \in \sigma_1(\mathcal{B})$).

Since $L^1(\lambda, \sigma_1(\mathcal{B})) \subseteq L^1(\lambda)$, (f_{k_n}) is weakly convergent in $L^1(\lambda)$; it means that (μ_{k_n}) is weakly convergent in $ca(\mathcal{A})$ (see Remark 1.43).

Therefore, $\mathcal{K}$ is weakly sequentially compact and then we can apply again Eberlein–Šmulian theorem ([62], V.6.1) to obtain that it is relatively weakly compact. □

Theorem 1.65 (Dunford–Pettis; see Theorem 2.54 of [85] and [57], p. 93). *Let $\lambda \in ca^+(\mathcal{A})$ and $K \subseteq L^1(\lambda)$.*

The following conditions are equivalent:

(i) *K is relatively weakly compact in $(L^1(\lambda), \|\cdot\|_1)$,*

(ii) *K is bounded and the integrals of members of K are uniformly σ-additive,*

(iii) *K is bounded and λ-equicontinuous.*

Proof. According to the notations introduced at the beginning of the Section 1.2, let $\Psi : L^1(\lambda) \to ca(\mathcal{A})$ defined by $\Psi(f) = \mu_f$, where, for every $f \in L^1(\lambda)$, μ_f is defined by $\mu_f(A) = \int_A f d\lambda$, for every $A \in \mathcal{A}$. Ψ is an isometric isomorphism between the normed spaces $L^1(\lambda)$ and $ca_\lambda(\mathcal{A}) \subseteq ca(\mathcal{A})$ (Theorem 1.43). Therefore the subset $K \subseteq L^1(\lambda)$ is relatively weakly compact in $(L^1(\lambda), \|\cdot\|_1)$ if and only if $\mathcal{K} = \Psi(K)$ is relatively weakly compact in $ca_\lambda(\mathcal{A})$; since $ca_\lambda(\mathcal{A})$ is a closed subset of $ca(\mathcal{A})$, this is equivalent with $\Psi(K)$ is relatively weakly compact in $ca(\mathcal{A})$.

Obviously, the condition (ii) of Theorem 1.64 on $\mathcal{K}$ is equivalent with the above condition (ii) on K.

For the conditions (iii) of these two theorems, we need to demonstrate the equivalence between the two following conditions:

(a) K is λ-equicontinuous;

(b) $\exists \lambda' \in ca^+(\mathcal{A})$ for which $\mathcal{K}$ is λ'-equicontinuous.

We can rewrite these conditions:

(a) For every $\varepsilon > 0$, there exists $\delta > 0$ such that $\int_A |f| d\lambda < \varepsilon$, for every $A \in \mathcal{A}$ with $\lambda(A) < \delta$ and every $f \in K$.

(b) There exists $\lambda' \in ca^+(\mathcal{A})$ such that, for every $\varepsilon > 0$, there exists $\delta > 0$ such that $\int_A |f| d\lambda < \varepsilon$, for every $A \in \mathcal{A}$ with $\lambda'(A) < \delta$ and every $f \in K$.

Obviously, (a)$\Longrightarrow$(b) (with $\lambda' = \lambda$).

(b)$\Longrightarrow$(a) Let $\lambda' = \lambda'_1 + \lambda'_2$ be the Lebesgue decomposition of λ' relatively to λ (Theorem 1.47) where $\lambda'_1, \lambda'_2 \in ca(\mathcal{A})$, $\lambda'_1 \ll \lambda$ and $\lambda'_2 \perp \lambda$; therefore there exists a set $A_0 \in \mathcal{A}$ such that $\lambda'_2(A_0) = 0 = \lambda(X \setminus A_0)$.

According to (b), for every $\varepsilon > 0$, there exists $\delta_1 > 0$ such that, for every $A \in \mathcal{A}$ with $\lambda'(A) < \delta_1$, we have $\int_A |f| d\lambda < \varepsilon$, for any $f \in K$. Since $\lambda'_1 \ll \lambda$, there exists $\delta > 0$ such that $\lambda(A) < \delta \Rightarrow \lambda'_1(A) < \delta_1$.

Then, for any $A \in \mathcal{A}$ with $\lambda(A) < \delta$,

$$\lambda'(A \cap A_0) = \lambda'_1(A \cap A_0) + \lambda'_2(A \cap A_0) \leq \lambda'_1(A) < \delta_1$$

and therefore

$$\int_{A\cap A_0} |f|d\lambda < \varepsilon, \quad \text{for every} \quad f \in K$$

or yet

$$\int_A |f|d\lambda = \int_{A\cap A_0} |f|d\lambda + \int_{A\setminus A_0} |f|d\lambda = \int_{A\cap A_0} |f|d\lambda < \varepsilon, \text{ for every } f \in K.$$

Which was to be demonstrated. □

Remark 1.66.

(i) If $\lambda : \mathcal{A} \to \overline{\mathbb{R}}_+$ is a σ-finite measure, then $K \subseteq L^1(\lambda)$ is relatively weakly compact if and only if

  (a) K is bounded,

  (b) K is λ-equicontinuous,

  (c) $\forall\varepsilon > 0, \exists A_\varepsilon \in \mathcal{A}$ with $\lambda(A_\varepsilon) < +\infty$ and $\int_{X\setminus A_\varepsilon} |f|d\lambda < \varepsilon$, for any $f \in K$.

(ii) If $(f_n)_{n\in\mathbb{N}} \subseteq L^1(\lambda)$ is weakly convergent, then $K = \{f_n : n \in \mathbb{N}\}$ is λ-equicontinuous. Indeed, K is relatively weakly compact in $(L^1(\lambda), \|\cdot\|_1)$ and therefore, by Theorem 1.65, K is λ-equicontinuous.

(iii) We can now restate the Vitali's theorem (see Theorem 1.54) in the following manner:

Let $\lambda \in ca^+(\mathcal{A})$; a sequence $(f_n)_n \subseteq L^1(\lambda)$ is strongly convergent to f if and only if (f_n) converges in measure and (f_n) is weakly convergent to f.

1.5 The Bidual of $L^1(\lambda)$

Let λ be a σ-additive, σ-finite, positive measure on the σ-algebra $\mathcal{A}$, let $L^1(\lambda)$ be the space of all λ-integrable mappings, let $L^\infty(\lambda)$ be the dual space of $L^1(\lambda)$ and, in accordance with Definition 1.28, let

$$ba_\lambda(\mathcal{A}) = \{\mu \in ba(\mathcal{A}) : \mu \ll \lambda\}.$$

In this section, we will identify $ba_\lambda(\mathcal{A})$ with the bidual of $L^1(\lambda)$. Then, we will give the characteristic properties of the measures of this bidual and we will identify a purely finite-additive measure of $ba(\mathcal{A})$ as being a measure orthogonal on every measure of $ca(\mathcal{A})$. We will conclude this part with the Hewitt–Yosida decomposition of every additive measure in a σ-additive part and a purely finite-additive one.

Theorem 1.67 (Theorem IV.8.16 of [62]). *The dual space of $L^\infty(\lambda)$ is isometrically isomorphic as normed space with $ba_\lambda(\mathcal{A})$.*

Proof. Let $x^* \in [L^\infty(\lambda)]^*$; then $x^* : L^\infty(\lambda) \to \mathbb{R}$ is a continuous linear mapping on $L^\infty(\lambda)$. Since, for all $A \in \mathcal{A}$, $\chi_A \in L^\infty(\lambda)$, we can define $\mu_{x^*} : \mathcal{A} \to \mathbb{R}$ by: $\mu_{x^*}(A) = x^*(\chi_A)$, for every $A \in \mathcal{A}$. Obviously, μ_{x^*} is additive.

$$|\mu_{x^*}(A)| \le \|x^*\| \cdot \|\chi_A\|_\infty \le \|x^*\| < +\infty, \quad \text{for every} \quad A \in \mathcal{A}.$$

μ_{x^*} is therefore bounded.

For every $A \in \mathcal{A}$ with $\lambda(A) = 0$, $\chi_A = 0$, λ-a.e. We have, therefore, $\mu_{x^*}(A) = x^*(\chi_A) = 0$; therefore $\mu_{x^*} \in ba_\lambda(\mathcal{A})$.

Let now $S : [L^\infty(\lambda)]^* \to ba_\lambda(\mathcal{A})$ defined by:

$$S(x^*) = \mu_{x^*}, \quad \text{for every} \quad x^* \in [L^\infty(\lambda)]^* .$$

We will show that S is an isometric isomorphism of normed spaces.

• It is evident that S is linear and that

$$\begin{aligned}\|S(x^*)\| &= \|\mu_{x^*}\| = |\mu_{x^*}|(X) \\ &= \sup\left\{\sum_{i=1}^n |x^*(\chi_{A_i})| : \{A_1, \dots, A_n\} = \mathcal{A} - \quad \text{partition of} \quad X\right\}.\end{aligned}$$

For every $\mathcal{A}$-partition $\{A_1, \dots, A_n\}$ of X, we can rewrite the sets in such a manner as

$$x^*(\chi_{A_1}), \dots, x^*(\chi_{A_p}) \ge 0 \quad \text{and} \quad x^*(\chi_{A_{p+1}}), \dots, x^*(\chi_{A_n}) < 0.$$

Then

$$\begin{aligned}\sum_1^n |x^*(\chi_{A_i})| &= \sum_1^p x^*(\chi_{A_i}) - \sum_{p+1}^n x^*(\chi_{A_i}) = x^*\left(\sum_1^p \chi_{A_i} - \sum_{p+1}^n \chi_{A_i}\right) \\ &\le \|x^*\| \cdot \left\|\sum_1^p \chi_{A_i} - \sum_{p+1}^n \chi_{A_i}\right\|_\infty = \|x^*\|,\end{aligned}$$

therefore

$$\|S(x^*)\| \le \|x^*\|, \quad \text{for every} \quad x^* \in [L^\infty(\lambda)]^* . \tag{1}$$

• S is surjective.

Indeed, for every $\mu \in ba_\lambda(\mathcal{A})$, we define $x^* : \mathcal{E}(\mathcal{A}) \to \mathbb{R}$ letting $x^*(f) = \sum_{i=1}^n a_i \mu(A_i)$, for every $f = \sum_1^n a_i \chi_{A_i} \in \mathcal{E}(\mathcal{A})$, where $\{a_1, \dots, a_n\} \subseteq \mathbb{R}$ and $\{A_1, \dots, A_n\}$ is a $\mathcal{A}$-partition of X. Then x^* is a linear mapping on $\mathcal{E}(\mathcal{A})$ and, since $\mu \ll \lambda$, for every $f \in \mathcal{E}(\mathcal{A})$,

$$\begin{aligned}|x^*(f)| = \left|\sum_1^n a_i \mu(A_i)\right| &\le \sum_1^n |a_i| \cdot |\mu(A_i)| \\ &\le \sup\{|a_i| : \mu(A_i) \neq 0\} \cdot \sum_1^n |\mu(A_i)| \\ &\le \sup\{|a_i| : \lambda(A_i) \neq 0\} \cdot \sum_1^n |\mu(A_i)| \le \|f\|_\infty \cdot \|\mu\|.\end{aligned}$$

We have shown that:

$$|x^*(f)| \le \|\mu\| \cdot \|f\|_\infty, \quad \text{for every} \quad f \in \mathcal{E}(\mathcal{A}). \tag{2}$$

Hence x^* is a continuous linear mapping on $\mathcal{E}(\mathcal{A})$. We can extend by continuity this mapping to $L^\infty(\lambda) = \overline{\mathcal{E}(\mathcal{A})}^\infty$. The extension, written again x^*, is a continuous linear real mapping on $(L^\infty(\lambda), \|\cdot\|_\infty)$, therefore $x^* \in [L^\infty(\lambda)]^*$. For every $g \in L^\infty(\lambda)$, let us denote

$$x^*(g) = \int_X g d\mu.$$

Then $S(x^*) = \mu_{x^*}$, where, for every $A \in \mathcal{A}$, $\mu_{x^*}(A) = x^*(\chi_A) = \mu(A)$.

Therefore, we obtain $S(x^*) = \mu$.

• Now we show that the map S preserves the norm. For every $f \in L^\infty(\lambda)$ with $\|f\|_\infty \le 1$, there exists a sequence $(f_n) \subseteq \mathcal{E}(\mathcal{A})$ such that $f_n \xrightarrow{\|\cdot\|_\infty} f$. Then, by (2),

$$|x^*(f)| = \lim_n |x^*(f_n)| \le \lim_n \|\mu\| \cdot \|f_n\|_\infty = \|\mu\| \cdot \|f\|_\infty \le \|\mu\| = \|S(x^*)\|$$

hence

$$\|x^*\| = \sup_{\|f\|_\infty \le 1} |x^*(f)| \le \|S(x^*)\|. \tag{3}$$

From (1) and (3), we obtain:

$$\|S(x^*)\| = \|x^*\|, \quad \text{for every} \quad x^* \in [L^\infty(\lambda)]^* .$$

S is therefore an isometric isomorphism between the normed spaces $[L^\infty(\lambda)]^*$ and $ba_\lambda(\mathcal{A})$. □

According to Theorem 1.49, we deduce the following corollary.

Corollary 1.68. *The space $(ba_\lambda(\mathcal{A}), \|\cdot\|)$ is the bidual of $(L^1(\lambda), \|\cdot\|_1)$.*

Remark 1.69. $ca_\lambda(\mathcal{A}) \subseteq ba_\lambda(\mathcal{A})$ and the mapping $\Psi : L^1(\lambda) \to ba_\lambda(\mathcal{A})$, defined by $\Psi(f) = f \cdot \lambda$ for every $f \in L^1(\lambda)$, is the natural embedding of $L^1(\lambda)$ in its bidual. As a consequence of Radon–Nikodym theorem (1.40), we have shown (Theorem 1.43) that $L^1(\lambda)$ is isometric isomorphic with $ca_\lambda(\mathcal{A})$. According to Alaoglu theorem [62, V.4.2] and its immediate consequence, the Goldstine's theorem ([62] V.4.5 and 6), $\Psi(L^1(\lambda))$ is $\sigma(L^\infty(\lambda), ba_\lambda(\mathcal{A}))$ -dense in $ba_\lambda(\mathcal{A})$. This will allow us to characterize the measures of $ba_\lambda(\mathcal{A})$ through nets (generalized sequences).

Theorem 1.70. *Let $\lambda : \mathcal{A} \to \overline{\mathbb{R}}_+$ be a σ-additive and σ- finite measure and let $\mu \in ba(\mathcal{A})$.*

$$\mu \in ba_\lambda(\mathcal{A}) \iff \text{there exists a net} \quad (f_i)_{i\in I} \subseteq L^1(\lambda) \quad \text{such that}$$
$$\mu(A) = \lim_i \int_A f_i \, d\lambda, \quad \text{for every} \quad A \in \mathcal{A}.$$

If this net is λ-equicontinuous, then $\mu \in ca_\lambda(\mathcal{A})$.

Proof. ($\Rightarrow$) According to Theorem 1.43 and with the notations already used, the mapping $\Psi : L^1(\lambda) \to ba_\lambda(\mathcal{A})$, defined by $\Psi(f) = \mu_f = f \cdot \lambda$, is an isometric isomorphism between the normed spaces $L^1(\lambda)$ and $ca_\lambda(\mathcal{A})$. Let again Ψ be the natural embedding of $L^1(\lambda)$ in its bidual $ba_\lambda(\mathcal{A})$ and let w* be the topology $\sigma(ba_\lambda(\mathcal{A}), L^\infty(\lambda))$ on $ba_\lambda(\mathcal{A})$; according to the previous remark, $ca_\lambda(\mathcal{A}) = \Psi(L^1(\lambda))$ is w*- dense in $ba_\lambda(\mathcal{A})$. Therefore, for all $\mu \in ba_\lambda(\mathcal{A})$, there exists a net $(f_i)_{i\in I} \subseteq L^1(\lambda)$ such that $\Psi(f_i) \xrightarrow{w*} \mu$.

Since, for all $A \in \mathcal{A}, \chi_A \in L^\infty(\lambda)$, $\Psi(f_i)(\chi_A) = \int_E f_i \, d\lambda \to \mu(A)$. Therefore $\mu(A) = \lim_i \int_A f_i \, d\lambda$, for every $A \in \mathcal{A}$.

($\Leftarrow$) Let now $(f_i)_{i\in I} \subseteq L^1(\lambda)$ be a net such that, for every $A \in \mathcal{A}$, $\mu(A) = \lim_i \int_A f_i \, d\lambda$. For every $A \in \mathcal{A}$ with $\lambda(A) = 0, \int_A f_i \, d\lambda = 0, \quad \forall i \in I$; therefore $\mu(A) = 0$, from where $\mu \ll \lambda$ and $\mu \in ba_\lambda(\mathcal{A})$.

Now, if $(f_i)_{i\in I}$ is λ-equicontinuous, then, by Dunford–Pettis' theorem (Theorem 1.65), $\{f_i : i \in I\}$ is relatively weakly compact. So there exists a subnet of $(f_i)_{i\in I}$ weakly convergent to a function $f \in L^1(\lambda)$. Then $\mu(A) = \int_A f d\lambda$, for every $A \in \mathcal{A}$; therefore $\mu = f.\lambda$ and, according to Proposition 1.31, $\mu \in ca_\lambda(\mathcal{A})$. □

Remark 1.71.

(i) In accordance with the notations of the previous theorem, let $\mu_i : \mathcal{A} \to \mathbb{R}$, $\mu_i = f_i \cdot \lambda$. Therefore $\mu_i(E) \to \mu(E)$, for all $E \in \mathcal{A}$ and μ does not necessarily belong to $ca(\mathcal{A})$. This shows that the Nikodym's theorem (1.37) is no longer true if we replace the sequences by nets.

(ii) Let X be a countable set, $\mathcal{P}(X)$ be the family of all subsets of X and $\lambda_0 : \mathcal{P}(X) \to [0, +\infty]$, $\lambda_0(A) = \sum_{x\in X} \delta_x(A)$, where δ_x is the unit mass concentrated in x ($\lambda_0(A)$ gives the number of elements of A).

Then λ_0 is a positive, σ-additive and σ-finite measure on $\mathcal{P}(X)$.

Let $B(X) = \{f : X \to \mathbb{R} : f \quad \text{is bounded on} \quad X\}$ and $\|\cdot\| : B(X) \to \mathbb{R}_+$, $\|f\| = \sup_{x\in X} |f(x)|$.

Then $(B(X), \|\cdot\|) \equiv (L^\infty(\lambda_0), \|\cdot\|_\infty)$ is a Banach space. According to Theorem 1.67, the dual space of $(B(X), \|\cdot\|)$ is $ba_{\lambda_0}(\mathcal{P}(X))$.

Because $\lambda_0(A) = 0 \Leftrightarrow A = \emptyset, ba_{\lambda_0}(\mathcal{P}(X)) = ba(\mathcal{P}(X))$ $(\mu \ll \lambda_0, \quad \forall \mu \in ba(\mathcal{P}(X)))$. Therefore $(ba(\mathcal{P}(X), \|\cdot\|)$ is the dual space of $(B(X), \|\cdot\|)$.

Definition 1.72. (Definition III.7.7 of [62]) A measure $\mu \in ba^+(\mathcal{A})$ is said to be *purely finitely additive* if the inequality $0 \le \lambda \le \mu$, with λ countably additive, imply $\lambda = 0$.

Let $\mu \in ba(\mathcal{A})$ and let $\mu^+, \mu^- \in ba^+(\mathcal{A})$ be the positive and the negative variation of μ (see Definition 1.19). μ is called *purely finitely additive* if μ^+ and μ^- are purely finitely additive.

Definition 1.73. Two additive measures μ, ν on $\mathcal{A}$ are called *orthogonal measures* if, for every $\varepsilon > 0$, there exists $A_\varepsilon \in \mathcal{A}$ such that $|\mu|(A_\varepsilon) < \varepsilon$ and $|\nu|(X \setminus A_\varepsilon) < \varepsilon$; this is denoted by $\mu \perp \nu$.

Remark 1.74.

(i) This definition is in according to Definition 1.45; indeed if μ and ν are countable additives ($\mu, \nu \in ca(\mathcal{A})$), then, according with Proposition 1.46, the following assertions are equivalent:

(a) There exists a set $D \in \mathcal{A}$ such that $|\mu|$ is concentrated on D and $|\nu|$ is concentrated on $D^c = X \setminus D$.

(b) For every $\varepsilon > 0$, there exists a set $A_\varepsilon \in \mathcal{A}$ such that $|\mu|(A_\varepsilon) < \varepsilon$ and $|\nu|(X \setminus A_\varepsilon) < \varepsilon$.

(ii) If μ and ν are only finitely additive, the previous properties are no longer equivalent. We have the following implication: $(a) \Rightarrow (b)$.

As shown in the following example, the implication $(b) \Rightarrow (a)$ is not true.

Example 1.75. Let X be an arbitrary set and let $\mu : \mathcal{P}(X) \to \{0,1\}$ be an additive measure such that $\mu(X) = 1$. Let us denote $\mathcal{U} = \{A \subseteq X : \mu(A) = 1\}$. It is easy to see that $\mathcal{U}$ is an ultrafilter on X.

Reciprocally, for every ultrafilter $\mathcal{U}$ on X, μ, defined by $\mu(A) = 1$ if $A \in \mathcal{U}$ and $\mu(A) = 0$ if $A \notin \mathcal{U}$, is an additive measure on X associated to the ultrafilter $\mathcal{U}$.

Let now $\mathcal{F} = \{A \subseteq \mathbb{N} : \mathbb{N} \setminus A \text{ is finite }\}$. $\mathcal{F}$ is a filter on $\mathbb{N}$; let $\mathcal{U}$ be the ultrafilter which contains $\mathcal{F}$ and let μ be the additive measure associated to $\mathcal{U}$. μ is purely finitely additive. Indeed, $\mu \in ba^+(\mathcal{P}(\mathbb{N}))$ and, for every countably additive measure $\lambda \in ca^+(\mathcal{P}(\mathbb{N}))$ with $0 \le \lambda \le \mu$, $\lambda(\mathbb{N}) = \sum_{n \in \mathbb{N}} \lambda(\{n\}) = 0$ (if there were a number $n_0 \in \mathbb{N}$ with $\lambda(\{n_0\}) > 0$ then $\mu(\{n_0\}) = 1$ and so $\{n_0\} \in \mathcal{U}$ what is impossible because $\mathbb{N} \setminus \{n_0\} \in \mathcal{F} \subseteq \mathcal{U}$).

Let $m : \mathcal{P}(\mathbb{N}) \to \mathbb{R}_+$ defined by $m(A) = \sum_{n \in A} \frac{1}{(n+1)^2}$ if $A \neq \emptyset$ and $m(\emptyset) = 0$. It is clear that $m \in ca^+(\mathcal{P}(\mathbb{N}))$. We show that $m \perp \mu$. For every $\varepsilon > 0$, there exists $n_0 \in \mathbb{N}$ such that $\sum_{n \ge n_0} \frac{1}{(n+1)^2} < \varepsilon$ and therefore $m(A_0) < \varepsilon$, where $A_0 = \{n_0, n_0 + 1, \dots\}$. $\mathbb{N} \setminus A_0$ is finite so that $A_0 \in \mathcal{U}$; then $\mathbb{N} \setminus A_0 \notin \mathcal{U}$ and so $\mu(\mathbb{N} \setminus A_0) = 0 < \varepsilon$. Therefore $m \perp \mu$.

On the other hand, if there were a set $B_0 \in \mathcal{P}(\mathbb{N})$ with $m(B_0) = 0 = \mu(\mathbb{N} \setminus B_0)$, then $\mathbb{N} \setminus B_0 \notin \mathcal{U}$ so that $B_0 \in \mathcal{U}$ and then $\mu(B_0) = 1$; because $m(B_0) = 0$, $B_0 = \emptyset$ and then $\mathbb{N} \setminus B_0 \in \mathcal{U}$ which is impossible.

So $(b) \not\Rightarrow (a)$.

We have seen in the previous example that the purely finitely additive measure μ is orthogonal to the σ-additive measure m.

The following theorem shows that this is a general property for all purely finitely additive measure (see [175]).

Theorem 1.76. *Let $\mu \in ba(\mathcal{A})$; then μ is purely finitely additive if and only if μ is orthogonal on every σ-additive measure $\lambda \in ca(\mathcal{A})$.*

Proof. We can notice that $\mu \perp \lambda$ if and only if $\mu \perp |\lambda|$ and then we can suppose that $\lambda \in ca^+(\mathcal{A})$.

($\Longrightarrow$)

a) Let $\mu \in ba^+(\mathcal{A})$ be a purely finitely additive measure and let $\lambda \in ca^+(\mathcal{A})$. We first define $\nu : \mathcal{A} \to \mathbb{R}_+$, letting $\nu(E) = \inf\{\mu(A) + \lambda(E \setminus A) : A \in \mathcal{A}, A \subseteq E\}$, for every $E \in \mathcal{A}$. Then

$$0 \leq \nu(E) \leq \lambda(E), \quad \text{for every} \quad E \in \mathcal{A}. \tag{1}$$

Therefore ν is bounded.

ν is an additive measure. Indeed, for every $E, F \in \mathcal{A}$, such that $E \cap F = \emptyset$ we note $G = E \cup F$. For every $\varepsilon > 0$, there exist $A, B \in \mathcal{A}, A \subseteq E, B \subseteq F$ such that $\mu(A) + \lambda(E \setminus A) < \nu(E) + \varepsilon$ and $\mu(B) + \lambda(F \setminus B) < \nu(F) + \varepsilon$. If $C = A \cup B, C \in \mathcal{A}, C \subseteq G$ and $\nu(G) \leq \mu(C) + \lambda(G \setminus C) = \mu(A) + \mu(B) + \lambda(E \setminus A) + \lambda(F \setminus B) < \nu(E) + \nu(F) + 2\varepsilon$ from where

$$\nu(G) \leq \nu(E) + \nu(F). \tag{2}$$

On the other hand, there exists $D \in \mathcal{A}$ with $D \subseteq G$ such that $\mu(D) + \lambda(G \setminus D) < \nu(G) + \varepsilon$. Then $\nu(G) + \varepsilon > \mu(D) + \lambda(G \setminus D) = \mu(D \cap E) + \mu(D \cap F) + \lambda(E \setminus (D \cap E)) + \lambda(F \setminus (D \cap F)) \geq \nu(E) + \nu(F)$ from where

$$\nu(E) + \nu(F) \leq \nu(G). \tag{3}$$

(1), (2) and (3) lead to $\nu(E \cup F) = \nu(E) + \nu(F)$ and so $\nu \in ba^+(\mathcal{A})$.

Let us show that ν is σ-additive.

For every sequence $(E_n)_{n\in\mathbb{N}} \subseteq \mathcal{A}$ of pairwise disjoint sets, let us denote $E = \cup_1^\infty E_n$. From (1), $0 \leq \nu(E \setminus \cup_1^n E_k) \leq \lambda(E \setminus \cup_1^n E_k)$, for every $n \in \mathbb{N}$. Because $\lambda \in ca^+(\mathcal{A})$, $\lambda(E \setminus \cup_1^n E_k) \to 0$ when $n \to \infty$. Then $\nu(E) = \lim_n \nu(\cup_1^n E_k) = \lim_n \sum_1^n \nu(E_k) = \sum_1^\infty \nu(E_n)$, and so, $\nu \in ca^+(\mathcal{A})$.

From the definition of ν, for every $E \in \mathcal{A}$, $0 \leq \nu(E) \leq \mu(E)$. Since μ is purely finitely additive, $\nu = 0$.

Therefore $\nu(X) = 0$ and then, for every $\varepsilon > 0$, there exists $A_\varepsilon \in \mathcal{A}$ such that $\mu(A_\varepsilon) + \lambda(X \setminus A_\varepsilon) < \varepsilon$. Then $\mu(A_\varepsilon) < \varepsilon$ and $\lambda(X \setminus A_\varepsilon) < \varepsilon$. So $\mu \perp \lambda$.

b) If $\mu \in ba(\mathcal{A})$ is purely finitely additive then μ^+, μ^- are purely finitely additive and positive. From the part (a), μ^+ and μ^- are orthogonal to λ. Hence, for every $\lambda \in ca^+(\mathcal{A})$, and for all $\varepsilon > 0$, there exist A_ε and $B_\varepsilon \in \mathcal{A}$ such that $\mu^+(A_\varepsilon) < \frac{\varepsilon}{2}$, $\lambda(X \setminus A_\varepsilon) < \frac{\varepsilon}{2}$ and $\mu^-(B_\varepsilon) < \frac{\varepsilon}{2}$, $\lambda(X \setminus B_\varepsilon) < \frac{\varepsilon}{2}$. Let $C_\varepsilon = A_\varepsilon \cap B_\varepsilon \in \mathcal{A}$; then $|\mu|(C_\varepsilon) = \mu^+(A_\varepsilon \cap B_\varepsilon) + \mu^-(A_\varepsilon \cap B_\varepsilon) \leq \mu^+(A_\varepsilon) + \mu^-(B_\varepsilon) < \varepsilon$ and $\lambda(X \setminus C_\varepsilon) \leq \lambda(X \setminus A_\varepsilon) + \lambda(X \setminus B_\varepsilon) < \varepsilon$.

Therefore $\mu \perp \lambda$.

($\Longleftarrow$) Suppose that, for every $\lambda \in ca^+(\mathcal{A})$, $\mu \perp \lambda$. Then, for every $\varepsilon > 0$, there exists $A_\varepsilon \in \mathcal{A}$ with $|\mu|(A_\varepsilon) < \varepsilon$, and so $\mu^+(A_\varepsilon) < \varepsilon, \mu^-(A_\varepsilon) < \varepsilon$ and $\lambda(X \setminus A_\varepsilon) < \varepsilon$ from where, for every $\lambda \in ca^+(\mathcal{A})$, $\mu^+ \perp \lambda$ and $\mu^- \perp \lambda$.

Let now $\lambda \in ca^+(\mathcal{A})$ with $0 \le \lambda \le \mu^+$; since $\mu^+ \perp \lambda$, for every $\varepsilon > 0$, there exists $A_\varepsilon \in \mathcal{A}$ such that $\mu^+(A_\varepsilon) < \varepsilon$ and $\lambda(X \setminus A_\varepsilon) < \varepsilon$. Hence, for every $\varepsilon > 0$, $\lambda(X) = \lambda(A_\varepsilon) + \lambda(X \setminus A_\varepsilon) \le \mu^+(A_\varepsilon) + \lambda(X \setminus A_\varepsilon) < 2\varepsilon$; then $\lambda(X) = 0$ and so $\lambda = 0$. Therefore μ^+ is purely finitely additive. Similarly we show that μ^- is purely finitely additive and so μ is purely finitely additive. □

Corollary 1.77. *The subset $ba_p(\mathcal{A}) \subseteq ba(\mathcal{A})$ of all purely finitely additive measures is a subspace of the real vector space $ba(\mathcal{A})$.*

Proof. For every $\mu_1, \mu_2 \in ba_p(\mathcal{A})$, $\mu = \mu_1 + \mu_2 \in ba(\mathcal{A})$. From the previous theorem, for any $\lambda \in ca(\mathcal{A})$, $\mu_1 \perp \lambda$ and $\mu_2 \perp \lambda$. Therefore, for every $\varepsilon > 0$, there exist $A_1, B_1 \in \mathcal{A}$ such that $|\mu_1|(A_1) < \frac{\varepsilon}{2}$ and $|\lambda|(X \setminus A_1) < \frac{\varepsilon}{2}$, $|\mu_2|(A_2) < \frac{\varepsilon}{2}$ and $|\lambda|(X \setminus A_2) < \frac{\varepsilon}{2}$. Since $|\mu| = |\mu_1 + \mu_2| \le |\mu_1| + |\mu_2|$, $|\mu|(A_1 \cap A_2) \le |\mu_1|(A_1) + |\mu_2|(A_2) < \varepsilon$ and $|\lambda|(X \setminus (A_1 \cap A_2)) \le |\lambda|(X \setminus A_1) + |\lambda|(X \setminus A_2) < \varepsilon$. Therefore, for every $\lambda \in ca(\mathcal{A})$, $\mu \perp \lambda$ and, by previous theorem, $\mu \in ba_p(\mathcal{A})$.

For every $\mu \in ba_p(\mathcal{A})$, for all $a \in \mathbb{R}$, let $\nu = a \cdot \mu$, $\nu \in ba(\mathcal{A})$. Then $|\nu| = |a| \cdot |\mu|$ is evidently orthogonal to every $\lambda \in ca^+(\mathcal{A})$, hence $\nu \in ba_p(\mathcal{A})$. □

Now we can present an important result of decomposition for the finitely additive measures. It was proved by Hewitt and Yosida in [175] (see also Theorem 1.12 in [85] and Theorem III.7.8 in [62]).

Theorem 1.78 (Hewitt–Yosida). *Each finitely additive measure $\mu \in ba(\mathcal{A})$ is uniquely decomposed as the sum of a countably additive and a purely finitely additive measure:*

$$\mu = \lambda_\mu + \nu_\mu,$$

where $\lambda_\mu \in ca(\mathcal{A})$ and $\nu_\mu \in ba_p(\mathcal{A})$.

Proof. (i) Firstly, we suppose that $\mu \in ba^+(\mathcal{A})$.

Let $\Gamma = \{\gamma \in ca^+(\mathcal{A}) : 0 \le \gamma \le \mu\}$ and let $\alpha = \sup_{\gamma \in \Gamma} \gamma(X)$; then

$$0 \le \alpha \le \mu(X) < +\infty.$$

Let $(\gamma_n)_{n \in \mathbb{N}} \subseteq \Gamma$ be a sequence such that $\lim_{n \to \infty} \gamma_n(X) = \alpha$. We associate an increasing sequence $(\lambda_n)_{n \in \mathbb{N}} \subseteq \Gamma$ so that $\lambda_n(X) \to \alpha$. Let $\lambda_1 = \gamma_1$ and $\lambda_2 : \mathcal{A} \to \mathbb{R}_+$ defined by:

$$\lambda_2(E) = \sup\{\gamma_1(A) + \gamma_2(E \setminus A) : A \in \mathcal{A}, A \subseteq E\}, \quad \text{for every} \quad A \in \mathcal{A}.$$

For every $E, A \in \mathcal{A}$ with $A \subseteq E$, $\gamma_1(A) + \gamma_2(E \setminus A) \le \mu(A) + \mu(E \setminus A) = \mu(E)$. Therefore $0 \le \lambda_1 \le \lambda_2 \le \mu$ and also, for every $E \in \mathcal{A}$, $\lambda_2(E) \in \mathbb{R}_+$.

(a) Let us show that λ_2 is additive. For every $E, F \in \mathcal{A}$ with $E \cap F = \emptyset$, let $G = E \cup F$. For any $\varepsilon > 0$, there exist $A, B \in \mathcal{A}$, $A \subseteq E$, $B \subseteq F$ so that

$$\gamma_1(A) + \gamma_2(E \setminus A) > \lambda_2(E) - \varepsilon,$$

$$\gamma_1(B) + \gamma_2(F \setminus B) > \lambda_2(F) - \varepsilon.$$

Then $C = A \cup B \subseteq G$ and $\lambda_2(G) \geq \gamma_1(C) + \gamma_2(G \setminus C) = \gamma_1(A) + \gamma_1(B) + \gamma_2(E \setminus A) + \gamma_2(F \setminus B) > \lambda_2(E) + \lambda_2(F) - 2\varepsilon$. Therefore

$$\lambda_2(E \cup F) \geq \lambda_2(E) + \lambda_2(F). \tag{1}$$

On the other hand, according to the definition of $\lambda_2(G) = \lambda_2(E \cup F)$, for every $\varepsilon > 0$, there exists $D \subseteq G$, $D \in \mathcal{A}$ such that $\lambda_2(G) - \varepsilon < \gamma_1(D) + \gamma_2(G \setminus D) = \gamma_1(D \cap E) + \gamma_1(D \cap F) + \gamma_2(E \setminus (D \cap E)) + \gamma_2(F \setminus (D \cap F)) \leq \lambda_2(E) + \lambda_2(F)$ and so

$$\lambda_2(E \cup F) \leq \lambda_2(E) + \lambda_2(F). \tag{2}$$

(1) and (2) show that λ_2 is finitely additive and therefore that $\lambda_2 \in ba^+(\mathcal{A})$.

(b) Let us show that λ_2 is σ-additive. Let $(E_n)_{n \in \mathbb{N}} \subseteq \mathcal{A}$ be a sequence of pairwise disjoint sets and let $E = \bigcup_{n=1}^{\infty} E_n$. $(\lambda_2(\cup_{k=1}^{n} E_k))_n$ is an increasing bounded sequence in $\mathbb{R}$. Therefore, because λ_2 is additive, there exists $\lim_n \lambda_2(E \setminus \bigcup_{k=1}^{n} E_k) = l \geq 0$. If $l > 0$ then, for every n, $\lambda_2(E \setminus \bigcup_{k=1}^{n} E_k) > l > 0$. Then, by definition of λ_2, there exist $A_n \in \mathcal{A}$, $A_n \subseteq E \setminus \bigcup_{k=1}^{n} E_k$ such that,

$$\gamma_1(A_n) + \gamma_2\left[\left(E \setminus \bigcup_{k=1}^{n} E_k\right) \setminus A_n\right] > l, \quad \text{for every} \quad n \in \mathbb{N}. \tag{3}$$

But $\gamma_1, \gamma_2 \in ca^+(\mathcal{A})$ and so,

$$\gamma_1(A_n) \leq \gamma_1\left(E \setminus \bigcup_{k=1}^{n} E_k\right) \xrightarrow[n \to \infty]{} 0 \quad \text{and}$$

$$\gamma_2\left[\left(E \setminus \bigcup_{k=1}^{n} E_k\right) \setminus A_n\right] \leq \gamma_2\left(E \setminus \bigcup_{k=1}^{n} E_k\right) \xrightarrow[n \to \infty]{} 0,$$

which is incompatible with (3). Then $l = 0$ and so λ_2 is σ-additive or $\lambda_2 \in ca^+(\mathcal{A})$. It is obvious that

$$0 \leq \lambda_1 \leq \lambda_2 \leq \mu \quad \text{and} \quad \gamma_2 \leq \lambda_2.$$

We construct then, by recurrence, the sequence $(\lambda_n)_n$, where, for every $n \in \mathbb{N}$, $\lambda_n : \mathcal{A} \to \mathbb{R}_+$ is defined by:

$$\lambda_n(E) = \sup\{\lambda_{n-1}(A) + \gamma_n(E \setminus A) : A \in \mathcal{A}, A \subseteq E\}, \quad \text{for every} \quad E \in \mathcal{A}.$$

With an analogous demonstration, we show that, for every n, $\lambda_n \in ca^+(\mathcal{A})$ and

$$0 \leq \lambda_1 \leq \ldots \leq \lambda_n \leq \ldots \leq \mu \quad \text{and} \quad \gamma_n \leq \lambda_n.$$

$(\lambda_n)_n \subseteq \Gamma$ and, since $\gamma_n(X) \leq \lambda_n(X) \leq \alpha$, we have:

$$\lim_n \lambda_n(X) = \alpha.$$

For every $E \in \mathcal{A}$, $(\lambda_n(E))_{n\in\mathbb{N}}$ is an increasing sequence, bounded by $\mu(E)$. So, for every $E \in \mathcal{A}$, there exists $\lambda_\mu(E) = \lim_n \lambda_n(E) \in \mathbb{R}$. According to Nikodym's theorem (1.37), we have $\lambda_\mu \in ca^+(\mathcal{A})$. It results also that $\lambda_\mu \leq \mu$ and therefore $\nu_\mu = \mu - \lambda_\mu \in ba^+(\mathcal{A})$.

(c) Let us show that ν_μ is purely finitely additive.

Let $\lambda \in ca(\mathcal{A})$ so that $0 \leq \lambda \leq \nu_\mu$. We have therefore: $\lambda + \lambda_\mu \leq \mu$ and $\lambda + \lambda_\mu \in ca^+(\mathcal{A})$; hence $\lambda + \lambda_\mu \in \Gamma$ and $(\lambda + \lambda_\mu)(X) \leq \alpha$.

On the other hand,

$$(\lambda + \lambda_\mu)(X) = \lambda(X) + \lambda_\mu(X) = \lambda(X) + \lim_n \lambda_n(X) = \lambda(X) + \alpha,$$

and so $\lambda(X) \leq 0$. Hence, there results that $\lambda(X) = 0$ and therefore $\lambda = 0$.

(ii) If $\mu \in ba(\mathcal{A})$, then let μ^+ be the positive variation and μ^- be the negative variation of μ. Then $\mu^+, \mu^- \in ba^+(\mathcal{A})$ and, from (i), there exist $\lambda_{\mu^+}, \lambda_{\mu^-} \in ca(\mathcal{A})$, $\nu_{\mu^+}, \nu_{\mu^-} \in ba_p(\mathcal{A})$ such that

$$\mu^+ = \lambda_{\mu^+} + \nu_{\mu^+} \quad \text{and} \quad \mu^- = \lambda_{\mu^-} + \nu_{\mu^-}.$$

We have therefore $\mu = \mu^+ - \mu^- = (\lambda_{\mu^+} - \lambda_{\mu^-}) + (\nu_{\mu^+} - \nu_{\mu^-})$. Let

$$\lambda_\mu = \lambda_{\mu^+} - \lambda_{\mu^-} \quad \text{and} \quad \nu_\mu = \nu_{\mu^+} - \nu_{\mu^-}.$$

Then $\mu = \lambda_\mu + \nu_\mu$ where $\lambda_\mu \in ca(\mathcal{A})$ and $\nu_\mu \in ba_p(\mathcal{A})$.

(iii) In order to demonstrate the uniqueness of the decomposition, let us suppose that there are two measures $\lambda \in ca(\mathcal{A})$ and $\nu \in ba_p(\mathcal{A})$ such that $\mu = \lambda_\mu + \nu_\mu = \lambda + \nu$. Then $\lambda_\mu - \lambda = \nu - \nu_\mu$ where $\lambda_\mu - \lambda \in ca(\mathcal{A})$ and $\nu - \nu_\mu \in ba_p(\mathcal{A})$. According to Theorem 1.76, $\nu - \nu_\mu \perp \lambda_\mu - \lambda$, therefore, for every $\varepsilon > 0$, there exists $A_\varepsilon \in \mathcal{A}$ such that

$$|\nu - \nu_\mu|(A_\varepsilon) < \varepsilon \quad \text{and} \quad |\lambda_\mu - \lambda|(X \setminus A_\varepsilon) < \varepsilon.$$

We have therefore: $|\lambda_\mu - \lambda|(X) = |\lambda_\mu - \lambda|(X \setminus A_\varepsilon) + |\lambda_\mu - \lambda|(A_\varepsilon) < \varepsilon + |\nu - \nu_\mu|(A_\varepsilon) < 2\varepsilon$, from where $|\lambda_\mu - \lambda|(X) = 0$ and so $\lambda_\mu = \lambda$ and $\nu = \nu_\mu$. □

We will now use the two results of decomposition (the theorem of Lebesgue decomposition 1.47 and the theorem of Hewitt–Yosida Theorem 1.78) to obtain an extension of Lebesgue decomposition theorem. We will first introduce a new notion (see [73]).

Definition 1.79. Let $\lambda \in ca^+(\mathcal{A})$ and let $c_\lambda : ba(\mathcal{A}) \to \mathbb{R}_+$ defined by:

$$c_\lambda(\mu) = \inf_{\varepsilon>0} \sup_{\lambda(A)<\varepsilon} |\mu|(A).$$

We say that c_λ is *the modulus of λ-continuity* and that $c_\lambda(\mu)$ is *the modulus of λ-continuity of measure μ.*

We can notice that $c_\lambda(\mu) = 0$ if and only if $\mu : (\mathcal{A}, d_\lambda) \to \mathbb{R}$ is d_λ-continuous (see Proposition 1.30) and, in the case where $\mu \in ca(\mathcal{A})$, iff μ is λ-continuous (see Definition 1.33).

The next sentence was taken from [73].

Proposition 1.80. *Let $\lambda \in ca^+(\mathcal{A})$ and let $c_\lambda : ba(\mathcal{A}) \to \mathbb{R}_+$ be the modulus of λ-continuity; then:*

(1) *c_λ is a semi-norm on the space $ba(\mathcal{A})$ and, for every $\mu \in ba(\mathcal{A})$, $0 \leq c_\lambda(\mu) \leq \|\mu\| = |\mu|(X)$.*

(2) *$c_\lambda(\mu) = \|\mu\|$ if and only if $\mu \perp \lambda$.*

Proof. (1) is an immediate consequence of the definition of c_λ.

(2) ($\Longrightarrow$) If we suppose that $c_\lambda(\mu) = \|\mu\| = |\mu|(X)$ then, for every $\varepsilon > 0$, $|\mu|(X) \leq \sup_{\lambda(A)<\varepsilon} |\mu|(A)$ and then there exists $A_\varepsilon \in \mathcal{A}$ such that $\lambda(A_\varepsilon) < \varepsilon$ and $|\mu|(A_\varepsilon) \geq |\mu|(X) - \varepsilon = c_\lambda(\mu) - \varepsilon$; hence $|\mu|(X \setminus A_\varepsilon) < \varepsilon$.

We have therefore $\mu \perp \lambda$.

($\Longleftarrow$) If $\mu \perp \lambda$ then, for every $\varepsilon > 0$, there exists $A_\varepsilon \in \mathcal{A}$ such that $|\mu|(X \setminus A_\varepsilon) < \varepsilon$ and $\lambda(A_\varepsilon) < \varepsilon$. Then $\sup_{\lambda(A)<\varepsilon} |\mu|(A) \geq |\mu|(A_\varepsilon) = |\mu|(X) - |\mu|(X \setminus A_\varepsilon) \geq \|\mu\| - \varepsilon$, for any $\varepsilon > 0$, from where $c_\lambda(\mu) \geq \|\mu\|$; since $c_\lambda(\mu) \leq \|\mu\|$, $c_\lambda = \|\mu\|$. □

The following is the above announced extension of Lebesgue decomposition theorem; its proof can be found in [73], Proposition 2.

Theorem 1.81. *Let $\lambda \in ca^+(\mathcal{A})$; for every $\mu \in ba(\mathcal{A})$, there exist $f \in L^1(\lambda)$ and $\theta \in ba(\mathcal{A})$, orthogonal on λ such that, for every $A \in \mathcal{A}$,*

$$\mu(A) = \int_A f d\lambda + \theta(A) \quad \textit{and} \quad c_\lambda(\mu) = \|\theta\|.$$

Proof. According to the Hewitt–Yosida decomposition theorem 1.78, for every $\mu \in ba(\mathcal{A})$, there exist $\lambda_\mu \in ca(\mathcal{A})$ and $\nu_\mu \in ba_p(\mathcal{A})$ such that $\mu = \lambda_\mu + \nu_\mu$.

By Lebesgue decomposition theorem 1.47, there exist $\mu_1, \mu_2 \in ca(\mathcal{A})$ such that $\lambda_\mu = \mu_1 + \mu_2$, where μ_1 is λ-continuous and $\mu_2 \perp \lambda$. Then $\mu = \mu_1 + (\mu_2 + \nu_\mu)$. Let $\theta = \mu_2 + \nu_\mu$. It is clear that $\theta \in ba(\mathcal{A})$. Since ν_μ is purely finitely additive and λ is σ- additive, after Theorem 1.76, it results that $\nu_\mu \perp \lambda$ and so $\theta \perp \lambda$.

Since μ_1 is λ-continuous, by Radon–Nikodym theorem 1.40, there exists $f \in L^1(\lambda)$ such that, for every $A \in \mathcal{A}$, $\mu_1(A) = \int_A f d\lambda$. Therefore, for every $A \in \mathcal{A}$,

$$\mu(A) = \int_A f d\lambda + \theta(A).$$

According to Proposition 1.80, c_λ is a semi-norm and therefore:

$$c_\lambda(\mu) = c_\lambda(\mu_1 + \theta) \le c_\lambda(\mu_1) + c_\lambda(\theta) = c_\lambda(\theta) \quad \text{and}$$
$$c_\lambda(\theta) = c_\lambda(\mu - \mu_1) \le c_\lambda(\mu) + c_\lambda(\mu_1) = c_\lambda(\mu).$$

Therefore $c_\lambda(\mu) = c_\lambda(\theta)$. Since $\theta \perp \lambda$, according to Proposition 1.80 (2), $c_\lambda(\theta) = \|\theta\|$ and so $c_\lambda(\mu) = \|\theta\|$. □

1.6 Extensions of Dunford–Pettis' Theorem

In the final paragraph of this chapter, we present some recent results of weak compactness in $L^1(\lambda)$. Part of these results use the modulus of uniform integrability introduced by H. P. Rosenthal [143] whose definition we will give as follows (see also [73]).

Definition 1.82. For any subset $H \subseteq L^1(\lambda)$,

$$\eta(H) = \inf_{\varepsilon>0} \left[\sup \left\{ \int_A |f| d\lambda : A \in \mathcal{A}, \lambda(A) < \varepsilon, f \in H \right\} \right] \in [0, +\infty].$$

We say that $\eta(H)$ is *the modulus of uniform integrability* of H. For every sequence $(f_n) \subseteq L^1(\lambda)$, we will note $\eta(\{f_n\}) = \eta((f_n))$.

According to Remark 1.53 (ii)), it results that $\eta(H) = 0$ if and only if H est λ-equicontinuous.

The following proposition and theorem motivate this denomination.

Proposition 1.83. *Let $H \subseteq L^1(\lambda)$ be a bounded set; then*

$$\eta(H) = \lim_{t\to\infty}\left[\sup_{f\in H}\int_{\{x\in X:|f(x)|>t\}} |f|\,d\lambda\right].$$

Proof. Let

$$\eta_1 = \lim_{t\to\infty}\left[\sup_{f\in H}\int_{(|f|>t)} |f|\,d\lambda\right] = \inf_{t>0}\left[\sup_{f\in H}\int_{(|f|>t)} |f|\,d\lambda\right]$$

and let $M = \sup\{\|f\|_1 : f \in H\}$.

Firstly, we show that $\eta(H) \geq \eta_1$. For every $\varepsilon > 0$, there exists $t_0 > 0$ such that $\frac{M}{t_0} < \varepsilon$. According to the definition of η_1, for every $\delta > 0$, there exists $f_0 \in H$ such that

$$\eta_1 - \delta < \int_{(|f_0|>t_0)} |f_0|\,d\lambda. \tag{1}$$

$$M \geq \int_X |f_0|\,d\lambda \geq \int_{(|f_0|>t_0)} |f_0|\,d\lambda > t_0\cdot\lambda(|f_0|>t_0); \quad \text{hence}$$

$$\lambda(|f_0|>t_0) < \frac{M}{t_0} < \varepsilon.$$

Therefore

$$\sup\left\{\int_E |f|\,d\lambda : E\in\mathcal{A}, \lambda(E)<\varepsilon, f\in H\right\} \geq \int_{(|f_0|>t_0)} |f_0|\,d\lambda > \eta_1-\delta,$$

for every $\delta > 0$ and then

$$\eta(H) \geq \eta_1. \tag{2}$$

On the other hand, let $\delta > 0$, $t > 0$ and let $\varepsilon > 0$ such that $\varepsilon < \frac{\delta}{t}$. According to the definition of $\eta(H)$, there exist $A_0 \in \mathcal{A}$ and $f_0 \in H$ such that $\lambda(A_0) < \varepsilon$ and

$$\eta(H) - \delta < \int_{A_0} |f_0|\,d\lambda. \tag{3}$$

Then

$$\begin{aligned}\int_{(|f_0|>t)} |f_0|\,d\lambda &\geq \int_{(|f_0|>t)\cap A_0} |f_0|\,d\lambda \geq \int_{A_0}|f_0|\,d\lambda - \int_{A_0\cap(|f_0|\leq t)}|f_0|\,d\lambda\\ &\geq \eta(H)-\delta-t\cdot\lambda(A_0\cap\{|f_0|\leq t\})\\ &\geq \eta(H)-\delta-t\lambda(A_0) > \eta(H)-\delta-t\varepsilon > \eta(H)-2\delta.\end{aligned}$$

So, for every $t > 0$ and every $\delta > 0$, $\sup_{f\in H}[\int_{(|f|>t)} |f| d\lambda] > \eta(H) - 2\delta$. We have therefore

$$\eta_1 \geq \eta(H). \tag{4}$$

By (2) and (4), $\eta(H) = \eta_1$. □

Theorem 1.84 (See [58]). *Let $\lambda \in ca^+(\mathcal{A})$ and let $H \subseteq L^1(\lambda)$. The following conditions are equivalent:*

(i) *H is relatively weakly compact in $(L^1(\lambda), \|\cdot\|_1)$.*

(ii) *H is bounded and λ-equicontinuous.*

(iii) *For every $\varepsilon > 0$, there exists $t_\varepsilon > 0$ such that, for every $f \in H$, $\int_{(|f|>t_\varepsilon)} |f| d\lambda < \varepsilon$.*

(iv) *There exists $\Phi : \mathbb{R}_+ \to \mathbb{R}_+$ an increasing function as $\Phi(0) = 0$, $\lim_{x\to\infty} \frac{1}{x}\cdot\Phi(x) = \infty$ and $\sup_{f\in H} \int_X \Phi(|f|) d\lambda < +\infty$.*

Proof. (i) $\Longleftrightarrow$ (ii) was demonstrated in the Dunford–Pettis theorem (1.65).

(ii) $\Longleftrightarrow$ (iii) is a consequence of Proposition 1.83.

(iii) $\Longrightarrow$ (iv) According to the hypothesis (iii), for every $k \in \mathbb{N}^*$, there exists $t_k > 0$ such that,

$$\int_{(|f|>t_k)} |f| d\lambda < \frac{1}{2^k}, \quad \text{for every} \quad f \in H. \tag{1}$$

Let $t_0 = 0$; we can choose the sequence $(t_k)_k$ such that, for every $k \in \mathbb{N}^*$, $t_k > 2t_{k-1}$.

Let $\Phi : [0, +\infty[\to [0, +\infty[$ be continuous and linear on every interval $[t_k, t_{k+1}]$ so that $\Phi(t_k) = k \cdot t_k$, for every $k \in \mathbb{N}$.

Then Φ is an increasing mapping and $\Phi(0) = 0$.

On every interval $[t_k, t_{k+1}]$,

$$\Phi(x) = \left(k + \frac{t_{k+1}}{t_{k+1} - t_k}\right) x - \frac{t_k \cdot t_{k+1}}{t_{k+1} - t_k}$$

and then

$$\frac{\Phi(x)}{x} \geq k + \frac{t_{k+1}}{t_{k+1} - t_k} - \frac{t_{k+1}}{t_{k+1} - t_k} = k.$$

Therefore $\lim_{x\to+\infty} \frac{\Phi(x)}{x} = +\infty$.

We also have, on every interval $[t_k, t_{k+1}]$, $\frac{\Phi(x)}{x} \leq (k + \frac{t_{k+1}}{t_{k+1}-t_k}) - \frac{t_k}{t_{k+1}-t_k} = k+1$, from where, for every $f \in H$,

$$\begin{aligned}\int_X \Phi(|f|)d\lambda &= \sum_{k=0}^{\infty} \int_{(t_k \leq |f| < t_{k+1})} \Phi(|f|)d\lambda \\ &= \int_{(|f|<t_1)} \Phi(|f|)d\lambda + \sum_{k=1}^{\infty} \int_{(t_k \leq |f| < t_{k+1})} \Phi(|f|)d\lambda \\ &\leq t_1 \cdot \lambda(\{|f| < t_1\}) + \sum_{k=1}^{\infty}(k+1) \int_{(t_k \leq |f| < t_{k+1})} |f|d\lambda \\ &\leq t_1\lambda(X) + \sum_{k=1}^{\infty} \frac{k+1}{2^k} < +\infty.\end{aligned}$$

Let $M = t_1\lambda(X) + \sum_{k=1}^{\infty} \frac{k+1}{2^k}$. Then, $\sup_{f\in H} \int_X \Phi(|f|)d\lambda \leq M < +\infty$.

(iv)$\Longrightarrow$ (ii) Let $\Phi : \mathbb{R}_+ \to \mathbb{R}_+$ be a function satisfying (iv) and let $M = \sup_{f\in H} \int_X \Phi(|f|)d\lambda$.

Since $\lim_{x\to\infty} \frac{\Phi(x)}{x} = +\infty$, there exists $x_0 \in]0, +\infty[$ such that, for every $x \geq x_0$, $\frac{\Phi(x)}{x} \geq 1$. Then, for every $f \in H$,

$$\int_{(|f|\geq x_0)} |f|d\lambda \leq \int_{(|f|\geq x_0)} \Phi(|f|)d\lambda \leq \sup_{g\in H} \int_X \Phi(|g|)d\lambda = M < +\infty.$$

Therefore, for every $f \in H$,

$$\|f\|_1 = \int_X |f|d\lambda \leq \int_{(|f|<x_0)} |f|d\lambda + M \leq x_0 \cdot \lambda(\{|f| < x_0\}) + M \leq x_0\lambda(X) + M.$$

Then H is bounded in $L^1(\lambda)$. Since $\lim_{x\to+\infty} \frac{\Phi(x)}{x} = +\infty$, for every $\varepsilon > 0$, there exists $x_1 \in]0, +\infty[$ such that, for every $x \geq x_1$,

$$\frac{\Phi(x)}{x} \geq \frac{2M}{\varepsilon}. \tag{2}$$

Then

$$|f| \geq x_1 \Longrightarrow \Phi(|f|) \geq \frac{2M}{\varepsilon} \cdot |f|, \quad \text{for every} \quad f \in H. \tag{3}$$

Therefore, for every $A \in \mathcal{A}$ and for every $f \in H$,

$$\begin{aligned}\int_A |f|d\lambda &= \int_{A\cap(|f|<x_1)} |f|d\lambda + \int_{A\cap(|f|\geq x_1)} |f|d\lambda \\ &\leq x_1\lambda(A) + \frac{\varepsilon}{2M}\int_{(|f|\geq x_1)} \Phi(|f|)d\lambda \geq x_1\lambda(A) + \frac{\varepsilon}{2M}\cdot M \\ &= x_1\lambda(A) + \frac{\varepsilon}{2}.\end{aligned}$$

Then there exists $\delta = \frac{\varepsilon}{2x_1}$ such that, for every $A \in \mathcal{A}$ with $\lambda(A) < \delta$,

$$\int_A |f|d\lambda < x_1\lambda(A) + \frac{\varepsilon}{2} < \varepsilon, \quad \text{for every} \quad f \in H.$$

Hence H is λ-equicontinuous. □

Definition 1.85. A set $H \subseteq L^1(\lambda)$ who satisfies the condition (iii) of previous theorem is called *uniformly integrable*. Therefore $H \subseteq L^1(\lambda)$ is uniformly integrable if, for every $\varepsilon > 0$, there exists $t_\varepsilon > 0$ such that, for every $f \in H$, $\int_{(|f|>t_\varepsilon)} |f|d\lambda < \varepsilon$.

According to Theorem 1.84, an uniformly integrable subset $H \subseteq L^1(\lambda)$ is relatively weakly compact and so it is bounded.

The equality from Proposition 1.83 justifies the name of modulus of uniform integrability given in Definition 1.82.

The following proposition is proved in [73], Proposition 4.

Proposition 1.86. *Let $(f_n)_{n\in\mathbb{N}} \subseteq L^1(\lambda)$ be a sequence with $f_n \geq 0$, for every $n \in \mathbb{N}$, and let η be the modulus of uniform integrability of (f_n); then*

$$\limsup_n \int_X f_n d\lambda \leq \int_X \limsup_n f_n d\lambda + \eta.$$

Proof. Let us denote $f = \limsup_n f_n$; it is clear that the inequality is satisfied in the both case where $f \notin L^1(\lambda)$ or if $\eta = +\infty$.

Let us suppose that $f \in L^1(\lambda)$ and $\eta < +\infty$. By the definition of η, for every $\varepsilon > 0$, there exists $\delta \in]0, \varepsilon[$ such that

$$\int_A f_n d\lambda < \eta + \varepsilon, \quad \text{for every } A \in \mathcal{A} \quad \text{with } \lambda(A) < \delta \quad \text{and for every } n \in \mathbb{N}. \tag{*}$$

Let $g_n = \sup_{k \geq n} f_k - f$ and $A_n = \{x \in X : g_n(x) \leq \delta\}$; then $A_n \in \mathcal{A}$. $(A_n)_{n \in \mathbb{N}}$ is an increasing sequence; hence $\lambda(\cup_1^\infty A_n) = \lim_n \lambda(A_n)$. Since $g_n \downarrow 0$, λ-a.e., $\lambda(\cup_1^\infty A_n) = \lambda(X)$. Therefore there exists $n_0 \in \mathbb{N}$ such that $\lambda(X \setminus A_{n_0}) < \delta$. Then, by $(*)$, for every $n \geq n_0$,

$$\int_X f_n d\lambda = \int_{X \setminus A_{n_0}} f_n d\lambda + \int_{A_{n_0}} f_n d\lambda < \eta + \varepsilon + \int_{A_{n_0}} \left[\sup_{k \geq n_0} (f_k - f) \right] d\lambda$$
$$+ \int_{A_{n_0}} f d\lambda \leq \eta + \varepsilon + \delta\lambda(A_{n_0}) + \int_{A_{n_0}} f d\lambda \leq \eta + \varepsilon(1 + \lambda(X)) + \int_X f d\lambda$$

and so we obtain

$$\limsup_n \int_X f_n d\lambda \leq \int_X f d\lambda + \eta. \qquad \square$$

Corollary 1.87. *If $(f_n) \subseteq L^1_+(\lambda)$ is a λ-equicontinuous sequence, then*

$$\limsup_n \int_X f_n d\lambda \leq \int_X \limsup_n f_n d\lambda.$$

Remark 1.88. The condition of λ-equicontinuity in the previous corollary is indispensable. Indeed, let $X = [0, 1]$ and let λ be the Lebesgue measure on X; for every $n \in \mathbb{N}^*$, let $f_n = n\chi_{[0, \frac{1}{n}]}$. Then $\limsup_n f_n = 0$, λ-a.e. and $\limsup_n \int_X f_n d\lambda = 1$.

We can easily demonstrate that $\eta((f_n)) = 1$.

Corollary 1.89 (Proposition 1 in [76]). *Let $(f_n) \subseteq L^1(\lambda)$ and let $f \in L^1(\lambda)$; then:*

(i) $\limsup_n \|f_n - f\|_1 \leq \eta((f_n)) + \int_X \limsup_n |f_n - f| d\lambda$.

(ii) *If $f_n \xrightarrow{\lambda} f$, then* $\limsup_n \|f_n - f\|_1 \leq \eta((f_n))$.

Proof. (i) It is sufficient to apply Proposition 1.86 to the sequence $(|f_n - f|)_{n \in \mathbb{N}}$ and to notice that $\eta((f_n - f)) = \eta((f_n))$.

(ii) Let $(f_n) \subseteq L^1(\lambda)$ be a sequence convergent in measure to $f \in L^1(\lambda)$. Let $(f_n')_n$ be a subsequence of (f_n) such that

$$\limsup_n \|f_n - f\|_1 = \lim_n \|f_n' - f\|_1.$$

Then there is a subsequence (f_n'') of (f_n') such that (f_n'') converges to f, λ-a.e.

Part (i) applied to the sequence (f_n'') allows us to write:

$$\begin{aligned}\limsup_n \|f_n - f\|_1 = \lim_n \|f_n'' - f\|_1 &\leq \eta((f_n'')) + \int_X \limsup_n |f_n'' - f| d\lambda \\ &= \eta((f_n'')) \leq \eta((f_n)).\end{aligned}$$ □

Remark 1.90. In the part (ii) of the previous corollary, if we suppose that (f_n) is λ-equicontinuous, then $f_n \xrightarrow{\|\cdot\|_1} f$. This gives another proof for the sufficiency of Vitali's theorem 1.54.

In [73], we introduced the a-convergence as a generalization of weak convergence.

Definition 1.91. Let $(f_i)_{i\in I} \subseteq L^1(\lambda)$ be a net, $f \in L^1(\lambda)$ and $a \geq 0$. The sequence $(f_i)_{i\in I}$ is called *a- convergent* to f if, for every $A \in \mathcal{A}$,

$$\limsup_{i\in I} \left| \int_A (f_i - f) d\lambda \right| \leq a,$$

which we write $f_i \xrightarrow{a} f$.

$H \subseteq L^1(\lambda)$ is said *relatively a-compact* if every net $(f_i)_{i\in I} \subseteq H$ admits at least one subnet a-convergent to a function $f \in L^1(\lambda)$.

Remark 1.92.

(i) Obviously, a net is 0-convergent if and only if it is weakly convergent in $L^1(\lambda)$. Similarly, a set $H \subseteq L^1(\lambda)$ is relatively 0-compact if and only if H is relatively weakly compact.

(ii) According to (ii) of Corollary 1.89, if $(f_n) \subseteq L^1(\lambda)$ is convergent in measure to f, then (f_n) is $\eta((f_n))$-convergent to f.

The following theorem is a first extension of the Dunford–Pettis theorem; it was demonstrated in [73] (see Theorem 1).

Theorem 1.93. *Let $H \subseteq L^1(\lambda)$ be a bounded set and let $\eta(H)$ be his modulus of uniform integrability.*

Then H is relatively $\eta(H)$-compact.

Proof. Let Ψ be the natural embedding of $L^1(\lambda)$ in its bidual $ba_\lambda(\mathcal{A})$ and let $(f_i)_{i\in I} \subseteq H$ be a net. Since H is bounded in $L^1(\lambda)$, $\Psi(H)$ it is bounded in

$ba_\lambda(\mathcal{A})$ also; therefore $\Psi(H)$ is relatively w*- compact, according to the notations already used in Remark 1.69 and in Theorem 1.70. So $(\Psi(f_i))_{i\in I}$ admits a subnet $(\Psi(f_j))_{j\in J}$ w*- convergent to a measure $\mu \in ba_\lambda(\mathcal{A})$. Therefore we have,

$$\int_A f_j d\lambda \xrightarrow[j\in J]{} \mu(A), \quad \text{for every} \quad A \in \mathcal{A}. \tag{1}$$

Thanks to Theorem 1.81, there exist $f \in L^1(\lambda)$ and $\theta \in ba(\mathcal{A})$, $\theta \perp \lambda$ such that,

$$\mu(A) = \int_A f d\lambda + \theta(A), \quad \text{for every} \quad A \in \mathcal{A} \quad \text{and} \tag{2}$$

$$c_\lambda(\mu) = \|\theta\|. \tag{3}$$

Let us show that $f_j \xrightarrow{\eta(H)} f$.

According to the definition of $\eta(H)$, for every $\varepsilon > 0$, there exists $\delta \in]0, \varepsilon[$ such that

$$\int_A |f| d\lambda < \eta(H) + \varepsilon, \;\text{ for every } A \in \mathcal{A} \text{ with } \lambda(A) < \delta \text{ and for every } f \in H. \tag{4}$$

After the definition of $c_\lambda(\mu)$, there exists $A_0 \in \mathcal{A}$ such that

$$c_\lambda(\mu) - \delta < |\mu|(A_0) \quad \text{and} \quad \lambda(A_0) < \delta. \tag{5}$$

According to the definition of $|\mu|$, there exists an $\mathcal{A}$-partition $\{A_1, \ldots, A_n\}$ of A_0 such that

$$|\mu|(A_0) - \varepsilon < \sum_{k=1}^n |\mu(A_k)|. \tag{6}$$

After (1), there exists $j_0 \in J$ such that

$$\left|\int_{A_k} f_{j_0} d\lambda - \mu(A_k)\right| < \frac{\varepsilon}{n}, \quad \text{for every} \quad k = 1, \ldots, n. \tag{7}$$

Therefore, since $\lambda(A_0) < \delta$, we obtain, from (7) and (4):

$$\begin{aligned}\sum_{k=1}^n |\mu(A_k)| &\leq \sum_{k=1}^n \left|\mu(A_k) - \int_{A_k} f_{j_0} d\lambda\right| + \sum_{k=1}^n \int_{A_k} |f_{j_0}| d\lambda \\ &< \varepsilon + \int_{A_0} |f_{j_0}| d\lambda < \varepsilon + \eta(H) + \varepsilon = \eta(H) + 2\varepsilon\end{aligned}$$

and, from (6) and (5),

$$c_\lambda(\mu) < \eta(H) + 4\varepsilon. \tag{8}$$

We have therefore

$$\|\theta\| = c_\lambda(\mu) \le \eta(H). \tag{9}$$

For every $A \in \mathcal{A}$, there exists $\varepsilon > 0$, and by (1), there exists $j_1 \in J$ such that

$$\left| \int_A f_j \, d\lambda - \mu(A) \right| < \varepsilon, \quad \text{for every} \quad j \ge j_1. \tag{10}$$

Then, from (2), (10) and (9), we have, for every $j \ge j_1$,

$$\begin{aligned} \left| \int_A (f_j - f)\, d\lambda \right| &= \left| \int_A f_j \, d\lambda - \mu(A) \right| + |\theta(A)| \\ &\le \varepsilon + |\theta|(A) \le \varepsilon + \|\theta\| < \varepsilon + \eta(H) \end{aligned}$$

from where,

$$\limsup_{j \in J} \left| \int_A (f_j - f)\, d\lambda \right| \le \eta(H), \quad \text{for every} \quad A \in \mathcal{A},$$

i.e. $f_j \xrightarrow{\eta(H)} f$. □

Remark 1.94. In the case where H is bounded and λ-equicontinuous, $\eta(H) = 0$ and, according to Remark 1.92(i), H is relatively weakly compact in $L^1(\lambda)$. We have then the implication (iii)$\Longrightarrow$(i) of the Dunford–Pettis theorem 1.65.

Now, we will give a new notion of convergence; it was introduced by J. K. Brooks and R. V. Chacon [37] and it will be used for a second extension of Dunford–Pettis theorem.

Definition 1.95. Let λ be a positive, σ-additive and σ-finite measure on the σ-algebra $\mathcal{A}$. A sequence $(f_n) \subseteq L^1(\lambda)$ is said *w^2-convergent* to $f \in L^1(\lambda)$ if the following two conditions are satisfied:

(a) there exists a decreasing sequence $(B_p)_{p \in \mathbb{N}} \subseteq \mathcal{A}$ such that $\lambda(B_p) \to 0$,

(b) for every $p \in \mathbb{N}$, $f_n \restriction_{X \setminus B_p} \xrightarrow[X \setminus B_p]{w} f \restriction_{X \setminus B_p}$.

We note this by $f_n \xrightarrow{w^2} f$.

We say that the sequence $(B_p)_{p \in \mathbb{N}} \subseteq \mathcal{A}$ proving the above conditions a) and b) *localizes the concentration of mass* of the w^2- convergent sequence (f_n).

Remark 1.96.

(i) If $(f_n) \subseteq L^1(\lambda)$ is w^2-convergent, then there exists only one function $f \in L^1(\lambda)$ so that $f_n \xrightarrow{w^2} f$.

Indeed, if we suppose that $f_n \xrightarrow{w^2} f$ and $f_n \xrightarrow{w^2} g$, then let (B_p) and (C_p) be two decreasing sequences of $\mathcal{A}$ such that $\lambda(B_p) \to 0, \lambda(C_p) \to 0$ and so that, for every $p \in \mathbb{N}$, $f_n \upharpoonright_{X \setminus B_p} \xrightarrow[X \setminus B_p]{w} f \upharpoonright_{X \setminus B_p}$ and $f_n \upharpoonright_{X \setminus C_p} \xrightarrow[X \setminus C_p]{w} g \upharpoonright_{X \setminus C_p}$.

Then, for all $p \in \mathbb{N}$, $\{x \in X : f(x) \neq g(x)\} \subseteq B_p \cup C_p$, from where, because of a), $f = g$ a.e.

(ii) From Theorem 1.57, it results that the condition b) of the previous definition is equivalent with:

$$\int_A f_n d\lambda \to \int_A f d\lambda, \quad \text{for every } p \in \mathbb{N}, \text{ and for every } A \in \mathcal{A} \text{ with } A \subseteq X \setminus B_p.$$

(iii) The sequence (B_p), which localizes the concentration of mass of a w^2- convergent sequence, is not uniquely determined.

If (B_p) and (C_p) are two sequences that localize the concentration of mass of $(f_n) \subseteq L^1(\lambda)$, then the sequence $(B_p \cap C_p)_{p \in \mathbb{N}}$ localizes the concentration of mass of (f_n), also. Indeed, $(B_p \cap C_p)$ is a decreasing sequence, $\lambda((B_p \cap C_p)) \to 0$ and, for every $A \in \mathcal{A}, A \subseteq X \setminus (B_p \cap C_p)$, there exist A_1 and $A_2 \in \mathcal{A}, A_1 \subseteq (X \setminus B_p), A_2 \subseteq (X \setminus C_p)$ such that $A = A_1 \cup A_2$. Then, as for all $p \in \mathbb{N}$,

$$f_n \upharpoonright_{X \setminus B_p} \xrightarrow[X \setminus B_p]{w} f \upharpoonright_{X \setminus B_p} \quad \text{and} \quad f_n \upharpoonright_{X \setminus C_p} \xrightarrow[X \setminus C_p]{w} f \upharpoonright_{X \setminus C_p},$$

$$\int_A f_n d\lambda = \int_{A_1} f_n d\lambda + \int_{A_2} f_n d\lambda \to \int_{A_1} f d\lambda + \int_{A_2} f d\lambda = \int_A f d\lambda$$

and, because of (ii), we have

$$f_n \upharpoonright_{X \setminus (B_p \cap C_p)} \xrightarrow[X \setminus (B_p \cap C_p)]{w} f \upharpoonright_{X \setminus (B_p \cap C_p)}, \quad \text{for every} \quad p \in \mathbb{N}.$$

(iv) We can notice that, if there exists $p \in \mathbb{N}$ with $B_p \cap C_p = \emptyset$, then $f_n \xrightarrow[L^1(\lambda)]{w} f$.

(v) If $f_n \xrightarrow[L^1(\lambda)]{w} f$, then $f_n \xrightarrow{w^2} f$.

As the following example shows, the inverse implication is not true.

Let λ be the Lebesgue measure on $[0,1]$ and let $f_n : [0,1] \to \mathbb{R}$, defined by $f_n = n\chi_{[0,\frac{1}{n}]}$. Then $f_n \xrightarrow{w^2} 0$ but $f_n \overset{w}{\nrightarrow} 0$. The sequence (B_p), $B_p = [0, \frac{1}{p}]$, localizes the concentration of mass of (f_n).

The following theorem will show that the w^2-convergence is not a topological one: there is no topology τ on $L^1(\lambda)$ for which to have the equivalence:

$$f_n \xrightarrow{w^2} f \Leftrightarrow f_n \xrightarrow{\tau} f.$$

(vi) A w^2-convergent sequence in $L^1(\lambda)$ is not necessarily bounded. Indeed, if λ is the Lebesgue measure on $[0,1]$ and if $f_n = n^2\chi_{[0,\frac{1}{n}]}$, then $f_n \xrightarrow{w^2} 0$ and $\|f_n\|_1 = n \to +\infty$.

(vii) If $f_n \xrightarrow{w^2} f$, then, for every subsequence $(g_k)_k$ of (f_n), $g_k \xrightarrow{w^2} f$. Moreover, the sequence (B_p) which qui localizes the concentration of mass of (f_n), localizes the concentration of mass of (g_k), also.

The following theorem says us that is no any topology τ on L^1 such that $f_n \xrightarrow{w^2} f \Leftrightarrow f_n \xrightarrow{\tau} f$ (see Theorem 2 in [38]).

Theorem 1.97. *Let λ be the Lebesgue measure on $[0,1]$; the w^2-convergence on $L^1(\lambda)$ is not a topological one.*

Proof. The statement is proved if we can construct a sequence $(f_n) \subseteq L^1(\lambda)$ which is not w^2-convergent to 0 but so that each of its subsequence (f_n') admits a subsequence (f_n'') w^2-convergent to 0.

Let $\mathbb{Q} \cap [0,1] = \{r_1, r_2, \dots, r_n\}$ be the set of rational numbers of $[0,1]$. For every $n \in \mathbb{N}$, let $r_n = \frac{p_n}{q_n}$, where p_n and $q_n \in \mathbb{N}$ are relatively prime.

For every $n \in \mathbb{N}$, let $f_n : (0,1) \to \mathbb{R}$, defined by

$$f_n(x) = \begin{cases} q_n & , x \in \left]\frac{p_n-2}{q_n}, \frac{p_n+2}{q_n}\right[\cap [0,1], \\ 0 & , x \notin \left]\frac{p_n-2}{q_n}, \frac{p_n+2}{q_n}\right[\cap [0,1]. \end{cases}$$

For every $n \in \mathbb{N}$, $(f_n) \subseteq L^1_+(\lambda)$ and

$$\|f_n\|_1 = \int_{[0,1]} f_n d\lambda = q_n \lambda \left(\left]\frac{p_n - 2}{q_n}, \frac{p_n + 2}{q_n}\right[\cap]0,1[\right) \le q_n \cdot \frac{4}{q_n} = 4.$$

Let us suppose that $f_n \xrightarrow{w^2} 0$. Then there exists a sequence $(B_p) \subseteq \mathcal{A}$, such that $\lambda(B_p) \downarrow 0$ and $f_n \restriction_{(0,1)\setminus B_p} \xrightarrow{w} 0$. Because $\lambda(B_p) \downarrow 0$, there exists $p_0 \in \mathbb{N}$ such that $\lambda(B_{p_0}) < \frac{1}{2}$. Then, if $E = [0,1] \setminus B_{p_0}$, $\lambda(E) > \frac{1}{2}$ and according to (ii) of Remark 1.96,

$$\lim_n \int_E f_n d\lambda = 0. \tag{1}$$

Since $\lambda(E) > \frac{1}{2} > 0$, there is at least a density point $x_0 \in E$, i.e. a point x_0 such that $\lim_{h\downarrow 0} \frac{1}{2h}\lambda(E \cap [x_0 - h, x_0 + h]) = 1$. Therefore, there exists $h_0 > 0$ such that,

$$\lambda\left(E \cap]x_0 - h, x_0 + h[\right) > h, \quad \text{for every} \quad h \in [0, h_0]. \tag{*}$$

For every $n \in \mathbb{N}$, let us denote $p_n = [nx_0]$, where $[a]$ is the largest integer not greater than a. Then $r'_n = \frac{p_n}{n} \in \mathbb{Q} \cap [0,1]$ and $(f_{r'_n})_{n\in\mathbb{N}}$ is a subsequence of (f_n). By (1),

$$\lim_n \int_E f_{r'_n} d\lambda = 0. \tag{2}$$

According to the definition of r'_n, for every $n \in \mathbb{N}^*$, $0 \le x_0 - \frac{p_n}{n} < \frac{1}{n}$ i.e.

$$|x_0 - r'_n| < \frac{1}{n}, \quad \text{for every} \quad n \in \mathbb{N}^*. \tag{3}$$

we have therefore,

$$\left]x_0 - \frac{1}{n}, x_0 + \frac{1}{n}\right[\subseteq \left]r'_n - \frac{2}{n}, r'_n + \frac{2}{n}\right[, \quad \text{for every} \quad n \in \mathbb{N}^*. \tag{4}$$

There exists $n_0 \in \mathbb{N}$ such that, for every $n \ge n_0$, $\frac{1}{n} < h_0$; by $(*)$

$$\lambda\left(E \cap \left]x_0 - \frac{1}{n}, x_0 + \frac{1}{n}\right[\right) > \frac{1}{n}, \quad \text{for every} \quad n \ge n_0 \tag{5}$$

and then, by (4) and (5), for every $n \geq n_0$,

$$\int_E f_{r'_n} d\lambda = n \cdot \lambda\left(E \cap \left]r'_n - \frac{2}{n}, r'_n + \frac{2}{n}\right[\right)$$
$$\geq n\lambda\left(E \cap \left]x_0 - \frac{1}{n}, x_0 + \frac{1}{n}\right[\right) > 1,$$

which contradicts (2).

Therefore (f_n) is not w^2-convergent to 0.

Let now $(f_{k_n})_{n\in\mathbb{N}}$ be any subsequence of (f_n); as (r_{k_n}) is bounded, it admits at least one convergent subsequence; let (r''_n) be a subsequence of (r_{k_n}) convergent to $x_1 \in [0,1]$. Then $(f_{r''_n})$ is a subsequence of $(f_{r_{k_n}})$, w^2-convergent to 0.

Indeed, for every $p \in \mathbb{N}$, let $B_p =]x_1 - \frac{1}{p}, x_1 + \frac{1}{p}[\cap]0,1[$; (B_p) is a decreasing sequence and $\lambda(B_p) \to 0$.

Since $r''_n = \frac{p''_n}{q''_n} \to x_1, q''_n \to +\infty$. Then, for every $p \in \mathbb{N}^*$, there exists $n_0 \in \mathbb{N}$ such that, for all $n \geq n_0$,

$$\frac{2}{q''_n} < \frac{1}{2p} \quad \text{and} \quad |r''_n - x_1| < \frac{1}{2p}. \tag{6}$$

We have $]r''_n - \frac{2}{q''_n}, r''_n + \frac{2}{q''_n}[\subseteq]x_1 - \frac{1}{p}, x_1 + \frac{1}{p}[$, for every $n \geq n_0$; therefore $f_{r''_n} \upharpoonright_{(0,1)\setminus B_p} = 0$, for every $n \geq n_0$, and so, $f_{r''_n} \upharpoonright_{(0,1)\setminus B_p} \xrightarrow{w} 0$, for every $p \in \mathbb{N}$. We have hence $f_{r''_n} \xrightarrow{w^2} 0$. □

The following results, proved in [76], Proposition 4, concern the calculation of the modulus of uniform integrability for the w^2-convergent sequences. They allow us to present a new demonstration for the subsequence splitting lemma, using biting lemma.

Theorem 1.98. *Let $(f_n) \subseteq L^1(\lambda)$ be a w^2-convergent sequence; if $(B_p) \subseteq \mathcal{A}$ is a sequence which localizes the concentration of mass of (f_n), then*

$$\eta((f_n)) = \lim_p \limsup_n \int_{B_p} |f_n| d\lambda.$$

Proof. Let

$$\eta = \eta((f_n)) = \inf_{\varepsilon>0} \sup_{\mu(A)<\varepsilon} \sup_n \int_A |f_n| d\lambda \quad \text{and}$$

$$\eta_1 = \lim_p \limsup_n \int_{B_p} |f_n| d\lambda = \inf_p \limsup_n \int_{B_p} |f_n| d\lambda.$$

(i) If $\eta = +\infty$, then, for every $n \in \mathbb{N}^*$, there exist $A_n \in \mathcal{A}$ and $k_n \geq n$ such that $\lambda(A_n) < \frac{1}{n}$ and $\int_{A_n} |f_{k_n}| d\lambda \geq n$.

For every $p \in \mathbb{N}$, $(f_{k_n} \upharpoonright_{X \setminus B_p})_{n \in \mathbb{N}}$ is weakly convergent and therefore, according to Remark 1.66(ii), $(f_{k_n} \upharpoonright_{X \setminus B_p})_{n \in \mathbb{N}}$ is λ-equicontinuous on $X \setminus B_p$.

Then $\lim_n \int_{A_n \cap B_p} |f_{k_n}| d\lambda = +\infty$, for every $p \in \mathbb{N}$ and so

$$\lim_n \int_{B_p} |f_{k_n}| d\lambda = \limsup_n \int_{B_p} |f_{k_n}| d\lambda = +\infty, \quad \text{for every} \quad p \in \mathbb{N}.$$

Therefore $\eta_1 = +\infty = \eta$.

(ii) Let now $\eta < +\infty$.

According to the definition of η, for every $\varepsilon > 0$, there exists $\delta > 0$, such that, for any $A \in \mathcal{A}$ with $\lambda(A) < \delta$ and for any $n \in \mathbb{N}$, $\int_A |f_n| d\lambda < \eta + \varepsilon$.

Since $\lambda(B_p) \to 0$, there exists $p_0 \in \mathbb{N}$ such that $\lambda(B_p) < \delta$, for every $p \geq p_0$ and therefore

$$\int_{B_p} |f_n| d\lambda < \eta + \varepsilon, \quad \text{for every } n \in \mathbb{N} \quad \text{and every } p \geq p_0.$$

Then $\limsup_n \int_{B_p} |f_n| d\lambda < \eta + \varepsilon$, for every $p \geq p_0$; hence, for every $\varepsilon > 0$, $\eta_1 < \eta + \varepsilon$ and so

$$\eta_1 \leq \eta. \tag{1}$$

If $\eta = 0$, $\eta_1 = \eta$.

If $0 < \eta < +\infty$, according to the definition of η_1, for every $\varepsilon \in]0, \frac{\eta}{2}[$, there exist $p_0 \in \mathbb{N}$ and $n_0 \in \mathbb{N}$ such that

$$\int_{B_{p_0}} |f_n| d\lambda < \eta_1 + \varepsilon, \quad \text{for every} \quad n \geq n_0. \tag{2}$$

Since $(f_n \upharpoonright_{X \setminus B_{p_0}})_{n \in \mathbb{N}}$ is weakly convergent on $X \setminus B_{p_0}$, $\{f_n \upharpoonright_{X \setminus B_{p_0}} : n \in \mathbb{N}\}$ is λ-equicontinuous on $X \setminus B_{p_0}$ (see (ii) of Remark 1.66). The finite set $\{f_0, \ldots, f_{n_0-1}\}$ is λ-equicontinuous on X. Then there exists $\delta \in]0, \varepsilon[$ such that, for every $n \in \mathbb{N}$,

$$\int_A |f_n| d\lambda < \varepsilon, \quad \text{for every} \quad A \in \mathcal{A}, A \subseteq X \setminus B_{p_0} \quad \text{with} \quad \lambda(A) < \delta \tag{3}$$

and, for any $i = 0, \ldots, n_0 - 1$,

$$\int_A |f_i| d\lambda < \varepsilon, \quad \text{for every} \quad A \in \mathcal{A} \quad \text{with} \quad \lambda(A) < \delta. \tag{4}$$

According to the definition of η, there exist $A_1 \in \mathcal{A}$ with $\lambda(A_1) < \delta$ and $n_1 \in \mathbb{N}$ such that

$$\eta - \delta < \int_{A_1} |f_{n_1}| d\lambda; \tag{5}$$

since $\delta < \varepsilon < \frac{\eta}{2}$,

$$\varepsilon < \frac{\eta}{2} < \eta - \delta < \int_{A_1} |f_{n_1}| d\lambda. \tag{6}$$

From (6) and (4), $n_1 \geq n_0$ and so, according to (5), (3) and (2), we have

$$\begin{aligned}\eta - \varepsilon < \eta - \delta < \int_{A_1} |f_{n_1}| d\lambda &= \int_{A_1 \setminus B_{p_0}} |f_{n_1}| d\lambda + \int_{A_1 \cap B_{p_0}} |f_{n_1}| d\lambda \\ &< \varepsilon + \int_{B_{p_0}} |f_{n_1}| d\lambda < \varepsilon + \eta_1 + \varepsilon,\end{aligned}$$

from where, for every $\varepsilon > 0$, $\eta < \eta_1 + 3\varepsilon$ and so

$$\eta \leq \eta_1. \tag{7}$$

By (1) and (7) $\eta = \eta_1$. □

Proposition 1.99. *Let $(f_n) \subseteq L^1(\lambda)$ be a sequence w^2-convergent. (f_n) is weakly convergent if and only if (f_n) is λ-equicontinuous.*

Proof. We have already seen (see (ii) of Remark 1.66) that every weakly convergent sequence is λ-equicontinuous.

Let now $(f_n)_{n \in \mathbb{N}}$ be a λ-equicontinuous sequence w^2-convergent to f. Then, for every $\varepsilon > 0$, there exists $\delta > 0$ such that,

$$\int_A |f| d\lambda < \frac{\varepsilon}{3} \quad \text{and} \quad \int_A |f_n| d\lambda < \frac{\varepsilon}{3}, \quad \forall A \in \mathcal{A}, \\ \text{with} \quad \lambda(A) < \delta, \quad \forall n \in \mathbb{N}. \tag{1}$$

Let $(B_p) \subseteq \mathcal{A}$ be a decreasing sequence which localizes the concentration of mass of (f_n). Then there exists $p_0 \in \mathbb{N}$ such that, for any $p \geq p_0$, $\lambda(B_p) < \delta$ and so,

$$\int_{B_p} |f_n| d\lambda < \frac{\varepsilon}{3} \text{ and } \int_{B_p} |f| d\lambda < \frac{\varepsilon}{3}, \text{ for every } n \in \mathbb{N} \text{ and every } p \geq p_0. \tag{2}$$

Because of w^2-convergence, for every p and every $A \in \mathcal{A}$, $A \subseteq X \setminus B_p$, $\int_A f_n d\lambda$ converges to $\int_A f d\lambda$. Therefore, for $p = p_0$, there exists $n_0 \in \mathbb{N}$ such that

$$\int_A (f_n - f) d\lambda < \frac{\varepsilon}{3}, \quad \text{for every} \quad n \geq n_0 \quad \text{and every} \quad A \in \mathcal{A}, A \subseteq X \setminus B_{p_0}. \tag{3}$$

According to (1), (2) and (3), for every $n \geq n_0$ and every $A \in \mathcal{A}$, we have

$$\begin{aligned} \left| \int_A (f_n - f)\, d\lambda \right| &\leq \left| \int_{A \setminus B_{p_0}} (f_n - f)\, d\lambda \right| + \int_{A \cap B_{p_0}} |f_n| d\lambda + \int_{A \cap B_{p_0}} |f| d\lambda \\ &\leq \left| \int_{A \setminus B_{p_0}} (f_n - f)\, d\lambda \right| + \int_{B_{p_0}} |f_n| d\lambda + \int_{B_{p_0}} |f| d\lambda \leq \varepsilon. \end{aligned}$$

Therefore, for every $A \in \mathcal{A}$, $\int_A f_n d\lambda \to \int_A f d\lambda$ i.e. $f_n \xrightarrow[L^1(\lambda)]{w} f$. □

Theorem 1.100 (Theorem 5 of [76]). *For every bounded sequence $(f_n)_n$, w^2-convergent in $L^1(\lambda)$, there exist a subsequence (g_n) and a sequence of pairwise disjoint sets $(A_n) \subseteq \mathcal{A}$ such that:*

(i) $\eta((f_n)) = \lim_n \int_{A_n} |g_n| d\lambda$ *and*

(ii) $\left(\chi_{X \setminus A_n} \cdot g_n\right)_{n \in \mathbb{N}}$ *is weakly convergent in $L^1(\lambda)$.*

Proof. Let $(B_p) \subseteq \mathcal{A}$ be a decreasing sequence which localizes the concentration of mass of (f_n). According to Theorem 1.98,

$$\eta = \eta((f_n)) = \lim_p \limsup_n \int_{B_p} |f_n| d\lambda.$$

Then, for every $n \in \mathbb{N}^*$, there exist $p_n \in \mathbb{N}$ and $k_n \geq n$ such that

$$\eta - \frac{1}{n} < \int_{B_{p_n}} |f_{k_n}| d\lambda < \eta + \frac{1}{n}. \tag{1}$$

As the sequence (B_p) is decreasing, we can select the sequence (p_n) to be increasing and so that $\lim p_n = +\infty$. For $n = i_0 = 1$, by (1), there exist $p_1 = p_{i_0}$ and $k_1 \geq 1$ such that

$$\eta - 1 < \int_{B_{p_{i_0}}} |f_{k_1}| d\lambda < \eta + 1. \tag{2}$$

Since $f_{k_1} \cdot \lambda \ll \lambda$, for $\varepsilon_1 = 1$, there exists $\delta_1 > 0$ such that

$$\int_A |f_{k_1}| d\lambda < 1 = \varepsilon_1, \quad \text{for every} \quad A \in \mathcal{A} \quad \text{with} \quad \lambda(A) < \delta_1. \tag{3}$$

Let $i_1 > i_0$ such that $\lambda(B_{p_{i_1}}) < \delta_1$; then, by (3), we have:

$$\int_{B_{p_{i_1}}} |f_{k_1}| d\lambda < 1. \tag{4}$$

and by (2) and (4),

$$\eta - 1 - 1 < \int_{B_{p_{i_0}}} |f_{k_1}| d\lambda - \int_{B_{p_{i_1}}} |f_{k_1}| d\lambda \le \int_{B_{p_{i_0}}} |f_{k_1}| d\lambda < \eta + 1.$$

Therefore

$$\eta - 2 < \int_{B_{p_{i_0}} - B_{p_{i_1}}} |f_{k_1}| d\lambda < \eta + 1. \tag{5}$$

By (1), for $n = i_1$, there exist $p_{i_1} \in \mathbb{N}$ and $k_{i_1} \ge i_1$ such that

$$\eta - \frac{1}{i_1} < \int_{B_{p_{i_1}}} |f_{k_{i_1}}| d\lambda < \eta + \frac{1}{i_1}. \tag{6}$$

Let $\varepsilon_2 = \frac{1}{2}$; there exists $\delta_2 > 0$ such that

$$\int_A |f_{k_{i_1}}| d\lambda < \frac{1}{2}, \quad \text{for every} \quad A \in \mathcal{A} \quad \text{with} \quad \lambda(A) < \delta_2. \tag{7}$$

Let $i_2 > i_1$ such that $\lambda(B_{p_{i_2}}) < \delta_2$; then, by (7), we have:

$$\int_{B_{p_{i_2}}} |f_{k_{i_1}}| d\lambda < \frac{1}{2} \tag{8}$$

and by (6) and (8),

$$\begin{aligned} \eta - \frac{1}{i_1} - \frac{1}{2} &< \int_{B_{p_{i_1}}} |f_{k_{i_1}}| d\lambda - \int_{B_{p_{i_2}}} |f_{k_{i_1}}| d\lambda = \int_{B_{p_{i_1}} \setminus B_{p_{i_2}}} |f_{k_{i_1}}| d\lambda \\ &\le \int_{B_{p_{i_1}}} |f_{k_{i_1}}| d\lambda < \eta + \frac{1}{i_1}, \end{aligned}$$

therefore

$$\eta - \frac{1}{i_1} - \frac{1}{2} < \int_{B_{p_{i_1}} \setminus B_{p_{i_2}}} |f_{k_{i_1}}| d\lambda < \eta + \frac{1}{i_1}. \tag{9}$$

Then we construct an increasing sequence $(i_n) \subseteq \mathbb{N}$, such that $(k_{i_n})_n$ is an increasing sequence, $k_{i_n} \geq i_n$, for every $n \in \mathbb{N}$ and

$$\eta - \frac{1}{i_n} - \frac{1}{n+1} < \int_{B_{p_{i_n}} \setminus B_{p_{i_{n+1}}}} |f_{k_{i_n}}| d\lambda < \eta + \frac{1}{i_n}. \tag{10}$$

For every $n \in \mathbb{N}$, let $h_n = f_{k_{i_n}}$. Then $(h_n)_n$ is a subsequence of (f_n).

Let $A_n = B_{p_{i_n}} \setminus B_{p_{i_{n+1}}}$. By (10), we have:

$$\eta = \eta((f_n)) = \lim_n \int_{A_n} |h_n| d\lambda. \tag{11}$$

We show that $\{\chi_{X \setminus A_n} h_n : n \in \mathbb{N}\}$ is λ-equicontinuous. Let

$$\begin{aligned} \eta_0 &= \eta(\{\chi_{X \setminus A_n} h_n : n \in \mathbb{N}\}) \\ &= \inf_{p>0} \sup \left\{ \int_E \chi_{X \setminus A_n} |h_n| d\lambda : \lambda(E) < \frac{1}{p} : n \in \mathbb{N} \right\}. \end{aligned}$$

If $\eta_0 > 0$, then, for every $0 < \alpha < \eta_0$ and every $p \in \mathbb{N}$, there exist $U_p \in \mathcal{A}$ with $\lambda(U_p) < \frac{1}{p}$ and $n_p \geq p$ such that

$$\int_{U_p \setminus A_{n_p}} |h_{n_p}| d\lambda > \alpha. \tag{12}$$

We remark that $U_p \setminus A_{n_p} = (U_p \cup A_{n_p}) \setminus A_{n_p}$ and so we obtain

$$\int_{U_p \cup A_{n_p}} |h_{n_p}| d\lambda - \int_{A_{n_p}} |h_{n_p}| d\lambda > \alpha.$$

According to (10), we have therefore

$$\eta - \frac{2}{n_p} < \int_{A_{n_p}} |h_{n_p}| d\lambda < \int_{U_p \cup A_{n_p}} |h_{n_p}| d\lambda - \alpha, \quad \text{for every} \quad p \in \mathbb{N}. \tag{13}$$

Since (h_{n_p}) is a subsequence of (f_n), $\eta((h_{n_p})) \leq \eta < \eta + \frac{1}{2} \cdot \alpha$. Then there exists $\delta > 0$ such that

$$\begin{aligned} &\int_E |h_{n_p}| d\lambda < \eta + \frac{1}{2} \cdot \alpha, \quad \text{for every} \quad E \in \mathcal{A} \\ &\qquad \text{with} \quad \lambda(E) < \delta, \quad \text{and} \quad p \in \mathbb{N}. \end{aligned} \tag{14}$$

$\lambda(A_{n_p}) \to 0$ and $\lambda(U_p \cup A_{n_p}) < \lambda(U_p) + \lambda(A_{n_p}) < \frac{1}{p} + \lambda(A_{n_p})$; therefore there exists $p_0 \in \mathbb{N}$ such that, for every $p \geq p_0$, $\lambda(U_p \cup A_{n_p}) < \delta$.

So, by (13) and (14), for every $0 < \alpha < \eta_0$ and for every $p \geq p_0$,

$$\eta - \frac{2}{n_p} < \int_{A_{n_p}} |h_{n_p}| d\lambda < \eta + \frac{1}{2} \cdot \alpha - \alpha = \eta - \frac{1}{2} \cdot \alpha$$

so that $\eta - \frac{2}{n_p} \leq \eta - \frac{1}{2}\eta_0$, for every $p \geq p_0$. When $p \to +\infty$, $\frac{2}{n_p} \to 0$ and, if $\eta_0 > 0$, then $\eta \leq \eta - \frac{1}{2}\eta_0$ which is impossible.

Therefore $\eta_0 = 0$ and hence $\{\chi_{X \setminus A_n} h_n : n \in \mathbb{N}\}$ is bounded and λ - equicontinuous in $L^1(\lambda)$. According to the Dunford–Pettis theorem 1.65, $\{\chi_{X \setminus A_n} \cdot h_n : n \in \mathbb{N}\}$ is relatively weakly compact and so it has a subsequence $(\chi_{X \setminus A_{n_k}} \cdot h_{n_k})_{k \in \mathbb{N}}$ weakly convergent. Let us note $g_k = h_{n_k}$; because (A_n) is decreasing and (f_n) is bounded in $L^1(\lambda)$, the sequence $(\chi_{X \setminus A_k} \cdot g_k)_{k \in \mathbb{N}}$ is also weakly convergent. By (11), we have $\eta((f_n)) = \lim_k \int_{A_k} |g_k| d\lambda$. □

Remark 1.101.

(i) The subsequence $(g_n)_n$ of $(f_n)_n$ in the previous theorem has the following property:

$$\eta((g_{n_k})_k) = \eta((f_n)_n), \quad \text{for every subsequence} \quad (g_{n_k})_k, \quad \text{of} \quad (g_n)_n.$$

Indeed, if $(B_p)_p$ localizes the concentration of mass for $(g_n)_n$, then, for every subsequence $(g_{n_k})_k$ of $(g_n)_n$, the sequence $(B_p)_p$ localizes again the concentration of mass of $(g_{n_k})_k$ (see (vii) of Remark 1.96). Then, by Theorem 1.98,

$$\eta((g_{n_k})) = \lim_p \limsup_k \int_{B_p} |g_{n_k}| d\lambda.$$

Therefore, for every $\varepsilon > 0$, there exist $p_0 \in \mathbb{N}$ and $k_0 \in \mathbb{N}$, such that

$$\int_{B_{p_0}} |g_{n_k}| d\lambda < \eta((g_{n_k})) + \varepsilon, \quad \text{for every} \quad k \geq k_0.$$

But, for k sufficiently large, through the definition of $A_n = B_{p_{i_n}} \setminus B_{p_{i_{n+1}}}$,

$$\int_{A_n} |g_{n_k}| d\lambda \leq \int_{B_{p_{i_n}}} |g_{n_k}| d\lambda \leq \int_{B_{p_0}} |g_{n_k}| d\lambda < \eta((g_{n_k})) + \varepsilon.$$

If we pass to the limit for $n \to \infty$, we obtain

$$\eta((f_n)) = \lim_n \int_{A_n} |g_{n_k}| d\lambda \leq \eta((g_{n_k})) + \varepsilon, \quad \text{for every} \quad \varepsilon > 0.$$

Then $\eta((f_n)) \leq \eta((g_{n_k}))$. Since (g_{n_k}) is a subsequence of (f_n), $\eta((g_{n_k})) \leq \eta((f_n))$. We have therefore $\eta((g_{n_k})) = \eta((f_n))$.

Let us remark that the concentration of mass for $(g_n)_n$ is the maximum between the concentration of mass for all subsequences of $(f_n)_n$.

(ii) As we show in the following example, the subsequence $(g_n)_n$ of $(f_n)_n$ does not generally coincide with the sequence $(f_n)_n$.

Let λ be the Lebesgue measure on $]0,1]$; for every $n \in \mathbb{N}$, let $f_n = n\chi_{]0,\frac{1}{n}]}$. Then $f_n \in L^1(\lambda)$ and $\|f_n\|_1 = 1$, for every $n \in \mathbb{N}$. Therefore (f_n) is a bounded sequence in $L^1(\lambda)$, $f_n \xrightarrow{w^2} 0$ and $\eta((f_n)) = 1$.

If we suppose that the sequence (g_n) coincides with the sequence (f_n), then there exists a sequence of pairwise disjoint subsets (A_n) of $]0,1]$ such that $\lim_n \int_{A_n} f_n d\lambda = 1 = \eta((f_n))$. Therefore, there exists $n_0 \in \mathbb{N}$ such that $\int_{A_n} f_n d\lambda > \frac{1}{2}$, for every $n \geq n_0$. We have therefore $\lambda(A_n) > \frac{1}{2n}$, for every $n \geq n_0$. Then the infinite series of general term $\lambda(A_n)$ is divergent. On the other hand, $\sum_{n=n_0}^{\infty} \lambda(A_n) = \lambda(\cup_{n=n_0}^{\infty} A_n) \leq 1$ which is a contradiction.

Therefore the sequence (g_n) is different from the sequence (f_n).

For every $n \in \mathbb{N}$, we take $g_n = f_{n^n}$ and $A_n =]\frac{1}{(n+1)^{n+1}}, \frac{1}{n^n}]$; then $\int_{A_n} g_n d\lambda = 1 - (\frac{n}{n+1})^n \cdot \frac{1}{n+1} \to 1 = \eta((f_n))$ and, for all $A \in \mathcal{A}$, $\int_A \chi_{(]0,1]\setminus A_n)} g_n d\lambda = n^n \lambda(A \cap]0, \frac{1}{(n+1)^{n+1}}]) \leq \frac{n^n}{(n+1)^{n+1}} \to 0$.

Therefore $\chi_{(]0,1]\setminus A_n)} g_n \xrightarrow[L^1(\lambda)]{w} \underline{0}$. Here $(g_n) = (f_{n^n})$ is the subsequence that suits to problem.

Proposition 1.102. *Let $(f_n) \subseteq L^1(\lambda)$ and $f \in L^1(\lambda)$ such that $f_n \xrightarrow{w^2} f$. Then,*

(i) *$(f_n)_n$ is $\eta((f_n))$-convergent to f;*

(ii) *If $(f_n)_n$ is convergent in measure, then (f_n) converges in measure to f.*

Proof. Let $(B_p) \subseteq \mathcal{A}$ be a decreasing sequence which localizes the concentration of mass of (f_n).

(i) According to Theorem 1.98, we have

$$\eta((f_n)) = \lim_p \limsup_n \int_{B_p} |f_n| d\lambda. \tag{1}$$

For every $A \in \mathcal{A}$, every $p \in \mathbb{N}$ and every $n \in \mathbb{N}$, one can write

$$\left|\int_A (f_n - f) d\lambda\right| \leq \left|\int_{(A \setminus B_p)} (f_n - f) d\lambda\right| + \int_{B_p} |f_n| d\lambda + \int_{B_p} |f| d\lambda. \tag{2}$$

Since, for every $p \in \mathbb{N}$, $f_n \restriction_{(X \setminus B_p)} \xrightarrow[X \setminus B_p]{w} f \restriction_{(X \setminus B_p)}$,

$$\lim_n \left|\int_{(A \setminus B_p)} (f_n - f) d\lambda\right| = 0, \quad \text{for every} \quad A \in \mathcal{A}, \quad \text{and every} \quad p \in \mathbb{N}. \tag{3}$$

By (2) and (3), we have

$$\limsup_n \left|\int_A (f_n - f) d\lambda\right| \leq \limsup_n \int_{B_p} |f_n| d\lambda + \int_{B_p} |f| d\lambda,$$
$$\text{for every} \quad p \in \mathbb{N}.$$

Hence, using (1),

$$\limsup_n \left|\int_A (f_n - f) d\lambda\right| \leq \eta((f_n)) + \lim_p \int_{B_p} |f| d\lambda = \eta((f_n)).$$

According to Definition 1.91, $f_n \xrightarrow{\eta((f_n))} f$. This proves the (i).

(ii) Let g be a measurable function such that $f_n \xrightarrow{\lambda} g$. Then, for all $p \in \mathbb{N}$, $\chi_{(X \setminus B_p)} f_n \xrightarrow{\lambda} \chi_{(X \setminus B_p)} g$ and since $(\chi_{(X \setminus B_p)} f_n)_{n \in \mathbb{N}}$ is λ-equicontinuous, for all $p \in \mathbb{N}$, it results from the Vitali's theorem (see Theorem 1.54) that

$$\chi_{(X \setminus B_p)} f_n \xrightarrow{\|\cdot\|_1} \chi_{(X \setminus B_p)} g, \quad \text{for every} \quad p \in \mathbb{N}. \tag{4}$$

Then, for every $p \in \mathbb{N}$ and every $A \in \mathcal{A}$ with $A \subseteq X \setminus B_p$,

$$\begin{aligned}\left|\int_A (f - g) d\lambda\right| &\leq \left|\int_A (f - f_n) d\lambda\right| + \left|\int_A (f_n - g) d\lambda\right| \\ &\leq \left|\int_{A \setminus B_p} (f - f_n) d\lambda\right| + \int_{X \setminus B_p} |f_n - g| \, d\lambda.\end{aligned}$$

By (3), $\lim_n \int_{A\setminus B_p}(f - f_n)d\lambda = 0$ and, by (4), $\lim_n \int_{X\setminus B_p} |f_n - g|\, d\lambda = 0$, for every $p \in \mathbb{N}$. Therefore, for every $A \in \mathcal{A}$ with $A \subseteq X\setminus B_p$, $\int_A (f - g)d\lambda = 0$. Since $\lambda(B_p) \to 0$, $f = g$, λ-a.e.; hence $f_n \overset{\lambda}{\to} f$. □

Now we are going to give another extension to the Dunford–Pettis' theorem of weak compactness. This extension, known under the name of "biting lemma", was demonstrated in 1980 by J. K. Brooks and R. V. Chacon ([37]). We can also mention other demonstrations of this result: M. Słaby [154], J. M. Ball and F. Murat [23], D. Q. Luu [115] and L. C. Florescu [75]. The demonstration given here is that of M. Słaby and was reproduced by M. Saadoune and M. Valadier in [149] (see also Theorem 6 in [76]).

Theorem 1.103 (Biting Lemma). *Every bounded sequence $(f_n)_n \subseteq L^1(\lambda)$ admits a subsequence $(g_n)_n$ w^2-convergent. Besides, for all subsequence $(g_{n_k})_k$ of $(g_n)_n$, $\eta((g_{n_k})_k) = \eta((f_n)_n)$.*

Proof. Let $(f_n)_n$ be a bounded sequence in $L^1(\lambda)$ and let $M = \sup_n \|f_n\|_1$; we have shown in Proposition 1.83 that

$$\eta = \eta((f_n)) = \lim_{t\to\infty}\left[\sup_n \int_{(|f_n|>t)} |f_n| d\lambda\right].$$

For every $i \in \mathbb{N}$, let $F_i : [0, +\infty[\to [0, +\infty[$,

$$F_i(t) = \sup_{n\ge i} \int_{(|f_n|>t)} |f_n| d\lambda.$$

It is clear that, for all i, F_i is a decreasing function on $[0, +\infty[$ and that $\lim_{t\to\infty} F_i(t) = \eta(\{f_n : n \ge i\})$.

As $\eta(\{f_n : n < i\}) = 0$, for every i, we have:

$$\lim_{t\to\infty} F_i(t) = \eta((f_n)) = \eta, \quad \text{for all} \quad i \in \mathbb{N}. \tag{1}$$

Thus, for $i = 0$, we can choose an increasing sequence $(t_q) \subseteq [0, +\infty[$, $t_q \uparrow +\infty$ such that

$$F_0(t_q) \le \eta + \frac{1}{q}, \quad \text{for every} \quad q \in \mathbb{N}^*. \tag{2}$$

Then, by (1) and (2), it results that, for every $i = q$

$$\eta = \inf_{t>0} F_q(t) \le F_q(t_q) = \sup_{n\ge q} \int_{(|f_n|>t_q)} |f_n| d\lambda.$$

Then, for all $q \in \mathbb{N}^*$

$$\eta - \frac{1}{q} < \sup_{n \geq q} \int_{(|f_n|>t_q)} |f_n| d\lambda.$$

Therefore, for all $q \in \mathbb{N}^*$, there exists $n_q \geq q$ such that

$$\eta - \frac{1}{q} < \int_{(|f_{n_q}|>t_q)} |f_{n_q}| d\lambda, \quad \text{for every} \quad q \in \mathbb{N}^*. \tag{3}$$

If $A_q = \{x \in X : |f_{n_q}(x)| > t_q\}$, then

$$M = \sup_n \|f_n\|_1 \geq \int_X |f_{n_q}| d\lambda \geq \int_{A_q} |f_{n_q}| d\lambda \geq t_q \lambda(A_q), \quad \text{for all} \quad q \in \mathbb{N}^*.$$

Hence $\lambda(A_q) \to 0$.

We will now show that the sequence $(\chi_{(X \setminus A_q)} f_{n_q})_q$ is λ-equicontinuous.

Let us define $F(t)$ by

$$F(t) = \sup_{q \in \mathbb{N}} \int_{(|f_{n_q}|>t) \setminus A_q} |f_{n_q}| d\lambda, \quad \text{for every} \quad t > 0.$$

From Proposition 1.83, we have $\eta((\chi_{(X \setminus A_q)} f_{n_q})) = \lim_{t \to \infty} F(t)$. Since F is a decreasing function, $\lim_{t \to \infty} F(t) = \lim_{j \to \infty} F(t_j)$, for every increasing sequence $(t_j)_j$ with $t_j \uparrow +\infty$. Then, by definition of A_q, for every $q \in \mathbb{N}$ and every $j \in \mathbb{N}$, $(|f_{n_q}| > t_j) \setminus A_q = (|f_{n_q}| > t_j) \cap (|f_{n_q}| \leq t_q) = (t_j < |f_{n_q}| \leq t_q)$. Thus, for all $j \geq q$, $(|f_{n_q}| > t_j) \setminus A_q = \emptyset$ and from (2) and (3), we obtain, for every $j \in \mathbb{N}$,

$$\begin{aligned} F(t_j) &= \sup_{q>j} \int_{(t_j < |f_{n_q}| \leq t_q)} |f_{n_q}| d\lambda \\ &= \sup_{q>j} \left[\int_{(|f_{n_q}|>t_j)} |f_{n_q}| d\lambda - \int_{(|f_{n_q}|>t_q)} |f_{n_q}| d\lambda \right] \\ &\leq F_0(t_j) - \inf_{q>j} \int_{(|f_{n_q}|>t_q)} |f_{n_q}| d\lambda < \eta + \frac{1}{q} - \left(\eta - \frac{1}{q}\right) = \frac{2}{q}. \end{aligned}$$

Therefore

$$\eta\left(\left(\chi_{(X \setminus A_q)} f_{n_q}\right)\right) = \lim_{j \to \infty} F(t_j) = 0$$

and then the sequence $(\chi_{(X \setminus A_q)} f_{n_q})_{q \in \mathbb{N}}$ is λ-equicontinuous.

We replace the sequence $(A_p)_p$ by a decreasing sequence $(B_p)_p$ in the following way:

Because $\lambda(A_p) \to 0$, for every $h \in \mathbb{N}$, there exists $l_h \geq h$ such that

$$\lambda(A_{l_h}) \leq \frac{1}{2^h}.$$

For every $p \in \mathbb{N}$, let $B_p = \bigcup_{h \geq p} A_{l_h}$; then the sequence (B_p) is decreasing. $\lambda(B_p) \leq \sum_{h=p}^{\infty} \lambda(A_{l_h}) < \frac{1}{2^{p-1}}$, hence $\lambda(B_p) \to 0$. Since, for every $p \in \mathbb{N}$, $B_p \supseteq A_{l_p}$,we have, for every $q \in \mathbb{N}$,

$$\begin{aligned}\eta\left(\left(\chi_{(X \setminus B_q)} \cdot f_{n_{l_q}}\right)\right) &= \lim_{t \to \infty} \sup_{q \in \mathbb{N}} \int_{(|f_{n_{l_q}}| > t) \setminus B_q} |f_{n_{l_q}}| d\lambda \\ &\leq \lim_{t \to \infty} \sup_{q \in \mathbb{N}} \int_{(|f_{n_{l_q}}| > t) \setminus A_{l_q}} |f_{n_{l_q}}| d\lambda \\ &\leq \lim_{t \to \infty} \sup_{q \in \mathbb{N}} \int_{(|f_{n_q}| > t) \setminus A_q} |f_{n_q}| d\lambda = \lim_{t \to \infty} F(t) = 0.\end{aligned}$$

The sequence $(\chi_{(X \setminus B_q)} f_{n_{l_q}})_{q \in \mathbb{N}}$ is therefore an equicontinuous and bounded one and, according to Dunford–Petis theorem (Theorem 1.65), there exists a subsequence $q_m \uparrow \infty$ such that $(\chi_{(X \setminus B_{q_m})} f_{n_{l_{q_m}}})_{m \in \mathbb{N}}$ is weakly convergent. We note, for every $m \in \mathbb{N}$, $g_m = f_{n_{l_{q_m}}}$ and $C_m = B_{q_m}$; the sequence $(\chi_{(X \setminus C_m)} g_m)_{m \in \mathbb{N}}$ is then weakly convergent in $L^1(\lambda)$ to a function f.

The sequence $(C_m)_m$ is a decreasing one and $\lambda(C_m) \downarrow 0$. For every $p \in \mathbb{N}$, every $A \in \mathcal{A}$ with $A \subseteq X \setminus C_p$ and every $n \geq p$, $C_n \subseteq C_p$ and so

$$\int_A g_n d\lambda = \int_{(A \cap (X \setminus C_n))} g_n d\lambda = \int_A \chi_{(X \setminus C_n)} g_n d\lambda \to \int_A f d\lambda.$$

Therefore the sequence $g_n \xrightarrow{w^2} f$.

We still need to demonstrate that, for every subsequence $(g_{n_k})_k$ of (g_n), we have $\eta((g_{n_k})) = \eta((f_n))$.

It is clear that

$$\eta((g_{n_k})) \leq \eta((f_n)) = \eta. \tag{4}$$

As $(g_{n_k})_k$ is a subsequence of $(g_n)_n$, there exists an increasing sequence $n_k \uparrow \infty$ such that $g_{n_k} = f_{n_{h_k}}$ where, for every $k \in \mathbb{N}$, $h_k = l_{q_{n_k}}$. By (3),

$$\eta - \frac{1}{l_{q_{n_k}}} = \eta - \frac{1}{h_k} < \int_{(|g_{n_k}| > t_{h_k})} |g_{n_k}| d\lambda, \quad \text{for all} \quad k \in \mathbb{N}.$$

k tending to infinity, we obtain:

$$\begin{aligned}\eta &\le \lim_{k\to\infty} \int_{(|g_{n_k}|>t_{h_k})} |g_{n_k}|d\lambda \le \lim_{k\to\infty} \sup_{n\in\mathbb{N}} \int_{(|g_{n_k}|>t_{h_k})} |g_{n_k}|d\lambda \\ &= \lim_{t\to\infty} \sup_k \int_{(|g_{n_k}|>t)} |g_{n_k}|d\lambda = \eta((g_{n_k})).\end{aligned} \tag{5}$$

According to (4) and (5), we have $\eta((g_{n_k})) = \eta$. Which was to be demonstrated. □

Remark 1.104.

(i) If $(f_n)_n$ is a bounded and λ-equicontinuous sequence, then, according to Theorem 1.103, there exists a subsequence $(g_n)_n$ of $(f_n)_n$, w^2-convergent such that $\eta((g_n)) = \eta((f_n)) = 0$. Then, by Proposition 1.99, $(g_n)_n$ is weakly convergent.

(ii) According to the previous theorem, let us remark that the concentration of mass for the subsequence $(g_n)_n$ (which is non null if and only if $\eta((f_n)) > 0$) is maximal among all the other subsequences of $(f_n)_n$.

Corollary 1.105. *Let $(f_n)_n$ be a bounded sequence of $L^1(\lambda)$ such that $f_n \xrightarrow{\lambda} f$; then there exists a subsequence $(g_n)_n$ of $(f_n)_n$ such that $g_n \xrightarrow{w^2} f$.*

Proof. By Theorem 1.103, there exists a subsequence $(g_n)_n$ of $(f_n)_n$ that is w^2-convergent. Since $g_n \xrightarrow{\lambda} f$, from (2) of Proposition 1.102, we have $g_n \xrightarrow{w^2} f$. □

The following result, known as "subsequence splitting lemma", was given by Rosenthal in 1979 ([143]) and published in 1980 ([33]).

Corollary 1.106 (Subsequence Splitting Lemma). *For every bounded sequence $(f_n)_n \subseteq L^1(\lambda)$, there exist a subsequence $(h_n)_n$ and a sequence of pairwise disjoint sets $(A_n)_n \subseteq \mathcal{A}$ such that*

(i) $\lim_n \int_{A_n} |h_n|d\lambda = \eta((f_n))$,

(ii) $(\chi_{(X\setminus A_n)} h_n)_n$ *is weakly convergent in* $L^1(\lambda)$.

Proof. According to Theorem 1.103, there exists a subsequence (g_n) of (f_n) w^2-convergent such that, for every subsequence (h_n) of (g_n), $\eta((h_n)) = \eta((f_n))$.

According to Theorem 1.100, there exist a subsequence (h_n) of (g_n) and a sequence of pairwise disjoint sets $(A_n) \subseteq \mathcal{A}$ such that $\eta((f_n)) = \eta((g_n)) = \lim_n \int_{A_n} |h_n| d\lambda$ and $(\chi_{(X \setminus A_n)} h_n)$ is weakly convergent. $\square$

Remark 1.107.

(i) According to Remark 1.101 (ii), $(h_n)_n$ does not generally coincide with (g_n).

(ii) Using the Rosenthal's lemma, H.-A. Klei demonstrated the Brooks–Chacon's lemma in [104] and [105]. Here, we adopt the reversed process: we have used the Brooks–Chacon's lemma to demonstrate the Rosenthal's lemma.

Chapter 2

Bounded Measures on Topological Spaces

If, in the first chapter, we studied measure spaces in an abstract frame, in this second chapter, we will present a number of properties of measures defined on topological spaces.

We will start by studying several types of regularities. After that, in the second paragraph, we present some properties of Polish and Suslin spaces, especially the property of being radonian. We will introduce then the narrow topology on the measure spaces and we will study the different properties of semicontinuity. In paragraph 4, we will present the Prohorov's compactness theorem and in paragraph 5, we will introduce the metrics of Dudley and of Levy–Prohorov on the space of positive Radon measures. Finally, in the last paragraph, as an application to the Prohorov's theorem, we will introduce the Wiener measure as limit of a sequence of measures. In writing this chapter, have been used, inter alia, the following monographs: [28, 30, 31, 32, 61, 70, 93, 152].

2.1 Regular Measures

In this section, without any further explanation,

- (T, τ) is a Hausdorff topological space.
- $D \in \tau$ means that D is an open set for the topology τ.
- $\mathcal{B}_T(\tau)$ (or simply $\mathcal{B}_T$ if is no ambiguity on the topology τ), is the Borel σ-algebra generated by all open sets for the topology τ.
- $\mathcal{K}_T(\tau)$ or $\mathcal{K}_T$ notes the family of all the compact sets of (T, τ).
- For all $t \in T$, $\delta_t : \mathcal{B}_T \to \mathbb{R}_+$ is the Dirac measure supported on the singleton t (the unit mass concentrated at t); it means that,

$$\delta_t(A) = \begin{cases} 1, & t \in A, \\ 0, & t \notin A. \end{cases}$$

  It is clear that, for every $t \in T$, $\delta_t \in ca(\mathcal{B}_T)$.
- For any σ-additive measure $\mu \in ca(\mathcal{B}_T(\tau))$, we note with $\hat{\mathcal{B}}_T(\tau, \mu)$ the completion of the σ-algebra $\mathcal{B}_T(\tau)$ with respect to μ.

We recall that

(i) $A \in \hat{\mathcal{B}}_T(\tau,\mu) \Leftrightarrow$ there exist $B \in \mathcal{B}_T(\tau)$ and $N \subseteq T$ such that $A = B \triangle N$ and $\mu^*(N) = 0$ where μ^* is the outer measure generated by the total variation, $|\mu|$, of μ (see Definition 1.19).

(ii) $\hat{\mathcal{B}}_T(\tau) = \bigcap\{\hat{\mathcal{B}}_T(\tau,\mu) : \mu \in ca(\mathcal{B}_T)\}$ (still noted $\hat{\mathcal{B}}_T$ if there is no ambiguity) is the σ–algebra of *universally measurable sets*.

(iii) Every measure $\mu \in ca(\mathcal{B}_T)$ admits an unique extension to a measure $\hat{\mu} \in ca(\hat{\mathcal{B}}_T)$.

Proposition 2.1. *Let $\mu \in ca(\mathcal{B}_T)$; then the following properties are equivalent:*

(i) $|\mu|(A) = \sup_{F=\overline{F}\subseteq A} |\mu|(F)$, *for every $A \in \mathcal{B}_T$.*

(ii) $|\mu|(A) = \inf_{D\in\tau, D\supseteq A} |\mu|(D)$, *for every $A \in \mathcal{B}_T$.*

(iii) *For every $\varepsilon > 0$ and every $A \in \mathcal{B}_T$, there exist $F = \overline{F} \subseteq A$ and $D \in \tau$, $D \supseteq A$ such that $|\mu|(D \setminus F) < \varepsilon$.*

Proof. (i)$\Rightarrow$(ii) For every $A \in \mathcal{B}_T$ and every $a > |\mu|(A)$, $|\mu|(T\setminus A) > |\mu|(T) - a$. By hypothesis, there exists $F = \overline{F} \subseteq T \setminus A$ such that $|\mu|(F) > |\mu|(T) - a$ or $a > |\mu|(T \setminus F)$. If $G = T \setminus F$, $G \in \tau$, then $G \supseteq A$ and $a > |\mu|(G)$. Therefore $|\mu|(A) \leq \inf_{D\in\tau, D\supseteq A} |\mu|(D)$ and so we have (ii).

(ii)$\Rightarrow$(iii) For every $A \in \mathcal{B}_T$, $T \setminus A \in \mathcal{B}_T$. By (ii), for every $\varepsilon > 0$, there exist $D, G \in \tau$ such that $A \subseteq D$, $T \setminus A \subseteq G$, $|\mu|(D) < |\mu|(A) + \frac{\varepsilon}{2}$ and $|\mu|(G) < |\mu|(T \setminus A) + \frac{\varepsilon}{2}$. If $F = T \setminus G$, then $F = \overline{F} \subseteq A$ and $|\mu|(D \setminus F) = |\mu|(D) - |\mu|(F) < |\mu|(A) + \frac{\varepsilon}{2} - |\mu|(A) + \frac{\varepsilon}{2} = \varepsilon$.

(iii)$\Rightarrow$(i) For every $A \in \mathcal{B}_T$ and every $a < |\mu|(A)$, let $\varepsilon = |\mu|(A) - a > 0$. Then, by (iii), there exist $F = \overline{F} \subseteq A$ and $D \in \tau$, $D \supseteq A$ such that $|\mu|(D \setminus F) < \varepsilon$. Therefore we have $|\mu|(F) > |\mu|(D) - \varepsilon \geq |\mu|(A) - \varepsilon = a$ and so $|\mu|(A) = \sup_{F=\overline{F}\subseteq A} |\mu|(F)$. □

Definition 2.2. A measure $\mu : \mathcal{B}_T \to \mathbb{R}$ is *regular* if it fulfils one of the equivalent conditions (i)–(iii) of the previous proposition. Let us note $rca(\mathcal{B}_T)$ the set of all regular measures of $ca(\mathcal{B}_T)$ and by $rca^+(\mathcal{B}_T)$ the cone of all positive measures of $rca(\mathcal{B}_T)$.

Obviously, $rca(\mathcal{B}_T)$ is a vector subspace of $ca(\mathcal{B}_T)$.

Remark 2.3.

(i) $rca(\mathcal{B}_T)$ is a closed vector subspace of $(ca(\mathcal{B}_T), \|\cdot\|)$; therefore it is a Banach space.

Indeed, let $(\mu_n)_n \subseteq rca(\mathcal{B}_T)$ be a sequence convergent in $(ca(\mathcal{B}_T), \|\cdot\|)$ to $\mu \in rca(\mathcal{B}_T)$; by Theorem 1.23, $|\mu_n - \mu|(T) \to 0$. Then, for every $\varepsilon > 0$, there exists $n_0 \in \mathbb{N}$ such that

$$|\mu_{n_0} - \mu|(T) < \frac{\varepsilon}{2}. \tag{1}$$

As μ_{n_0} is a regular measure, for every $A \in \mathcal{B}_T$, there exists $F_\varepsilon = \overline{F}_\varepsilon \subseteq A$ such that

$$|\mu_{n_0}|(A \setminus F_\varepsilon) < \frac{\varepsilon}{2}. \tag{2}$$

By (1) and (2), we obtain

$$|\mu|(A \setminus F_\varepsilon) \le |\mu - \mu_{n_0}|(A \setminus F_\varepsilon) + |\mu_{n_0}|(A \setminus F_\varepsilon) < \frac{\varepsilon}{2} + \frac{\varepsilon}{2} = \varepsilon.$$

Therefore $\mu \in rca(\mathcal{B}_T)$.

(ii) A measure μ is regular if and only if $|\mu|$ is regular.

(iii) μ is regular if and only if μ^+ and μ^- are regular measures.

(iv) In Proposition 2.1, the properties are equivalent as well, if we only suppose that $\mu \in ba(\mathcal{B}_T)$. We can therefore define, in an analogous way, the regular measures in $ba(\mathcal{B}_T)$.

(v) We can show (see Alexandroff's theorem in [62], III.5.13) that, if T is a compact space and if $\mu \in ba(\mathcal{B}_T)$ is a regular measure, then $\mu \in ca(\mathcal{B}_T)$.

The following theorem says that any Borel measure on a metric space is regular; we can find a proof in [28], Theorem 1.1 (see also Theorem 7.1.3 in [61] and Theorem 7.1.7 in [30]).

Theorem 2.4. *If (T, τ) is a metrizable space, then $rca(\mathcal{B}_T) = ca(\mathcal{B}_T)$.*

Proof. Let $\mu \in ca(\mathcal{B}_T)$; by replacing μ by $|\mu|$, we can suppose that $\mu \in ca^+(\mathcal{B}_T)$.

Let $\mathcal{A} = \{A \in \mathcal{B}_T : \forall \varepsilon > 0, \exists F_\varepsilon = \overline{F}_\varepsilon \subseteq A \subseteq D_\varepsilon \in \tau \text{ and } \mu(D_\varepsilon \setminus F_\varepsilon) < \varepsilon\}$. Firstly, let us show that $\mathcal{A}$ is a σ-algebra. Indeed, for every $A \in \mathcal{A}$ and every $\varepsilon > 0$, there exist $F_\varepsilon = \overline{F}_\varepsilon \subseteq A$ and $D_\varepsilon \in \tau$, $A \subseteq D_\varepsilon$ such that $\mu(D_\varepsilon \setminus F_\varepsilon) < \varepsilon$. Let $D'_\varepsilon = T \setminus F_\varepsilon$ and $F'_\varepsilon = T \setminus D_\varepsilon$; then $D'_\varepsilon \in \tau$, $D'_\varepsilon \supseteq T \setminus A$, $F'_\varepsilon = \overline{F'_\varepsilon} \subseteq T \setminus A$ and $\mu(D'_\varepsilon \setminus F'_\varepsilon) = \mu(D_\varepsilon \setminus F_\varepsilon) < \varepsilon$; therefore $T \setminus A \in \mathcal{A}$.

Let now $(A_n)_{n\in\mathbb{N}} \subseteq \mathcal{A}$ and $A = \bigcup_1^\infty A_n \in \mathcal{B}_T$; for every $n \in \mathbb{N}$ and every $\varepsilon > 0$, there exist $F_n = \overline{F}_n, D_n \in \tau$ such that $F_n \subseteq A \subseteq D_n$ and $\mu(D_n \setminus F_n) < \frac{\varepsilon}{2^{n+1}}$. The sequence $(\bigcup_{k=1}^n F_k)_{n\in\mathbb{N}}$ converges to $\bigcup_1^\infty F_n$ and so there exists $n_0 \in \mathbb{N}$ such that $\mu(\bigcup_{k=n_0+1}^\infty F_k) < \frac{\varepsilon}{2}$.

Let $F_\varepsilon = \bigcup_{k=1}^{n_0} F_k$ and $D_\varepsilon = \bigcup_{n=1}^\infty D_n$; then $F_\varepsilon = \overline{F}_\varepsilon \subseteq A, D_\varepsilon \in \tau$, $A \subseteq D_\varepsilon$ and

$$\begin{aligned}\mu(D_\varepsilon \setminus F_\varepsilon) &= \mu\left(\bigcup_{n=1}^\infty D_n \setminus \bigcup_{k=1}^{n_0} F_k\right) = \mu\left[\left(\bigcup_{n=1}^\infty D_n \setminus \bigcup_{k=1}^\infty F_k\right) \cup \bigcup_{k=n_0+1}^\infty F_k\right] \\ &\le \mu\left[\bigcup_{n=1}^\infty (D_n \setminus F_n)\right] + \mu\left[\bigcup_{k=n_0+1}^\infty F_k\right] \le \sum_{n=1}^\infty \mu(D_n \setminus F_n) + \frac{\varepsilon}{2} \\ &< \sum_{n=1}^\infty \frac{\varepsilon}{2^{n+1}} + \frac{\varepsilon}{2} = \varepsilon.\end{aligned}$$

Therefore $A \in \mathcal{A}$. Obviously, $T \in \mathcal{A}$ and so $\mathcal{A}$ is a σ–algebra.

Now we show that $\mathcal{A}$ contains the τ–closed sets.

Let d be a compatible metric on T ($\tau = \tau_d$). $\forall F = \overline{F} \subseteq T, F = \bigcap_{n=1}^\infty F^{\frac{1}{n}}$, where F^a is the open ball of center F and radius a. $F^{\frac{1}{n}} \downarrow F$ and so, for every $\varepsilon > 0$, there exists n_0 with $\mu(F^{\frac{1}{n_0}} \setminus F) < \varepsilon$. Because $F^{\frac{1}{n_0}} \in \tau$, we have $F \in \mathcal{A}$.

Therefore $\mathcal{A} \subseteq \mathcal{B}_T$ is a σ-algebra that contains the closed sets, so that $\mathcal{A} = \mathcal{B}_T$ and then μ is regular on $\mathcal{B}_T$. □

Remark 2.5. The proof of the previous theorem allows us to give the following result:

For every Hausdorff topological space T for which the closed sets are G_δ-set (a countable intersection of open sets), every σ-additive measure is regular, which means $rca(\mathcal{B}_T) = ca(\mathcal{B}_T)$.

Definition 2.6 (see [61], p. 224). A measure $\mu \in ca(\mathcal{B}_T)$ is called *tight* if, for every $\varepsilon > 0$, there exists $K_\varepsilon \in \mathcal{K}_T$ such that $|\mu|(T \setminus K_\varepsilon) < \varepsilon$. We will note by $tca(\mathcal{B}_T)$ the set of all tight measures in $ca(\mathcal{B}_T)$.

By using an analogous demonstration to that of Remark 2.3 (i), it easy to see that $tca(\mathcal{B}_T)$ is a closed vector subspace of $(ca(\mathcal{B}_T), \|\cdot\|)$ and so, a Banach subspace.

Definition 2.7 (see [152] and [32]). A measure $\mu \in ca(\mathcal{B}_T)$ is called a *Radon measure* if for every $A \in \mathcal{B}_T$,

$$|\mu|(A) = \sup_{K \in \mathcal{K}_T, K \subseteq A} |\mu|(K).$$

We will note the set of all Radon measures on $\mathcal{B}_T$ by $Rca(\mathcal{B}_T)$.

Obviously, $Rca(\mathcal{B}_T) \subseteq tca(\mathcal{B}_T)$, $Rca(\mathcal{B}_T) \subseteq rca(\mathcal{B}_T)$; if T is compact, then $rca(\mathcal{B}_T) = Rca(\mathcal{B}_T)$.

Proposition 2.8. $rca(\mathcal{B}_T) \cap tca(\mathcal{B}_T) = Rca(\mathcal{B}_T)$.

Proof. Let $\mu \in rca(\mathcal{B}_T) \cap tca(\mathcal{B}_T)$. As μ is regular, for every $A \in \mathcal{B}_T$ and every $\varepsilon > 0$, there exists $F_\varepsilon = \overline{F}_\varepsilon \subseteq A$ such that $|\mu|(A \setminus F_\varepsilon) < \frac{\varepsilon}{2}$. As μ is tight, there exists $K_\varepsilon \in \mathcal{K}_T$ such that $|\mu|(T \setminus K_\varepsilon) < \frac{\varepsilon}{2}$. Let $K = F_\varepsilon \cap K_\varepsilon \in \mathcal{K}_T$; then $K \subseteq A$ and $|\mu|(A \setminus K) = |\mu|[(A \setminus F_\varepsilon) \cup (A \setminus K_\varepsilon)] \leq |\mu|(A \setminus F_\varepsilon) + |\mu|(T \setminus K_\varepsilon) < \varepsilon$. Therefore $|\mu|(A) = \sup_{K \in \mathcal{K}_T, K \subseteq A} |\mu|(K)$ and $\mu \in Rca(\mathcal{B}_T)$. □

Definition 2.9 (Definition 8 in [152] and Definition 2, §3, [32]). A topological space T is said *Radon space* if every measure $\mu \in ca(\mathcal{B}_T)$ is a Radon measure.

For a Radon space T, $ca(\mathcal{B}_T) = rca(\mathcal{B}_T) = tca(\mathcal{B}_T) = Rca(\mathcal{B}_T)$.

In the following section, we will specify several types of Radon spaces. However, according to Proposition 2.1, we can now give the next proposition that early will be generalized (see Proposition 6 in [152]):

Proposition 2.10. *Every metrizable compact space is a Radon space.*

Remark 2.11. It is easy to see that

(i) $\mu \in Rca(\mathcal{B}_T) \Leftrightarrow \mu^+$ and $\mu^- \in Rca(\mathcal{B}_T)$.

(ii) For every $t \in T$, $\delta_t \in Rca(\mathcal{B}_T)$.

(iii) $Rca(\mathcal{B}_T)$ is a closed vector subspace of $(ca(\mathcal{B}_T), \|\cdot\|)$ and so, a Banach subspace.

We now recall the notion of support of a measure that be will used in the following sections.

Definition 2.12. Let $\mu \in ca(\mathcal{B}_T)$ non identically null; we call the *support* of μ the smallest non-void closed set of T on which is concentrated the measure $|\mu|$ (see Definition 1.11). It is noted $\operatorname{supp} \mu$.

In accordance with Definition 1.11, the support of μ is

$$\operatorname{supp} \mu = \cap \left\{F : F = \bar{F} \subseteq T, |\mu|(T \setminus F) = 0\right\}.$$

Proposition 2.13. *For every measure $\mu \in Rca(\mathcal{B}_T)$, $|\mu|(T \setminus \operatorname{supp} \mu) = 0$ (which means that every Radon measure is concentrated on its support).*

Proof. For every $K \in \mathcal{K}_T$ with $K \subseteq T \setminus \operatorname{supp} \mu$, we have

$$K \subseteq \cup \{D \in \tau, |\mu|(D) = 0\}.$$

From this cover of K, we can extract then a finite subcover $D_1, \dots, D_n \in \tau$ such that $K \subseteq \bigcup_{i=1}^{n} D_i$ and, for $i \leq n$, $|\mu|(D_i) = 0$; therefore $|\mu|(K) = 0$. As μ is Radon,

$$|\mu|(T \setminus \operatorname{supp} \mu) = \sup_{K \in \mathcal{K}_T, K \subseteq T \setminus \operatorname{supp} \mu} |\mu|(K) = 0. \qquad \square$$

Remark 2.14. The previous remark is not generally true for the measures of $ca(\mathcal{B}_T)$ that are not Radon measures. We can find in [59] an example of Borel measure on a compact space taking only the 0 and 1 values and without support. This example is mentioned also in [152] , p. 45.

2.2 Polish Spaces. Suslin Spaces

The aim of this section is to present several important classes of Radon spaces: Polish spaces and, more general, Suslin spaces. In this paragraph, we use the monographs [31, 68, 152].

We will first recall the definitions to be used in the following.

Definition 2.15. A topological space T is said

(i) *second-countable* if T has a countable base,

(ii) *separable* if T contains a countable dense set,

(iii) *Lindelöf space* if every open cover of T has a countable subcover,

(iv) *hereditarily Lindelöf* if every subspace of T is a Lindelöf space.

It is well known that a second-countable space is separable (see Proposition 12 of §2 in [31]); moreover:
In a metrizable space T, the following properties are equivalent:

(i) *T is second-countable,*

(ii) *T is separable,*

(iii) *T is homeomorphic to a subset of the Hilbert cube $[0, 1]^{\mathbb{N}}$.*

In the following, we present the definitions of Polish and Suslin spaces; see Definitions 1 and 2, §6 in [31] and Definitions 1, p. 93 and 3, p. 96 in [152] .

Definition 2.16. A topological space (P, τ) is said to be a *Polish space* if it is separable and if there exists a metric d on P compatible with τ (the topology induced by d coincides with τ) for which (P, d) is complete. We can always choose $d \leq 1$.

A Hausdorff topological space S is said to be a *Suslin space* if there exists a Polish space P and a continuous surjection from P to S.

It is clear that, if (S, τ) is a Suslin space, then (S, σ) is Suslin also, for every Hausdorff topology $\sigma \subseteq \tau$.

Example 2.17. As examples of Polish spaces, we can mention the compact metrizable spaces as well as the separable Banach spaces.

Every Polish space is a Suslin space, but there are important examples of Suslin spaces that are not Polish spaces, as the one to follow. Let $(X, \|\cdot\|)$ be a separable Banach space of infinite dimension; then the weak topology $\sigma(X, X^*)$ is not metrizable and, as $(X, \sigma(X, X^*))$ is the continuous image of the Polish space $(X, \tau_{\|\cdot\|})$, $(X, \sigma(X, X^*))$ is a Suslin space which is not a Polish one.

In the following theorems, we will present some important families and properties of the Polish and the Suslin spaces (see Propositions 1 and 2, §6, in [31] and Proposition 3, p. 104, in [152]).

Theorem 2.18.

(i) *Every space homeomorphic to a Polish space is Polish.*

(ii) *Every finite or countable product of Polish spaces is Polish.*

Every finite or countable sum of Polish spaces is a Polish space.

(iii) *Every closed (open) subspace of a Polish space is Polish.*

(iv) *Every Polish space is hereditarily Lindelöf.*

Proof. (i) Let P be a Polish space and let d be a complete metric compatible with the topology of P. If Q is a topological space and if f is a homeomorphism of P on Q, then $g : Q \times Q \to \mathbb{R}_+$, defined by

$$g(x, y) = d(f^{-1}(x), f^{-1}(y)),$$

is a metric on Q compatible with the topology of Q and (Q, g) is complete; then Q is a Polish space.

(ii) It is easy to see that, if $\{(P_n, \tau_n) : n \in I\}$ is a finite or countable family of Polish spaces, then $(\prod_{n\in I} P_n, \prod_{n\in I} \tau_n)$ is a Polish space .

For the sum, we suppose that the sets P_n are pairwise disjoint (if it is not the case, then one can replace the P_n by $P'_n = P_n \times \{n\}$ and τ_n with $\tau'_n = \tau_n \times \{\emptyset, \{n\}\}$).

Let $P = \bigcup_{n=0}^{\infty} P_n$ and $\tau = \{D \subseteq P : D \cap P_n \in \tau_n, \quad \text{for every} \quad n \in \mathbb{N}\}$. Let us show that $(P, \tau) = \bigoplus_{n=0}^{\infty}(P_n, \tau_n)$ is a Polish space.

Indeed, if, for every $n \in \mathbb{N}$, A_n is a countable and dense set in (P_n, τ_n), then $A = \bigcup_{n=0}^{\infty} A_n$ is countable and dense in (P, τ). Moreover, if $d_n \leq 1$ is a metric on P_n such that $\tau_n = \tau_{d_n}$ and that (P_n, d_n) is complete, then $d : P \times P \to \mathbb{R}_+$,

$$d(x, y) = \begin{cases} d_n(x, y), & x, y \in P_n, \\ 1, & x \in P_n, y \in P_m, n \neq m \end{cases},$$

is a metric on P with $\tau = \tau_d$ and (P, d) is complete (if $(x_k)_k$ is a Cauchy sequence in P, then there exists a n such that, from a certain rank, the sequence $(x_k)_k$ is contained in P_n).

(iii) Because every closed subspace of a complete space is complete, it is clear that every closed subspace of a Polish is Polish.

Let us now show the property for the open subspace. Let P be a Polish space, d be a complete metric compatible with the topology of P and Q be an open subset of P. Then $\mathbb{R} \times P$ is Polish and therefore

$$G = \{(x, p) \in \mathbb{R} \times P : x \cdot d(p, P \setminus Q) = 1\}$$

is Polish, also (G is a closed subspace of $\mathbb{R} \times P$).

The projection $\pi : G \to Q, \pi((x, p)) = p$ is a homeomorphism of G on Q.

By (i), Q is Polish.

(iv) Every Polish space P is a second-countable and metrizable space; then the subspaces of P are metrizable and second-countable spaces and, according to Proposition 13 of §2, n° 8 in [31], they are Lindelöf spaces. □

Theorem 2.19 (Corollary p. 94 in [152]). *A topological space is Polish if and only if it is homeomorphic to a G_δ subset of Hilbert cube $[0, 1]^{\mathbb{N}}$ (a countable intersection of open sets in $[0, 1]^{\mathbb{N}}$).*

Proof. By (iii) and (ii) of Theorem 2.18, $H = [0,1]^{\mathbb{N}}$ is a Polish space.

Firstly, let us show that a G_δ subset of $[0,1]^{\mathbb{N}}$ is a Polish space.

Let $(D_n)_{n\in\mathbb{N}}$ be a sequence of open subsets of $[0,1]^{\mathbb{N}}$ and $Q = \bigcap_{n\in\mathbb{N}} D_n$. By (iii) of 2.18, for every $n \in \mathbb{N}$, D_n is Polish and by (ii) of 2.18, $\prod_{n\in\mathbb{N}} D_n$ is Polish, also. The diagonal Δ of the product $\prod_{n\in\mathbb{N}} D_n$ is a closed subset of $\prod_{n\in\mathbb{N}} D_n$, therefore it is Polish. For every $x \in Q$, let $\tilde{x} = (x_n)_{n\in\mathbb{N}}$, where, for every n, $x_n = x$; then the mapping $x \longrightarrow \tilde{x}$ is a homeomorphism of Q on Δ so that Q is Polish.

Conversely, let (P,τ) be a Polish space, let d be a complete metric compatible with τ such that $d \le 1$ and let $A = \{p_n : n \in \mathbb{N}\}$ be a countable dense subset of P. The mapping $f : P \to H$, defined by $f(p) = (d(p,p_n))_{p\in\mathbb{N}}$ is a homeomorphism of P on $f(P) = Q$. Indeed, because A is dense in P, it is clear that f is a continuous injective mapping. Let now $B = B_d(p,r)$ and let n_0 such that $d(p,p_{n_0}) < \frac{r}{3}$. If $V = \{x \in P : |d(x,p_{n_0}) - d(p,p_{n_0})| < \frac{r}{3}\}$, then V is a neighborhood of p whose image by f is a neighborhood of $f(p)$ in $Q = f(P)$.

By (i) of Theorem 2.18, Q is Polish.

Now let g be a complete metric compatible with the topology of Q. Then $\tau_g = \tau_0 \restriction_Q$, where τ_0 is the product topology on $H = [0,1]^{\mathbb{N}}$. For every $n \in \mathbb{N}^*$, let

$$U_n = \bigcup\left\{U \in \tau_0 : \delta_g(U \cap Q) < \frac{1}{n}\right\},$$

where $\delta_g(E)$ is the diameter of E in the metric space (Q,g).

We first notice that, for every $n \in \mathbb{N}, Q \subseteq U_n$. Indeed, if $y \in Q$, then the ball $B_g(y,\frac{1}{2n})$ is an open subset of Q; hence there exists $U \in \tau_0$ such that $B_g(y,\frac{1}{2n}) = U \cap Q$ and therefore $y \in U_n$. Then we have $Q \subseteq \bigcap_{n=1}^{\infty}(U_n \cap \overline{Q})$. We will show that

$$Q = \bigcap_{n=1}^{\infty}(U_n \cap \overline{Q}). \tag{*}$$

which was to be demonstrated.

For every $x \in \bigcap_{n=1}^{\infty}(U_n \cap \overline{Q})$, there exists a sequence $(x_k) \subseteq Q$ such that $x_k \xrightarrow{\tau_0} x$. For every $\varepsilon > 0$ let $n_0 \in \mathbb{N}^*$ with $\frac{1}{n_0} < \varepsilon$. As $x \in U_{n_0}$, there exists $U \in \tau_0$ such that $x \in U$ and $\delta_g(U \cap Q) < \frac{1}{n_0}$. Therefore, there exists $k_0 \in \mathbb{N}$ such that, for every $k \ge k_0$, $x_k \in U$ and, for every $k,l \ge k_0$, $g(x_k,x_l) < \frac{1}{n_0} < \varepsilon$; then (x_n) is a Cauchy sequence in the complete metric space (Q,g). So, there exists $y \in Q$ such that $x_k \xrightarrow{g} y$. We have therefore $x_k \xrightarrow{\tau_0} y$ and $x = y \in Q$ and so $(*)$ is demonstrated.

As $\overline{Q}$ is a closed subset in the metrizable space $H = [0,1]^{\mathbb{N}}$, it is a countable intersection of open subsets of H from where, $Q = \bigcap_{n=1}^{\infty}(U_n \cap \overline{Q})$ is a G_δ subset of H. □

The following proposition gives a family of Polish spaces which will be used later.

Proposition 2.20. *A locally compact space is Polish if and only if it is second-countable.*

Proof. Every Polish space is second-countable.

If T is a locally compact space which is second-countable, then its Alexandroff's compactification $\widehat{T}$ is metrizable (see corollary of Proposition 16 of §2 in [31]) and second-countable. Therefore $\widehat{T}$ is Polish and, as T is open in $\widehat{T}$, by (iii) of Theorem 2.18, T is Polish. □

Let us now give some properties of Suslin spaces (see Proposition 4, Section 6, in [31]).

Theorem 2.21.

(i) *Every Suslin space is separable.*

(ii) *Every Suslin space is hereditarily Lindelöf.*

(iii) *Every regular Suslin space is a normal space.*

Proof. Let (S,τ) be a Suslin space; there exist a Polish space P and a surjective continuous mapping $f : P \to S$.

(i) If A is a countable dense subset of P, then $f(A)$ is countable and dense in S, hence S is separable.

(ii) To show that S is hereditarly Lindelöf it is sufficient to show that every open subset $D \subseteq S$ is Lindelöf. Let $\{D_\gamma : \gamma \in \Gamma\}$ be an open cover of D. Then $\{f^{-1}(D_\gamma) : \gamma \in \Gamma\}$ is an open cover of $f^{-1}(D)$ in P. As P is hereditarily Lindelöf (see (iv) of Theorem 2.18), there exists a countable family $\Gamma_0 \subseteq \Gamma$ such that $\{f^{-1}(D_\gamma : \gamma \in \Gamma_0\}$ is a cover of $f^{-1}(D)$. Then $\{D_\gamma : \gamma \in \Gamma_0\}$ is a countable cover of D.

(iii) Let (S,τ) be a regular Suslin space and let A and B two closed and disjoint subsets of S. For every $x \in A$ and every $y \in B$, there exist $U_x, V_y \in \tau$ such that $x \in U_x \subseteq \overline{U}_x \subseteq S \setminus B$ and $y \in V_y \subseteq \overline{V}_y \subseteq S \setminus A$. The family $\{U_x : x \in A\}$ is an open cover of A and the family $\{V_y : y \in B\}$ an open cover of B.

As S is hereditarly Lindelöf, there exist $(x_n)_{n\in\mathbb{N}} \subseteq A$ and $(y_n)_{n\in\mathbb{N}} \subseteq B$ such that $A \subseteq \bigcup_{n=1}^{\infty} U_{x_n}$ and $B \subseteq \bigcup_{n=1}^{\infty} V_{y_n}$.

For every $n \in \mathbb{N}$, let $D_n = U_{x_n} \setminus \bigcup_{m \leq n} \overline{V}_{y_m}$ and $G_n = V_{y_n} \setminus \bigcup_{m \leq n} \overline{U}_{x_m}$.

$D_n, G_n \in \tau$ and if $D = \bigcup_{n=1}^{\infty} D_n$ and $G = \bigcup_{n=1}^{\infty} G_n \in \tau$, then $A \subseteq D$, $B \subseteq G$ and $D \cap G = \bigcup_{n=1}^{\infty} \bigcup_{m=1}^{\infty} (D_n \cap G_m)$.

Let $n, m \in \mathbb{N}$; if we suppose that $m \leq n$, then $D_n \subseteq S \setminus \overline{V}_{y_m} \subseteq S \setminus V_{y_m} \subseteq S \setminus G_m$ and so $D_n \cap G_m = \emptyset$. Similarly in the case where $n \geq m$.

Therefore $D_n \cap G_m = \emptyset$, for every $n, m \in \mathbb{N}$ and then $D \cap G = \emptyset$. Hence (S, τ) is a normal space. □

In the following sections, the property of a space to be completely regular will be indispensable in order to obtain significant results; then we will give the next corollary:

Corollary 2.22. *Every regular Suslin space is a completely regular space.*

Proof. Since any normal space is completely regular, the corollary results immediately from (iii) of previous theorem. □

The following theorem lists some important properties of Suslin spaces (see Propositions 5, 7 and 8, §6, in [31]).

Theorem 2.23.

(i) *Every space homeomorphic to a Suslin space is Suslin.*

(ii) *Every closed or open subspace of a Suslin space is Suslin.*

(iii) *A continuous image of a Suslin space in a Hausdorff space is Suslin.*

(iv) *Every finite or countable product of Suslin spaces is Suslin. The sum of a finite or countable family of Suslin spaces is Suslin.*

(v) *Every union and every intersection of a finite or countable family of Suslin subspaces of a Hausdorff space is Suslin.*

Proof. The proofs of (i), (ii) and (iii) are obvious.

(iv) Let $\{(S_n, \tau_n) : n \in \mathbb{N}\}$ be a countable family of Suslin spaces and let, for every $n \in \mathbb{N}$, (P_n, σ_n) be a Polish space and $f_n : P_n \to S_n$ be a surjective continuous mapping; then, according to Theorem 2.18 (ii), $(P, \sigma) = \prod_{n=0}^{\infty} (P_n, \sigma_n)$ is a Polish space and, if $f : P \to S = \prod_0^{\infty} S_n$ is defined by: $f((x_n)_{n \in \mathbb{N}}) = (f_n(x_n))_{n \in \mathbb{N}}$ and if τ the product topology $\prod_n \tau_n$, then f is a continuous surjection from (P, σ) on (S, τ). Therefore (S, τ) is Suslin.

Let now suppose that S_n are pairwise disjoint and let $(S, \tau) = \bigoplus_{n=0}^{\infty} (S_n, \tau_n)$. We can suppose also that P_n are pairwise disjoint. According to Theorem 2.18 (ii), $(P, \sigma) = \bigoplus_{n=0}^{\infty} (P_n, \sigma_n)$ is Polish. Then the mapping $f : P \to S$, defined by: $f(x) = f_n(x)$, if $x \in P_n$, is a continuous surjection and therefore (S, τ) is Suslin.

(v) Let $(S_n, \tau_n)_{n \in \mathbb{N}}$ be a sequence of Suslin subspaces of the Hausdorff space X and let $S = \cup_{n=0}^{\infty} S_n$. For every $n \in \mathbb{N}$, let $A_n = S_n \times \{n\}$ and $\sigma_n = \tau_n \bigotimes \{\emptyset, \{n\}\}$. Then, for every $n \in \mathbb{N}$, σ_n is a topology on A_n and (A_n, σ_n) is a Suslin space. By (iv), $\bigoplus_{n=0}^{\infty}(A_n, \sigma_n) = (A, \sigma)$ is Suslin. The mapping $\varphi : A \to S$, defined by: $\varphi((x, n)) = x$, for every $(x, n) \in A_n$, is a continuous surjection and therefore S is Suslin.

Let now (X, τ) be a Hausdorff space; for every $n \in \mathbb{N}$, let (S_n, τ_n) be a Suslin subspace, where $\tau_n = \tau \restriction_{S_n}$. Then, by (iv), $S = \prod_{n=0}^{\infty} S_n$ is a Suslin subspace of $X^{\mathbb{N}}$. The diagonal $\Delta = \{(x_n)_{n \in \mathbb{N}} :$ where $x_n = x \in X,$ for every $n \in \mathbb{N}\}$, is a closed subset of the product space $X^{\mathbb{N}}$; then $\Delta \cap S$ is a closed subset of S and so, by (ii), $\Delta \cap S$ is Suslin. The mapping $\varphi : \bigcap_{n=0}^{\infty} S_n \to \Delta \cap S$, that, to every x associates $(x_n)_{n \in \mathbb{N}}$ where $x_n = x$, for every $n \in \mathbb{N}$, is a homeomorphism; by (i), $\bigcap_{n=0}^{\infty} S_n$ is Suslin. □

Theorem 2.24 (Proposition 11, §6, in [31] and Corollary 1, p. 101, in [152]).

(i) *Every Borel subspace of a Suslin space is Suslin.*

(ii) *If (S, τ) is a Suslin space then $B \in \mathcal{B}_S(\tau) \Leftrightarrow B$ and $S \setminus B$ are Suslin subspaces of S.*

Proof. (i) Let (S, τ) be a Suslin space and let $\mathcal{A} = \{A \subseteq S : A$ and $S \setminus A$ are Suslin$\}$. Thanks to (v) of Theorem 2.23 and from the definition of $\mathcal{A}$, it is clear that $\mathcal{A}$ is a σ-algebra. By (ii) of Theorem 2.23, $\tau \subseteq \mathcal{A}$ and so $\mathcal{B}_S(\tau) \subseteq \mathcal{A}$.

(ii) The $\Rightarrow$ is a consequence of (i). The implication $\Leftarrow$ has a long demonstration and we omit it; one can found it in Corollary 1, p. 101 of [152] . □

Theorem 2.25.

(i) *For every Suslin space (S, τ), there exists a topology $\sigma \subseteq \tau$ such that (S, σ) is a second-countable Suslin space.*

(ii) *If (S, τ) is a regular Suslin space, then there exists a metric d on S such that:*

  (a) $\tau_d \subseteq \tau$,

  (b) τ_d *is second-countable and*

  (c) $\mathcal{B}_S(\tau_d) = \mathcal{B}_S(\tau)$.

Proof. (i) Let (S, τ) be a Suslin space; since S is a Hausdorff space, $\Delta = \{(x, x) : x \in S\}$ is a closed subset in $S \times S$ and so $D = (S \times S) \setminus \Delta$ is open in $S \times S$. For every $(x, y) \in D$, $x \neq y$; therefore there exist $U_{(x,y)}$ and $V_{(x,y)} \in \tau$ such that $x \in U_{(x,y)}$, $y \in V_{(x,y)}$, $U_{(x,y)} \cap V_{(x,y)} = \emptyset$ and $D = \bigcup_{(x,y)\in D}[U_{(x,y)} \times V_{(x,y)}]$. According to Theorems 2.21 (ii) and 2.23 (iii), $S \times S$ is hereditarily Lindelöf and then there exist $(x_n, y_n)_{n\in\mathbb{N}} \subseteq D$ such that

$$D = \bigcup_{n=1}^{\infty} \left[U_{(x_n,y_n)} \times V_{(x_n,y_n)}\right]. \tag{*}$$

Let now $\mathcal{S} = \{U_{(x_n,y_n)}, V_{(x_n,y_n)} : n \in \mathbb{N}\} \subseteq \tau$. By (*), for every $(x, y) \in D$, there exists $n \in \mathbb{N}$ such that $(x, y) \in U_{(x_n,y_n)} \times V_{(x_n,y_n)}$. Then, for every $x \in S$, there is a $n \in \mathbb{N}$ with $x \in U_{(x_n,y_n)}$. Therefore $\mathcal{S}$ is a countable open cover of S.

Let σ be the topology on S generated by the subbase $\mathcal{S}$. Then σ is a Hausdorff topology, $\sigma \subseteq \tau$ and σ has a countable base. (S, σ) is a Suslin space, also.

(ii) If (S, τ) is a regular Suslin space, then, by (iii) of 2.21, (S, τ) is normal space and so a completely regular one. Therefore, for every $(x, y) \in D = (S \times S) \setminus \Delta$, there exists a continuous mapping $f : (S, \tau) \to [0, 1]$ such that $f(x) \neq f(y)$. Let $\{f_i : i \in I\}$ be a family of continuous mappings from S in $[0, 1]$ which separates the points of S and, for any $i \in I$, let $D_i = \{(x, y) \in S \times S : f_i(x) \neq f_i(y)\}$. The sets D_i are open in $S \times S$ and $D = \bigcup_{i\in I} D_i$. As $S \times S$ is hereditarily Lindelöf, there exist $(i_n)_{n\in\mathbb{N}} \subseteq I$ such that $D = \bigcup_{n=1}^{\infty} D_{i_n}$.

Let now $d : S \times S \to \mathbb{R}_+, d(x, y) = \sum_{n=1}^{\infty} \frac{1}{2^n}|f_{i_n}(x) - f_{i_n}(y)|$. d is a metric on S and, as the mappings $\{f_{i_n} : n \in \mathbb{N}\}$ are τ–continuous, $\tau_d \subseteq \tau$. Since a continuous image of a Lindelöf space is again a Lindelöf one, (S, τ_d) is Lindelöf. If $B_d(t, r)$ designates the open d-ball of center t and ray r, then, for every $p \in \mathbb{N}^*, \{B_d(x, \frac{1}{p}) : x \in S\}$ is a τ_d–open cover of S; then there exists a countable set $A_p \subseteq S$ such that

$$S = \bigcup_{x\in A_p} B_d\left(x, \frac{1}{p}\right).$$

The set $A = \bigcup_{p=1}^{\infty} A_p \subseteq S$ is countable τ_d–dense in S and so τ_d has a countable base.

Now we show (c). Since $\tau_d \subseteq \tau$, $\mathcal{B}_S(\tau_d) \subseteq \mathcal{B}_S(\tau)$. Let $B \in \mathcal{B}_S(\tau)$; by (i) of Theorem 2.24, B and $S \setminus B$ are Suslin sets with respect to the topology τ. As $\tau_d \subseteq \tau$ and τ_d is a Hausdorff topology, B and $S \setminus B$ are Suslin with respect to the topology τ_d. According to (ii) of Theorem 2.24, $B \in \mathcal{B}_S(\tau_d)$. Therefore $\mathcal{B}_S(\tau_d) = \mathcal{B}_S(\tau)$. □

Corollary 2.26 (Corollary 2, p. 106 and Theorem 6, p. 111, in [152]). *Every locally compact Suslin space is a Polish space. Particularly, every compact subspace of a Suslin space is metrizable and so it is sequentially compact.*

Proof. 1. If (S,τ) is a compact Suslin space, then it is regular and, according to (ii) of Theorem 2.25, there exists a metric d on S such that $\tau_d \subseteq \tau$ and τ_d has a countable base. Therefore, the identity map of (S,τ) on (S,τ_d) is a continuous bijection of compact space (S,τ) on the Hausdorff space (S,τ_d) and so it is a homeomorphism. Then $\tau = \tau_d$. As the metrizable space (S,τ_d) is compact, (S,d) is complete and then (S,τ) is a compact Polish space.

2. If (S,τ) is a locally compact Suslin space, then, by (i) of Theorem 2.21, it is separable. Let $\{x_n : n \in \mathbb{N}\}$ be a dense subset of S; for every $n \in \mathbb{N}$, let V_n be an open and relatively compact neighborhood of x_n. $\overline{V_n}$ is a compact Suslin space and, after the first part of the demonstration, a Polish space. Let now $\{U_n^m : m \in \mathbb{N}\}$ be a countable base of τ-open sets of the topology of V_n. Then $\{U_n^m : n,m \in \mathbb{N}\}$ is a countable base of S, so that S is second-countable locally compact space; according to Proposition 2.20, it is a Polish space. □

In the following we will also show that the Polish spaces and, more generally, the Suslin spaces are Radon spaces. The next preliminary result was established by A. D. Alexandrov (1940) and was rediscovered by Yu. Prohorov (1956).

Theorem 2.27 (Theorem 1.4 in [28]). *Let (P,τ) be a Polish space; then every measure $\mu \in ca(\mathcal{B}_P(\tau))$ is tight.*

Proof. Let d be a complete metric which generates the topology of P and let $A = \{t_n : n \in \mathbb{N}^*\}$ be a dense subset of (P,τ_d). For every $n \in \mathbb{N}^*$, $P = \bigcup_{k=1}^{\infty} B_d(t_k, \frac{1}{n})$ and so, for every $\varepsilon > 0$ and every $n \in \mathbb{N}^*$, there exists $k_n \in \mathbb{N}^*$ such that:

$$|\mu|\left(P \setminus \bigcup_{k=1}^{k_n} B_d\left(t_k, \frac{1}{n}\right)\right) < \frac{\varepsilon}{2^n}. \tag{*}$$

Let $A_\varepsilon = \bigcap_{n=1}^{\infty} \bigcup_{k=1}^{k_n} B_d(t_k, \frac{1}{n}) \subseteq P$; as A_ε is totally bounded, $K_\varepsilon = \overline{A_\varepsilon}$ is totally bounded also and, (P,d) being complete, K_ε is compact. Then, by $(*)$,

$$\begin{aligned} |\mu|(P \setminus K_\varepsilon) \le |\mu|(P \setminus A_\varepsilon) &= |\mu|\left[\bigcup_{n=1}^{\infty}\left(P \setminus \bigcup_{k=1}^{k_n} B_d(t_k, \frac{1}{n})\right)\right] \\ &\le \sum_{n=1}^{\infty} |\mu|\left(P \setminus \bigcup_{k=1}^{k_n} B_d(t_k, \frac{1}{n})\right) < \sum_{n=1}^{\infty} \frac{\varepsilon}{2^n} = \varepsilon, \end{aligned}$$

which finishes the proof. □

We recall that a Radon space is a topological space T for which $ca(\mathcal{B}_T) = Rca(\mathcal{B}_T)$ (see Definition 2.9).

The following corollary was first proved by J. C. Oxtoby and S. M. Ulam.

Corollary 2.28 (Theorem 9, p. 122, in [152]). *Any Polish space is a Radon space.*

Proof. Let (P, τ) be a Polish space; according to 2.27, $ca(\mathcal{B}_P(\tau)) = tca(\mathcal{B}_P(\tau))$ and by Theorem 2.4, $ca(\mathcal{B}_P(\tau)) = rca(\mathcal{B}_P(\tau))$. Then the result is a consequence of Proposition 2.8. □

Theorem 2.29 (Theorem 10, p. 122, in [152] and Proposition 3, §3, in [32]). *Any Suslin space is Radon space.*

Proof. Let P be a Polish space and let $f : P \to S$ be a continuous surjection on S. Let us consider the multivalued mappings $F : S \to \mathcal{P}_0(P)$ defined by: $F(s) = f^{-1}(s)$, for every $s \in S$, where $\mathcal{P}_0(P)$ designates the family of all not empty subsets of P; then the graph of F, $G_F = \{(s, p) \in S \times P : p \in F(s)\} = \{(s, p) : s = f(p)\} = \{(f(p), p) : p \in P\}$, is a closed subset of $S \times P$ and therefore $G_F \in \mathcal{B}_{S\times P} = \mathcal{B}_S \otimes \mathcal{B}_P$. According to Aumann–Sainte Beuve theorem ([42], T. III.22), there exists a mapping $g : S \to P$, $(\hat{\mathcal{B}}_S - \mathcal{B}_P)$–measurable such that, for every $s \in S$, $g(s) \in F(s)$ (g is a measurable selection of F). Then, for every $s \in S$, $g(s) \in f^{-1}(s)$ or, equivalently, $f(g(s)) = s$, for every $s \in S$.

Let now $\mu \in ca(\mathcal{B}_S)$ and let $\hat{\mu} \in ca(\hat{\mathcal{B}}_S)$ be the unique extension of μ on $\hat{\mathcal{B}}_S$.

For every $A \in \mathcal{B}_P$, $g^{-1}(A) \in \hat{\mathcal{B}}_S$. If we note $\lambda(A) = \hat{\mu}(g^{-1}(A))$, it is clear that $\lambda : \mathcal{B}_P \to \mathbb{R}$ belongs to $ca(\mathcal{B}_P)$ and that, for any $C \in \mathcal{B}_S$, $f^{-1}(C) \in \mathcal{B}_P$ and $\lambda(f^{-1}(C)) = \hat{\mu}(g^{-1}(f^{-1}(C))) = \hat{\mu}(C) = \mu(C)$. Therefore $\mu = \lambda \circ f^{-1}$ is the image of λ under the mapping f. According to Corollary 2.28, $\lambda \in ca(\mathcal{B}_P) = Rca(\mathcal{B}_P)$. We have therefore

$$\lambda(A) = \sup_{K\in\mathcal{K}_P, K\subseteq A} \lambda(K), \quad \text{for every} \quad A \in \mathcal{B}_P.$$

Then, for each $C \in \mathcal{B}_S$ and each $a < \mu(C) = \lambda(f^{-1}(C))$, there exists $K \in \mathcal{K}_P$, $K \subseteq f^{-1}(C)$ such that $a < \lambda(K) = \hat{\mu}(g^{-1}(K)) \leq \hat{\mu}(f(K)) = \mu(f(K))$, because $f(K) \in \mathcal{K}_S \subseteq \mathcal{B}_S$. Thus, for any $a < \mu(C)$, there exists a compact set $f(K) \in \mathcal{K}_S$, $f(K) \subseteq C$ with $a < \mu(f(K))$ from where

$$\mu(C) = \sup_{K\in\mathcal{K}_S, K\subseteq C} \mu(K), \quad \text{for every} \quad C \in \mathcal{B}_S.$$

Which means that $\mu \in Rca(\mathcal{B}_S)$. □

Remark 2.30. Let S be a Suslin space, P be a Polish space and f be a continuous surjection of P on S; let us define the mapping

$$\bar{f} : Rca(\mathcal{B}_P) = ca(\mathcal{B}_P) \to Rca(\mathcal{B}_S) = ca(\mathcal{B}_S) \quad \text{by:}$$

$$\bar{f}(\lambda)(A) = \lambda(f^{-1}(A)), \quad \text{for every} \quad \lambda \in ca(\mathcal{B}_P) \quad \text{and every} \quad A \in \mathcal{B}_S.$$

We show that $\bar{f}$ is a surjection of $Rca(\mathcal{B}_P)$ on $Rca(\mathcal{B}_S)$. In fact, according to the proof of Theorem 2.29, for all $\mu \in ca(\mathcal{B}_S)$, there exists $\lambda \in ca(\mathcal{B}_P)$ such that $\lambda(A) = \hat{\mu}(g^{-1}(A))$ (where $\hat{\mu}$ is the extension of μ on $\hat{\mathcal{B}}_S$ and g is a $(\hat{\mathcal{B}}_S - \mathcal{B}_P)$-measurable selection of multivalued mapping f^{-1}). $\bar{f}(\lambda) = \mu$. The restriction of $\bar{f}$ to $Rca^+(\mathcal{B}_P)$ is again a surjection of $Rca^+(\mathcal{B}_P)$ on $Rca^+(\mathcal{B}_S)$. This remark will be used later to show that, if S is a Suslin space, $Rca^+(\mathcal{B}_S)$ is also Suslin.

2.3 Narrow Topology

In the following, we intend to introduce on $ca(\mathcal{B}_T)$ a topology weaker than that induced by the norm.

Let us first recall that $ca(\mathcal{B}_T)$ (as well as $rca(\mathcal{B}_T)$, $Rca(\mathcal{B}_T)$ and $tca(\mathcal{B}_T)$), provided with the norm $\|\cdot\|$, given by $\|\mu\| = |\mu|(T)$ - where $|\mu|$ is variation of μ, is a Banach space (see Theorem 1.23). The notation $(ca(\mathcal{B}_T))^*$ refers to the topological dual of $(ca(\mathcal{B}_T), \|\cdot\|)$.

Notations and recalls. In the following, T designates a Hausdorff topological space, $\mathcal{B}_T$ the σ–algebra of all Borel sets of T and $\mathcal{K}_T$ the family of all compacts of T. We will note:

$C(T) = \{f : T \to \mathbb{R} : f \quad \text{continuous on} \quad T\}$,

$C_b(T) = \{f \in C(T) : f \quad \text{bounded on} \quad T\}$,

$C_c(T) = \{f \in C(T) : \operatorname{supp} f \in \mathcal{K}_T\}$—the subspace of all functions with compact support,

$C_0(T) = \{f \in C(T) : \forall \varepsilon > 0, \exists K \in \mathcal{K}_T \quad \text{s.t.} \quad |f(x)| < \varepsilon, \forall x \in T \setminus K\}$—the subspace of all functions vanishing at infinity,

$C_\infty(T) = \{f \in C(T) : \exists a \in \mathbb{R} \quad \text{such that} f - \underline{a} \in C_0(T)\}$—the subspace of all functions having a limit at infinity.

Here, $\underline{a}$ designates the constant function equal to a on T. If $f - \underline{a} \in C_0(T)$, then a is said to be the limit of f at infinity.

Obviously,

$$C_c(T) \subseteq C_0(T) \subseteq C_\infty(T) \subseteq C_b(T) \subseteq C(T)$$

and, if T is a compact space, then the previous inclusions become equalities.

Let T be a locally compact space and let $\hat{T} = T \cup \{\omega\}$ be the Alexandroff compactification of T, obtained by addition of a point to the infinity; then $f \in C_\infty(T)$ if and only if there exists $g \in C(\hat{T})$ such that $f = g\restriction_T$. In this case, the limit of f at infinity is $g(\omega)$.

Let now T be a Hausdorff space; $C_b(T)$ (and its subspaces) can be endowed with the uniform convergence norm $\|\cdot\|_\infty$: if $f \in C_b(T)$, then $\|f\|_\infty = \sup_{x\in T}|f(x)|$. $(C_b(T), \|\cdot\|_\infty)$ is a Banach space; $C_0(T)$ and $C_\infty(T)$ are closed subspaces of $C_b(T)$ and therefore they are Banach spaces, also.

Generally, $\overline{C_c(T)} \subseteq C_0(T)$. If T is locally compact, then $\overline{C_c(T)} = C_0(T)$ and so $C_0(T)$ is the completion of $C_c(T)$; in this case the topological dual of $C_c(T)$ with respect to uniform convergence norm is identical with the dual space of $C_0(T)$ and coincides with $(Rca(\mathcal{B}_T), \|\cdot\|)$.

Definition 2.31 (Definition 1, §5, in [32]). If $f \in C_b(T)$ and $\mu \in ca(\mathcal{B}_T)$, then $f \in L^1(\mu)$ (see Definition 1.1.22). In this part, $\mu(f)$ will designate $\mu_f(T) = \int_T f d\mu$ (cf. the notations of Chapter 1). For every $f \in C_b(T)$, let $I_f : ca(\mathcal{B}_T) \to \mathbb{R}$ defined by $I_f(\mu) = \int_T f d\mu = \mu(f)$; obviously, I_f is a linear map on $ca(\mathcal{B}_T)$.

The weakest topology on $ca(\mathcal{B}_T)$ making continuous the mappings $\{I_f : f \in C_b(T)\}$ is said *narrow topology* on $ca(\mathcal{B}_T)$; this topology is noted by $\mathcal{T}(T)$ or $\mathcal{T}$ if there is no ambiguity on the space T. We also say that $\mathcal{T}$ is generated by $C_b(T)$.

A net $(\mu_i)_{i\in I} \subseteq ca(\mathcal{B}_T)$ is narrowly convergent (or $\mathcal{T}$–convergent) to $\mu \in ca(\mathcal{B}_T)$ if and only if $\mu_i(f) \to \mu(f)$, for every $f \in C_b(T)$; we note this by $\mu_i \xrightarrow{\mathcal{T}} \mu$. Particularly, since $\underline{1} \in C_b(T)$, if $\mu_i \xrightarrow{\mathcal{T}} \mu$, then $\mu_i(T) \to \mu(T)$.

We will note $\mathcal{T}_c(T), \mathcal{T}_0(T)$ and $\mathcal{T}_\infty(T)$ the topologies generated on $ca(\mathcal{B}_T)$ by the families of mappings $C_c(T), C_0(T)$ and $C_\infty(T)$, respectively.

The topology $\mathcal{T}_c(T)$ is called the *vague topology* on $ca(\mathcal{B}_T)$.

Remark 2.32.

(i) In general case, the mapping $\mu \mapsto |\mu|$ of $ca(\mathcal{B}_T)$ on $ca^+(\mathcal{B}_T)$ is not narrowly continuous. Indeed, let $\mu_n : \mathcal{B}_\mathbb{R} \to \mathbb{R}, \mu_n = \delta_{\frac{1}{n}} - \delta_{-\frac{1}{n}}$, for every $n \in N^*$; then the sequence $(\mu_n)_n$ is narrowly convergent to the measure $\underline{0}$ but $|\mu_n| \xrightarrow{\mathcal{T}} 2\delta_0$.

(ii) Because of the inclusions between the generating families, we have

$$\mathcal{T}_c(T) \subseteq \mathcal{T}_0(T) \subseteq \mathcal{T}_\infty(T) \subseteq \mathcal{T}(T).$$

If T is compact, it is obvious that these five topologies coincide. Without any supplementary hypotheses on T, the above-mentioned inclusions are, generally, strict. For example, the sequence $(\delta_n)_{n\in\mathbb{N}}$ is $\mathcal{T}_0(\mathbb{R})$–convergent to the null measure and it is not $\mathcal{T}_\infty(\mathbb{R})$–convergent. In this section, we will study cases in which these inclusions are equalities.

(iii) For every $f \in C_b(T)$ and every $\mu \in ca(\mathcal{B}_T)$,

$$\left|I_f(\mu)\right| \leq \int_T |f|d|\mu| \leq \|f\|_\infty \cdot |\mu|(T) = \|f\|_\infty \cdot \|\mu\|.$$

Then, for all $f \in C_b(T)$, I_f is a linear continuous functional on the norm space $(ca(\mathcal{B}_T), \|\cdot\|)$; hence $I_f \in (ca(\mathcal{B}_T))^*$. We have therefore

$$\mathcal{T}(T) \subseteq \sigma(ca(\mathcal{B}_T), (ca(\mathcal{B}_T))^*) \subseteq \tau_{\|\cdot\|}$$

We will see later (see Remark 2.39) that, in general, these inclusions are strict.

(iv) The family of applications $\{\|\cdot\|_f : f \in C_b(T)\}$, where $\|\mu\|_f = |\int_T f d\mu|$, is a family of semi-norms on $ca(\mathcal{B}_T)$ that generates the narrow topology; therefore $(ca(\mathcal{B}_T), \mathcal{T}(T))$ is a locally convex vector space. We can make the same statement about the spaces $(ca(\mathcal{B}_T), \mathcal{T}_\infty(T))$, $(ca(\mathcal{B}_T), \mathcal{T}_0(T))$ and $(ca(\mathcal{B}_T), \mathcal{T}_c(T))$.

(v) If T is compact, then $\mathcal{T}_c(T) = \mathcal{T}_0(T) = \mathcal{T}_\infty(T) = \mathcal{T}(T)$; then we have $rca(\mathcal{B}_T) = Rca(\mathcal{B}_T) = [C(T)]^*$ and $\mathcal{T}(T)$ is the weak* topology on $[C(T)]^*$.

Therefore, we have the following property:

For every $\alpha \in \mathbb{R}$, let $rca_\alpha(\mathcal{B}_T) = \{\mu \in rca(\mathcal{B}_T) : |\mu|(T) \leq \alpha\}$; if T is compact, then $rca_\alpha(\mathcal{B}_T)$ is a compact subset of $(ca(\mathcal{B}_T), \mathcal{T})$.

Indeed, according to the Banach–Alaoglu theorem, $rca_1(\mathcal{B}_T)$ (the closed unit ball in the normed space $(rca(\mathcal{B}_T), \|\cdot\|) = ([C(T)]^*, \|\cdot\|)$) is weak*-compact and therefore it is $\mathcal{T}$-compact. Since $rca_\alpha(\mathcal{B}_T) = \alpha \cdot rca_1(\mathcal{B}_T)$, it is $\mathcal{T}$-compact, too.

We also have the following property:

If T is compact, then $\mathcal{P}_T$ (the set of all probabilities on T) is $\mathcal{T}$- compact. To demonstrate it, it is sufficient to prove that $\mathcal{P}_T = \{\mu \in rca^+(\mathcal{B}_T) : \mu(T) = 1\}$ is a closed subset in the compact space $rca_1(\mathcal{B}_T)$. Indeed, for every net $(\mu_i)_{i\in I} \subseteq \mathcal{P}_T$ with $\mu_i \xrightarrow{\mathcal{T}} \mu$, $\mu_i(T) \to \mu(T)$, so that μ is a probability on T.

For every $\alpha \in \mathbb{R}_+$, let us note by $rca_\alpha^+(\mathcal{B}_T) = rca_\alpha(\mathcal{B}_T) \cap rca^+(\mathcal{B}_T) = \{\mu \in rca^+(\mathcal{B}_T) : \mu(T) \leq \alpha\}$.

Proposition 2.33. $\mathcal{T}_0(T) = \mathcal{T}(T)$ *if and only if* T *is compact.*

Proof. ($\Leftarrow$) This is evident because, if T is compact, then the generating families of these two topologies coincide.

($\Rightarrow$) The mapping $\underline{1} \in C_b(T)$; hence $I_{\underline{1}} : ca(\mathcal{B}_T) \to \mathbb{R}$, $I_{\underline{1}}(\mu) = \int_T \underline{1} d\mu = \mu(T)$, is $\mathcal{T}$–continuous and so it is $\mathcal{T}_0$–continuous. For $\varepsilon = \frac{1}{2}$, there exist $\delta > 0$ and $f_1, \ldots, f_n \in C_0(T)$ such that, for every $i \leq n$ and every $\mu \in ca(\mathcal{B}_T)$ with $|\int_T f_i d\mu| < \delta$, we have $|I_{\underline{1}}(\mu)| = |\mu(T)| < \frac{1}{2}$. For every i, the mapping f_i vanish at infinity; there exists $K_i \in \mathcal{K}_T$ such that, $|f_i(t)| < \delta$, for every $i \leq n$ and every $t \in T \setminus K_i$. Let now $K = \cup_{i=1}^n K_i \in \mathcal{K}_T$; we will demonstrate by the reduction to absurd that $T = K$. If we suppose that there exists $t_0 \in T \setminus K$, then $\delta_{t_0} \in ca(\mathcal{B}_T)$. Besides, for all $i = 1, \ldots, n$, $|\int_T f_i d\delta_{t_0}| = |f_i(t_0)| < \delta$ ($t_0 \notin K_i$); therefore $\delta_{t_0}(T) < \frac{1}{2} < 1$, what is impossible. Therefore $T = K$ is compact. □

Corollary 2.34. *$\mathcal{T}_c(T) = \mathcal{T}(T)$ if and only if T is compact.*

Proof. If T is compact, then $\mathcal{T}_c(T) = \mathcal{T}(T)$.

Conversely, from (ii) of Remark 2.32, immediately results that, if $\mathcal{T}_c(T) = \mathcal{T}(T)$, then $\mathcal{T}_0(T) = \mathcal{T}(T)$ and the previous proposition allows us to finish the demonstration. □

In order to demonstrate the next theorem, we need the following lemma:

Lemma 2.35. *Let (T, d) be a metric space; for every lower semicontinuous mapping $f : T \to \mathbb{R}$, bounded from below and for every $n \in \mathbb{N}$, let $f_n : T \to \mathbb{R}$ be the Yosida's transform of f: $f_n(t) = \inf\{f(s) + n \cdot d(t, s) : s \in T\}$, for any $t \in T$.*

Then the sequence $(f_n)_{n\in\mathbb{N}}$ has the following properties:

(i) *$\inf\{f(s) : s \in T\} \leq f_n(t) \leq f(t)$, for every $t \in T$, and every $n \in \mathbb{N}$.*

(ii) *$f_n(t) \leq f_{n+1}(t)$, for every $t \in T$ and every $n \in \mathbb{N}$.*

(iii) *$|f_n(t) - f_n(s)| \leq n \cdot d(t, s)$, for every $t, s \in T$ and every $n \in \mathbb{N}$.*

(iv) *$f_n(t) \uparrow f(t)$, for every $t \in T$.*

Proof. (i) Let $m = \inf\{f(s) : s \in T\} \in \mathbb{R}$; for every $t, s \in T$ and every $n \in \mathbb{N}$, $m \leq f(s) \leq f(s) + n \cdot d(t, s)$, from where $m \leq f_n(t)$. By the definition of f_n, $f_n(t) \leq f(t)$, for every $t \in T$.

(ii) follows immediately.

(iii) For every $t, s \in T$ and every $\varepsilon > 0$, there exist $t_1, s_1 \in T$ such that,

$$f(t_1) + n \cdot d(t_1, t) < f_n(t) + \varepsilon,$$
$$f(s_1) + n \cdot d(s_1, s) < f_n(s) + \varepsilon.$$

We have therefore

$$\begin{aligned} f_n(t) - f_n(s) &< f(s_1) + n \cdot d(t, s_1) - f(s_1) - n \cdot d(s, s_1) + \varepsilon \\ &\leq n \cdot d(t, s) + \varepsilon, \\ f_n(s) - f_n(t) &< f(t_1) + n \cdot d(t_1, s) - f(t_1) - n \cdot d(t_1, t) + \varepsilon \\ &\leq n \cdot d(t, s) + \varepsilon. \end{aligned}$$

Then, for every $\varepsilon > 0$,

$$|f_n(t) - f_n(s)| < n \cdot d(t, s) + \varepsilon$$

what demonstrates (iii).

(iv) By the definition of Yosida's transform results that, for every $t \in T$ and every $n \in \mathbb{N}^*$, there exists $s_n \in T$ such that

$$f(s_n) + n \cdot d(t, s_n) < f_n(t) + \frac{1}{n} \leq f(t) + \frac{1}{n}. \tag{1}$$

Let us suppose that $s_n \nrightarrow t$; with perhaps extracting a subsequence noted again $(s_n)_n$, there exists $\varepsilon_0 > 0$ such that $d(s_n, t) \geq \varepsilon_0$, for any $n \in \mathbb{N}^*$. Therefore, we obtain from (1)

$$m + n \cdot \varepsilon_0 \leq f(s_n) + n \cdot \varepsilon_0 < f_n(t) + \frac{1}{n} \leq f(t) + \frac{1}{n}, \quad \text{for all} \quad n \in \mathbb{N}^*.$$

Then $f(t) = +\infty$; this contradicts the hypothesis that $f(T) \subseteq \mathbb{R}$. Therefore $s_n \to t$ and, as f is lower semicontinuous, $f(t) \leq \liminf_n f(s_n)$. By using (1), we obtain

$$\limsup_n n \cdot d(t, s_n) \leq \limsup_n (f(t) - f(s_n)) \leq 0, \tag{2}$$

from where $\lim_n n \cdot d(t, s_n) = 0$. Therefore, by making in (1) n tend to ∞,

$$f(t) \leq \liminf_n f(s_n) \leq \lim_n f_n(t) \leq f(t).$$

Therefore, for every $t \in T, \lim_n f_n(t) = f(t)$. □

Theorem 2.36. *Let T be a locally compact space whose topology is second-countable; then:*

(i) $\mathcal{T}(T)\restriction_{ca^+(\mathcal{B}_T)} = \mathcal{T}_\infty(T)\restriction_{ca^+(\mathcal{B}_T)}$ *and* $\mathcal{T}(T)\restriction_{rca^+(\mathcal{B}_T)} = \mathcal{T}_\infty(T)\restriction_{rca^+(\mathcal{B}_T)}$

(ii) $\mathcal{T}(T)\restriction_{\mathcal{P}_T} = \mathcal{T}_c(T)\restriction_{\mathcal{P}_T}$.

Proof. (i) If T is a second-countable locally compact space, by Proposition 2.20, T is a Polish space. Let $\hat{d}$ be a compatible metric on the Alexandroff one-point compactification $\hat{T} = T \cup \{\omega\}$ of T; as $\hat{T}$ is compact, $\hat{d}$ is a bounded metric and his restriction to $T \times T$ is a bounded metric on T compatible with its topology.

Let $f \in C_b(T)$ and $m = \inf\{f(t) : t \in T\}$; the mapping $\hat{f} : \hat{T} \to \mathbb{R}$, defined by

$$\hat{f}(t) = \begin{cases} f(t), & t \in T \\ m, & t = \omega \end{cases},$$

is bounded from below and lower semicontinuous on $\hat{T}$. Let $(\hat{f}_n)_{n\in\mathbb{N}}$ be the sequence of Yosida's transforms of $\hat{f}$; by (iii) of the previous lemma, $(\hat{f}_n)_{n\in\mathbb{N}} \subseteq C(\hat{T})$. Therefore, if we note $f_n = \hat{f}_n \restriction_T$, then $(f_n)_n \subseteq C_\infty(T)$ (the limit of f_n at infinity is $\hat{f}_n(\omega)$). So, we have proved that, for every $f \in C_b(T)$, there exists a sequence $(f_n)_n \subseteq C_\infty(T)$ such that $f_n \uparrow f$. $(f_n - \underline{m})_{n\in\mathbb{N}}$ is, therefore, an increasing sequence of continuous and positive mappings converging to $f - \underline{m}$; so, for every $\mu \in ca^+(\mathcal{B}_T)$, $\lim_n \int_T (f_n - \underline{m}) d\mu = \int_T (f - \underline{m}) d\mu$, from where $\lim_n \int_T f_n d\mu = \int_T f d\mu$ or $I_f(\mu) = \sup_n I_{f_n}(\mu)$. Because, for every $n \in \mathbb{N}$, I_{f_n} is $\mathcal{T}_\infty(T)$–continuous on $ca^+(\mathcal{B}_T)$, I_f is $\mathcal{T}_\infty(T)$-l.s.c. on $ca^+(\mathcal{B}_T)$, for every $f \in C_b(T)$. We have then the same property for $-f$ and, as $I_{-f} = -I_f$, I_f is $\mathcal{T}_\infty(T)$– continuous on $ca^+(\mathcal{B}_T)$. According to Theorem 2.4, the second relation is identical with the first.

(ii) From (ii) of Remark 2.32, there results immediately that $\mathcal{T}_c \restriction_{\mathcal{P}_T} \subseteq \mathcal{T} \restriction_{\mathcal{P}_T}$. For every $f \in C_b(T)$, let $I_f : \mathcal{P}(T) \to \mathbb{R}$, $I_f(\mu) = \int_T f d\mu$; according to the definition of narrow topology $\mathcal{T}$, I_f is $\mathcal{T}$–continuous. In order to demonstrate that $\mathcal{T} \restriction_{\mathcal{P}_T} \subseteq \mathcal{T}_c \restriction_{\mathcal{P}_T}$, let us show that I_f is $\mathcal{T}_c$–continuous. Let $\lambda_0 \in \mathcal{P}(T)$ and let $(\lambda_i)_{i\in I} \subseteq \mathcal{P}(T)$ be a net with $\lambda_i \xrightarrow{\mathcal{T}_c} \lambda_0$. Since T is Polish, for every $\varepsilon > 0$, there exists $K_\varepsilon \in \mathcal{K}_T$ such that $\lambda_0(T \setminus K_\varepsilon) < \varepsilon$ (see 2.28). Then there exists $g \in C_c(T)$ with $g(T) \subseteq [0, 1]$ and $g \restriction_{K_\varepsilon} = 1$. As $\lambda_i(g) \to \lambda_0(g)$, there exists $i_1 \in I$ such that, for every $i \geq i_1$, $|\lambda_i(g) - \lambda_0(g)| < \varepsilon$. Let $S_\varepsilon = \operatorname{supp} g \supseteq K_\varepsilon$. As $g \leq \chi_{S_\varepsilon}$ and $\chi_{K_\varepsilon} \leq g$, we have:

$$\begin{aligned}\lambda_i(T \setminus S_\varepsilon) &= 1 - \lambda_i(S_\varepsilon) \leq 1 - \lambda_i(g) \\ &= 1 + \lambda_0(g) - \lambda_i(g) - \lambda_0(g) \leq 1 - \lambda_0(g) + \varepsilon \\ &\leq 1 - \lambda_0(K_\varepsilon) + \varepsilon = \lambda_0(T \setminus K_\varepsilon) + \varepsilon < 2\varepsilon.\end{aligned}$$

Therefore, for $i \geq i_1$,

$$\lambda_i(T \setminus S_\varepsilon) < 2\varepsilon. \tag{3}$$

Since S_ε is compact, there exists $h \in C_c(T)$ such that $h(T) \subseteq [0, 1], h \restriction_{S_\varepsilon} = 1$. Let now $f_0 = h \cdot f$; because $\operatorname{supp} f_0 \subseteq \operatorname{supp} h$, $f_0 \in C_c(T)$. As $\lambda_i \xrightarrow{\mathcal{T}_c} \lambda_0$,

$\lambda_i(f_0) \to \lambda_0(f_0)$; so there exists $i_2 \in I$ such that, for every $i \geq i_2$,

$$|\lambda_i(f_0) - \lambda_0(f_0)| < \varepsilon. \tag{4}$$

Let $i_0 \geq i_1, i_0 \geq i_2$ and $i \geq i_0$. Then

$$\begin{aligned}
\left|I_f(\lambda_i) - I_f(\lambda_0)\right| \leq & \underbrace{\left|I_f(\lambda_i) - I_{f_0}(\lambda_i)\right|}_{T_1} + \underbrace{\left|I_{f_0}(\lambda_i) - I_{f_0}(\lambda_0)\right|}_{T_2} \\
& + \underbrace{\left|I_{f_0}(\lambda_0) - I_f(\lambda_0)\right|}_{T_3}
\end{aligned}$$

$$|f - f_0|(x) = \begin{cases} 0, & x \in S_\varepsilon, \\ \leq \|f\|, & x \in T \setminus S_\varepsilon, \end{cases} \quad \text{hence}$$

$$\begin{aligned}
T_1 &= \left|I_f(\lambda_i) - I_{f_0}(\lambda_i)\right| \\
&\leq \int_T |f - f_0| d\lambda_i \leq \|f\|\lambda_i(T \setminus S_\varepsilon) \overset{(3)}{\leq} 2\varepsilon\|f\|. \\
T_2 &= \left|I_{f_0}(\lambda_i) - I_{f_0}(\lambda_0)\right| \overset{(4)}{<} \varepsilon. \\
T_3 &= \left|I_{f_0}(\lambda_0) - I_f(\lambda_0)\right| \leq \int_T |f - f_0| d\lambda_0 \leq \|f\|\lambda_0(T \setminus S_\varepsilon) \\
&\leq \|f\|\lambda_0(T \setminus K_\varepsilon) < \varepsilon\|f\|.
\end{aligned}$$

So, for all $i \geq i_0$,

$$\left|I_f(\lambda_i) - I_f(\lambda_0)\right| < \varepsilon(3\|f\| + 1)$$

and then $I_f(\lambda_i) \to I_f(\lambda_0)$. Therefore $I_f \restriction_{\mathcal{P}_T}$ is $\mathcal{T}_c$–continuous, for every $f \in C_b(T)$. The proof is finished if we remark that $\mathcal{T}\restriction_{\mathcal{P}_T}$ is the weakest topology making continuous all mappings of family $\{I_f : f \in C_b(T)\}$; therefore $\mathcal{T}_c \restriction_{\mathcal{P}_T} \supseteq \mathcal{T}\restriction_{\mathcal{P}_T}$. □

Remark 2.37.

(i) Let $(\lambda_i)_{i \in I} \subseteq rca^+(\mathcal{B}_T)$ be a net and let $\lambda_0 \in rca^+(\mathcal{B}_T)$; if T is a second-countable locally compact space, then (i) of Theorem 2.36 can be rewritten:

$$\lambda_i \overset{\mathcal{T}}{\to} \lambda_0 \Leftrightarrow \lambda_i(f) \to \lambda_0(f), \quad \text{for every} \quad f \in C_\infty(T).$$

We remark that $f \in C_\infty(T)$ if and only if there exists $a \in \mathbb{R}$ such that $f - \underline{a} \in C_0(T)$.

Therefore, if T is a second-countable locally compact space, then

$$\lambda_i \overset{\mathcal{T}}{\to} \lambda_0 \Longleftrightarrow \begin{cases} \lambda_i(T) \to \lambda_0(T) & \text{and} \\ \lambda_i(f) \to \lambda_0(f), & \text{for every} \quad f \in C_0(T). \end{cases}$$

(ii) We can rewrite the second part of the previous theorem in the following manner: *If T is a second-countable locally compact space, then a net of probabilities is narrowly convergent if and only if it is vaguely convergent.*

Let us note that this result, which is true for the probabilities, cannot be applied for $ca^+(\mathcal{B}_T) = rca^+(\mathcal{B}_T)$; in fact, the sequence $(\delta_n)_n$ converges vaguely to 0, but it is not narrowly convergent.

(iii) We will show later (see 2.52) that, in the case where T is only locally compact (that is, without being second-countable), then, for every net $(\lambda_i)_{i\in I} \subseteq Rca^+(\mathcal{B}_T)$ and every measure $\lambda \in Rca^+(\mathcal{B}_T)$,

$$\lambda_i \xrightarrow{\mathcal{T}} \lambda \iff \lambda_i \to \lambda \quad \text{vaguely and} \quad \lambda_i(T) \to \lambda(T)$$

(even though, in this case, $Rca(\mathcal{B}_T) \neq ca(\mathcal{B}_T)$).

We will now mention some properties of the Radon measure spaces on a completely regular space. Let us remind that a Hausdorff space T is completely regular (or Tychonoff space) if, for every $x \in T$ and every closed set F with $x \notin F$, there exists a continuous mapping $f : T \to [0,1]$ such that $f(x) = 0$ and $f(y) = 1$, for any $y \in F$.

Proposition 2.38. *Let (T,τ) be a completely regular space and let $\Psi : T \to Rca(\mathcal{B}_T)$ be the application defined by: $\Psi(t) = \delta_t$, for every $t \in T$. Then Ψ is a homeomorphism of (T,τ) on $(\Psi(T), \mathcal{T}\restriction_{\Psi(T)})$.*

Proof. Since T is completely regular, for every pair $(s,t) \in T \times T$ with $s \neq t$, there exists $f \in C_b(T)$ such that $f(s) = 0$ and $f(t) = 1$ what implies that $\delta_s \neq \delta_t$. Then, Ψ is injective and therefore a bijection of T on $\Psi(T)$.

Let us show that Ψ is bicontinuous. Let $(t_i)_{i\in I}$ be a net of (T,τ) which τ-converges to t. For every $\varepsilon > 0$ and every $f \in C_b(T)$, $U_{f,\varepsilon}(\delta_t) = \{\mu \in Rca(\mathcal{B}_T) : |\mu(f) - \delta_t(f)| < \varepsilon\}$ is a $\mathcal{T}$-neighborhood of $\Psi(t) = \delta_t$. Since f is continuous, $f(t_i) \to f(t)$ and so there exists $i_0 \in I$ such that

$$|\delta_{t_i}(f) - \delta_t(f)| = |f(t_i) - f(t)| < \varepsilon, \quad \text{for every} \quad i \geq i_0.$$

Then, for every $i \geq i_0$, $\delta_{t_i} \in U_{f,\varepsilon}(\delta_t)$ and so $\Psi(t_i) \xrightarrow{\mathcal{T}} \Psi(t)$; therefore Ψ is continuous.

Now suppose that $\Psi(t_i) \xrightarrow{\mathcal{T}} \Psi(t)$; for every $f \in C_b(T)$, $\delta_{t_i}(f) \to \delta_t(f)$ or, equivalently, $f(t_i) \to f(t)$. For every τ–neighborhood V of t, let $f_0 : T \to [0,1]$ be a continuous mapping such that $f_0(t) = 0$ and $f_0\restriction_{T\setminus V} = 1$.

As $f_0(t_i) \to f_0(t)$, there exists $i_0 \in I$ such that, for every $i \geq i_0$, $|f_0(t_i) - f_0(t)| = |f_0(t_i)| < \frac{1}{2}$. Then, for every $i \geq i_0$, $t_i \in V$ which means that the sequence $(t_i)_{i\in I}$ is τ-convergent to t. Therefore Ψ is a homeomorphism of T on $\Psi(T)$. $\square$

Remark 2.39. If T is completely regular and if $t_n \underset{T}{\longrightarrow} t$, with $t_n \neq t$, for every $n \in \mathbb{N}$, then, by Proposition 2.38, $\delta_{t_n} \overset{\mathcal{T}}{\to} \delta_t$. But $\delta_{t_n}(\{t\}) = 0 \nrightarrow \delta_t(\{t\}) = 1$. So, according to Corollary 1.58, $(\delta_{t_n})_{n\in\mathbb{N}}$ is not weakly convergent to δ_t in $ca(\mathcal{B}_T)$. Hence, we infer that *when T is completely regular, the narrow topology on $ca(\mathcal{B}_T)$ is strictly less fine than weak topology of $ca(\mathcal{B}_T)$.*

In the following, we will present some topological properties of Radon measures space $(Rca(\mathcal{B}_T), \mathcal{T})$, where we will still note $\mathcal{T}$ instead of $\mathcal{T}\restriction_{Rca(\mathcal{B}_T)}$, the trace of the narrow topology on $Rca(\mathcal{B}_T)$.

Proposition 2.40. *Let T be a completely regular space and let μ be a positive Radon measure on T; then:*

(i) *for every lower semi-continuous (l.s.c.) mapping $f : T \to \overline{\mathbb{R}}_+$, $\mu(f) = \sup\{\mu(g) : g \in C_b(T), 0 \leq g \leq f\}$,*

(ii) *for every bounded from above upper semicontinuous mapping $f : T \to \overline{\mathbb{R}}$, $\mu(f) = \inf\{\mu(g) : g \in C_b(T), g \geq f\}$.*

Proof. (i) Let $G = \{g \in C_b(T) : 0 \leq g \leq f\}$. Let us first show that

$$\sup_{g\in G} g = f. \tag{*}$$

It is obvious that $\sup_{g\in G} g \leq f$. Suppose that there is $t_0 \in T$ such that $a = \sup_{g\in G} g(t_0) < f(t_0)$; if $0 < \delta < f(t_0) - a$, then, as f is l.s.c., there exists an open neighborhood V of t_0, such that $a + \delta < f(t)$, for every $t \in V$. As T is completely regular, there exists a continuous mapping $h : T \to [0, 1]$ such that $h(t_0) = 1$ and $h\restriction_{T\setminus V} = 0$. Let now $g = (a + \delta) \cdot h$; then $g \in C_b(T)$ and $g \leq f$ so that $g \in G$. Therefore, we would have $a \geq g(t_0) = a + \delta$ which is impossible. Therefore the property $(*)$ is demonstrated.

Now, let us show that $\mu(f) = \int_T f d\mu = \sup_{g\in G} \int_T g d\mu = \sup_{g\in G} \mu(g)$. For every $g \in G$, $\int_T g d\mu \leq \int_T f d\mu$ and so

$$\sup_{g\in G} \mu(g) \leq \mu(f). \tag{1}$$

Let φ be a positive $\mathcal{B}_T$-simple function such that $\varphi \leq f$. φ can be written $\varphi = \sum_{i=1}^n a_i \chi_{A_i}$, where, for every $i = 1, \dots, n$, $a_i \neq 0$ and $\{A_i : i = 1, \dots, n\} \subseteq \mathcal{B}_T$ are pairwise disjoint nonempty sets. As $\mu \in Rca^+(\mathcal{B}_T)$, for every $\varepsilon > 0$ and every $i \leq n$, there exists $K_i \in \mathcal{K}_T$, $K_i \subseteq A_i$ such that

$$\mu(A_i \setminus K_i) < \frac{\varepsilon}{a_i \cdot n}. \tag{2}$$

By $(*)$, we have $a_i - \varepsilon = \varphi(t) - \varepsilon < f(t) = \sup_{g \in G} g(t)$, for every $t \in K_i \subseteq A_i$. Therefore, there exists $g_t \in G$ such that $a_i - \varepsilon < g_t(t)$ and, as g_t is continuous, there is an open neighborhood V_t of t, such that

$$a_i - \varepsilon < g_t(s), \quad \text{for every} \quad s \in V_t. \tag{3}$$

$K_i \in \mathcal{K}_T$ and $K_i \subseteq \bigcup_{t \in K_i} V_t$; therefore, there exist $t_1, \dots, t_p \in K_i$ such that $K_i \subseteq \bigcup_{j=1}^p V_{t_j}$. Let now $g_i = \max\{g_{t_1}, \dots, g_{t_p}\}$; then $g_i \in G$ and, by (3), we have

$$a_i - \varepsilon < g_i(t), \quad \text{for every} \quad t \in K_i. \tag{4}$$

If $g = \max\{g_1, \dots, g_n\}$, then $g \in G$ and, by (3), we have:

$$a_i - \varepsilon < g(t), \quad \text{for every} \quad i = 1, \dots, n \quad \text{and every} \quad t \in K_i. \tag{5}$$

Therefore, $(a_i - \varepsilon)\mu(K_i) \leq \int_{K_i} g d\mu$, for every $i = 1, \dots, n$; hence

$$\sum_{i=1}^n (a_i - \varepsilon)\mu(K_i) \leq \int_{\bigcup_1^n K_i} g d\mu \leq \int_T g d\mu. \tag{6}$$

Then

$$\begin{aligned}\int_T \varphi d\mu &= \sum_{i=1}^n a_i \mu(A_i) \overset{(2)}{<} \sum_{i=1}^n a_i \mu(K_i) + \varepsilon \overset{(6)}{<} \varepsilon\mu(T) + \int_T g d\mu + \varepsilon \\ &< \varepsilon(1 + \mu(T)) + \sup_{g \in G} \int_T g d\mu.\end{aligned}$$

Since ε is arbitrary, we have that

$$\int_T \varphi d\mu \leq \sup_{g \in G} \mu(g), \quad \text{for every positive and} \quad \mathcal{B}_T - \text{simple mapping} \quad \varphi \leq f.$$

Hence $\mu(f) \leq \sup_{g \in G} \mu(g)$ and (i) is demonstrated.

(ii) Let $f : T \to \overline{\mathbb{R}}$ be an u.s.c. mapping bounded from above and let $M > 0$ with $M > \sup_{t \in T} f(t)$. Then $\underline{M} - f$ is a positive and l.s.c. mapping. By (i), we have

$$\begin{aligned} \mu(\underline{M} - f) &= \sup\{\mu(g) : g \in C_b^+(T), g \le \underline{M} - f\} \\ &= \sup\{M \cdot \mu(T) - \mu(\underline{M} - g) : f \le \underline{M} - g, g \in C_b(T)\} \\ &= M \cdot \mu(T) - \inf\{\mu(h) : f \le h, h \in C_b(T)\}. \end{aligned}$$

As $\mu(\underline{M} - f) = M \cdot \mu(T) - \mu(f)$, (ii) is demonstrated. □

Corollary 2.41 (Proposition 6, §5, in [32]). *Let T be a completely regular space.*

(i) *For every l.s.c. function $f : T \to \overline{\mathbb{R}}_+$, the mapping $\mu \mapsto \mu(f)$ is l.s.c. on $(Rca^+(\mathcal{B}_T), \mathcal{T})$.*

(ii) *For every bounded from above and u.s.c. function $f : T \to \overline{\mathbb{R}}$, the mapping $\mu \mapsto \mu(f)$ is u.s.c. on $(Rca^+(\mathcal{B}_T), \mathcal{T})$.*

Proof. (i) From previous proposition, $I_f(\mu) = \mu(f) = \sup_{g \in G} I_g(\mu)$; since the mappings I_g are continuous, I_f is lower semicontinuous.

The proof of (ii) is analogous. □

Remark 2.42. According to previous corollary, if $(\mu_i)_{i \in I}$ is a net $\mathcal{T}$-convergent to μ in $Rca^+(\mathcal{B}_T)$, then:

(i) $\int_T f d\mu \le \liminf_i \int_T f d\mu_i$, for every l.s.c. function $f : T \to \overline{\mathbb{R}}_+$.

(ii) $\int_T f d\mu \ge \limsup_i \int_T f d\mu_i$, for every bounded from above and u.s.c. function $f : T \to \overline{\mathbb{R}}$.

Proposition 2.43. *If T is a completely regular space, then the narrow topology $\mathcal{T}$ on $Rca^+(\mathcal{B}_T)$ is the weakest of all topologies on $Rca^+(\mathcal{B}_T)$ for which the application: $\mu \mapsto \mu(f)$ is u.s.c., for every bounded and u.s.c. function $f : T \to \mathbb{R}$.*

Proof. Let $\mathcal{T}'$ be the weakest topology on $Rca^+(\mathcal{B}_T)$ under which every mapping $\mu \mapsto \mu(f)$ is u.s.c., for every bounded and u.s.c. function $f : T \to \mathbb{R}$. According to (ii) of Corollary 2.41, $\mathcal{T}' \subseteq \mathcal{T}$.

On the other hand, if $\mu \mapsto \mu(f)$ is u.s.c., for every bounded and u.s.c. function $f : T \to \mathbb{R}$, then the mappings $\mu \mapsto \mu(f)$ and $\mu \mapsto \mu(-f) = -\mu(f)$ are u.s.c., for every $f \in C_b(T)$. Therefore the mapping $\mu \mapsto \mu(f)$ is continuous. Hence $\mathcal{T} \subseteq \mathcal{T}'$ and then the two topologies are indentical. □

In the following proposition, we extend the result of (i) in Proposition 2.40 (see Proposition 1, §5, in [32]).

Proposition 2.44. *Let T be a completely regular space and let μ be a Radon measure on T; for every lower semicontinuous function $f : T \to \overline{\mathbb{R}}_+$,*

$$|\mu|(f) = \sup\{|\mu(g)| : g \in C_b(T), |g| \leq f\}.$$

Proof. For every function $g \in C_b(T)$ with $|g| \leq f, |\mu(g)| \leq |\mu|(|g|) \leq |\mu|(f)$; therefore

$$\sup\{|\mu(g)| : g \in C_b(T), |g| \leq f\} \leq |\mu|(f). \tag{1}$$

For every $a < |\mu|(f)$, by (i) of Proposition 2.40, there exists $h \in C_b(T)$ such that $0 \leq h \leq f$ and

$$a < |\mu|(h). \tag{2}$$

According to Theorem 1.13 and its Corollary 1.21, there exists $A_M \in \mathcal{B}_T$ such that, for every $A \in \mathcal{B}_T, A \subseteq A_M, \mu(A) \geq 0$ and, for every $A \in \mathcal{B}_T, A \subseteq T \backslash A_M$, $\mu(A) \leq 0, \forall A \in \mathcal{B}_T$. We have therefore

$$|\mu|(h) = \int_{A_M} h d\mu - \int_{T \backslash A_M} h d\mu. \tag{3}$$

Since μ is a Radon measure, for every $n \in \mathbb{N}^*$, there exist the compacts K_n and K'_n, and the open sets D_n and D'_n with $K_n \subseteq A_M \subseteq D_n$, $K'_n \subseteq T \setminus A_M \subseteq D'_n$ satisfying $|\mu|(D_n \setminus K_n) < \frac{1}{n}$ and $|\mu|(D'_n \setminus K'_n) < \frac{1}{n}$. Because T is completely regular, there exist the continuous mappings $\varphi_n, \varphi'_n : T \to [0,1]$ such that $\varphi_n \restriction_{K_n} = 1, \varphi_n \restriction_{T \backslash D_n} = 0$ and $\varphi'_n \restriction_{K'_n} = 1, \varphi'_n \restriction_{T \backslash D'_n} = 0$. Then

$$\int_T h\varphi_n d\mu = \int_{A_M} h d\mu - \int_{A_M \backslash K_n} h d\mu + \int_{D_n \backslash K_n} h\varphi_n d\mu. \tag{4}$$

Because $h \in L^1(\mu)$, $|h\varphi_n| \leq h$, $\mu(A_M \setminus K_n) \xrightarrow[n \to \infty]{} 0$ and $\mu(D_n \setminus K_n) \xrightarrow[n \to \infty]{} 0$, the passage to the limit in (4) leads to:

$$\lim_n \int_T h\varphi_n d\mu = \int_{A_M} h d\mu. \tag{5}$$

Similarly, we obtain:

$$\lim_n \int_T h\varphi'_n d\mu = \int_{T \backslash A_M} h d\mu. \tag{6}$$

According to (3), (5) and (6), we have then:

$$|\mu|(h) = \lim_n \int_T h\varphi_n d\mu - \lim_n \int_T h\varphi'_n d\mu.$$

By (2), there exists $n_0 \in \mathbb{N}$ such that

$$\int_T h(\varphi_{n_0} - \varphi'_{n_0}) d\mu > a.$$

As $|\varphi_{n_0} - \varphi'_{n_0}| \leq 1$, $g = h(\varphi_{n_0} - \varphi'_{n_0}) \in C_b(T)$ and $|g| = |h| \cdot |\varphi_{n_0} - \varphi'_{n_0}| \leq f$; then $a < \int_T g d\mu \leq |\int_T g d\mu| = |\mu(g)|$.

Therefore, we have shown that, for every $a < |\mu|(f)$, there exists $g \in C_b(T)$, $|g| \leq f$ such that $a < |\mu(g)|$. According to the inequality (1), it means that $|\mu|(f) = \sup\{|\mu(g)| : g \in C_b(T), |g| \leq f\}$. □

Corollary 2.45. *Let T be a completely regular space. For every lower semicontinuous function $f : T \to \bar{\mathbb{R}}_+$,*

(i) *the mapping $\mu \mapsto I_f(|\mu|) = \int_T f d|\mu|$ of $Rca(\mathcal{B}_T)$ in $\mathbb{R}_+$, is narrowly lower semicontinuous.*

(ii) *for every net $(\mu_i)_{i \in I} \subseteq Rca(\mathcal{B}_T)$, $\mathcal{T}$-convergent to $\mu \in Rca(\mathcal{B}_T)$,*

$$\int_T f d|\mu| \leq \liminf_i \int_T f d|\mu_i|.$$

The following proposition is an important consequence of Proposition 2.44.

Proposition 2.46. *If T is completely regular, then $Rca(\mathcal{B}_T)$ provided with the narrow topology is a completely regular space.*

Proof. $Rca(\mathcal{B}_T)$ is a vector subspace of $ca(\mathcal{B}_T)$, therefore, according to (iv) of Remark 2.32, $(Rca(\mathcal{B}_T), \mathcal{T})$ is a locally convex space. Let us show that $(Rca(\mathcal{B}_T), \mathcal{T})$ is a Hausdorff space. Since the topology $\mathcal{T}$ is generated by the family of applications $\mathcal{F} = \{I_f : f \in C_b(T)\}$, $\mathcal{T}$ is separated if and only if $\mathcal{F}$ separates the points of $Rca(\mathcal{B}_T)$.

Let $\mu, \nu \in Rca(\mathcal{B}_T)$ such that, for every $f \in C_b(T)$, $I_f(\mu) = I_f(\nu)$ and let $\theta = \mu - \nu \in Rca(\mathcal{B}_T)$. Then, for every $f \in C_b(T)$, $I_f(\theta) = \int_T f d\theta = 0$. For every $D \in \tau_T$, χ_D is l.s.c. and positive; therefore, according to Proposition 2.44, $|\theta|(D) = \sup\{|\theta(g)| : g \in C_b(T), |g| \leq \chi_D\} = 0$. As $|\theta|$ is a regular measure, for every $A \in \mathcal{B}(T)$, $|\theta|(A) = \inf_{D \in \tau_T, A \subseteq D} |\theta|(D) = 0$. Therefore, $|\theta| = 0$ and so $\theta = 0$ or $\mu = \nu$. □

Remark 2.47. According to Theorem 2.29, every Suslin space T is a Radon space ($ca(\mathcal{B}_T) = rca(\mathcal{B}_T)$); therefore, from (iii) of Theorem 2.21 and from the previous proposition it results that $(ca(\mathcal{B}_T), \mathcal{T})$ is a completely regular space, if T is a regular Suslin space.

Proposition 2.48. *Let T be a completely regular space; $Rca^+(\mathcal{B}_T)$ is a narrowly closed subset of $Rca(\mathcal{B}_T)$.*

Proof. Let $(\mu_i)_{i\in I} \subseteq Rca^+(\mathcal{B}_T)$ be a net such that $\mu_i \stackrel{\mathcal{T}}{\to} \mu \in Rca(\mathcal{B}_T)$; for every $f \in C_b^+(T)$ and every $i \in I$, $\mu_i(f) \geq 0$; therefore we have: $\mu(f) \geq 0$, for every $f \in C_b^+(T)$.

Let now $K \in \mathcal{K}_T$; for every $\varepsilon > 0$, there exists $D \in \tau_T$ such that $K \subseteq D$ and $|\mu|(D \setminus K) < \varepsilon$. Let $f : T \to [0,1]$ be a continuous function such that $f \restriction_K = 1$ and $f \restriction_{T\setminus D} = 0$. Then $\mu(f) = \int_T f d\mu = \int_D f d\mu = \mu(K) + \int_{D\setminus K} f d\mu$. Therefore $\mu(K) = \mu(f) - \int_{D\setminus K} f d\mu \geq 0 - |\mu|(D \setminus K) > -\varepsilon$, for every $\varepsilon > 0$. Then $\mu(K) \geq 0$, for every $K \in \mathcal{K}_T$. As μ is a Radon measure, it is inner regular with respect to the family of compact sets so that $\mu \in Rca^+(\mathcal{B}_T)$. □

Corollary 2.49. *If T is a completely regular space, then $(Rca^+(\mathcal{B}_T), \mathcal{T})$ is also a completely regular space.*

Proposition 2.50. *Let T be a completely regular space, let $\mu \in Rca^+(\mathcal{B}_T)$ and let $(\mu_i)_i \subseteq Rca^+(\mathcal{B}_T)$ be a net; then $\mu_i \stackrel{\mathcal{T}}{\to} \mu$ if and only if $\mu(f) \leq \liminf_i \mu_i(f)$, for every function $f : T \to \overline{\mathbb{R}}$, bounded from below and l.s.c.*

Proof. ($\Longrightarrow$) If $\mu_i \stackrel{\mathcal{T}}{\to} \mu$ and if $f : T \to \overline{\mathbb{R}}$ is bounded from below and lower semicontinuous, then there exists $m \in \mathbb{R}$ such that $\underline{m} + f$ is a positive and l.s.c. function. According to (i) of Remark 2.42, $\mu(\underline{m} + f) \leq \liminf_i \mu_i(\underline{m} + f)$. As $\mu_i(\underline{m}) = m \cdot \mu_i(T) \to m \cdot \mu(T)$, we have $\mu(f) \leq \liminf_i \mu_i(f)$, for every bounded from below and l.s.c. function $f : T \to \overline{\mathbb{R}}$.

($\Longleftarrow$) For every $f \in C_b(T)$, f is bounded and l.s.c. and so

$$\mu(f) \leq \liminf_i \mu_i(f). \tag{1}$$

On the other hand, as $-f$ is also bounded and l.s.c., $\mu(-f) \leq \liminf_i \mu_i(-f)$, from where

$$\limsup_i \mu_i(f) \leq \mu(f). \tag{2}$$

By (1) and (2), we have $\lim_i \mu_i(f) = \mu(f)$ and, therefore, $(\mu_i)_i$ is $\mathcal{T}$-convergent to μ. □

Let (T,d) be a metric space; we note by $BL(T,d)$ the set of all bounded, real-valued Lipschitz functions on (T,d); the following proposition is a consequence of the previous proposition and of Lemma 2.35.

Proposition 2.51 (Theorem 11.3.3 in [61]). *Let (T,d) be a metric space, let $(\mu_i)_i \subseteq Rca^+(\mathcal{B}_T)$ be a net and let $\mu \in Rca^+(\mathcal{B}_T)$; then*

$$\mu_i \xrightarrow{\mathcal{T}} \mu \Leftrightarrow \mu_i(f) \to \mu(f), \quad \textit{for every} \quad f \in BL(T,d).$$

Proof. The implication $\Rightarrow$ is a consequence of the inclusion $BL(T,d) \subseteq C_b(T)$.

($\Leftarrow$) Let us suppose that $\mu_i(f) \to \mu(f)$, for every $f \in BL(T,d)$ and let $f : T \to \mathbb{R}$ be a bounded l.s.c. function. Thanks to Lemma 2.35, there exists a sequence $(f_n)_n$ of Lipschitz functions such that $f_n(t) \uparrow f(t)$, for every $t \in T$. According to (i) of the same lemma, f_n is bounded, for every $n \in \mathbb{N}$ and therefore $(f_n) \subseteq BL(T,d)$. As $\mu(f_n) \uparrow \mu(f)$, for every $\varepsilon > 0$, there exists $n_0 \in \mathbb{N}$ such that

$$\mu(f) - \varepsilon < \mu(f_{n_0}) = \lim_i \mu_i(f_{n_0}) \leq \liminf_i \mu_i(f).$$

Proposition 2.50 allows us to conclude that $\mu_i \xrightarrow{\mathcal{T}} \mu$. □

Proposition 2.52. *Let T be a locally compact space, let $\mu \in Rca^+(\mathcal{B}_T)$ and let $(\mu_i)_{i\in I} \subseteq Rca^+(\mathcal{B}_T)$ be a net; then*

$$\mu_i \xrightarrow{\mathcal{T}} \mu \Longleftrightarrow \begin{cases} 1. & \mu_i \to \mu \quad \textit{vaguely and} \\ 2. & \mu_i(T) \to \mu(T). \end{cases}$$

Proof. ($\Longrightarrow$) If $\mu_i \xrightarrow{\mathcal{T}} \mu$, then, according to (ii) of Remark 2.32, $(\mu_i)_i$ is vaguely convergent to μ and $\mu_i(T) \to \mu(T)$.

($\Longleftarrow$) Let us suppose that the net $(\mu_i)_i$ is vaguely convergent to μ and $\mu_i(T) \to \mu(T)$. For every bounded l.s.c. function $f : T \to \mathbb{R}$, there exists $m \in \mathbb{R}$ such that $\underline{m} + f \geq 0$. Therefore, through an analogous demonstration to that of Proposition 2.40, we have

$$\mu(\underline{m} + f) = \sup\{\mu(g) : g \in C_c(T), 0 \leq g \leq \underline{m} + f\}.$$

So that, for every $\varepsilon > 0$, there exists $g \in C_c(T), 0 \leq g \leq \underline{m} + f$ such that

$$\mu(\underline{m} + f) - \varepsilon \leq \mu(g) = \lim_i \mu_i(g) \leq \liminf_i \mu_i(\underline{m} + f).$$

As ε is arbitrary and $\mu_i(\underline{m}) = m \cdot \mu_i(T) \to m \cdot \mu(T) = \mu(\underline{m})$, $\mu(f) \leq \liminf_i \mu_i(f)$. Proposition 2.50 allows us to conclude. □

The following theorem characterizes the narrow topology on $Rca^+(\mathcal{B}_T)$; the theorem can be find in the literature as Portmanteau theorem (see Theorem 2.1 in [28]).

Theorem 2.53. *Let (T, τ) be a completely regular space, let $\mu \in Rca^+(\mathcal{B}_T)$ and let $(\mu_i)_{i\in I} \subseteq Rca^+(\mathcal{B}_T)$ be a net; the following properties are equivalent:*

(i) *$(\mu_i)_{i\in I}$ is narrowly convergent to μ.*

(ii) *$\lim_i \mu_i(T) = \mu(T)$ and $\mu(D) \leq \liminf_i \mu_i(D)$, for every $D \in \tau_T$.*

(iii) *$\lim_i \mu_i(T) = \mu(T)$ and $\mu(F) \geq \limsup_i \mu_i(F)$, for every $F = \overline{F}^{\tau}$.*

Each of these conditions leads to:

(iv) *For every $A \in \mathcal{B}_T$ such that $\mu(\bar{A} \setminus A^{\circ}) = 0$, $\lim_i \mu_i(A) = \mu(A)$.*

If T is metrizable, then (iv) is equivalent to the previous conditions.

Proof. (i) $\Rightarrow$ (ii) Since $(\mu_i)_{i\in I}$ is narrowly convergent to μ, $\mu_i(T) = \mu_i(\underline{1}) \to \mu(\underline{1}) = \mu(T)$. On the other hand, for every $D \in \tau, \chi_D$ is l.s.c. and positive; therefore, following (i) in Remark 2.42, $\int_T \chi_D d\mu \leq \liminf_i \int_T \chi_D d\mu_i$ or $\mu(D) \leq \liminf_i \mu_i(D)$.

(ii) $\Rightarrow$ (iii) For every $F = \bar{F} \subseteq T, D = T \setminus F \in \tau$ and so, by (ii), $\mu(T \setminus F) \leq \liminf_i \mu_i(T \setminus F)$ and, as $\mu_i(T) \to \mu(T)$, we have (iii).

(iii) $\Rightarrow$ (i) Let $f \in C_b(T)$ and let $m, M \in \mathbb{R}$ such that $m < f(x) < M$, for every $x \in T$. If $g : T \to]0,1[$ is defined by $g(x) = \frac{1}{M-m}[f(x) - m]$, for every $x \in T$, then $g \in C_b(T)$ and so, for every $n \in \mathbb{N}^*$ and every $k = 0, \dots, n$, $F_k^n = \{x \in T : \frac{k}{n} \leq g(x)\}$ is a closed subset of T. As $T = g^{-1}(]0,1[) = \bigcup_{k=1}^n g^{-1}([\frac{k-1}{n}, \frac{k}{n}[) = \bigcup_{k=1}^n (F_{k-1}^n \setminus F_k^n)$, we have $\int_T g d\nu = \sum_{k=1}^n \int_{F_{k-1}^n \setminus F_k^n} g d\nu$, for every $\nu \in Rca^+(\mathcal{B}_T)$.

Since, for every $k = 1, \dots, n$,

$$\frac{k-1}{n} \cdot \nu(F_{k-1}^n \setminus F_k^n) \leq \int_{F_{k-1}^n \setminus F_k^n} g d\nu \leq \frac{k}{n} \cdot \nu(F_{k-1}^n \setminus F_k^n),$$

we have

$$\sum_{k=1}^n \frac{k-1}{n} \cdot \nu(F_{k-1}^n \setminus F_k^n) \leq \int_T g d\nu \leq \sum_{k=1}^n \frac{k}{n} \cdot \nu(F_{k-1}^n \setminus F_k^n). \tag{1}$$

As $F_0^n = T$ and $F_n^n = \emptyset$,

$$\sum_{k=1}^{n} \frac{k}{n} \cdot \nu(F_{k-1}^n \setminus F_k^n) = \frac{1}{n} \left[\nu(F_0^n \setminus F_1^n) + 2\nu(F_1^n \setminus F_2^n) + \cdots + n\nu(F_{n-1}^n \setminus F_n^n)\right]$$
$$= \frac{1}{n} \left[\nu(T) + \nu(F_1^n) + \cdots + \nu(F_{n-1}^n) - n\nu(F_n^n)\right]$$
$$= \frac{1}{n} \sum_{k=1}^{n} \nu(F_k^n) + \frac{1}{n}\nu(T).$$

Similarly, we obtain:

$$\sum_{k=1}^{n} \frac{k-1}{n} \cdot \nu(F_{k-1}^n \setminus F_k^n) = \frac{1}{n} \sum_{k=1}^{n} \nu(F_k^n).$$

Therefore, (1) leads to:

$$\frac{1}{n} \sum_{k=1}^{n} \nu(F_k^n) \leq \int_T g d\nu \leq \frac{1}{n}\nu(T) + \frac{1}{n} \sum_{k=1}^{n} \nu(F_k^n). \tag{2}$$

If we use (2) for μ_i and for μ, due to the hypothesis (iii), then we obtain, for every $n \in \mathbb{N}^*$:

$$\limsup_i \int_T g d\mu_i \leq \limsup_i \left[\frac{1}{n} \cdot \mu_i(T) + \frac{1}{n} \sum_{k=1}^{n} \mu_i(F_k^n)\right]$$
$$\leq \frac{1}{n} \cdot \mu(T) + \frac{1}{n} \sum_{k=1}^{n} \limsup_i \mu_i(F_k^n)$$
$$\leq \frac{1}{n}\mu(T) + \frac{1}{n} \sum_{k=1}^{n} \mu(F_k^n) \leq \frac{1}{n}\mu(T) + \int_T g d\mu.$$

If this takes place for all $n \in \mathbb{N}^*$, we have

$$\limsup_i \mu_i(g) \leq \mu(g). \tag{3}$$

Returning to $f = (M - m)g + \underline{m}$, we obtain

$$\limsup_i \mu_i(f) \leq \mu(f), \quad \text{for every} \quad f \in C_b(T). \tag{4}$$

By replacing $f \in C_b(T)$ with $-f$ in (4), we obtain

$$\liminf_i \mu_i(f) \geq \mu(f). \tag{5}$$

Finally, by (4) and (5), $\mu(f) = \lim_i \mu_i(f)$, for every $f \in C_b(T)$; therefore $(\mu_i)_i$ is narrowly convergent to μ.

(iii)$\Rightarrow$ (iv) In the hypothesis (iii), the properties (i) and (ii) are satisfied; then, for every $A \in \mathcal{B}_T$ such that $\mu(\bar{A} \setminus A^\circ) = 0, \mu(\bar{A}) = \mu(A^\circ) = \mu(A)$. Since μ_i and μ are positive measures, due to properties (iii) and (ii),

$$\begin{aligned}\mu(A) = \mu(\bar{A}) &\geq \limsup_i \mu_i(\bar{A}) \geq \limsup_i \mu_i(A) \geq \liminf_i \mu_i(A)\\ &\geq \liminf_i \mu_i(A^\circ) \geq \mu(A^\circ) = \mu(A).\end{aligned}$$

Therefore $\mu(A) = \lim_i \mu_i(A)$.

Let us suppose that $(T, \mathcal{T})$ is a metrizable space and let d be a metric on T such that $\tau = \tau_d$. In this case, we show that (iv)$\Rightarrow$ (iii).

For every $F = \bar{F} \subseteq T, \forall a > 0$, let $F^a = \{x \in T : d(x, F) < a\}$; then

$$\begin{aligned}H_a &= \overline{F^a} \setminus (F^a)^\circ \subseteq \{x \in T : d(x, F) \leq a\} \setminus \{x \in T : d(x, F) < a\}\\ &= \{x \in T : d(x, F) = a\}.\end{aligned}$$

It is obvious that $H_a \cap H_b = \emptyset$, if $a \neq b$. By reducing to the absurd, we show that there exists a decreasing sequence

$$(a_n)_n \subseteq \mathbb{R}, a_n \downarrow 0 \quad \text{such that} \quad \mu(H_{a_n}) = 0, \quad \text{for every} \quad n \in \mathbb{N}. \tag{6}$$

In fact, if there is $a > 0$ such that, $\mu(H_b) > 0$, for every $b \in]0, a[$, then $]0, a[= \bigcup_{n=1}^\infty A_n$, where $A_n = \{b \in]0, a[: \mu(H_b) \geq \frac{1}{n}\}$; there exists $n_0 \in \mathbb{N}$ such that A_{n_0} is infinite. We can, therefore, find a sequence $\{b_p : p \in \mathbb{N}\} \subseteq A_{n_0}$ such that $b_p \neq b_q$, for $p \neq q$ and, since μ is a measure positive and finite, if $H = \bigcup_{p=0}^\infty H_{b_p} \in \mathcal{B}_T$, then

$$+\infty > \mu(T) \geq \mu(H) = \sum_{p=1}^\infty \mu(H_{b_p}) \geq +\infty \cdot \frac{1}{n_0} = +\infty$$

which is impossible. Therefore the property (6) is proved.

We have, therefore, for all $n \in \mathbb{N}, \mu(\overline{F^{a_n}} \setminus (F^{a_n})^\circ) = 0$ and, due to hypothesis (iv),

$$\lim_i \mu_i(F^{a_n}) = \mu(F^{a_n}), \quad \text{for every} \quad n \in \mathbb{N}. \tag{7}$$

From (7), there results that $\limsup_i \mu_i(F) \le \lim_i \mu_i(F^{a_n}) = \mu(F^{a_n})$, for every $n \in \mathbb{N}$. As $\lim_n \mu(F^{a_n}) = \mu(F)$, we have shown that $\limsup_i \mu_i(F) \le \mu(F)$, for every $F = \bar{F} \subseteq T$. Finally, since $\mu(\bar{T} \setminus T^{\circ}) = 0$, from hypothesis (iv), we immediately obtain that $\lim_i \mu_i(T) = \mu(T)$. (iii) is then proved. □

Remark 2.54. The properties of the previous theorem do not remain equivalent on $Rca(\mathcal{B}_T)$. Indeed, let τ_u be the usual topology on $\mathbb{R}$ and let $\mathcal{B}_u$ be the Borel σ-algebra on $(\mathbb{R}, \tau_u)$. For every $n \in \mathbb{N}$, let us note $\mu_n = \delta_{\frac{1}{n}} - \delta_{-\frac{1}{n}}$; then $(\mu_n)_n \subseteq Rca(\mathcal{B}_u)$, and $\mu_n \xrightarrow{\mathcal{T}} \mu = \underline{0} \in Rca(\mathcal{B}_u)$. Therefore the condition (i) is satisfied. On the other hand, $D = (-\infty, 0) \in \tau_u$, $\mu(D) = 0$ and $\liminf_n \mu_n(D) = -1$, therefore the condition (ii) is not satisfied.

Proposition 2.55. *Let (T, τ) be a completely regular space, let $\mu \in Rca^+(\mathcal{B}_T)$ and let $(\mu_n)_n \subseteq Rca^+(\mathcal{B}_T)$. If $\mu_n \xrightarrow{\mathcal{T}} \mu$, then* $\operatorname{supp} \mu \subseteq \operatorname{Ls}_n(\operatorname{supp} \mu_n)$ *(where $\operatorname{Ls}_n(A_n) = \cap_{p=1}^{\infty} \overline{\cup_{n=p}^{\infty} A_n}$ is the Kuratowski upper limit of $(A_n)_{n \in \mathbb{N}}$).*

Proof. According to Definition 2.12, $\operatorname{supp} \mu = \cap\{F = \bar{F} : \mu(T \setminus F) = 0\}$ (the support of μ).

Suppose that there is $t \in \operatorname{supp} \mu$ and $t \notin \operatorname{Ls}_n(\operatorname{supp} \mu_n)$.

We have, therefore, $t \in T \setminus \operatorname{Ls}_n(\operatorname{supp} \mu_n) = \bigcup_{p=1}^{\infty} (T \setminus \overline{\bigcup_{n=p}^{\infty} \operatorname{supp} \mu_n})$. Let $p_0 \in \mathbb{N}$ and $D \in \tau_T$ such that $t \in D$ and $D \cap \bigcup_{n=p_0}^{\infty} (\operatorname{supp} \mu_n) = \emptyset$.

By (ii) of Theorem 2.53, it results that:

$$\mu(D) \le \liminf_n \mu_n(D) \le \liminf_n \mu_n(T \setminus \operatorname{supp} \mu_n) = 0.$$

According to Proposition 2.13, μ_n are concentrated on their supports; so that $\mu_n(T \setminus \operatorname{supp}\mu_n) = 0$, for every $n \in \mathbb{N}$. Then $F = T \setminus D$ is a closed set and $\mu(T \setminus F) = 0$; since $t \in \operatorname{supp} \mu, t \in F = T \setminus D$ which is a contradiction. □

Remark 2.56. From the previous proposition, it results that, if $\mu_n \xrightarrow{\mathcal{T}} \mu$, then μ is concentrated on the Kuratowski upper limit $\operatorname{Ls}_n(\operatorname{supp} \mu_n)$.

Proposition 2.57 (Proposition 8, §5, in [32]). *Let (X, τ_X) be a completely regular space, let T be a set and let $i : T \to X$ be an injection. Let us note by:*

τ_T $= i^{-1}(\tau_X)$*—the preimage topology of τ_X by i,*
$\mathcal{B}_T$ *—the Borel sets of (T, τ_T),*
$\mathcal{B}_X$ *—the Borel sets of (X, τ_X) and*
$\mathcal{M}_T$ $= \{\nu \in Rca(\mathcal{B}_X) : |\nu|^*(X \setminus i(T)) = 0\}$.

Then:

(i) $\mathcal{B}_T = i^{-1}(\mathcal{B}_X) = \{i^{-1}(A) : A \in \mathcal{B}_X\}$.

(ii) *For every measure* $\mu \in Rca(\mathcal{B}_T)$, *let* $I(\mu) : \mathcal{B}_X \to \mathbb{R}$ *defined by* $I(\mu)(A) = \mu(i^{-1}(A))$, *for every* $A \in \mathcal{B}_X$; *then* $I(\mu)$ *is a Radon measure on* $\mathcal{B}_X$, $|I(\mu)| = I(|\mu|)$ *and* $I(\mu) \in \mathcal{M}_T$.

Let $I : Rca(\mathcal{B}_T) \to \mathcal{M}_T \subseteq Rca(\mathcal{B}_X)$, $\mu \mapsto I(\mu)$; *then:*

(iii) I *is a bijection,*

(iv) $\|I(\mu)\| = \|\mu\|$, *for every* $\mu \in Rca(\mathcal{B}_T)$ *and*

(v) I *is a homeomorphism between the subspaces* $Rca^+(\mathcal{B}_T)$ *and* $\mathcal{M}_T^+$ *provided with their narrow topology.*

Before the demonstration, let us remind that

- $|\nu|$ is the total variation of measure ν (see Definition 1.19).
- $|\nu|^* : \mathcal{P}(X) \to \mathbb{R}_+$ indicates the outer measure generated by $|\nu|$, defined by $|\nu|^*(A) = \inf\{|\nu|(B) : B \in \mathcal{B}_X, A \subseteq B\}$.
- $\mathcal{M}_T \subseteq Rca(\mathcal{B}_X)$ is the subset of all measures on X concentrated on T (according to the terminology of [32], Chap. 9).

Proof. (i) Let $\mathcal{B}'_T = i^{-1}(\mathcal{B}_X)$. Since $\tau_T = i^{-1}(\tau_X) = \{i^{-1}(A) : A \in \tau_X\}$, $\mathcal{B}_T = \mathcal{B}(i^{-1}(\tau_X))$. On the other hand, $\mathcal{B}'_T$ is a σ-algebra on T generated by $i^{-1}(\tau_X)$ since τ_X generates $\mathcal{B}_X$. Therefore, we have $\mathcal{B}'_T = \mathcal{B}_T$.

(ii) It is obvious that $I(\mu)$ is a positive measure on X and that $I(\mu)(X) = \mu(T)$. Therefore $I(\mu) \in ca(\mathcal{B}_X)$.

According to Theorem 1.17, for every $A \in \mathcal{B}_X$, $|I(\mu)|(A) =$

$$= \sup\left\{\sum_{j=1}^{n} \left|I(\mu)(A_j)\right| : n \in \mathbb{N}^*, \{A_j\}_{j=1}^n = \mathcal{B}_X - \text{partition of} \quad A\right\}$$

$$= \sup\left\{\sum_{j=1}^{n} |\mu(i^{-1}(A_j))| : n \in \mathbb{N}^*, \{A_j\}_{j=1}^n = \mathcal{B}_X - \text{partition of} \quad A\right\}$$

$$= \sup\left\{\sum_{j=1}^{n} |\mu(B_j)| : n \in \mathbb{N}^*, \{B_j\}_{j=1}^n = \mathcal{B}_T - \text{partition of} \quad B = i^{-1}(A)\right\}$$

$$= |\mu|(B) = |\mu|(i^{-1}(A)) = I(|\mu|)(A).$$

As $\mu \in Rca(\mathcal{B}_T)$, for every $A \in \mathcal{B}_X$ and every $\varepsilon > 0$, there exists $K \in \mathcal{K}_T$ with $K \subseteq i^{-1}(A)$ such that $|\mu|(K) > |\mu|(i^{-1}(A)) - \varepsilon$. If $K' = i(K)$, then $K' \in \mathcal{K}_X$. So, we have $|I(\mu)|(K') = I(|\mu|)(K') = |\mu|(K) > I(|\mu|)(A) - \varepsilon = |I(\mu)|(A) - \varepsilon$. Then, for every $A \in \mathcal{B}_X$, $|I(\mu)|(A) = \sup\{|I(\mu)|(K') : K' \in \mathcal{K}_X, K' \subseteq A\}$ and so $I(\mu) \in Rca(\mathcal{B}_X)$.

We have $\mu \in Rca(\mathcal{B}_T)$; therefore, for every $\varepsilon > 0$, there exists $K_1 \in \mathcal{K}_T$ such that $|\mu|(K_1) > |\mu|(T) - \varepsilon$. If $K_1' = i(K_1) \in \mathcal{K}_X$, then $|I(\mu)|(K_1') = I(|\mu|)(K_1') = |\mu|(K_1) > |\mu|(i^{-1}(X)) - \varepsilon = I(|\mu|)(X) - \varepsilon = |I(\mu)|(X) - \varepsilon$. So $|I(\mu)|(X \setminus K_1') < \varepsilon$ and, as $X \setminus K_1' \supseteq X \setminus i(T)$, $|I(\mu)|^*(X \setminus i(T)) < \varepsilon$. Hence $|I(\mu)|^*(X \setminus i(T)) = 0$ and therefore $I(\mu) \in \mathcal{M}_T$.

(iii) The mapping $I : Rca(\mathcal{B}_T) \to \mathcal{M}_T$, which associates to every μ, $I(\mu)$, is well defined and is injective. Now, let us show that I is a surjection of $Rca(\mathcal{B}_T)$ on $\mathcal{M}_T$. If $\nu \in \mathcal{M}_T$, then $\nu \in Rca(\mathcal{B}_X)$ and $|\nu|^*(X \setminus i(T)) = 0$. Let $\mu : \mathcal{B}_T \to \mathbb{R}$ defined by $\mu(B) = \nu(A)$, where $A \in \mathcal{B}_X$ is so that $B = i^{-1}(A)$. μ is correctly defined since if $A, C \in \mathcal{B}_X$ and $i^{-1}(A) = i^{-1}(C)$, then $A \Delta C \subseteq X \setminus i(T)$. Hence $|\nu|(A \triangle C) = 0$; as $|\nu(A) - \nu(C)| \leq |\nu|(A \triangle C)$, $\nu(A) = \nu(C)$.

Firstly, we will show that μ is a measure on $\mathcal{B}_T$ and that $I(\mu) = \nu$. It is obvious that $\mu(\emptyset) = \nu(\emptyset) = 0$. Let $(B_n)_{n \in \mathbb{N}} \subseteq \mathcal{B}_T$ be a pairwise disjoint sequence of sets and let $B = \bigcup_1^\infty B_n \in \mathcal{B}_T$; then, by (i), there exists $(A_n)_n \subseteq \mathcal{B}_X$ such that, $B_n = i^{-1}(A_n)$, for every $n \in \mathbb{N}$. We note $A_1' = A_1$ and $A_n' = A_n \setminus \bigcup_{k=1}^{n-1} A_k$, for every $n \geq 2$; we obtain recursively that, for every $n \in \mathbb{N}$, $B_n = i^{-1}(A_n')$, where $(A_n')_n \subseteq \mathcal{B}_X$ is a pairwise disjoint sequence of sets. Therefore, we have $B = \bigcup_1^\infty i^{-1}(A_n') = i^{-1}(\bigcup_1^\infty A_n')$ and $\mu(B) = \nu(\bigcup_1^\infty A_n') = \sum_1^\infty \nu(A_n') = \sum_1^\infty \mu(B_n)$. So μ is a measure on $\mathcal{B}_T$ and $\mu(T) = \nu(X)$.

For every $B \in \mathcal{B}_T$, there exists $A \in \mathcal{B}_X$ such that $B = i^{-1}(A)$; because $A \setminus i(B) \subseteq X \setminus i(T)$, we have $|\nu|^*(A \setminus i(B)) \leq |\nu|^*(X \setminus i(T)) = 0$. Therefore $i(B)$ is ν–measurable and there exist $C \in \mathcal{B}_X$ and $N \subseteq X$ such that $|\nu|^*(N) = 0$, $C \cap N = \emptyset$ and $i(B) = C \cup N \subseteq i(T)$. We have then $|\nu|^*(A \setminus C) \leq |\nu|^*(A \setminus i(B)) + |\nu|^*(N) = 0$ and so $\nu(A) = \nu(C)$. Since $\nu \in Rca(\mathcal{B}_X)$, for every $\varepsilon > 0$, there exists $K \in \mathcal{K}_X, K \subseteq C$ such that $|\nu|(K) > |\nu|(C) - \varepsilon$. Then $|\mu|(B) = |\nu|(A) = |\nu(C)| < |\nu|(K) + \varepsilon$. As i is a homeomorphism of T on $i(T) \subseteq X$ and as $K \subseteq i(T)$, $H = i^{-1}(K) \in \mathcal{K}_T$. Then $H = i^{-1}(K) \subseteq B$ and $|\nu|(K) = |\mu|(H)$. For every $B \in \mathcal{B}_T$ and every $\varepsilon > 0$, we find $H \in \mathcal{K}_T$, $H \subseteq B$ such that $|\mu|(B) < |\mu|(H) + \varepsilon$. Therefore $\mu \in Rca(\mathcal{B}_T)$.

For every $A \in \mathcal{B}_X$, $I(\mu)(A) = \mu(i^{-1}(A)) = \nu(A)$ which means that $I(\mu) = \nu$. Therefore, I is a surjection of $Rca(\mathcal{B}_T)$ on $\mathcal{M}_T$.

(iv) $\|I(\mu)\| = |I(\mu)|(X) = I(|\mu|)(X) = |\mu|(T) = \|\mu\|$.

(v) Let us show that I is continuous on $Rca^+(\mathcal{B}_T)$. Let $(\mu^j)_{j \in J} \subseteq Rca^+(\mathcal{B}_T)$ be a net narrowly convergent to $\mu \in Rca^+(\mathcal{B}_T)$; for every $D \in \tau_X$, $i^{-1}(D) \in \tau_T$

and so, by (ii) of Theorem 2.53,

$$\mu(i^{-1}(D)) \leq \liminf_j \mu^j(i^{-1}(D)) \quad \text{from where}$$

$$I(\mu)(D) \leq \liminf_j I(\mu^j)(D).$$

Moreover, $I(\mu^j)(X) = \mu^j(T) \to \mu(T) = I(\mu)(X)$ and, still according to Theorem 2.53, $(I(\mu^j))_{j\in J}$ is narrowly convergent to $I(\mu)$ in $\mathcal{M}_T^+$. Therefore, I is a continuous mapping.

We have to show yet that I^{-1} is continuous. Let $(\nu^j)_{i\in J} \subseteq \mathcal{M}_T^+$ be a net narrowly convergent to $\nu \in \mathcal{M}_T^+$; for every $D \in \tau_T$, there exists $G \in \tau_X$ such that $D = i^{-1}(G)$. Therefore $I^{-1}(\nu)(D) = \nu(G) \leq \liminf_j \nu^j(G) = \liminf_j I^{-1}(\nu^j)(D)$. According to (ii) of Theorem 2.53, $(I^{-1}(\nu^j))_j$ is narrowly convergent to $I^{-1}(\nu)$ in $Rca^+(\mathcal{B}_T)$ and, therefore, I is a homeomorphism of $Rca^+(\mathcal{B}_T)$ on $\mathcal{M}_T^+$. □

In the rest of this section, we present several properties that can be transferred from (T, τ) towards $(Rca^+(\mathcal{B}_T), \mathcal{T})$.

Theorem 2.58 (Proposition 10 and corollaries 1 and 2, §5, in [32]).

a) *If (T, τ) is a Polish space, then $(Rca^+(\mathcal{B}_T), \mathcal{T})$ is also Polish.*

b) *If (T, τ) is a metrizable second-countable space, then $(Rca^+(\mathcal{B}_T), \mathcal{T})$ is also metrizable and second-countable.*

c) *If (T, τ) is a Suslin space, then $(Rca^+(\mathcal{B}_T), \mathcal{T})$ is also Suslin space.*

Proof. (a) (i) We first assume that (T, τ) is a metrizable compact space. By (v) of Remark 2.32 and according to Theorem 2.4, we obtain: $ca(\mathcal{B}_T) = rca(\mathcal{B}_T) = Rca(\mathcal{B}_T) = [C(T)]^*$ and then $\mathcal{T} = \sigma(Rca(\mathcal{B}_T), C(T))$ is the weak* topology on $Rca(\mathcal{B}_T)$. Still according to (v) of Remark 2.32, the closed unit ball $rca_1(\mathcal{B}_T)$ of $Rca(\mathcal{B}_T)$ is, therefore, $\mathcal{T}$–compact. As T is metrizable compact, it is separable and so the closed unit ball $rca_1(\mathcal{B}_T)$ provided with the weak* topology $\mathcal{T}$ is metrizable; hence $(rca_1(\mathcal{B}_T), \mathcal{T})$ is a metrizable compact space. According to Proposition 2.48, $Rca^+(\mathcal{B}_T) = rca^+(\mathcal{B}_T)$ is a $\mathcal{T}$–closed subset of $rca(\mathcal{B}_T)$; then $rca_1^+(\mathcal{B}_T) = \{\mu \in Rca^+(\mathcal{B}_T) : \mu(T) \leq 1\}$ is also a metrizable compact space and then a Polish one.

Let $\underline{1}$ be the constant function equal to 1 on T; the mapping $I_1 : rca_1^+(\mathcal{B}_T) \to \mathbb{R}$, defined by: $I_1(\mu) = \mu(\underline{1}) = \mu(T)$, is $\mathcal{T}$–continuous and, therefore,

$$M = \{\mu \in rca_1^+(\mathcal{B}_T) : \mu(T) < 1\} = I_1^{-1}(]-\infty, 1[)$$

is an open subset of $(rca_1^+(\mathcal{B}_T), \mathcal{T})$.

From (iii) of Theorem 2.18 it results that M is a $\mathcal{T}$–Polish space.

Finally, the application $\varphi : Rca^+(\mathcal{B}_T) \to M, \varphi(\mu) = \frac{1}{1+\mu(T)} \cdot \mu$, is a homeomorphism and so, following (i) of Theorem 2.18, $Rca^+(\mathcal{B}_T)$ is Polish.

(ii) Let now (T, τ) be an arbitrary Polish space; according to Theorem 2.19, T is homeomorphic to a G_δ set $P \subseteq H = [0, 1]^{\mathbb{N}}$.

Therefore there exist the τ_H–open sets $\{G_n : n \in \mathbb{N}\}$, such that $P = \bigcap_{n=0}^{\infty} G_n$. According to (i), $rca^+(\mathcal{B}_H) = Rca^+(\mathcal{B}_H)$ is a Polish space. For every $n \in \mathbb{N}$ and every $p \in \mathbb{N}^*$, let us note $D_{np} = \{\mu \in rca^+(\mathcal{B}_H) : \mu(H \setminus G_n) < \frac{1}{p}\}$. As G_n is open in H, $\chi_{H \setminus G_n}$ is a bounded l.s.c. function; then, by (ii) of Corollary 2.41, for every $n \in \mathbb{N}$, $I_{\chi_{H\setminus G_n}}$ is $\mathcal{T}$–u.s.c. on $rca^+(\mathcal{B}_H)$.

As $D_{np} = I^{-1}_{\chi_{H\setminus G_n}}(]-\infty, \frac{1}{p}[)$, D_{np} is $\mathcal{T}$–open in $rca^+(\mathcal{B}_H)$, for every $n \in \mathbb{N}$ and every $p \in \mathbb{N}^*$.

Then $A = \bigcap_{n=0}^{\infty} \bigcap_{p=1}^{\infty} D_{np}$ is a G_δ subset of Polish space $rca^+(\mathcal{B}_H)$. By (ii) and (iii) of Theorem 2.18, A is Polish space.

We remark that $A = \{\mu \in rca^+(\mathcal{B}_H) : \mu(H \setminus P) = 0\}$ (the set of all measures concentrated on P). According to Proposition 2.57, $Rca^+(\mathcal{B}_P)$ is homeomorphic to A, therefore $Rca^+(\mathcal{B}_P)$ is Polish. As T is homeomorphic to P, $Rca^+(\mathcal{B}_T)$ is homeomorphic to $Rca^+(\mathcal{B}_P)$ and $Rca^+(\mathcal{B}_P)$ is also a Polish space.

(b) Let (T, τ) be a metrizable second-countable space and let d be a metric on T compatible with τ (i.e. $\tau_d = \tau$).

Let $(\hat{T}, \hat{d})$ be a completion of the metric space (T, d); then $(\hat{T}, \tau_{\hat{d}})$ is a Polish space and, by a), $Rca^+(\mathcal{B}_{\hat{T}})$ is also Polish. According to Proposition 2.57, $Rca^+(\mathcal{B}_T)$ is homeomorphic to $\mathcal{M}_T^+ = \{\mu \in Rca^+(\mathcal{B}_{\hat{T}}) : \mu^*(\hat{T} \setminus T) = 0\}$. As $\mathcal{M}_T^+$ is a subspace of $Rca^+(\mathcal{B}_{\hat{T}})$, $\mathcal{M}_T^+$ is a metrizable separable space and then $Rca^+(\mathcal{B}_T)$ is also separable and metrizable.

(c) Let (T, τ) be a Suslin space, let P be a Polish space and let $f : P \to T$ be a continuous surjection on T. Just as in Remark 2.30, let $\bar{f} : Rca^+(\mathcal{B}_P) \to Rca^+(\mathcal{B}_T)$ be the mapping defined by $\bar{f}(\lambda) : \mathcal{B}_T E \to \mathbb{R}_+, \bar{f}(\lambda)(A) = \lambda(f^{-1}(A))$. Then, for every net $(\mu^i)_{i\in I} \subseteq Rca^+(\mathcal{B}_P)$ with $\mu^i \to \mu \in Rca^+(\mathcal{B}_P)$ and every $D \in \tau_T, f^{-1}(D) \in \tau_P$ and so $\mu(f^{-1}(D)) \le \liminf_i \mu^i(f^{-1}(D))$, from where $\bar{f}(\mu^i) \to \bar{f}(\mu)$. Therefore $\bar{f}$ is continuous on $Rca^+(\mathcal{B}_P)$. $Rca^+(\mathcal{B}_P)$ is Polish and, according to Remark 2.30, $\bar{f}$ is a surjection; then $Rca^+(\mathcal{B}_T)$ is a Suslin space. □

Corollary 2.59. *If T is a Suslin space and if $H \subseteq Rca^+(\mathcal{B}_T)$ is $\mathcal{T}$– compact, then H is sequentially $\mathcal{T}$–compact.*

Proof. According to (c) of Theorem 2.58, $(Rca^+(\mathcal{B}_T), \mathcal{T})$ is a Suslin space; the result is an immediately consequence of Corollary 2.26. □

Remark 2.60. According to Proposition 2.38, the converse of b) of Theorem 2.58 is also true. Therefore we can present the following result:

(T, τ) is a metrizable second-countable space if and only if $(Rca^+(\mathcal{B}(T)), \mathcal{T})$ is metrizable and second-countable.

2.4 Compactness Results

Definition 2.61 (Definition 2, §5, in [32]). Let T be a completely regular space and let $H \subseteq ca(\mathcal{B}_T)$; H is *tight* if satisfies:

(a) $\sup_{\mu \in H} |\mu|(T) < +\infty$.

(b) For every $\varepsilon > 0$, there exists $K_\varepsilon \in \mathcal{K}_T$ such that $|\mu|(T \setminus K_\varepsilon) \le \varepsilon$, for every, $\mu \in H$.

A sequence $(\mu_n)_n \subseteq ca(\mathcal{B}_T)$ is said be tight if $\{\mu_n : n \in \mathbb{N}\}$ is tight.

Remark 2.62.

(i) This definition is coherently with Definition 2.1.6.

(ii) Let $A \subseteq T$ and $H = \Psi(A)$, where Ψ is the application defined in Proposition 2.38; then $H = \{\delta_x : x \in A\} \subseteq Rca^+(\mathcal{B}_T)$. H is tight if and only if A is relatively compact in T.

Proposition 2.63 (Exercise 10, §5, in [32]). *Let H be a bounded subset of $ca(\mathcal{B}_T)$. H is tight if and only if there exists an inf-compact application $\varphi : T \to [0, +\infty]$ such that*

$$\sup_{\mu \in H} \int_T \varphi d|\mu| < +\infty.$$

Proof. We recall that an application $\varphi : T \to [0, +\infty]$ is inf-compact if, for every $a \in \mathbb{R}_+^*$, the a-level set $\varphi^{-1}([0, a])$ is a compact of T.

Let $H \subseteq ca(\mathcal{B}_T)$ be a bounded set.

If H is tight, then, according to condition b) of the previous definition, there exists an increasing sequence $(K_n)_n$ of compact subsets of T such that, for every $n \in \mathbb{N}$, $\sup_{\mu \in H} |\mu|(T \setminus K_n) \le \frac{1}{2^n}$. Let $\varphi = \sum_{n=0}^{\infty} \chi_{T \setminus K_n}$. If $[a]$ note the integer part of a, $\varphi^{-1}([0, a]) = K_{[a]+1} \in \mathcal{K}_T$, for every $a \in [0, +\infty[$. Therefore φ is inf-compact. Obviously, $\sup_{\mu \in H} \int_T \varphi d|\mu| = \sup_{\mu \in H} \sum_{n=0}^{\infty} |\mu|(T \setminus K_n) \le \sum_{n=0}^{\infty} \sup_{\mu \in H} |\mu|(T \setminus K_n) \le \sum_{n=0}^{\infty} \frac{1}{2^n} < +\infty$.

Conversely, if we suppose that $\varphi : T \to [0, +\infty]$ is an inf-compact application such that $M = \sup_{\mu \in H} \int_T \varphi d|\mu| < +\infty$, then $K_\varepsilon = \varphi^{-1}([0, \frac{M}{\varepsilon}]) \in \mathcal{K}_T$, for every $\varepsilon > 0$; therefore, for every $\mu \in H$,

$$|\mu|(T \setminus K_\varepsilon) = |\mu|\left(\varphi > \frac{M}{\varepsilon}\right) \leq \frac{\varepsilon}{M} \int_T \varphi d|\mu| \leq \varepsilon.$$

Hence H is tight. □

Proposition 2.64 (Proposition 11, §5, in [32]). *If $H \subseteq ca(\mathcal{B}_T)$ is tight, then the $\mathcal{T}$-closure of H, $\bar{H}$, is also tight.*

Proof. Let H be a tight set and let $M = \sup_{\mu \in H} |\mu|(T) < +\infty$. For every $\mu \in \bar{H}$, there exists $(\mu_i)_{i \in I} \subseteq H$ such that $\mu_i \overset{\mathcal{T}}{\to} \mu$. As $\underline{1}$ is l.s.c. and positive, according to Corollary 2.45,

$$|\mu|(T) \leq \liminf_i |\mu_i|(T) \leq M.$$

For every $\varepsilon > 0$, there exists $K \in \mathcal{K}_T$ such that $|\mu|(T \setminus K) < \varepsilon$, for every $\mu \in H$.

The mapping $\chi_{T \setminus K}$ is l.s.c. and positive; therefore, according to the same Corollary 2.45,

$$|\mu|(T \setminus K) \leq \liminf_i |\mu_i|(T \setminus K) \leq \varepsilon.$$

Hence, for every $\mu \in \bar{H}$, $|\mu|(T \setminus K) < \varepsilon$; therefore $\bar{H}$ is tight. □

We introduced the following notions in [78], Definition 3.7.

Definition 2.65. For every bounded set $H \subseteq ca(\mathcal{B}_T)$, we note

$$t(H) = \inf_{K \in \mathcal{K}_T} \sup_{\mu \in H} |\mu|(T \setminus K).$$

$t(H) \in \mathbb{R}_+$ is called *the modulus of narrow compactness* of H.

It is clear that H is tight if and only if $t(H) = 0$.

In what will follow, we present the compactness results related to weaker convergences than the narrow convergence.

Definition 2.66. Let $a \geq 0$; a net $(\mu_i)_{i \in I} \subseteq ca(\mathcal{B}_T)$ is said *a–convergent* to $\mu \in ca(\mathcal{B}_T)$ if

$$\limsup_i |\mu_i(f) - \mu(f)| \leq a \cdot \|f\|, \quad \text{for every} \quad f \in C_b(T),$$

which will be noted $\mu_i \overset{a}{\to} \mu$.

A set $H \subseteq ca(\mathcal{B}_T)$ is said *relatively a-compact* if every net $(\mu_i)_{i\in I} \subseteq H$ has at least one subnet a–convergent to a measure $\mu \in ca(\mathcal{B}_T)$.

Remark 2.67. In 1.91, we have introduced a notion of a-convergence for the sequences of integrable functions: $\lambda \in ca^+(\mathcal{B}_T)$ and $a \geq 0$ being fixed, a net $(f_i)_{i\in I} \subseteq L^1(\lambda)$ is said a–convergent to $f \in L^1(\lambda)$ if

$$\limsup_{i\in I} \left| \int_A (f_i - f)d\lambda \right| \leq a, \quad \text{for every} \quad A \in \mathcal{B}_T.$$

If $\mu_i, \mu : \mathcal{B}_T \to \mathbb{R}$ are defined by $\mu_i(A) = \int_A f_i d\lambda, \mu(A) = \int_A f d\lambda$, then $(\mu_i) \subseteq ca(\mathcal{B}_T), \mu \in ca(\mathcal{B}_T)$ and $f_i \overset{a}{\to} f$ if and only if

$$\limsup_{i\in I} \left|\mu_i(\chi_A) - \mu(\chi_A)\right| \leq a \cdot \|\chi_A\|, \quad \text{for every} \quad A \in \mathcal{B}_T.$$

There is, therefore, no inconvenience in using the same notion for the two convergences.

Remark 2.68.

(i) $(\mu_i)_{i\in I}$ is 0–convergent to μ if and only if $(\mu_i)_{i\in I}$ is narrowly convergent to μ. Hence, a set $H \subseteq ca(\mathcal{B}_T)$ is relatively 0-compact if and only if H is relatively narrowly compact.

(ii) We can interpret $\inf\{a \geq 0 : (\mu_i)_{i\in I} \quad \text{is} \quad a-\text{convergent}\}$ as indicating the deficit of narrow convergence of the net $(\mu_i)_i$.

The following theorem generalizes the Prohorov's theorem (see Theorem 3.8 of [78]).

Theorem 2.69. *Let T be a completely regular space, let H be a bounded subset of $Rca(\mathcal{B}_T)$ and let $t(H)$ be its modulus of narrow compactness; then H is relatively $t(H)$-compact.*

Proof. Let X be the Čech–Stone compactification of T, let i be a homeomorphic embedding of T in X $(\overline{i(T)} = X)$ and let I be the injection of $Rca(\mathcal{B}_T)$ in $rca(\mathcal{B}_X) = Rca(\mathcal{B}_X)$ associated to injection i (see Proposition 2.57). Then I is a isometric injection; moreover, for every $F \in C(X)$, $f = F \circ i \in C_b(T)$ and, for every $\mu \in Rca(\mathcal{B}_T)$, $I(\mu)(F) = \int_X Fd(I(\mu)) = \int_T fd\mu = \mu(f)$. Since $H \subseteq Rca(\mathcal{B}_T)$ is bounded, $I(H)$ is also bounded in $rca(\mathcal{B}_X)$. Therefore, the set $I(H)$ is relatively weak* compact, which means that it is relatively $\mathcal{T}(X)$-compact (see (v) of Remark 2.32).

Let $(\mu_j)_{j\in J}$ be a net in H and, for every $j \in J$, let $\nu_j = I(\mu_j)$. Then $(\nu_j)_{j\in J}$ is a net in $I(H)$ so that there exists a subnet $(\nu_{j_l})_{l\in L}$, $\mathcal{T}(X)$–convergent to a measure $\nu \in Rca(\mathcal{B}_X)$. By the definition of $t(H)$, for every $\mu \in H$ and every $n \in \mathbb{N}^*$, there exists $K_n \subseteq \mathcal{K}_T$ such that

$$|\mu|(T \setminus K_n) < t(H) + \frac{1}{n}. \tag{1}$$

Let $T_0 = \bigcup_{n=1}^{\infty} K_n$ and

$$X_0 = \bigcup_{n=1}^{\infty} i(K_n) = i(T_0). \tag{2}$$

For every $n \in \mathbb{N}, i(K_n) \in \mathcal{K}_X$ hence $X_0 \in \mathcal{B}_X$ and $X \setminus i(K_n)$ is an open subset of X. Therefore, according to Corollary 2.45, the mapping $\lambda \mapsto |\lambda|(X \setminus i(K_n))$ of $rca(\mathcal{B}_X)$ in $\mathbb{R}^+$ is $\mathcal{T}(X)$–l.s.c. on $rca(\mathcal{B}_X)$ and so,

$$|\nu|(X \setminus i(K_n)) \leq \liminf_l |\nu_{j_l}|(X \setminus i(K_n)), \quad \text{for every} \quad n \in \mathbb{N}^*. \tag{3}$$

By (1), for every $l \in L$ and every $n \in \mathbb{N}^*$, $|\nu_{j_l}|(X \setminus i(K_n)) = |I(\mu_{j_l})|(X \setminus i(K_n)) = |\mu_{j_l}|(T \setminus K_n) < t(H) + \frac{1}{n}$. Then, by (3), we obtain:

$$|\nu|(X \setminus i(K_n)) \leq t(H) + \frac{1}{n}, \quad \text{for every} \quad n \in \mathbb{N}^*. \tag{4}$$

Therefore, according to (2),

$$|\nu|(X \setminus X_0) \leq t(H). \tag{5}$$

As X is the Čech–Stone compactification of T, every continuous mapping $f \in C_b(T)$, is extendable to a mapping $F \in C(X)$ $(F \circ i = f)$; moreover, $\|F\| = \|f\|$ (see Theorem 3.6.1. of [68]).

Let now $J : C_b(T) \to \mathbb{R}$ defined by

$$J(f) = \nu(F \cdot \chi_{X_0}) = \int_{X_0} F d\nu.$$

Obviously, J is a continuous linear mapping on $C_b(T)$. Since ν is a Radon measure on X, i is a homeomorphism of T on $i(T) \subseteq X$ and $X_0 = i(T_0) \in \mathcal{B}_X$, then, for every $\varepsilon > 0$, there exists $K \in \mathcal{K}_T, K \subseteq T_0$ (and so $i(K) \subseteq X_0$) such that $|\nu|(X_0 \setminus i(K)) < \varepsilon$.

For every $g \in C_b(T)$ such that $|g| \leq 1$ and $g\restriction_K = 0$, let $G \in C(X)$ such that $G \circ i = g$ and $\|G\| = \|g\| \leq 1$. Then

$$J(g) = \nu(G \cdot \chi_{X_0}) = \int_{X_0} G d\nu = \int_{X_0 \setminus i(K)} G d\nu + \int_{i(K)} G d\nu = \int_{X_0 \setminus i(K)} G d\nu \text{ and}$$

$$|J(g)| \leq \left| \int_{X_0 \setminus i(K)} G d\nu \right| \leq |\nu|(X_0 \setminus i(K)) < \varepsilon.$$

Then, according to Proposition 5 of Chap. 9, §5 of [32] , there exists a measure $\mu \in Rca(\mathcal{B}_T)$ such that $\mu(f) = J(f)$, for every $f \in C_b(T)$.

For every $f \in C_b(T)$, let us note as previously $F \in C(X)$, the extension of f; $F \circ i = f$ and $\|F\| = \|f\|$. Then $|\mu_{j_l}(f) - \mu(f)| = |\mu_{j_l}(F \circ i) - \nu(F \cdot \chi_{X_0})| \leq |\nu_{j_l}(F) - \nu(F)| + |\nu(F \cdot \chi_{X \setminus X_0})|$.

As $(\nu_{j_l})_{l \in L}$ is $\mathcal{T}(X)$–convergent to ν, by passing to limit, we obtain:

$$\limsup_l |\mu_{j_l}(f) - \mu(f)| \leq \|F\| \cdot |\nu|(X \setminus X_0)$$

and then, using the inequality (5),

$$\limsup_l |\mu_{j_l}(f) - \mu(f)| \leq t(H) \cdot \|f\|, \quad \text{for every} \quad f \in C_b(T).$$

Therefore, $(\mu_{j_l})_l$ is $t(H)$–convergent to μ and so H is $t(H)$–compact. □

Remark 2.70. From previous theorem, it results that: *For every bounded net $(\mu_i)_{i \in I} \subseteq Rca(\mathcal{B}_T)$, there exist a subnet $(\mu_{i_j})_{j \in J}$ and a measure $\mu \in Rca(\mathcal{B}_T)$ such that*

$$\limsup_j |\mu_{i_j}(f) - \mu(f)| \leq t(\{\mu_i : i \in I\}) \cdot \|f\|, \quad \text{for every} \quad f \in C_b(T).$$

Now, we obtain as corollary an important result of compactness—Prohorov's theorem (see Theorem 1, §5, in [32] and Theorem 11.5.4 of [61]).

Corollary 2.71 (Prohorov's Theorem). *If T is a completely regular space, then every tight of $Rca(\mathcal{B}_T)$ is relatively $\mathcal{T}$-compact.*

Proof. If H is tight, then H is bounded and $t(H) = 0$. □

From Corollaries 2.59 and 2.71, we obtain:

Corollary 2.72. *Let T be a Suslin space; every tight $(\mu_n)_{n \in \mathbb{N}} \subseteq Rca^+(\mathcal{B}_T)$ admits one subsequence $(\mu_{k_n})_{n \in \mathbb{N}}$ $\mathcal{T}$–convergent to a measure $\mu \in Rca^+(\mathcal{B}_T)$.*

We can formulate a reciprocal of the Prohorov's theorem in the case of locally compact spaces or of Polish spaces but, only for the positive measures (see Theorem 2, §5, in [32]).

Theorem 2.73. *Let T be a locally compact space and let $H \subseteq Rca^+(\mathcal{B}_T)$ be a relatively $\mathcal{T}(T)$-compact set; then H is tight.*

Proof. The mapping $\mu \mapsto \mu(T)$ is continuous on $(Rca^+(\mathcal{B}_T), \mathcal{T}(T))$. Hence, since $\bar{H}$ is $\mathcal{T}(T)$-compact, $\sup_{\mu\in H} \mu(T) < +\infty$.

For every $\varepsilon > 0$ and every $\mu \in \bar{H}$, there exists $K_\mu \in \mathcal{K}_T$ such that $\mu(T \setminus K_\mu) < \varepsilon$. Now let U_μ be a $\mathcal{T}(T)$-open relatively compact neighborhood of K_μ. Since $\chi_{T\setminus U_\mu}$ is u.s.c. and bounded, in the light of (ii) of Corollary 2.41, the mapping $\lambda \mapsto \lambda(T \setminus U_\mu)$ is $\mathcal{T}(T)$-u.s.c. Therefore $V_\mu = \{\lambda \in \bar{H} : \lambda(T \setminus U_\mu) < \varepsilon\}$ is $\mathcal{T}(T) \upharpoonright_{\bar{H}}$-open and, as $\mu \in V_\mu$, V_μ is a neighborhood $\mathcal{T}(T)$-open of μ. $\{V_\mu : \mu \in \bar{H}\}$ is a $\mathcal{T}(T)$-open cover of the compact set $\bar{H}$; then there exist $\mu_1, \ldots, \mu_n \in \bar{H}$ such that $\bar{H} = \cup_{i=1}^n V_{\mu_i}$.

Then $K = \bigcup_{i=1}^n \bar{U}_{\mu_i} \in \mathcal{K}_T$ and, for every $\lambda \in H$, there exists i so that $\lambda \in V_{\mu_i}$ and so $\lambda(T \setminus U_{\mu_i}) < \varepsilon$. Then, we have $\lambda(T \setminus K) < \varepsilon$ which leads to the fact that H is tight. □

Theorem 2.74. *Let T be a Polish space and let $H \subseteq Rca^+(\mathcal{B}_T)$ be a relatively $\mathcal{T}(T)$-compact set; then H is tight.*

Proof. Just as in the proof of the previous theorem, $\sup_{\mu\in H} \mu(T) < +\infty$. As T is Polish, according to Theorem 2.19, T is homeomorphic to a G_δ subset of $[0,1]^{\mathbb{N}}$. For every $p \in \mathbb{N}$, let T_p be an open subset of $[0,1]^{\mathbb{N}}$ and let $i : T \to \cap_{p=0}^\infty T_p$ be a homeomorphism. For every $p \in \mathbb{N}$, let $i_p : T \to T_p$ be the application defined by $i_p(t) = i(t)$, for any $t \in T$. Then, we define $I_p : Rca^+(\mathcal{B}_T) \to Rca^+(\mathcal{B}_{T_p})$ by $I_p(\mu)(A) = \mu(i_p^{-1}(A))$, for every $\mu \in Rca^+(\mathcal{B}_T)$ and every $A \in \mathcal{B}_{T_p}$. According to (v) of Proposition 2.57, I_p is continuous and $H_p = I_p(\bar{H})$ is a compact set in $(Rca^+(\mathcal{B}_{T_p}), \mathcal{T}(T_p))$. Since T_p is an open set in the compact space $[0,1]^{\mathbb{N}}$, T_p is locally compact. From Theorem 2.73 it results that H_p is tight. Therefore, for every $p \in \mathbb{N}$ and every $\varepsilon > 0$, there exists $C_p \in \mathcal{K}_{T_p} \subseteq \mathcal{K}_{[0,1]^{\mathbb{N}}}$ such that

$$\nu(T_p \setminus C_p) < \frac{\varepsilon}{2^p}, \quad \text{for every} \quad \nu \in H_p.$$

Let $C = \bigcap_0^\infty C_p$. C is a compact of $\bigcap_{p=0}^\infty T_p$. Since i is a homeomorphism, $K = i^{-1}(C)$ is a compact of T and, for every $\mu \in H$ and every p, $I_p(\mu) \in H_p$.

Hence we have

$$\mu(T\setminus K) \le \sum_{p=1}^{\infty} \mu(T\setminus i_p^{-1}(C_p)) \le \sum_{p=1}^{\infty} \mu(i_p^{-1}(T_p\setminus C_p))$$
$$= \sum_{p=1}^{\infty} I_p(\mu)(T_p\setminus C_p) < \sum_{p=1}^{\infty} \frac{\varepsilon}{2^p} = \varepsilon$$

which proves that H is tight. □

Remark 2.75.

(i) The demonstration of Theorem 2.74 does not explicitly use the fact that T is homeomorphic to a G_δ-subset $[0,1]^{\mathbb{N}}$, only that T is homeomorphic to a G_δ-subset of a compact space. A completely regular space homeomorphic to a G_δ-subset of a compact space is called a *Čech-complete* space; the locally compact spaces and the Polish spaces are Čech-complete spaces. According to this remark, we can formulate the following theorem which generalizes Theorems 2.73 and 2.74:

Theorem. If T is a Čech-complete space, every relatively $\mathcal{T}(T)$-compact subset $H \subseteq Rca^+(\mathcal{B}_T)$ is tight.

A space T is called a *Prohorov space* if every relatively $\mathcal{T}(T)$-compact subset of $Rca^+(\mathcal{B}_T)$ is tight (see Definition 8.10.8 in [30]).

The previous theorem asserts that any Čech-complete space (particularly, any locally compact space and any Polish space) is a Prohorov space (see also Corollary 8.10.11 of [30]).

(ii) If T is not a Čech-complete space, the condition of being tight is not generally obtained from narrow compactness. The following example (Chap. 9, Exercise 5 of § 5 of [32]), in which T is a Suslin space (but not a Polish one !), highlights a sequence of narrowly convergent probabilities which is not tight.

Example 2.76. Let $T = \ell^2 = \{(x_n)_n \subseteq \mathbb{R} : \sum_0^\infty x_n^2 < +\infty\}$ equipped with its usual norm $\|\cdot\|_2$. ℓ^2 is a Hilbert space and its dual is itself; let τ be the weak topology on ℓ^2. Then, by 2.17, $(T,\tau_T) = (\ell^2,\tau)$ is a completely regular Suslin space not-metrizable; let $\mathcal{B}_T(\tau) = \mathcal{B}_T$ the σ-algebra of Borel sets relatively to the weak topology τ.

Let $\{u^p : p \in \mathbb{N}\}$ be the classical orthonormal basis of ℓ^2: for every $p \in \mathbb{N}$, $u^p = (u_n^p)_{n\in\mathbb{N}} \in \ell^2$, where, for every n and every $p \in \mathbb{N}$,

$$u_n^p = \begin{cases} 0, & n \ne p, \\ 1, & n = p. \end{cases}$$

For every $n \in \mathbb{N}$, let $\mu_n = \sum_{p=0}^{\infty} e^{-\frac{p}{n^3}}(1 - e^{-\frac{1}{n^3}})\delta_{n\cdot u^p}$ where $\delta_{n\cdot u^p}$ is the Dirac measure concentrated at $n \cdot u^p \in \ell^2$.

Then $(\mu_n)_n \subseteq Rca^+(\mathcal{B}_T)$ and, for every $n \in \mathbb{N}$, $\mu_n(T) = 1$.

We will show that $(\mu_n)_n$ is $\mathcal{T}$-convergent to δ_0 and that the sequence $(\mu_n)_n$ is not tight.

• Firstly, let us show that (μ_n) is $\mathcal{T}$-convergent to δ_0.

As $\mu_n(T) = 1 \to \delta_0(T)$, by (ii) of Theorem 2.53, it is sufficient to show that, for every $D \in \tau$, $\delta_0(D) \leq \liminf_n \mu_n(D)$. For the open sets not containing 0, the inequality is obvious. Among the open sets which contain 0, it is sufficient to show the inequality for the sets of type $U_a = \{x \in \ell^2 : |(x,a)| < 1\}$, where $a = (a_n)_n \in \ell^2$. Therefore it is sufficient to show that, for every $a = (a_n)_n \in \ell^2$, $\lim_n \mu_n(U_a) = 1$.

$$\mu_n(U_a) = \sum_{|a_p|<\frac{1}{n}} e^{-\frac{p}{n^3}}(1 - e^{-\frac{1}{n^3}}).$$

For every $n \in \mathbb{N}$, let $\{p_1, \ldots, p_{q_n}\}$ be the set of all $p \in \mathbb{N}$ with $|a_p| \geq \frac{1}{n}$. Then, for all $p \in \{p_1, \ldots, p_{q_n}\}$, $a_p^2 \geq \frac{1}{n^2}$ from where, for all $n \in \mathbb{N}$, $\frac{q_n}{n^2} \leq \sum_{i=1}^{q_n} a_{p_i}^2 \leq \|a\|_2^2$. We have so:

$$\lim_n \frac{q_n}{n^3} = 0. \tag{*}$$

Then

$$\begin{aligned}
\mu_n(U_a) &= \sum_{|a_p|<\frac{1}{n}} e^{-\frac{p}{n^3}}(1 - e^{-\frac{1}{n^3}}) \\
&= \sum_{p\in\mathbb{N}} e^{-\frac{p}{n^3}}(1 - e^{-\frac{1}{n^3}}) - \sum_{|a_p|\geq\frac{1}{n}} e^{-\frac{p}{n^3}}(1 - e^{-\frac{1}{n^3}}) \\
&= 1 - (1 - e^{-\frac{1}{n^3}}) \sum_{i=1}^{q_n} e^{-\frac{p_i}{n^3}} \geq 1 - (1 - e^{-\frac{1}{n^3}}) \sum_{p=0}^{q_n-1} e^{-\frac{p}{n^3}} = e^{-\frac{q_n}{n^3}}
\end{aligned}$$

and, by $(*)$, $\lim_n \mu_n(U_a) = 1$.

• The sequence $(\mu_n)_n$ is not tight. Indeed, every weakly compact set $K \subseteq T$ is bounded; therefore, for every K, there exists $M > 0$ such that $\|x\| \leq M$, for every $x \in K$. Then, for any $n > M$, $n \cdot u^p \notin K$, for all $p \in \mathbb{N}$ and $\mu_n(K) = 0$ so that $H = \{\mu_n : n \in \mathbb{N}\}$ is not tight (in fact we have here $t(H) = 1$).

Let us remark that, according to c) of Theorem 2.58, $Rca^+(\mathcal{B}_T)$ is a Suslin space and H is a compact subset of $Rca^+(\mathcal{B}_T)$. Therefore, on the grounds of Corollary 2.26, H is a Polish subspace. Let us mention again that, in this case, T is neither Polish, nor locally compact and nor Čech-complete.

2.5 Metrics on the Space $(Rca^+(\mathcal{B}_T), \mathfrak{T})$

We have demonstrated in a) of Theorem 2.58 that, if (T, τ) is a Polish space, then $(Rca^+(\mathcal{B}_T), \mathfrak{T})$ is also a Polish space; in b) of the same theorem we proved that, if (T, τ) is a metrizable second-countable space, then $(Rca^+(\mathcal{B}_T), \mathfrak{T})$ is also metrizable and second-countable. In this section, we will effectively construct possible distances compatible on $(Rca^+(\mathcal{B}_T), \mathfrak{T})$.

2.5.1 Dudley's Metric

Let T be a metrizable and second-countable space and let d be a metric generating the topology of T; then $(Rca^+(\mathcal{B}_T), \mathfrak{T})$ is metrizable and second-countable. In achieving an embedding of $Rca^+(\mathcal{B}_T)$ in the dual of a Banach space, R. M. Dudley showed that the narrow topology on this space is a trace of a topology of a normed space [60, 61].

Let $BL(T, d)$, or simply $BL(T)$, be the set of all bounded, real-valued Lipschitz functions on T. $BL(T)$ is a Banach space with respect to the norm $\|\cdot\|_{BL}$ defined by:

$$\|f\|_{BL} = \|f\|_\infty + \|f\|_L = \sup_{t\in T}|f(t)| + \sup_{t,s\in T, t\neq s}\left\{\frac{|f(t)-f(s)|}{d(t,s)}\right\}.$$

Let us note by $BL(T)^*$ the dual space of $(BL(T), \|\cdot\|_{BL})$.

Proposition 2.77. *The mapping $I : Rca^+(\mathcal{B}_T) \to BL(T)^*$ defined by*

$$I(\mu)(f) = \int_T f d\mu, \quad \textit{for every} \quad f \in BL(T) \quad \textit{and every} \quad \mu \in Rca^+(\mathcal{B}_T),$$

is an injection.

Proof. It is obvious that $I(\mu) \in BL(T)^*$ and that

$$\|I(\mu)\| = \sup_{\|f\|_{BL}\leq 1} |I(\mu)(f)| = \mu(T), \quad \text{for every} \quad \mu \in Rca^+(\mathcal{B}_T).$$

Let us show that I is an injection. Let $\mu, \nu \in Rca^+(\mathcal{B}_T)$ such that $I(\mu) = I(\nu)$; then $\int_T f d\mu = \int_T f d\nu$, for every $f \in BL(T)$.

Let $f \in C_b(T)$; according to Lemma 2.35, there exists a sequence $(f_n) \subseteq BL(T)$ such that $f_n(t) \uparrow f(t)$, for every $t \in T$. Then $\int_T f d\mu = \lim_n \int_T f_n d\mu = \lim_n \int_T f_n d\nu = \int_T f d\nu$ and therefore $\mu(f) = \nu(f)$, for every $f \in C_b(T)$.

The proof of Proposition 2.46 allows us, then, to conclude. □

Remark 2.78.

(i) Owing to the embedding of the previous proposition, we can consider $Rca^+(\mathcal{B}_T)$ as a subset of $BL(T)^*$.

(ii) According to Proposition 2.51, it results that, for every net $(\mu_i)_i \subseteq Rca^+(\mathcal{B}_T) \subseteq BL(T)^*$ and every $\mu \in Rca^+(\mathcal{B}_T) \subseteq BL(T)^*$, $\mu_i \xrightarrow{\mathcal{T}} \mu \Longleftrightarrow$ $(\mu_i)_i$ is weak*-convergent to μ (convergent with respect to the topology $\sigma(BL(T)^*, BL(T))$.

Then, the narrow topology $\mathcal{T}$ on $Rca^+(\mathcal{B}_T)$ is the trace of weak* topology on $BL(T)^*$.

In what follows, we will show that $\mathcal{T}$ is also the trace of the strong topology on $BL(T)^*$. For this, we need the following lemma:

Lemma 2.79. *Let T be a Polish space, let $(\mu_n)_{n\geq 1} \subseteq Rca^+(\mathcal{B}_T)$ be a sequence narrowly convergent to $\mu_0 \in Rca^+(\mathcal{B}_T)$ and let $(f_n)_{n\geq 1} \subseteq C_b(T)$ be an uniformly bounded sequence, uniformly convergent on the compact subsets of T to $f_0 \in C_b(T)$; then*

$$\int_T f_n d\mu_n \to \int_T f_0 d\mu_0.$$

Proof. Let $N > 0$ such that

$$\|f_n\|_\infty = \sup_{t\in T} |f_n(t)| \leq N, \quad \text{for every} \quad n \in \mathbb{N}. \tag{1}$$

As T is Polish, $(Rca^+(\mathcal{B}_T), \mathcal{T})$ is also Polish (see a) of Theorem 2.58). Then $\{\mu_n : n \in \mathbb{N}\}$ is narrowly compact and hence it is tight (see Theorem 2.74).

Therefore

$$M = \sup_{n\in\mathbb{N}} \mu_n(T) < +\infty. \tag{2}$$

For every $\varepsilon > 0$, there exists $K \in \mathcal{K}_T$ such that

$$\mu_n(T \setminus K) < \varepsilon_1 = \frac{\varepsilon}{6N}, \quad \text{for every} \quad n \in \mathbb{N}. \tag{3}$$

As $f_n \xrightarrow[K]{u} f_0$ and $\mu_n(f_0) \to \mu_0(f_0)$, there exists $n_0 \in \mathbb{N}$ such that

$$|f_n(t) - f_0(t)| < \varepsilon_2 = \frac{\varepsilon}{3M}, \quad \text{for every } n \geq n_0 \quad \text{and every } t \in K \quad \text{and} \tag{4}$$

$$|\mu_n(f_0) - \mu_0(f_0)| < \varepsilon_3 = \frac{\varepsilon}{3}, \text{ for every } n \geq n_0. \tag{5}$$

Then, for every $n \geq n_0$,

$$\begin{aligned}
|\mu_n(f_n) - \mu_0(f_0)| &\leq |\mu_n(f_n) - \mu_n(f_0)| + |\mu_n(f_0) - \mu_0(f_0)| \\
&\overset{(5)}{<} \left|\int_K (f_n - f_0) d\mu_n\right| + \left|\int_{T\setminus K} (f_n - f_0) d\mu_n\right| + \varepsilon_3 \\
&\overset{(4)+(1)}{<} \varepsilon_2 \mu_n(K) + 2N\mu_n(T \setminus K) + \varepsilon_3 \\
&\overset{(2)+(3)}{<} \varepsilon_2 \cdot M + 2N\varepsilon_1 + \varepsilon_3 = \varepsilon.
\end{aligned}$$

□

Definition 2.80. Let $\|\cdot\| : Rca^+(\mathcal{B}_T) \to \mathbb{R}_+$, defined by

$$\|\mu\| = \sup_{\|f\|_{BL} \leq 1} |\mu(f)|, \text{ for every } \mu \in Rca^+(\mathcal{B}_T),$$

the trace on $Rca^+(\mathcal{B}_T)$ of the norm on the dual space $BL(T)^*$.

The mappping $D : Rca^+(\mathcal{B}_T) \times Rca^+(\mathcal{B}_T) \to \mathbb{R}_+$, defined by

$$D(\mu, \nu) = \|\mu - \nu\| = \sup_{\|f\|_{BL} \leq 1} |\mu(f) - \nu(f)|, \text{ for every } \mu, \nu \in Rca^+(\mathcal{B}_T),$$

is a metric on $Rca^+(\mathcal{B}_T)$. This metric is called *the Dudley's metric* on $Rca^+(\mathcal{B}_T)$ (see [61], p. 394).

Theorem 2.81 (Theorem 11.3.3 in [61]). *Let (T, d) be a second-countable metric space; for every sequence $(\mu_n)_{n\in\mathbb{N}} \subseteq Rca^+(\mathcal{B}_T)$ and every $\mu \in Rca^+(\mathcal{B}_T)$,*

$$\mu_n \overset{\mathcal{T}}{\to} \mu \iff D(\mu_n, \mu) \to 0.$$

Proof. ($\Longrightarrow$) Let us suppose that $\mu_n \overset{\mathcal{T}}{\to} \mu$.

Let (T_1, d_1) be a completion of (T, d) and let $i : T \to T_1$ be the natural embedding of T in T_1; then, according to Proposition 2.57, the sequence $(\bar{\mu}_n)_{n\in\mathbb{N}} = (I(\mu_n))_{n\in\mathbb{N}}$ is narrowly convergent to $\bar{\mu} = I(\mu)$ in $Rca^+(\mathcal{B}_{T_1})$. Moreover, the space $BL(T, d)$ being isometric with $BL(T_1, d_1)$, $D(\mu_n, \mu) \to 0$ if and only if $D(\bar{\mu}_n, \bar{\mu}) \to 0$. Therefore, without loss of generality, we can take (T, d) as a complete space and therefore a Polish one.

Let us suppose that $D(\mu_n, \mu) = \sup_{\|f\|_{BL} \leq 1} |\mu_n(f) - \mu(f)| \not\to 0$.

With perhaps extracting a subsequence, still noted (μ_n), we can find $\varepsilon_0 > 0$ and $(f_n) \subseteq BL(T) \subseteq C_b(T)$ such that

$$\left|\int_T f_n d\mu_n - \int_T f_n d\mu\right| \geq \varepsilon_0 \quad \text{and} \quad \|f_n\|_{BL} \leq 1, \quad \text{for every} \quad n \in \mathbb{N}. \quad (1)$$

As T is second-countable, by passing perhaps to a subsequence, still noted (f_n), we can suppose that there exists a countable dense subset $A \subseteq T$ such that $(f_n(t))_{n\in\mathbb{N}}$ is convergent, for every $t \in A$.

Let $g : A \to \mathbb{R}, g(t) = \lim_n f_n(t)$, for every $t \in A$.

As $\|f_n\|_L \leq \|f_n\|_{BL} \leq 1$, for every $n \in \mathbb{N}$, $|f_n(t) - f_n(s)| \leq d(t,s)$, from where $|g(t) - g(s)| \leq d(t,s)$, for every $t, s \in A$. This means that g is a Lipschitz function on A.

We can extend this application on T by writing, for every $t \in T$,

$$f(t) = \inf\{g(s) + d(t,s) : s \in A\}.$$

The function f is Lipschitz on T (see Lemma 2.35) and then, as A is dense in T,

$$f_n(t) \to f(t), \quad \text{for every} \quad t \in T.$$

For every $t, s, u \in T$ and every $n \in \mathbb{N}$,

$$|f_n(t)| + \frac{|f_n(s) - f_n(u)|}{d(s,u)} \leq 1$$

and then

$$|f(t)| + \frac{|f(s) - f(u)|}{d(s,u)} \leq 1$$

from where $\|f\|_{BL} \leq 1$ and therefore $f \in BL(T) \subseteq C_b(T)$.

Let us suppose that (f_n) is not uniformly convergent to f on the compacts of T; then, there exists $K \in \mathcal{K}_T$ such that (f_n) is not uniformly convergent to f on K. With eventually extracting an other subsequence, still noted (f_n), there exist $\varepsilon_1 > 0$ and $(t_n) \subseteq K$ such that

$$|f_n(t_n) - f(t_n)| \geq \varepsilon_1, \quad \text{for every} \quad n \in \mathbb{N}. \quad (2)$$

As K is compact, we can assume that (t_n) converges to $t \in K$ and, as $f_n(t) \to f(t)$, there exists $n_0 \in \mathbb{N}$ such that

$$d(t_n, t) < \frac{\varepsilon_1}{3}, \quad \text{for every} \quad n \geq n_0 \quad \text{and} \quad (3)$$

$$|f_n(t) - f(t)| < \frac{\varepsilon_1}{3}, \quad \text{for every} \quad n \geq n_0. \quad (4)$$

Let $n \in \mathbb{N}, n \geq n_0$; then, by (2), (3) and (4),

$$\begin{aligned}\varepsilon_1 &\leq |f_n(t_n) - f(t_n)| \leq |f_n(t_n) - f_n(t)| + |f_n(t) - f(t)| + |f(t) - f(t_n)| \\ &\leq d(t_n, t) + \frac{\varepsilon_1}{3} + d(t_n, t) < \varepsilon_1.\end{aligned}$$

Therefore (f_n) is uniformly bounded and uniformly convergent to f on every compact of T. According to Lemma 2.79, $\mu_n(f_n) \to \mu(f)$. Then, we can find $n_1 \in \mathbb{N}$ such that

$$|\mu_n(f_n) - \mu(f)| < \varepsilon_2 = \frac{\varepsilon_0}{3 + \mu(T)}, \quad \text{for every} \quad n \geq n_1. \tag{5}$$

Since $\mu \in Rca^+(\mathcal{B}_T)$ is tight, there exists $C \in \mathcal{K}_T$ such that

$$\mu(T \setminus C) < \varepsilon_2. \tag{6}$$

As $f_n \xrightarrow[C]{u} f$, there exists $n_2 \in \mathbb{N}$ such that

$$|f_n(t) - f(t)| < \varepsilon_2, \quad \text{for every} \quad n \geq n_2 \quad \text{and every} \quad t \in C. \tag{7}$$

By (6) and (7), we obtain

$$\begin{aligned}|\mu(f_n) - \mu(f)| &\leq \int_{T \setminus C} |f_n - f| d\mu + \int_C |f_n - f| d\mu \\ &\leq 2\mu(T \setminus C) + \varepsilon_2 \mu(C) \leq \varepsilon_2(2 + \mu(T)).\end{aligned} \tag{8}$$

By (1), (5) and (8), for every $n \geq \max\{n_1, n_2\}$,

$$\begin{aligned}\varepsilon_0 &\leq |\mu_n(f_n) - \mu(f_n)| \leq |\mu_n(f_n) - \mu(f)| + |\mu(f) - \mu(f_n)| \\ &< \varepsilon_2 + \varepsilon_2(2 + \mu(T)) = \varepsilon_0.\end{aligned}$$

($\Longleftarrow$) Now we suppose that

$$D(\mu_n, \mu) = \sup_{\|f\|_{BL} \leq 1} |\mu_n(f) - \mu(f)| \to 0.$$

As $\underline{1} \in BL(T)$, $\mu_n(T) = \mu_n(\underline{1}) \to \mu(\underline{1}) = \mu(T)$.

Thanks to (ii) of Theorem 2.53, in order to conclude, it is sufficient to show that $\mu(D) \leq \liminf_n \mu_n(D)$, for every open subset $D \subseteq T$.

Therefore, let D be an open subset of T and, for all $n \in \mathbb{N}^*$, let

$$F_n = T \setminus (T \setminus D)^{\frac{1}{n}} = \left\{t \in T : d(t, T \setminus U) \geq \frac{1}{n}\right\}.$$

Then $F_n = \overline{F}_n \subseteq D$ is a closed subset of D.

Moreover, $(F_n)_{n \in \mathbb{N}}$ is an increasing sequence and $\cup_{n=1}^{\infty} F_n = D$.

The function $f_n : F_n \cup (T \setminus D) \to [0, 1]$, defined by

$$f_n(t) = \begin{cases} 1, & t \in F_n \\ 0, & t \in T \setminus U \end{cases},$$

is a Lipschitz one. Indeed, for every $t, s \in F_n \cup (T \setminus D)$,

$$|f_n(t) - f_n(s)| \leq n \cdot d(t, s).$$

We can therefore extend this application to T by stating that, for all $t \in T$,

$$g_n(t) = \inf\{f_n(s) + n \cdot d(t, s) : s \in F_n \cup (T \setminus D)\}.$$

The application $g_n : T \to [0, 1]$ is Lipschitz and $\|g_n\|_L = n$, $\|g_n\|_\infty = 1$ (see Lemma 2.35). Then $(g_n) \subseteq BL(T)$ and $\|g_n\|_{BL} = n + 1$.

As $F_n \uparrow D$, for every $\varepsilon > 0$, there exists $n_0 \in \mathbb{N}$ such that $\mu(F_{n_0}) > \mu(D) - \frac{\varepsilon}{2}$ and then

$$\mu(g_{n_0}) \geq \mu(F_{n_0}) > \mu(D) - \frac{\varepsilon}{2}. \tag{1}$$

As $D(\mu_n, \mu) = \sup_{\|f\|_{BL} \leq 1} |\mu_n(f) - \mu(f)| \to 0$, there exists $n_1 \in \mathbb{N}$ such that, for every $n \geq n_1$ and every $f \in BL(T)$ with $\|f\| \leq 1$,

$$|\mu_n(f) - \mu(f)| < \frac{\varepsilon}{2(n_0 + 1)}. \tag{2}$$

Let now $f_0 = \frac{1}{n_0+1} \cdot g_{n_0}$; then $\|f_0\|_{BL} = 1$ and, by (2),

$$|\mu_n(g_{n_0}) - \mu(g_{n_0})| < \frac{\varepsilon}{2}, \quad \text{for every} \quad n \geq n_1. \tag{3}$$

Therefore, as $g_{n_0} \restriction_{T \setminus D} = 0$, by (3) and (1), we obtain

$$\mu_n(D) \geq \mu_n(g_{n_0}) \geq \mu(g_{n_0}) - \frac{\varepsilon}{2} > \mu(D) - \varepsilon, \quad \text{for every} \quad n \geq n_1,$$

from where $\liminf_n \mu_n(D) \geq \mu(D)$. □

Remark 2.82. $(Rca^+(\mathcal{B}_T), \mathcal{T})$ is a metrizable space and D is a metric compatible with the narrow topology $\mathcal{T}$; so $\mathcal{T}$ is the trace of strong topology of $BL(T)^*$. Therefore, for every net $(\mu_i)_i \subseteq Rca^+(\mathcal{B}_T)$ and every $\mu \in Rca^+(\mathcal{B}_T)$, $\mu_i \xrightarrow{\mathcal{T}} \mu \iff D(\mu_i, \mu) \to 0$.

2.5.2 Lévy–Prohorov's Metric

According to Theorem 2.58, if (T, τ) is a Polish space, then $(Rca^+(\mathcal{B}_T), \mathcal{T})$ is also Polish. In the following theorem, we present a complete metric compatible with the narrow topology on $Rca^+(\mathcal{B}_T)$ (see Exercise 8, §5, in [32]).

Theorem 2.83. *Let (T, τ) be a Polish space and let d be a metric on T compatible with τ. For every $\varepsilon > 0$ and every $F \subseteq T$ let $F^\varepsilon = \{x \in T : d(x, F) < \varepsilon\}$ be the open ball of radius ε centred at F.*

Let $L : Rca^+(\mathcal{B}_T) \times Rca^+(\mathcal{B}_T) \to \mathbb{R}_+$ be a mapping defined by:

$$L(\mu, \nu) = \inf \left\{ \varepsilon > 0 : \begin{array}{l} \mu(F) \le \nu(F^\varepsilon) + \varepsilon \\ \nu(F) \le \mu(F^\varepsilon) + \varepsilon \end{array}, \quad \text{for every} \quad F = \overline{F} \subseteq T \right\}.$$

Then L is a complete metric on $Rca^+(\mathcal{B}_T)$ that generates the narrow topology on $Rca^+(\mathcal{B}_T)$.

The metric L is called the Lévy–Prohorov metric.

Proof. • Firstly, we will show that L is a metric on $Rca^+(\mathcal{B}_T)$.

For every $\mu, \nu \in Rca^+(\mathcal{B}_T)$, $L(\mu, \nu) \le \max[\mu(T), \nu(T)] < +\infty$ (μ and ν are bounded measures); therefore L is real valued.

If $\mu = \nu$, then, for every $\varepsilon > 0$ and every $F = \bar{F}$, $\mu(F) \le \nu(F^\varepsilon) + \varepsilon$ and $\nu(F) \le \mu(F^\varepsilon) + \varepsilon$, from where $L(\mu, \nu) = 0$. Conversely, if $L(\mu, \nu) = 0$, then there exists a decreasing sequence $\varepsilon_n \downarrow 0$ such that, for every $n \in \mathbb{N}$ and every $F = \bar{F}$, $\mu(F) \le \nu(F^{\varepsilon_n}) + \varepsilon_n$ and $\nu(F) \le \mu(F^{\varepsilon_n}) + \varepsilon_n$. Then the sequence $(F^{\varepsilon_n})_n$ is decreasing and $F = \cap_{n=1}^{\infty} F^{\varepsilon_n}$. Therefore, $\mu(F) = \lim_n \mu(F^{\varepsilon_n})$ and $\nu(F) = \lim_n \nu(F^{\varepsilon_n})$. We have then, for every $F = \bar{F}$, $\mu(F) = \nu(F)$ what assures that $\mu = \nu$ since μ and ν are regular measures.

The equality $L(\mu, \nu) = L(\nu, \mu)$ is obvious.

Let $\lambda, \mu, \nu \in Rca^+(\mathcal{B}_T)$; for every $\varepsilon > 0$, there exist $\delta_1 > 0$ and $\delta_2 > 0$ such that $\delta_1 < L(\lambda, \mu) + \varepsilon$, $\delta_2 < L(\mu, \nu) + \varepsilon$ and, for every $F = \bar{F}$,

$$\begin{cases} \lambda(F) \le \mu(F^{\delta_1}) + \delta_1, \\ \mu(F) \le \lambda(F^{\delta_1}) + \delta_1. \end{cases} \tag{1}$$

$$\begin{cases} \mu(F) \le \nu(F^{\delta_2}) + \delta_2, \\ \nu(F) \le \mu(F^{\delta_2}) + \delta_2. \end{cases} \tag{2}$$

An elementary calculation leads to

$$\overline{F^{\delta_1}}^{\delta_2} \subseteq F^{\delta_1 + \delta_2}. \tag{3}$$

Then, by (1) - (3), we obtain, for every $F = \bar{F}$,

$$\begin{aligned}\lambda(F) &\leq \mu(F^{\delta_1}) + \delta_1 \leq \mu(\overline{F^{\delta_1}}) + \delta_1 \\ &\leq \nu(\overline{F^{\delta_1}}^{\delta_2}) + \delta_1 + \delta_2 \leq \nu(F^{\delta_1+\delta_2}) + \delta_1 + \delta_2\end{aligned}$$

and similarly $\nu(F) \leq \lambda(F^{\delta_1+\delta_2}) + \delta_1 + \delta_2$.

These last relations leads us to: $L(\lambda, \nu) \leq \delta_1 + \delta_2 < L(\lambda, \mu) + L(\mu, \nu) + 2\varepsilon$ from where, $L(\lambda, \nu) \leq L(\lambda, \mu) + L(\mu, \nu)$.

Therefore, L is a metric on $Rca^+(\mathcal{B}_T)$.

Let us show that $\mathcal{T}_L = \mathcal{T}$. Since $\mathcal{T}$ is a metrizable topology, it is sufficient to show that, for every sequence $(\mu_n)_n \subseteq Rca^+(\mathcal{B}_T)$ and every $\mu \in Rca^+(\mathcal{B}_T)$,

$$\mu_n \xrightarrow{\mathcal{T}} \mu \Longleftrightarrow L(\mu_n, \mu) \to 0.$$

• Let us suppose that $\mu_n \xrightarrow{\mathcal{T}} \mu$ and let $r > 0$. The open balls of T having the same center t and different radii have their boundaries disjoint. Then, since μ is bounded, the μ-measure of the boundaries of these balls is null, except perhaps for at most countable family of balls.

Since T is separable, there exists a family $\{B_n : n \in \mathbb{N}^*\}$ of open balls B_n of radius $r_n < \frac{r}{6}$ such that $\mu(\partial B_n) = 0$ and $T = \bigcup_{n=1}^{\infty} B_n$.

Let $K \in \mathcal{K}_T$ such that $\mu(T \setminus K) < \frac{r}{3}$. As $K \subseteq \bigcup_{n=1}^{\infty} B_n$, there exists $p \in \mathbb{N}$ such that $K \subseteq \bigcup_{n=1}^{p} B_n = A_0$ and

$$\mu(T \setminus A_0) < \frac{r}{3}. \tag{4}$$

Let $\mathcal{A}$ be the family of all finite union of balls of the set $\{B_1, \dots, B_p\}$. $\mathcal{A}$ is finite and, if $A \in \mathcal{A}$, then $A = \cup_{k=1}^{j} B_{i_k}$, $\partial A = \bar{A} \setminus A^\circ \subseteq \cup_{k=1}^{j}(\bar{B}_{i_k} \setminus B_{i_k}^\circ) = \cup_{k=1}^{j} \partial B_{i_k}$, from where $\mu(\partial A) = 0$.

Let $U = \bigcap_{A \in \mathcal{A}} \{\nu \in Rca^+(\mathcal{B}_T) : |\nu(A) - \mu(A)| < \frac{r}{6}\} \cap \{\nu \in Rca^+(\mathcal{B}_T) : |\nu(T) - \mu(T)| < \frac{r}{6}\}$. According to (iv) of Theorem 2.53, for every $A \in \mathcal{A}$, $\mu_n(A) \to \mu(A)$ and $\mu_n(T) \to \mu(T)$; then, there exists $n_0 \in \mathbb{N}$ such that,

$$\mu_n \in U, \quad \text{for every} \quad n \geq n_0. \tag{5}$$

Let us show that

$$U \subseteq T(\mu, r) = \{\nu : L(\mu, \nu) \leq r\}. \tag{6}$$

For every $F = \bar{F}$, let $A_1 = \cup\{B_n : n = 1, \dots, p, B_n \cap F \neq \emptyset\}$. Then $A_1 \in \mathcal{A}$, $F \subseteq A_1 \cup (T \setminus A_0)$ and $A_1 \subseteq F^{\frac{r}{3}}$. Therefore, for every $\nu \in U$, we have

$$\begin{aligned}\nu(F) &\leq \nu(A_1) + \nu(T \setminus A_0) = \nu(A_1) + \nu(T) - \nu(A_0)\\ &< \mu(A_1) + \frac{r}{6} + \mu(T) + \frac{r}{6} - \mu(A_0) + \frac{r}{6} = \mu(A_1) + \mu(T \setminus A_0) + \frac{r}{2}\\ &< \mu(A_1) + \frac{r}{3} + \frac{r}{2} < \mu(A_1) + r < \mu(F^{\frac{r}{3}}) + r \leq \mu(F^r) + r.\end{aligned}$$

Therefore, we showed that, for every $\nu \in U$, $\nu(F) < \mu(F^r) + r$. By applying an analogous method, we obtain $\mu(F) < \nu(F^r) + r$ from where $L(\mu, \nu) \leq r$ and so $\nu \in T(\mu, r)$. Hence we have $U \subseteq T(\mu, r)$.

Therefore, by (5), for every $r > 0$, there exists n_0 such that, for any $n \geq n_0$, $L(\mu_n, \mu) \leq r$, which means that $L(\mu_n, \mu) \to 0$.

• Now, let us suppose that $L(\mu_n, \mu) \to 0$. We show that, for every $F = \bar{F}$, $\limsup_n \mu_n(F) \leq \mu(F)$ and that $\mu_n(T) \to \mu(T)$ which will ensure, according to (iii) of Theorem 2.53, that $\mu_n \xrightarrow{\mathcal{T}} \mu$.

Let $F = \bar{F} \subseteq T$; since $\lim_{r \downarrow 0} \mu(F^r) = \mu(F)$, there exist $\varepsilon > 0$ and $r > 0$ such that

$$r < \frac{\varepsilon}{2} \quad \text{and} \quad \mu(F^r) < \mu(F) + \frac{\varepsilon}{2}. \tag{7}$$

Let $n_0 \in \mathbb{N}$ such that,

$$L(\mu_n, \mu) < r, \quad \text{for every} \quad n \geq n_0. \tag{8}$$

By (8), for every $n \geq n_0$, there exists $\delta_n < r$ such that

$$\begin{cases} \mu_n(F) < \mu(F^{\delta_n}) + \delta_n, \\ \mu(F) < \mu_n(F^{\delta_n}) + \delta_n. \end{cases} \tag{9}$$

By (7) and (9), we obtain: $\mu_n(F) < \mu(F^{\delta_n}) + \delta_n \leq \mu(F^r) + r < \mu(F) + \frac{\varepsilon}{2} + \frac{\varepsilon}{2} = \mu(F) + \varepsilon$, for every $n \geq n_0$, from where

$$\limsup_n \mu_n(F) \leq \mu(F). \tag{10}$$

For every $\varepsilon > 0$, there exists $n_0 \in \mathbb{N}$ such that, for every $n \geq n_0$, $L(\mu_n, \mu) < \varepsilon$. Hence, for every $n \geq n_0$, there exists $\delta_n < \varepsilon$ such that $\mu_n(T) < \mu(T^{\delta_n}) + \delta_n = \mu(T) + \delta_n$ and $\mu(T) < \mu_n(T^{\delta_n}) + \delta_n = \mu_n(T) + \delta_n$, from where

$$|\mu_n(T) - \mu(T)| < \delta_n < \varepsilon, \quad \text{for every} \quad n \geq n_0. \tag{11}$$

(10) and (11) lead to $\mu_n \xrightarrow{\mathcal{T}} \mu$.

Therefore, we have shown the equality of topologies: $\mathcal{T}_L = \mathcal{T}$.

In the rest of the proof, we will show directly that $(Rca^+(\mathcal{B}_T), L)$ is a Polish space. Since $\mathcal{T}_L = \mathcal{T}$, the space $(Rca^+(\mathcal{B}_T), L)$ is separable.

• We only have to show that $(Rca^+(\mathcal{B}_T), L)$ is complete.

Let $(\mu_n)_{n\in\mathbb{N}} \subseteq (Rca^+(\mathcal{B}_T), L)$ be a Cauchy sequence; then $(\mu_n)_n$ is tight.

Indeed, since $(\mu_n)_n$ is Cauchy, for every $\varepsilon > 0$ and every $a > 0$, there exists $n_0 \in \mathbb{N}$ such that $L(\mu_n, \mu_{n_0}) < \frac{\varepsilon}{2}$, for every $n \geq n_0$. Therefore, for every $n \geq n_0$, there exists $\delta_n < \frac{\varepsilon}{2}, \delta_n < \frac{a}{2}$ such that

$$\mu_n(F) < \mu_{n_0}(F^{\delta_n}) + \delta_n, \quad \text{for every} \quad F = \bar{F} \subseteq T. \tag{12}$$

Let $K \in \mathcal{K}_T$ such that

$$\mu_n(T \setminus K) < \frac{\varepsilon}{2}, \quad \text{for every} \quad n = 1, \dots, n_0. \tag{13}$$

According to (12), for every $n > n_0, \mu_n(T) < \mu_{n_0}(T^{\delta_n}) + \delta_n = \mu_{n_0}(T) + \delta_n$ and $\mu_{n_0}(T) < \mu_n(T) + \delta_n$, from where

$$\sup_{n\geq n_0} |\mu_n(T) - \mu_{n_0}(T)| \leq \frac{\varepsilon}{2} \tag{14}$$

and then $\sup_{n\in\mathbb{N}} |\mu_n|(T) = \sup_n \mu_n(T) < +\infty$.

By (12) and (13), we obtain:

$$\mu_n(T \setminus K^{\frac{a}{2}}) < \mu_{n_0}\left((T \setminus K^{\frac{a}{2}})^{\delta_n}\right) + \delta_n \leq \mu_{n_0}(T \setminus K) + \delta_n < \varepsilon, \ \forall n \geq n_0. \tag{15}$$

As K is compact, there exists a finite set $F \subseteq T$ such that $K^{\frac{a}{2}} \subseteq F^a$ and so, by (15), $\mu_n(T \setminus F^a) < \varepsilon$, for every $n \geq n_0$. For any $n < n_0$, $\mu_n(T \setminus F^a) \leq \mu_n(T \setminus K) < \varepsilon$ and then, by (13) and (15), for every $\varepsilon > 0$ and every $a > 0$, there exists a finite set $F \subseteq T$ such that

$$\mu_n(T \setminus F^a) < \varepsilon, \quad \text{for every} \quad n \in \mathbb{N}. \tag{16}$$

Let $\varepsilon > 0$; by (16), to every $p \in \mathbb{N}$, we associate an $a = \frac{1}{2^p}$ and a finite set $F_p \subseteq T$ such that

$$\mu_n\left(T \setminus F_p^{\frac{1}{2^n}}\right) < \frac{\varepsilon}{2^p}, \quad \text{for every} \quad n \in \mathbb{N}. \tag{17}$$

Therefore, the set $K_\varepsilon = \bigcap_{p=1}^{\infty} \overline{F_p^{\frac{1}{2^p}}}$ is a closed and totally bounded set in the complete metric space (T, d), hence a compact one: $K_\varepsilon \in \mathcal{K}_T$. By (17) it follows that, for every $n \in \mathbb{N}$,

$$\mu_n(T \setminus K_\varepsilon) = \mu_n\left(\bigcup_{p=1}^{\infty}\left(T \setminus \overline{F_p^{\frac{1}{2^p}}}\right)\right) \le \sum_{p=1}^{\infty} \mu_n\left(T \setminus F_p^{\frac{1}{2^p}}\right) < \varepsilon.$$

Therefore, the sequence $(\mu_n)_n$ is tight; according to Corollary 2.71, there exists $k_n \uparrow +\infty$ such that $(\mu_{k_n})_n$ to be narrowly convergent to a measure $\mu \in Rca^+(\mathcal{B}_T)$.

Since $(\mu_n)_n$ is L–Cauchy, $\mu_n \xrightarrow{\mathcal{T}} \mu$. □

Remark 2.84.

(i) Let $\Psi : T \to Rca^+(\mathcal{B}_T)$, $\Psi(t) = \delta_t$ (see Proposition 2.38). It is obvious that, for all $s, t \in T$, $L(\Psi(t), \Psi(s)) = d(t,s)$. Therefore, Ψ is an isometry of T on its image $\Psi(T)$.

(ii) The completness of (T,d) was used only to show that $(Rca^+(\mathcal{B}_T), L)$ is complete. Therefore, if (T,d) is a second-countable metric space, L remains a compatible metric for $(Rca^+(\mathcal{B}_T), \mathcal{T})$.

We have noticed in Example 2.76, that a narrowly convergent sequence of measures on a Suslin space is not necessarily tight.

As we will show in the following theorem, in the case of a metrizable space, every narrowly convergent sequence is tight.

Theorem 2.85 (Le Cam, see [61], 11.5.3). *Let (T,d) be a metric space, let $(\mu_n)_{n\in\mathbb{N}} \subseteq Rca^+(\mathcal{B}_T)$ and let $\mu \in Rca^+(\mathcal{B}_T)$ such that $\mu_n \xrightarrow{\mathcal{T}} \mu$; then $(\mu_n)_n$ is tight.*

Proof. Every μ_n is tight; so it is concentrated on a σ - compact set S_n which is separable (S_n is a countable union of separable sets). Then $S = \overline{\cup_{n=0}^{\infty} S_n}$ is also separable. Since every bounded continuous function on S can be extended to a bounded continuous function on T, $(\mu_n \restriction_S)_{n\in\mathbb{N}}$ remains narrowly convergent to $\mu\restriction_S$.

Therefore, we can suppose that T is second-countable itself.

According to (ii) of the previous remark, $L(\mu_n, \mu) \to 0$ and $\mu_n(T) \to \mu(T)$.

Since $\mu \in Rca^+(\mathcal{B}_T)$, for each $\varepsilon > 0$, there exists $K \in \mathcal{K}_T$ such that $\mu(T \setminus K) < \frac{\varepsilon}{4}$ or

$$\mu(K) > \mu(T) - \frac{\varepsilon}{4}. \tag{1}$$

For every $n \in \mathbb{N}$,

$$\begin{aligned}&\{\delta > 0 : \mu(F) \le \mu_n(F^\delta) + \delta, \mu_n(F) \le \mu(F^\delta) + \delta, \text{ for every } F = \overline{F} \subseteq T\}\\ &\quad\subseteq \{\delta > 0 : \mu(F) \le \mu_n(F^\delta) + \delta, \text{ for every } F = \overline{F} \subseteq T\}\\ &\quad\subseteq \{\delta > 0 : \mu(K) \le \mu_n(K^\delta) + \delta\}.\end{aligned}$$

Then

$$l_n = \inf\{\delta > 0 : \mu(K) \le \mu_n(K^\delta) + \delta\} \le L(\mu_n, \mu), \quad \text{for all} \quad n \in \mathbb{N}. \tag{2}$$

Since $L(\mu_n, \mu) \to 0$, there exist $n_0 \in \mathbb{N}$ and a sequence $(\delta_n)_{n\in\mathbb{N}} \subseteq]0, +\infty[$ such that

$$\mu(K) \le \mu_n(K^{\delta_n}) + \delta_n, \delta_n < \frac{\varepsilon}{4} \quad \text{and} \quad \delta_n < l_n + \frac{1}{n}, \quad \text{for every} \quad n \ge n_0. \tag{3}$$

For every $n \in \mathbb{N}, n \ge n_0$, as $\mu_n \in Rca^+(\mathcal{B}_T)$, there exists a compact $K_n \subseteq K^{\delta_n}$ such that

$$\mu_n(K_n) > \mu_n(K^{\delta_n}) - \frac{\varepsilon}{4}. \tag{4}$$

Let $n_1 \in \mathbb{N}$ such that, for every $n \ge n_1$,

$$\mu(T) > \mu_n(T) - \frac{\varepsilon}{4}. \tag{5}$$

According to (4), (3), (1) and (5)

$$\mu_n(T \setminus K_n) < \varepsilon, \quad \text{for every} \quad n \ge \max\{n_0, n_1\} = n_2. \tag{6}$$

Since $\{\mu_0, \dots, \mu_{n_2-1}\}$ is tight, we can suppose that (6) is satisfied for all $n \in \mathbb{N}$.

Let $C = K \cup \bigcup_{n=0}^{\infty} K_n$; then $\mu_n(T \setminus C) < \varepsilon$, for every $n \in \mathbb{N}$.

We only have to show that C is compact.

Let $(x_p)_{p\in\mathbb{N}} \subseteq C$. If $(x_p)_{p\in\mathbb{N}}$ has a subsequence with all the terms in the same compact K or K_n, then it also has a subsequence convergent in C.

If no, let $(p_k)_{k\in\mathbb{N}}$ and $(n_k)_{k\in\mathbb{N}}$ two sequence in $\mathbb{N}$ such that $x_{p_k} \in K_{n_k}$, for every $k \in \mathbb{N}$. Since $K_{n_k} \subseteq K^{\delta_{n_k}}$, for every $k \in \mathbb{N}$, there exists $y_k \in K$ with $d(x_{p_k}, y_k) < \delta_{n_k}$. Then $(y_k)_k$ has a subsequence, still noted $(y_k)_k$, convergent to $y \in K$. Since $\delta_{n_k} \to 0$ (see (3)), we have $x_{p_k} \to y \in C$. □

Remark 2.86. Generally speaking, a metrizable space is not a Prohorov space (see Remark 2.75) but, according to the previous theorem, it is sequentially Prohorov.

2.6 Wiener Measure

In this section, we will present an important application of compactness theorem of Prohorov in the construction of the Wiener measure on $C[0,1]$.

The presentation of Wiener measure represents a processing of Paragraphs 8 and 9 of Chapter 2 of Billingsley's monograph [28].

Let $\mathbf{C} = C[0,1]$ be the space of all real-valued continuous functions on $[0,1]$ equipped with norm of uniform convergence, noted $\|.\|$.

If $\tau_{\|\cdot\|}$ is the norm topology, then $(\mathbf{C}, \tau_{\|\cdot\|})$ is a Polish space.

Before establishing the existence of Wiener measure, we begin this section by studying the narrow convergence of measures on $\mathbf{C}$ in connection to the narrow convergence of the image of this measures on the finite dimensional subspaces.

In the following, $\mathcal{B}$ is the σ-algebra of all Borel sets of $\mathbf{C}$ and, for every $n \in \mathbb{N}^*$, $\mathcal{B}_n$ indicates the σ-algebra of Borel sets of $\mathbb{R}^n$.

Definition 2.87. For every $t_1, t_2, \dots, t_n \in [0,1]$ with $t_1 < t_2 \cdots < t_n$, let $\pi_{t_1,\dots,t_n} : \mathbf{C} \to \mathbb{R}^n$ be *the natural projection* from $\mathbf{C}$ to $\mathbb{R}^n$ defined by:

$$\pi_{t_1,\dots,t_n}(x) = (x(t_1), \dots, x(t_n)), \quad \text{for every} \quad x \in \mathbf{C}.$$

A set $C \subseteq \mathbf{C}$ is said to be a *cylinder set* if there are $t_1, t_2, \dots, t_n \in [0,1]$ with $t_1 < t_2 < \cdots < t_n$ and $H \in \mathcal{B}_n$, such that $C = \pi^{-1}_{t_1,\dots,t_n}(H)$. By $\mathcal{C}$ we denote the family of all cylinder sets:

$$\mathcal{C} = \{\pi^{-1}_{t_1,\dots,t_n}(H) : n \in \mathbb{N}^*, 0 \le t_1 < \cdots < t_n \le 1, H \in \mathcal{B}_n\}.$$

Proposition 2.88. *$\mathcal{C}$ is an algebra that generates $\mathcal{B}$. $\mathcal{C}$ is said the algebra of cylinder sets.*

Proof. Since every projection $\pi_{t_1,\dots,t_n}$ is continuous, $\pi^{-1}_{t_1,\dots,t_n}(H) \in \mathcal{B}$, for every $H \in \mathcal{B}$ therefore $\mathcal{C} \subseteq \mathcal{B}$

• For every $I, J \in \mathcal{C}$, there exist $n, m \in \mathbb{N}$, $0 \le t_1 < \cdots < t_n \le 1$, $0 \le s_1 < \cdots < s_m \le 1$ and $B \in \mathcal{B}_n$, $C \in \mathcal{B}_m$ such that $I = \pi^{-1}_{t_1,\dots,t_n}(B)$ and $J = \pi^{-1}_{s_1,\dots,s_m}(C)$. Let $\{r_1, \dots, r_p\} = \{t_1, \dots, t_n\} \cup \{s_1, \dots, s_m\}$ where $\max\{n,m\} \le p \le n+m$ and $0 \le r_1 < r_2 < \cdots < r_p \le 1$. We suppose that , for every $k = 1, \dots, n$, $r_{i_k} = t_k$ and let $\varphi_{n,p} : \mathbb{R}^n \to \mathbb{R}^p$ be the embedding defined by:

$$\varphi_{n,p}(u_1, \dots, u_n) = (v_1, \dots, v_p), \text{ where } v_j = \begin{cases} u_k, & j = i_k, k = 1, \dots, n, \\ 0, & j \notin \{i_1, \dots, i_n\}. \end{cases}$$

According to (i) of Proposition 2.57, $\mathcal{B}_n = \varphi^{-1}_{n,p}(\mathcal{B}_p)$, from where $\varphi_{n,p}(B) \in \mathcal{B}_p$ and $I = \pi^{-1}_{r_1,\dots,r_p}(\varphi_{n,p}(B))$.

Similarly, we define the embedding $\varphi_{m,p} : \mathbb{R}^m \to \mathbb{R}^p$ so that $\varphi_{m,p}(C) \in \mathcal{B}_p$ and $J = \pi^{-1}_{r_1,\dots,r_p}(\varphi_{m,p}(C))$.

$I \cup J = \pi^{-1}_{r_1,\dots,r_p}(\varphi_{n,p}(B) \cup \varphi_{m,p}(C))$ and $I \setminus J = \pi^{-1}_{r_1,\dots,r_p}(\varphi_{n,p}(B) \setminus \varphi_{m,p}(C))$.

Therefore $I \cup J, I \setminus J \in \mathcal{C}$.

Since $\mathbf{C} = \pi_t^{-1}(\mathbb{R})$, for every $t \in [0,1]$, $\mathbf{C} \in \mathcal{C}$, and therefore $\mathcal{C}$ is an algebra on $\mathbf{C}$.

• Let $\mathcal{A}$ be the σ-algebra generated by $\mathcal{C}$; then $\mathcal{A} \subseteq \mathcal{B}$. For every $x \in \mathbf{C}$ and every $\varepsilon > 0$, let $T(x,\varepsilon) = \{y \in \mathbf{C} : \|x - y\| \le \varepsilon\}$ be the closed ball of radius ε centered at x and let

$$H_n = \prod_{i=1}^{n} \left[x\left(\frac{i}{n}\right) - \varepsilon, x\left(\frac{i}{n}\right) + \varepsilon \right].$$

H_n is a closed subset of $\mathbb{R}^n$; therefore, for every $n \in \mathbb{N}$, $H_n \in \mathcal{B}_n$. Let us show that

$$T(x,\varepsilon) = \bigcap_{n=1}^{\infty} \pi^{-1}_{\frac{1}{n},\frac{2}{n},\dots,\frac{n}{n}}(H_n). \tag{*}$$

If $y \in \bigcap_{n=1}^{\infty} \pi^{-1}_{\frac{1}{n},\frac{2}{n},\dots,\frac{n}{n}}(H_n)$, then, for every $n \in \mathbb{N}^*$, $(y(\frac{1}{n}),\dots,y(\frac{n}{n})) \in H_n$, form where, for every $n \in \mathbb{N}^*$ and every $i = 1,\dots,n$, $|y(\frac{i}{n}) - x(\frac{i}{n})| \le \varepsilon$. Since x and y are uniformly continuous on $[0,1]$, for every $\eta > 0$, there exists $\delta > 0$ such that $|t - s| < \delta$ implies $|x(t) - x(s)| < \eta$ and $|y(t) - y(s)| < \eta$. Let now $n \in \mathbb{N}^*$ with $\frac{1}{n} < \delta$; for every $t \in [0,1]$, there exists $i \in \{1,\dots,n\}$ such that $\frac{i-1}{n} \le t < \frac{i}{n}$. So we have $|x(t) - y(t)| \le |x(t) - x(\frac{i}{n})| + |x(\frac{i}{n}) - y(\frac{i}{n})| + |y(\frac{i}{n}) - y(t)| < 2\eta + \varepsilon$ from where $\|x - y\| = \sup_{t\in[0,1]} |x(t) - y(t)| \le 2\eta + \varepsilon$, for any $\eta > 0$. Therefore $\|x - y\| \le \varepsilon$ and $y \in T(x,\varepsilon)$.

Conversely, if $y \in T(x,\varepsilon)$, then, for every $n \in \mathbb{N}^*$ and every $i = 1,\dots,n$, $|x(\frac{i}{n}) - y(\frac{i}{n})| \le \|x - y\| \le \varepsilon$ and so, $y \in \pi^{-1}_{\frac{1}{n},\frac{2}{n},\dots,\frac{n}{n}}(H_n)$, for every $n \in \mathbb{N}^*$.

By (*), for every $x \in \mathbf{C}$ and every $\varepsilon > 0$, $T(x,\varepsilon) \in \mathcal{A}$. Since $\mathbf{C}$ is separable, every open set of $\mathbf{C}$ is a countable union of closed balls and, therefore, every open set belongs to $\mathcal{A}$ and, consequently, $\mathcal{A} = \mathcal{B}$. □

Remark 2.89. According to the previous proposition, if the measures μ and $\nu \in Rca^+(\mathcal{B})$ coincide on $\mathcal{C}$, then $\mu = \nu$.

We will now characterize the narrow convergence on $Rca^+(\mathcal{B})$ with the help of the narrow convergence of the image measures on the finite dimensional subspaces.

Let us remember that, according to Theorem 2.58, $(Rca^+(\mathcal{B}), \mathcal{T})$ is a Polish space.

For every $\mu \in Rca^+(\mathcal{B})$ and every $t_1, \ldots, t_p \in [0,1]$, let $\pi_{t_1,\ldots,t_p}(\mu) = \mu \circ \pi^{-1}_{t_1,\ldots,t_p} : \mathcal{B}_p \to \mathbb{R}_+$, defined by: $(\mu \circ \pi^{-1}_{t_1,\ldots,t_p})(A) = \mu(\pi^{-1}_{t_1,\ldots,t_p}(A))$, for every $A \in \mathcal{B}_p$. It is obvious that $\pi_{t_1,\ldots,t_p}(\mu) \in Rca^+(\mathcal{B}_p)$. Let us denote by $\mathcal{T}_p$ the narrow topology on $Rca^+(\mathcal{B}_p)$.

Theorem 2.90. *For every $(\mu_n)_{n\in\mathbb{N}} \subseteq Rca^+(\mathcal{B})$ and every $\mu \in Rca^+(\mathcal{B})$;*

$$\mu_n \xrightarrow{\mathcal{T}} \mu \Longleftrightarrow \begin{cases} 1. \quad \pi_{t_1,\ldots,t_p}(\mu_n) \xrightarrow{\mathcal{T}_p} \pi_{t_1,\ldots,t_p}(\mu), \\ \qquad \textit{for every} \quad p \in \mathbb{N}^* \textit{ and every } 0 \le t_1 < \cdots < t_p \le 1, \\ 2. \quad (\mu_n)_{n\in\mathbb{N}} \quad \textit{is tight} \quad . \end{cases}$$

Proof. ($\Longrightarrow$) Let $\mu_n \xrightarrow{\mathcal{T}} \mu$; for every $p \in \mathbb{N}^*$, every $0 \le t_1 < \cdots < t_p \le 1$ and every $n \in \mathbb{N}$, let $\nu_n = \pi_{t_1,\ldots,t_p}(\mu_n)$ and let $\nu = \pi_{t_1,\ldots,t_p}(\mu)$. According to (iv) of Theorem 2.53,

$$\nu_n \xrightarrow{\mathcal{T}_p} \nu \Leftrightarrow \lim_n \nu_n(A) = \nu(A), \text{ for every } A \in \mathcal{B}_p \text{ such that } \nu(\bar{A} \setminus A^\circ) = 0.$$

Let $A \in \mathcal{B}_p$ with $\nu(\bar{A} \setminus A^\circ) = 0$ and let $B = \pi^{-1}_{t_1,\ldots,t_p}(A) \in \mathcal{B}$.

Since $\overline{\pi^{-1}_{t_1,\ldots,t_p}(A)} \setminus (\pi^{-1}_{t_1,\ldots,t_p}(A))^\circ = \bar{B} \setminus B^\circ \subseteq \pi^{-1}_{t_1,\ldots,t_p}(\bar{A} \setminus A^\circ)$, we obtain $\mu(\bar{B} \setminus B^\circ) = 0$.

As $\mu_n \xrightarrow{\mathcal{T}} \mu$, $\lim_n \mu_n(B) = \mu(B)$, from where, for every $A \in \mathcal{B}_p$ such that $\nu(\bar{A} \setminus A^\circ) = 0$, $\lim_n \nu_n(A) = \nu(A)$ and so $(\nu_n)_{n\in\mathbb{N}}$ is $\mathcal{T}_p$-convergent to ν. Which demonstrates 1.

Since $(Rca^+(\mathcal{B}), \mathcal{T})$ is Polish space and $\mu_n \xrightarrow{\mathcal{T}} \mu$, $(\mu_n)_n$ is relatively $\mathcal{T}$–compact; according to Theorem 2.73, $(\mu_n)_n$ is tight.

($\Longleftarrow$) According to Corollary 2.72, as $(\mu_n)_n$ is tight and $(Rca^+(\mathcal{B}), \mathcal{T})$ is a Polish space, $(\mu_n)_n$ is relatively $\mathcal{T}$-compact. Therefore, every subsequence $(\mu'_n)_n$ of $(\mu_n)_n$ has a subsequence $(\mu''_n)_n$ $\mathcal{T}$–convergent to $\lambda \in Rca^+(\mathcal{B})$.

For every $C \in \mathcal{C} \subseteq \mathcal{B}$, there exist $p \in \mathbb{N}^*, 0 \le t_1 < \cdots < t_p \le 1$ and $A \in \mathcal{B}_p$ such that $C = \pi^{-1}_{t_1,\ldots,t_p}(A)$.

Since $\pi_{t_1,\ldots,t_p}(\mu''_n) \xrightarrow{\mathcal{T}_p} \pi_{t_1,\ldots,t_p}(\mu)$ and $\pi_{t_1,\ldots,t_p}(\mu''_n) \xrightarrow{\mathcal{T}_p} \pi_{t_1,\ldots,t_p}(\lambda)$, $\pi_{t_1,\ldots,t_p}(\mu) = \pi_{t_1,\ldots,t_p}(\lambda)$ ($(Rca^+(\mathcal{B}_p), \mathcal{T}_p)$ is a Hausdorff space). Therefore, for every $C \in \mathcal{C}$, $\mu(C) = \lambda(C)$. Then, according to Remark 2.89, $\mu = \lambda$. Therefore, every subsequence of $(\mu_n)_n$ admits a subsequence $\mathcal{T}$-convergent to μ; what assures that $\mu_n \xrightarrow{\mathcal{T}} \mu$. □

Remark 2.91. In Theorem 2.90, the quality of being tight is essential and cannot be removed, as it is demonstrated in the following example:

Let $x_n : [0,1] \to \mathbb{R}$, defined by $x_n(t) = \begin{cases} n \cdot t, & 0 \le t \le \frac{1}{n}, \\ 2 - n \cdot t, & \frac{1}{n} < t \le \frac{2}{n}, \\ 0, & \frac{2}{n} < t \le 1 \end{cases}$

The sequence $(x_n)_n$ is pointwise convergent to $\underline{0}$ but it is not uniformly convergent therefore, $\|x_n\| \overset{\|\cdot\|}{\nrightarrow} \underline{0}$; according to Proposition 2.38, $\delta_{x_n} \overset{\tau}{\nrightarrow} \delta_{\underline{0}}$.

However, for every $p \in \mathbb{N}^*$, every $0 \le t_1 < \cdots < t_p \le 1$ and every $A \in \mathcal{B}_p$ such that $\pi_{t_1,\dots,t_p}(\delta_{\underline{0}})(\bar{A} \setminus A^\circ) = 0$, $\underline{0} \notin \pi^{-1}_{t_1,\dots,t_p}(\partial A)$. We have then either $0_p = (0,\dots,0) \in A^\circ$ or $0_p \notin \bar{A}$.

If $0_p \in A^\circ$ and if p_0 is chosen as $\frac{2}{p_0} < \min\{t_i : i = 1,\dots,n, t_i > 0\}$, then $\pi_{t_1,\dots,t_p}(\delta_{x_n})(A) = 1 = \delta_{\underline{0}}\pi_{t_1,\dots,t_p}(\delta_{\underline{0}})(A)$, for every $p \ge p_0$.

If $0_p \notin \bar{A}$, then $\pi_{t_1,\dots,t_p}(\delta_{x_n})(A) = 0 = \delta_{\underline{0}}\pi_{t_1,\dots,t_p}(\delta_{\underline{0}})(A)$.

Therefore, we have $\pi_{t_1,\dots,t_p}(\delta_{x_n}) \overset{\tau_p}{\longrightarrow} \pi_{t_1,\dots,t_p}(\delta_{\underline{0}})$, for every $p \in \mathbb{N}^*$ and every $0 \le t_1 < \cdots < t_p \le 1$.

It is obvious that the sequence $(\delta_{x_n})_{n\in\mathbb{N}}$ is not tight.

Now, we will give some characterizations of tightness for a sequence of positive Radon measures on $\mathbf{C}$.

Definition 2.92. For every $x \in \mathbf{C}$, let $w_x :]0,1] \to \mathbb{R}_+$, defined by: $w_x(\delta) = \sup_{|s-t|<\delta} |x(s) - x(t)|$, for every $\delta \in]0,1]$.

The mapping w_x is said *the modulus of uniform continuity* of x.

Remark 2.93.

(i) For every $x \in \mathbf{C}$, $\lim_{\delta\to 0} w_x(\delta) = 0$.

(ii) For every $x, y \in \mathbf{C}$ and every $\delta \in]0,1]$, $|w_x(\delta) - w_y(\delta)| \le 2\|x - y\|$.

Indeed, by definition of $w_x(\delta)$, for every $\varepsilon > 0$, there exist $s, t \in [0,1]$ with $|s - t| < \delta$ such that $w_x(\delta) < |x(s) - x(t)| + \varepsilon$. Then $w_x(\delta) - w_y(\delta) < |x(s) - x(t)| + \varepsilon - |y(s) - y(t)| \le |x(s) - x(t) - y(s) + y(t)| + \varepsilon \le 2\|x - y\| + \varepsilon$. Therefore we have $w_x(\delta) - w_y(\delta) \le 2\|x - y\|$. Analogously, we show that $w_y(\delta) - w_x(\delta) \le 2\|x - y\|$, from where $|w_x(\delta) - w_y(\delta)| \le 2\|x - y\|$.

Proposition 2.94. *Let $\mu \in Rca^+(\mathcal{B})$; then*

(i) *for every $\varepsilon > 0$, there exists $K \in \mathcal{K}_{\mathbf{C}}$ with $\mu(\mathbf{C} \setminus K) < \varepsilon$, such that* $\lim_{\delta \to 0} w_x(\delta) = 0$, *uniformly on K* ($\lim_{\delta \to 0} w_x(\delta) = 0$, *$\mu$-almost uniformly on $\mathbf{C}$).*

(ii) *there exists $N \subseteq \mathbf{C}$ such that $\mu^*(N) = 0$ and* $\lim_{\delta \to 0} w_x(\delta) = 0$, *for every $x \in \mathbf{C} \setminus N$ (*$\lim_{\delta \to 0} w_x(\delta) = 0$, *$\mu$-a.e. on $\mathbf{C}$).*

(iii) *for every $\varepsilon > 0$ and every $\eta > 0$, there exists $\delta \in]0, 1]$ such that $\mu(\{x \in \mathbf{C} : w_x(\delta) \geq \varepsilon\}) < \eta$* ($\lim_{\delta \to 0} w_x(\delta) = 0$ *in μ-measure).*

Proof. (i) Since $\mu \in Rca^+(\mathcal{B})$, for every $\varepsilon > 0$, there exists $K \in \mathcal{K}_{\mathbf{C}}$ such that $\mu(\mathbf{C} \setminus K) < \varepsilon$. By the Arzelà–Ascoli theorem, as K is compact in $\mathbf{C}$, K is uniformly equicontinuous; therefore there exists $\delta_0 \in]0, 1]$ such that, for every $s, t \in [0, 1]$ with $|s - t| < \delta_0$ and for every $x \in K$, $|x(s) - x(t)| < \varepsilon$, which means that $\lim_{\delta \to 0} w_x(\delta) = 0$, uniformly on K.

(ii) and (iii) are immediate consequences of (i). According to Remark 2.93 (ii), we can see that $w_{.}(\delta)$ is uniformly continuous, for every $\delta > 0$, hence $\{x \in \mathbf{C} : w_x(\delta) \geq \varepsilon\}$ is a closed subset of $\mathbf{C}$, and, therefore, a Borel subset of $\mathbf{C}$. □

Theorem 2.95. *The sequence $(\mu_n) \subseteq Rca^+(\mathcal{B})$ is tight if and only if*

(i) *$(\pi_0(\mu_n))_{n \in \mathbb{N}}$ is tight on $\mathbb{R}$ and*

(ii) *$\forall \varepsilon, \eta > 0, \exists \delta \in]0, 1]$ such that $\mu_n(\{x \in \mathbf{C} : w_x(\delta) \geq \varepsilon\}) < \eta, \forall n \in \mathbb{N}$.*

Proof. (i) Firstly, let us show that the sequence $(\pi_0(\mu_n))_{n \in \mathbb{N}}$ is tight on $\mathbb{R}$ if and only if:

(1) $\sup_n \mu_n(\mathbf{C}) < +\infty$ and

(2) $\forall \varepsilon > 0, \exists a > 0$ such that $\mu_n(\{x \in \mathbf{C} : |x(0)| > a\}) < \varepsilon, \forall n \in \mathbb{N}$.

According to Definition 2.61, $(\pi_0(\mu_n))_{n \in \mathbb{N}}$ is tight on $\mathbb{R}$ if and only if:

(a) $\sup_n (\pi_0(\mu_n))(\mathbb{R}) < +\infty$ and,

(b) $\forall \varepsilon > 0, \exists K \in \mathcal{K}_{\mathbb{R}}$ such that $(\pi_0(\mu_n))(\mathbb{R} \setminus K) < \varepsilon, \forall n \in \mathbb{N}$.

(1) $\Longleftrightarrow$ (a) since $(\pi_0(\mu_n))(\mathbb{R}) = \mu_n(\{x \in \mathbf{C} : x(0) \in \mathbb{R}\}) = \mu_n(\mathbf{C})$ and, therefore, $(\pi_0(\mu_n))_{n \in \mathbb{N}}$ is bounded if and only if $(\mu_n)_{n \in \mathbb{N}}$ is bounded.

(2) $\Longleftrightarrow$ (b) Indeed, it is clear that (b) $\Longleftrightarrow$ for every $\varepsilon > 0$, there exists $a > 0$ such that, for every $n \in \mathbb{N}, (\pi_0(\mu_n))(\mathbb{R} \setminus [-a, a]) = \mu_n(\{x \in \mathbf{C} : |x(0)| > a\}) < \varepsilon$.

(ii) According to (ii) of Remark 2.93, for every $\delta \in (0, 1]$, $w_{.}(\delta)$ is an uniformly continuous function, hence $\{x \in \mathbf{C} : w_x(\delta) \geq \varepsilon\}$ is a closed subset of $\mathbf{C}$ and so a Borel subset. Therefore, we can consider the measure $\mu_n(\{x \in \mathbf{C} : w_x(\delta) \geq \varepsilon\})$.

($\Longrightarrow$) Let us suppose that the sequence $(\mu_n)_n$ is tight. Then $\sup_n \mu_n(\mathbf{C}) < +\infty$ and (1) is proved. By (b), for every $\eta > 0$, there exists $K \in \mathcal{K}_{\mathbf{C}}$ such that, for all $n \in \mathbb{N}$, $\mu_n(\mathbf{C} \setminus K) < \eta$. By Arzelà–Ascoli theorem, K is bounded and uniformly equicontinuous.

Since K is bounded, there exists $a > 0$ such that, for every $x \in K$, $\|x\|_0 \leq a$ from where $\{x \in \mathbf{C} : |x(0)| > a\} \subseteq \mathbf{C} \setminus K$ and so, for every $n \in \mathbb{N}$, $\mu_n(\{x \in \mathbf{C} : |x(0)| > a\}) < \eta$. (1) and (2) being satisfied $(\pi_0(\mu_n))_{n\in\mathbb{N}}$ is tight on $\mathbb{R}$.

Since K is uniformly equicontinuous, there exists $\delta \in]0,1]$ such that, for every $s, t \in [0,1]$ with $|s-t| < \delta$ and for every $x \in K$, $|x(t) - x(s)| < \varepsilon$, i.e., for every $x \in K$, $w_x(\delta) < \varepsilon$. Therefore, we have $\{x \in \mathbf{C} : w_x(\delta) \geq \varepsilon\} \subseteq \mathbf{C} \setminus K$ and, for every $n \in \mathbb{N}$, $\mu_n(\{x \in \mathbf{C} : w_x(\delta) \geq \varepsilon\}) < \eta$. (ii) is then satisfied.

($\Longleftarrow$) Let us suppose that $(\mu_n)_{n\in\mathbb{N}}$ satisfies the conditions (i) and (ii). Since $(\pi_0(\mu_n))_{n\in\mathbb{N}}$ is tight, (1) is satisfied: $\sup_n \mu_n(\mathbf{C}) < +\infty$. Also we have:

For every $\eta > 0$, there exists $a > 0$ such that

$$\mu_n(\{x \in \mathbf{C} : |x(0)| > a\}) < \frac{1}{2} \cdot \eta, \quad \text{for every} \quad n \in \mathbb{N}. \tag{1}$$

According to the condition (ii), we have: for every $k \in \mathbb{N}^*$, there exists $\delta_k \in]0,1]$ such that

$$\mu_n\left(\left\{x \in \mathbf{C} : w_x(\delta_k) \geq \frac{1}{k}\right\}\right) < \frac{\eta}{2^{k+1}}, \quad \text{for every} \quad n \in \mathbb{N}. \tag{2}$$

Let $A = \{x \in \mathbf{C} : |x(0)| \leq a\}$, $A_k = \{x \in \mathbf{C} : w_x(\delta_k) < \frac{1}{k}\}$ and $B = A \cap (\bigcap_{k=1}^{\infty} A_k)$.

It is clear that $B \in \mathcal{B}$, that $|x(0)| \leq a$, for every $x \in B$, and that, for every $k \in \mathbb{N}^*$, $w_x(\delta_k) < \frac{1}{k}$. Therefore B is uniformly equicontinuous.

Let $0 = t_0 < t_1 < \cdots < t_p = 1$ such that $t_{i+1} - t_i < \delta_1$, for every $i = 0, \ldots, p-1$; then, for all $t \in [0,1]$, there exists an unique i such that $t \in [t_i, t_{i+1}]$. Therefore we have $|x(t)| \leq |x(t) - x(t_i)| + |x(t_i) - x(t_{i-1})| + \cdots + |x(t_1) - x(t_0)| + |x(t_0)| \leq p \cdot 1 + a$. So, for every $x \in B$, $\|x\| \leq p + a$.

Then B is bounded and uniformly equicontinuous hence relatively compact in $\mathbf{C}$. If $K = \bar{B} \in \mathcal{K}_{\mathbf{C}}$, by (1) and (2), we have, for every $n \in \mathbb{N}$,

$$\mu_n(\mathbf{C} \setminus K) \leq \mu_n(\mathbf{C} \setminus B) \leq \mu_n(\mathbf{C} \setminus A) + \sum_{k=1}^{\infty} \mu_n(\mathbf{C} \setminus A_k) < \eta.$$

Therefore the sequence $(\mu_n)_{n\in\mathbb{N}}$ is tight. □

Remark 2.96.

(i) Condition (ii) from the previous theorem is equivalent to:

For every $\varepsilon > 0, \lim_{\delta \to 0} \mu_n(\{x \in \mathbf{C} : w_x(\delta) \geq \varepsilon\}) = 0$, uniformly in $n \in \mathbb{N}$.

(ii) In Theorem 2.95, we can substitute condition (ii) with the weaker condition (ii'):

(ii') For every $\varepsilon > 0$, there exist $\eta > 0$, $\delta_0 \in]0, 1]$ and $n_0 \in \mathbb{N}$ such that,

$$\mu_n (\{x \in \mathbf{C} : w_x(\delta_0) > \varepsilon\}) < \eta, \quad \text{for every} \quad n \geq n_0.$$

In fact, if we suppose (ii'), for every $n = 1, \ldots, n_0 - 1$, by (iii) of Proposition 2.94, there exists $\delta_n \in]0, 1]$ such that

$$\mu_n \left(\left\{ x \in \mathbf{C} : w_x(\delta_n) \geq \frac{\varepsilon}{2} \right\} \right) < \eta.$$

If $\delta = \min\{\delta_0, \delta_1, \ldots, \delta_{n_0 - 1}\} \in (0, 1]$, then, for every $n \geq n_0$, $\mu_n(\{x \in \mathbf{C} : w_x(\delta) > \varepsilon\}) \leq \mu_n(\{x \in \mathbf{C} : w_x(\delta_0) > \varepsilon\}) < \eta$ and, for every $n \leq n_0 - 1$, $\mu_n(\{x \in \mathbf{C} : w_x(\delta) > \varepsilon\}) \leq \mu_n(\{x \in \mathbf{C} : w_x(\delta) \geq \frac{\varepsilon}{2}\}) \leq \mu_n(\{x \in \mathbf{C} : w_x(\delta_n) \geq \frac{\varepsilon}{2}\}) < \eta$ what gives (ii).

Proposition 2.97. *Let $(\mu_n)_n \subseteq Rca^+(\mathcal{B})$ be a sequence satisfying the following two conditions:*

(a) *$(\pi_0(\mu_n))_{n \in \mathbb{N}}$ is tight on $\mathbb{R}$,*

(b) *for every $\varepsilon > 0$ and every $\eta > 0$, there exist $\delta \in]0, 1]$ and $n_0 \in \mathbb{N}$ such that, for every $n \geq n_0$, and every $t \in [0, 1]$,*
$\frac{1}{\delta} \cdot \mu_n(\{x \in \mathbf{C} : \sup_{t \leq s \leq t+\delta} |x(s) - x(t)| \geq \varepsilon\}) \leq \eta.$

Then $(\mu_n)_n$ is tight.

Proof. According to Theorem 2.95 and to (ii) of Remark 2.96, it is sufficient to show that (b)$\Longrightarrow$(ii').

By (b), for every $\varepsilon > 0$, there exist $\eta > 0$, $\delta_0 \in]0, 1]$ and $n_0 \in \mathbb{N}$ such that, for every $n \geq n_0$ and every $t \in [0, 1]$,

$$\mu_n \left(\left\{ x \in \mathbf{C} : \sup_{t \leq s \leq t+\delta_0} |x(s) - x(t)| \geq \frac{\varepsilon}{3} \right\} \right) \leq \frac{\eta . \delta_0}{2}. \qquad (*)$$

For every $t \in [0,1)$, let $A_t = \{x \in \mathbf{C} : \sup_{t \le s \le t+\delta_0} |x(s) - x(t)| \ge \frac{\varepsilon}{3}\}$. Then

$$\{x \in \mathbf{C} : w_x(\delta_0) > \varepsilon\} \subseteq \bigcup_{\substack{i \in \mathbb{N} \\ i \cdot \delta_0 < 1}} A_{i \cdot \delta_0}. \tag{**}$$

Indeed, for all $x \in \mathbf{C}$ with $w_x(\delta_0) > \varepsilon$, there exist $s, t \in [0,1]$ with $|s - t| < \delta_0$ and $|x(s) - x(t)| > \varepsilon$. We can suppose that $t < s \le 1$ and $t < s < t + \delta_0$. Let $p \in \mathbb{N}$ such that $p \cdot \delta_0 < 1 \le (p+1) \cdot \delta_0$.

Since $t \in [0, 1[\subseteq \bigcup_{i=1}^{p+1} [(i-1)\delta_0, i\delta_0[$, there exists $i_0 \in \{1, \ldots, p+1\}$ such that $t \in [(i_0 - 1)\delta_0, i_0\delta_0[$.

Let us show that:

(1) If $i_0 \le p$, then $x \in A_{(i_0-1)\delta_0} \cup A_{i_0\delta_0}$.

(2) If $i_0 = p+1$, then $x \in A_{p\delta_0}$.

Indeed, in the case (1), if we suppose that $x \notin A_{(i_0-1)\delta_0} \cup A_{i_0\delta_0}$, then

$$\sup_{(i_0-1)\delta_0 \le u \le i_0\delta_0} |x((i_0-1)\delta_0) - x(u)| < \frac{\varepsilon}{3} \quad \text{and} \tag{α}$$

$$\sup_{i_0\delta_0 \le u \le (i_0+1)\delta_0} |x(i_0\delta_0) - x(u)| < \frac{\varepsilon}{3}. \tag{β}$$

Since $(i_0 - 1)\delta_0 \le t < s < t + \delta_0 < (i_0 + 1)\delta_0$, by ($\alpha$) and ($\beta$)

$$|x((i_0-1)\delta_0) - x(s)| < \frac{\varepsilon}{3} \quad \text{or} \quad |x(i_0\delta_0) - x(s)| < \frac{\varepsilon}{3}.$$

By $|x((i_0-1)\delta_0) - x(t)| < \frac{\varepsilon}{3}$ and by $|x((i_0-1)\delta_0) - x(i_0\delta_0)| < \frac{\varepsilon}{3}$, we have then $|x(t) - x(s)| < \varepsilon$ which contradicts the choice of s and t.

In the case (2), if we suppose that $x \notin A_{p\delta_0}$, then we have

$$\sup_{p\delta_0 \le u \le 1} |x(p\delta_0) - x(u)| < \frac{\varepsilon}{3}.$$

Since $p\delta_0 \le t < s < 1$,

$$|x(p\delta_0) - x(t)| < \frac{\varepsilon}{3} \quad \text{and} \quad |x(p\delta_0) - x(s)| < \frac{\varepsilon}{3},$$

from where $|x(t) - x(s)| < \frac{2\varepsilon}{3} < \varepsilon$ which is a contradiction.

Then, by (1) and (2), we have $x \in \bigcup_{\substack{i \in \mathbb{N} \\ i \cdot \delta_0 < 1}} A_{i\delta_0}$ which proves the inclusion $(**)$.

From $(**)$ and $(*)$, we obtain:

$$\mu_n(\{x \in \mathbf{C} : w_x(\delta_0) > \varepsilon\}) \leq \sum_{i<\frac{1}{\delta_0}} \mu_n(A_{i\delta_0}) < \sum_{i<\frac{1}{\delta_0}} \frac{\eta.\delta_0}{2} = \left(1 + \left[\frac{1}{\delta_0}\right]\right)\frac{\eta.\delta_0}{2}$$

$$\leq \left(1 + \frac{1}{\delta_0}\right)\frac{\eta.\delta_0}{2} < \frac{(1+\delta_0).\eta}{2} \leq \eta, \text{ for every } n \geq n_0$$

which show (ii′). Therefore, the sequence $(\mu_n)_n$ is tight. □

Remark 2.98. The previous proposition gives only a sufficient condition for a sequence $(\mu_n)_n$ to be tight.

In the rest of this section, we will rewrite the conditions of Proposition 2.97 for a particular sequence of measures connected to the definition of Wiener measure. Before construction of this sequence, we will recall some classical notions and results of probabilities that we will use.

Definition 2.99. Let $(\Omega, \mathcal{A}, P)$ be a probability space.

A *random variable* is a real-valued $(\mathcal{A} - \mathcal{B}_{\mathbb{R}})$-measurable function $\xi : \Omega \to \mathbb{R}$.

If $\xi \in L^1(P)$, then $\int_\Omega \xi dP$ is said to be *the mean* or *expectation value* of ξ; it will be noted $E(\xi)$ or $\overline{\xi}$.

For a random variable, we can also define its k-th moments. Therefore, if $k \in \mathbb{N}^*$ and if there exists $E(\xi^k)$, then $E(\xi^k)$ is said to be the *k-th moment* of the random variable ξ and it will be noted with $m_k(\xi)$.

$E((\xi - \overline{\xi})^k) = m_k(\xi - \overline{\xi})$ is said the *k-th central moment* of ξ. Obviously, $m_1(\xi)$ is the mean $E(\xi)$ of ξ. If $\xi \in L^2(P)$, then there exists $m_2(\xi - \overline{\xi})$; it is called the *variance* of ξ and wil be noted by $V(\xi) = \sigma(\xi)^2$.

$\sigma(\xi) = \sqrt{V(\xi)}$ is said the *standard deviation* of ξ. We can see $\sigma(\xi)$ as a useful measure of spread values of ξ around its mean value $E(\xi)$.

The *distribution function* F of random variable ξ, also called the *cumulative distribution function* or *cumulative frequency function*, describes the probability that ξ takes on a value less than or equal to a number x. Therefore $F : \mathbb{R} \to [0, 1]$ is defined by

$$F(x) = P(\{\omega \in \Omega : \xi(\omega) < x\}) = P(\xi < x).$$

The distribution function is therefore related to a continuous probability density function f by $F(x) = P(\xi < x) = \int_{-\infty}^{x} f(x)dx$; so f (when it exists) is simply the derivative a.e. of the distribution function F ($f = F'$ a.e.).

A random variable ξ is *normal distributed* $\mathcal{N}(m, \sigma^2)$ (with mean m and variance σ^2) if its distribution function F admits as density $f(x) = \frac{1}{\sigma\sqrt{2\pi}} e^{-\frac{(x-m)^2}{2\sigma^2}}$. Then we have,

$$P(\xi^{-1}(A)) = \frac{1}{\sigma\sqrt{2\pi}} \int_A e^{-\frac{(x-m)2}{2\sigma^2}} dx, \quad \text{for every} \quad A \in \mathcal{B}_{\mathbb{R}}.$$

If $m = 0$ and $\sigma^2 = 1$ we say that ξ has a *standard normal distribution* $\mathcal{N}(0, 1)$; its distribution function F is defined by $F(x) = \frac{1}{\sqrt{2\pi}} \int_{-\infty}^{x} e^{-\frac{t^2}{2}} dt$, where dt denote the Lebesgue measure on $\mathbb{R}$.

An arbitrary set of random variables $\{\xi_i : i \in I\}$ is said *independent* if, for every $n \in \mathbb{N}^*$, every $J = \{j_1, \dots, j_n\} \subseteq I$ and every $\{B_1, \dots, B_n\} \subseteq \mathcal{B}_{\mathbb{R}}$,

$$P\left(\bigcap_{i=1}^{n} \xi_{j_i}^{-1}(B_i)\right) = \prod_{i=1}^{n} P\left(\xi_{j_i}^{-1}(B_i)\right).$$

Two random variables ξ and η are *independent* if the set $\{\xi, \eta\}$ is independent. It is clear that the independence of a set of random variables depends to the probability space on which they are defined. Where it will be necessary, we will specify the independence in connection to the probability space taken into consideration or in connection to the probability of this space.

Proposition 2.100.

(i) *If* $\xi = \sum_{i=1}^{n} a_i \chi_{A_i}$ *and* $\eta = \sum_{j=1}^{m} b_j \chi_{B_j}$ *are two simple random variables, then* ξ *and* η *are independent if and only if*

$$P(A_i \cap B_j) = P(A_i) \cdot P(B_j), \quad \textit{for all} \quad i = 1, \dots, n \quad \textit{and} \quad j = 1, \dots, m.$$

(ii) *If* ξ *and* η *are independent, then the couples* $\{|\xi|, \eta\}$, $\{\xi^+, \eta\}$ *and* $\{\xi^-, \eta\}$ *are also independent.*

(iii) ξ *and* η *are independent if and only if, for every* $\alpha, \beta \in \mathbb{R}$,

$$\begin{aligned} &P\left(\xi^{-1}(]-\infty, \alpha]) \cap \eta^{-1}(]-\infty, \beta]\right) \\ &= P\left(\xi^{-1}(]-\infty, \alpha])\right) \cdot P\left(\eta^{-1}(]-\infty, \beta])\right). \end{aligned}$$

Proof. (i) ($\Leftarrow$) For every $U, V \in \mathcal{B}_{\mathbb{R}}$,

$$\begin{aligned} P\left(\xi^{-1}(U) \cap \eta^{-1}(V)\right) &= P\left[\left(\cup_{a_i \in U} A_i\right) \cap \left(\cup_{b_j \in V} B_j\right)\right] \\ &= \sum_{a_i \in U} \sum_{b_j \in V} P(A_i \cap B_j) = \sum_{a_i \in U} \sum_{b_j \in V} P(A_i) \cdot P(B_j) \\ &= P\left(\cup_{a_i \in U} A_i\right) \cdot P\left(\cup_{b_j \in V} B_j\right) \\ &= P\left(\xi^{-1}(U)\right) \cdot P\left(\eta^{-1}(V)\right). \end{aligned}$$

($\Rightarrow$) For all $i = 1, \dots, n$, $j = 1, \dots, m$, $P(A_i \cap B_j) = P(\xi^{-1}(a_i) \cap \eta^{-1}(b_j)) =$

$$= P\left(\xi^{-1}(a_i)\right) \cdot P\left(\eta^{-1}(b_j)\right) = P(A_i) \cdot P(B_j).$$

(ii) Let $\varphi : \mathbb{R} \to \mathbb{R}$ be a Borel function and let $B, C \in \mathcal{B}_{\mathbb{R}}$; then

$$\begin{aligned} P\left((\varphi \circ \xi)^{-1}(B) \cap \eta^{-1}(C)\right) &= P\left(\xi^{-1}(\varphi^{-1}(B)) \cap \eta^{-1}(C)\right) \\ &= P\left(\xi^{-1}(\varphi^{-1}(B))\right) \cdot P\left(\eta^{-1}(C)\right) \\ &= P\left((\varphi \circ \xi)^{-1}(B)\right) \cdot P\left(\eta^{-1}(C)\right). \end{aligned}$$

Therefore $\varphi \circ \xi$ and η are independent.

By taking successively $\varphi(x) = |x|$, $\max\{x, 0\}$ and $\max\{-x, 0\}$, $\varphi \circ \xi = |\xi|, \xi^+$ and ξ^-, respectively and we come to the required conclusion.

(iii) The implication ($\Rightarrow$) is obvious.

($\Leftarrow$): Let $\alpha \in \mathbb{R}$ and let $\mathfrak{M}_\alpha = \{A \subseteq \mathbb{R} : \eta^{-1}(A) \in \mathcal{A}, P[(\xi \le \alpha) \cap (\eta \in A)] = P(\xi \le \alpha) \cdot P(\eta \in A)\}$.

We notice that $\mathfrak{M}_\alpha$ is a monotone class and that, for every $\beta \in \mathbb{R},]-\infty, \beta] \in \mathfrak{M}_{\alpha},$. Then $\mathcal{B}_{\mathbb{R}} \subseteq \mathfrak{M}_\alpha$; therefore, $\mathcal{B}_{\mathbb{R}} \subseteq \bigcap_{\alpha \in \mathbb{R}} \mathfrak{M}_\alpha$.

For every $A \in \mathcal{B}_{\mathbb{R}}$, let $\mathfrak{M}_A = \{B \subseteq \mathbb{R} : \xi^{-1}(B) \in \mathcal{A}, P((\xi \in B) \cap (\eta \in A)) = P(\xi \in B) \cdot P(\eta \in A)\}$. It is clear that $\mathfrak{M}_A$ is a monotone class and, as for all $\alpha \in \mathbb{R}$, $A \in \mathcal{B}_{\mathbb{R}} \subseteq \mathfrak{M}_\alpha$,

$$P\left[(\xi \le \alpha) \cap (\eta \in A)\right] = P\left((\xi \le \alpha)\right) \cdot P\left(\eta \in A\right)$$

from where $]-\infty, \alpha] \in \mathfrak{M}_A$; therefore $\mathcal{B}_{\mathbb{R}} \subseteq \mathfrak{M}_A$ what finishes the proof of (iii). □

Proposition 2.101.

(i) *For every $n \in \mathbb{N}$, let $\{\xi_n, \eta_n\}$ be a couple of independent random variables; if $\xi_n \to \xi$ and $\eta_n \to \eta$, P-almost everywhere, then ξ and η are independent.*

(ii) *If $\{\xi, \eta, \theta\}$ is a set of independent random variables, then, for every $a \in \mathbb{R}$, $a\xi + \eta$ and θ are independent and* $\max\{\xi, \eta\}$ *and θ are also independent.*

Proof. (i) Let $N = \{\omega \in \Omega : \xi_n(\omega) \nrightarrow \xi(\omega)\}$ and let $M = \{\omega \in \Omega : \eta_n(\omega) \nrightarrow \eta(\omega)\}$. Then $P(N) = P(M) = 0$. So, we can suppose that $\xi_n \to \xi$ and $\eta_n \to \eta$ pointwise on Ω. For every α and $\beta \in \mathbb{R}$,

$$\begin{aligned}(\xi \leq \alpha) \cap (\eta \leq \beta) &= \bigcap_{n=1}^{\infty} \liminf_m \left[\left(\xi_m < \alpha + \frac{1}{n}\right) \cap \left(\eta_m < \beta + \frac{1}{n}\right)\right] \\ &= \bigcap_{n=1}^{\infty} \limsup_m \left[\left(\xi_m < \alpha + \frac{1}{n}\right) \cap \left(\eta_m < \beta + \frac{1}{n}\right)\right].\end{aligned}$$

Then

$$\begin{aligned}P\left((\xi \leq \alpha) \cap (\eta \leq \beta)\right) &= \lim_n \lim_m P\left(\left(\xi_m < \alpha + \frac{1}{n}\right) \cap \left(\eta_m < \beta + \frac{1}{n}\right)\right) \\ &= \lim_n \lim_m P\left(\xi_m < \alpha + \frac{1}{n}\right) \cdot P\left(\eta_m < \beta + \frac{1}{n}\right) \\ &= \lim_n P\left(\xi \leq \alpha + \frac{1}{n}\right) \cdot P\left(\eta \leq \beta + \frac{1}{n}\right) \\ &= P\left(\xi \leq \alpha\right) \cdot P\left(\eta \leq \beta\right).\end{aligned}$$

Therefore, from (iii) of Proposition 2.100, we can infer that ξ and η are independent.

(ii) We first suppose that

$$\xi = \sum_{i=1}^{n} a_i \chi_{A_i}, \eta = \sum_{j=1}^{m} b_j \chi_{B_j} \quad \text{and} \quad \theta = \sum_{k=1}^{p} c_k \chi_{C_k}.$$

Then we have, $a\xi + \eta = \sum_{i=1}^{n} \sum_{j=1}^{m} (a \cdot a_i + b_j)\chi_{A_i \cap B_j}$ and $\max\{\xi, \eta\} = \sum_{i=1}^{n} \sum_{j=1}^{m} \max\{a_i, b_j\}\chi_{A_i \cap B_j}$. By (i) of Proposition 2.100, for every $i \leq n$, $j \leq m$ and every $k \leq p$, $P(A_i \cap B_j \cap C_k) = P(A_i) \cdot P(B_j) \cdot P(C_k) = P(A_i \cap B_j) \cdot P(C_k)$ and therefore $a \cdot \xi + \eta$ and θ are independent as well as $\max\{\xi, \eta\}$ and θ.

Now let ξ, η and θ be random variables and, for every $n \in \mathbb{N}$, let

$$\xi_n = \sum_{i=-n\cdot 2^n}^{n\cdot 2^n} \frac{i}{2^n} \chi_{A_i^n}, \quad \eta_n = \sum_{i=-n\cdot 2^n}^{n\cdot 2^n} \frac{i}{2^n} \chi_{B_i^n}, \quad \theta_n = \sum_{i=-n\cdot 2^n}^{n\cdot 2^n} \frac{i}{2^n} \chi_{C_i^n}, \text{ where}$$

$$A_i^n = \xi^{-1}\left(\left[\frac{i}{2^n}, \frac{i+1}{2^n}\right[\right), \ B_i^n = \eta^{-1}\left(\left[\frac{i}{2^n}, \frac{i+1}{2^n}\right[\right), \ C_i^n = \theta^{-1}\left(\left[\frac{i}{2^n}, \frac{i+1}{2^n}\right[\right).$$

Then $\xi_n \to \xi$, $\eta_n \to \eta$ and $\theta_n \to \theta$. We have also, $P(A_i^n \cap B_j^n \cap C_l^n) =$

$$= P\left(\xi \in \left[\frac{i}{2^n}, \frac{i+1}{2^n}\right[, \eta \in \left[\frac{j}{2^n}, \frac{j+1}{2^n}\right[, \theta \in \left[\frac{l}{2^n}, \frac{l+1}{2^n}\right[\right)$$
$$= P\left(\xi \in \left[\frac{i}{2^n}, \frac{i+1}{2^n}\right[\right) \cdot P\left(\eta \in \left[\frac{j}{2^n}, \frac{j+1}{2^n}\right[\right) \cdot P\left(\theta \in \left[\frac{l}{2^n}, \frac{l+1}{2^n}\right[\right).$$

From (i) of Proposition 2.100, for every $n \in \mathbb{N}$, $\{\xi_n, \eta_n, \theta_n\}$ is an independent set of simple random variables. According to the first part of this demonstration, for every $n \in \mathbb{N}$, $a \cdot \xi_n + \eta_n$ and θ_n are independent as well as $\max\{\xi_n, \eta_n\}$ and θ_n. By (i), $a \cdot \xi + \eta$ are therefore independent as well as $\max\{\xi, \eta\}$ and θ. □

Remark 2.102.

(i) If $\{\xi, \theta\}$ and $\{\eta, \theta\}$ are two independent sets, then we cannot conclude that the set $\{\xi + \eta, \theta\}$ is independent, as shown in the following example:

Let $\xi = \chi_{[0,\frac{1}{2}[}$, $\eta = \chi_{[\frac{1}{4},\frac{3}{4}]}$ and $\theta = \chi_{[0,\frac{1}{4}[\cup[\frac{1}{2},\frac{3}{4})[}$. Then, it is easy to see that $\{\xi, \theta\}$ and $\{\eta, \theta\}$ are independent sets but $\xi + \eta = \chi_{[0,\frac{1}{4}[\cup[\frac{1}{2},\frac{3}{4}[} + 2 \cdot \chi_{[\frac{1}{4},\frac{1}{2}[}$ and θ are not. We notice also that ξ and η are not independent.

(ii) We can extend the properties of Propositions 2.100 and 2.101 to every finite set of random variables.

The following proposition is classical and is given without demonstration.

Proposition 2.103.

(i) *Let $\xi, \eta : \Omega \to \mathbb{R}$ two random variables. If $\xi, \eta \in L^1(P)$, then their means exist and, for every $a \in \mathbb{R}$, $E(a \cdot \xi + \eta) = a \cdot E(\xi) + E(\eta)$.*

(ii) *Let $\xi, \eta : \Omega \to \mathbb{R}$ two random variables independent. If $\xi, \eta \in L^1(P)$, then $\xi \cdot \eta \in L^1(P)$ and $E(\xi \cdot \eta) = E(\xi) \cdot E(\eta)$.*

If $\xi, \eta \in L^2(P)$, then their variances exist and, for every $a \in \mathbb{R}$, $V(a \cdot \xi + \eta) = a^2 \cdot V(\xi) + V(\eta)$.

We will now give the last preliminary result, necessary for the definition of a sequence that will allow the construction of the Wiener measure.

Theorem 2.104. *Let $(\Omega, \mathcal{A}, P)$ be a probability space and let $X : \Omega \to \mathbf{C}$. X is $(\mathcal{A} - \mathcal{B})$-measurable if and only if, for every $t \in [0, 1]$, the mapping $X_t : \Omega \to \mathbb{R}$, defined by: $X_t(\omega) = X(\omega)(t)$, for every $\omega \in \Omega$, is a random variable.*

Proof. ($\Rightarrow$) Let us suppose that X is $(\mathcal{A} - \mathcal{B})$-measurable; for each $A \in \mathcal{B}, X^{-1}(A) \in \mathcal{A}$. For every $t \in [0,1], \pi_t : \mathbf{C} \to \mathbb{R}, \pi_t(x) = x(t)$, is a continuous mapping; hence, for every $\alpha \in \mathbb{R}$,

$$A_t = \{x \in \mathbf{C} : x(t) \leq \alpha\} = \pi_t^{-1}(-\infty, \alpha]$$

is a closed subset of $\mathbf{C}$. Therefore, $A_t \in \mathcal{B}$. Then we have

$$X^{-1}(A_t) = \{\omega \in \Omega : X(\omega) \in A_t\} = \{\omega \in \Omega : X_t(\omega) = X(\omega)(t) \leq \alpha\}$$

which means that

$$X^{-1}(A_t) = X_t^{-1}(]-\infty, \alpha]) \in \mathcal{A}, \quad \text{for every} \quad \alpha \in \mathbb{R}.$$

So, for every $t \in [0,1]$, X_t is $(\mathcal{A} - \mathcal{B})$-measurable.

($\Leftarrow$) Conversely, let us suppose that, for every $t \in [0,1]$, X_t is a random variable. To show that X is $(\mathcal{A} - \mathcal{B})$-measurable, it is enough to show that, for any open subset $D \subseteq \mathbf{C}$, $X^{-1}(D) \in \mathcal{A}$. Since $\mathbf{C}$ is separable, every open subset $D \subseteq \mathbf{C}$ is a countable union of closed balls. Therefore, it is sufficient to show that the inverse image by X of every closed ball of $\mathbf{C}$ belongs to $\mathcal{A}$.
For every $x \in \mathbf{C}$ and every $r > 0$, let us denote $T(x,r) = \{y \in \mathbf{C} : \|x - y\| \leq r\}$ the closed sphere with center x and radius r. Then

$$\begin{aligned} X^{-1}(T(x,r)) &= \bigcap_{q \in \mathbb{Q} \cap [0,1]} \{\omega \in \Omega : x(q) - r \leq X(\omega)(q) \leq x(q) + r\} \\ &= \bigcap_{q \in \mathbb{Q} \cap [0,1]} X_q^{-1}([x(q) - r, x(q) + r]) \in \mathcal{A}. \end{aligned}$$ □

Definition 2.105. Let $(\xi_n)_{n \in \mathbb{N}}$ be a sequence of independent random variables such that every one is normal distributed $\mathcal{N}(0,1)$. We associate to this:

(i) The sequence $(X_n)_{n \in \mathbb{N}}$ of mappings $X_n : \Omega \to \mathbf{C}$, defined, for each $n \in \mathbb{N}$, by:

(1) $X_n(\omega)(0) = 0$.

(2) $X_n(\omega)(\frac{i}{n}) = \frac{1}{\sqrt{n}}(\xi_1(\omega) + \cdots + \xi_i(\omega))$, for every $i = 1, \ldots, n$.

(3) $X_n(\omega)$ is continuous and affine on every interval $[\frac{i-1}{n}, \frac{i}{n}]$.

(ii) The sequence $(P_n)_{n \in \mathbb{N}}$ of probabilities $P_n : \mathcal{B} \to \mathbb{R}_+$, defined, for every $n \in \mathbb{N}$, by: $P_n(A) = P(X_n^{-1}(A))$, for every $A \in \mathcal{B}$.

Remark 2.106.

(i) For every $t \in [\frac{i-1}{n}, \frac{i}{n}[$ and every $i = 1, \ldots, n$,

$$X_n(\omega)(t) = \frac{1}{\sqrt{n}}(\xi_1(\omega) + \cdots + \xi_{i-1}(\omega)) + \left(t - \frac{i-1}{n}\right)\sqrt{n} \cdot \xi_i(\omega). \tag{*}$$

Let us remark that, for every $t \in [\frac{i-1}{n}, \frac{i}{n}[$, the integer part of nt is $[nt] = i - 1$; then, for every $t \in [0, 1]$ and every $\omega \in \Omega$,

$$X_n(\omega)(t) = \frac{1}{\sqrt{n}}\left[\xi_1(\omega) + \cdots + \xi_{[nt]}(\omega)\right] + \frac{1}{\sqrt{n}}(nt - [nt]) \cdot \xi_{[nt]+1}(\omega). \tag{**}$$

For every $t \in [0, 1]$, $X_n(.)(t)$ is so a random variable; according to Theorem 2.104, $X_n : \Omega \to \mathbf{C}$ is an $(\mathcal{A} - \mathcal{B})$-measurable mapping and, therefore, for every $n \in \mathbb{N}$, P_n is the image of P under X_n and $(P_n)_{n\in\mathbb{N}} \subseteq \mathcal{P}_{\mathbf{C}} \subseteq rca_1^+(\mathcal{B}) \subseteq Rca^+(\mathcal{B})$ (let us remember that, according to Remark 2.32 (v), $\mathcal{P}_{\mathbf{C}}$ is the family of all probabilities on $\mathbf{C}$).

(ii) For every $t \in [0, 1]$ and every $n \in \mathbb{N}$, the random variable $X_n(.)(t)$ is normally distributed $\mathcal{N}(0, \sigma_n^2(t))$, where

$$\sigma_n^2(t) = \frac{[nt]}{n} + \frac{(nt - [nt])^2}{n}.$$

Indeed, since $(\xi_n)_{n\in\mathbb{N}}$ is independent and normally distributed $\mathcal{N}(0, 1)$, it results from Proposition 2.103 that

$$E\left(X_n(.)(t)\right) = \frac{1}{\sqrt{n}}\sum_{j=1}^{i-1} E(\xi_j) + \left(t - \frac{i-1}{n}\right)\sqrt{n}E(\xi_i) = 0 \quad \text{and}$$

$$\begin{aligned} V\left(X_n(.)(t)\right) &= \frac{1}{n}\sum_{j=1}^{i-1} V(\xi_j) + \left(t - \frac{i-1}{n}\right)^2 \cdot nV(\xi_i) \\ &= \frac{i-1}{n} + \frac{(nt-(i-1))^2}{n} = \frac{[nt]}{n} + \frac{(nt-[nt])^2}{n}. \end{aligned}$$

(iii) For every $n \in \mathbb{N}$, $\pi_0(P_n) = \delta_0$ and, therefore, the sequence $(\pi_0(P_n))_{n\in\mathbb{N}}$ is tight. Indeed, for every $A \in \mathcal{B}_{\mathbb{R}}$, $P_n(\pi_0^{-1}(A)) = P(X_n^{-1}(\pi_0^{-1}(A)))$ where

$$X_n^{-1}(\pi_0^{-1}(A)) = \begin{cases} \Omega, & 0 \in A, \\ \emptyset, & 0 \notin A \end{cases}$$

Since P is a probability, $\pi_0(P_n) = \delta_0$.

From Proposition 2.97 and from (iii) of Remark 2.106, we obtain the following corollary:

Corollary 2.107. *If $(P_n)_{n\in\mathbb{N}}$ satisfies the following condition:*
For every $\varepsilon > 0$ and every $\eta > 0$, there exist $\delta \in]0,1]$ and $n_0 \in \mathbb{N}$ such that, for all $n \geq n_0$ and all $t \in [0,1]$,

$$\frac{1}{\delta}\cdot P\left(\left\{\omega \in \Omega : \sup_{t\leq s\leq t+\delta} |X_n(\omega)(t) - X_n(\omega)(s)| \geq \varepsilon\right\}\right) \leq \eta, \qquad (T_1)$$

then $(P_n)_{n\in\mathbb{N}}$ is tight.

Now, we transform the condition (T_1) to one easier to handle.

Theorem 2.108. *Let $(X_n)_{n\in\mathbb{N}}$ be the sequence of measurable applications and let $(P_n)_{n\in\mathbb{N}}$ be the sequence of probabilities defined in 2.105 and specified in Remark 2.106.*

If $(P_n)_n$ satisfies the condition:
For every $\varepsilon > 0$, there exist $\lambda > 1$ and $n_0 \in \mathbb{N}$ such that, for every $n \geq n_0$ and every $k \in \mathbb{N}$,

$$P\left(\left\{\omega \in \Omega : \max_{i\leq n} |\xi_{k+1}(\omega) + \cdots + \xi_{k+i}(\omega)| \geq \lambda\sqrt{n}\right\}\right) \leq \frac{\varepsilon}{\lambda^2}, \qquad (T_2)$$

then the sequence $(P_n)_n$ is tight.

Proof. For every $i = 1,\ldots,n$, let us note $S_i = \xi_1 + \xi_2 + \cdots + \xi_i$. For every $t \in [0,1]$ and every $n \in \mathbb{N}$, there exists a single $i \in \{0,\ldots,n\}$ such that $\frac{i}{n} \leq t < \frac{i+1}{n}$. Let $\delta > 0$ and $s \in [t, t+\delta]$; then there exists $j \geq i$ such that $\frac{j}{n} \leq t+\delta < \frac{j+1}{n}$ and so $s \in [t, t+\delta] \subseteq \bigcup_{k=i}^{j} [\frac{k}{n}, \frac{k+1}{n}[$. Since $X_n(\omega)$ is affine on every interval $[\frac{i}{n}, \frac{i+1}{n}]$,

$$\begin{aligned}
|X_n(\omega)(t) - X_n(\omega)(s)| &\leq \left|X_n(\omega)(t) - X_n(\omega)\left(\frac{i}{n}\right)\right| + \left|X_n(\omega)\left(\frac{i}{n}\right) - X_n(\omega)(s)\right| \\
&\leq \left|X_n(\omega)\left(\frac{i+1}{n}\right) - X_n(\omega)\left(\frac{i}{n}\right)\right| \\
&\quad + \max_{i+1\leq k\leq j+1} \left|X_n(\omega)\left(\frac{i}{n}\right) - X_n(\omega)\left(\frac{k}{n}\right)\right| \\
&\leq 2 \max_{0\leq l\leq j-1} \left|X_n(\omega)\left(\frac{i}{n}\right) - X_n(\omega)\left(\frac{l+i+1}{n}\right)\right| \\
&= 2 \max_{0\leq l\leq j-i} \frac{1}{\sqrt{n}} |S_{i+l+1}(\omega) - S_i(\omega)|.
\end{aligned}$$

As $j - i < n\delta + 1 < j - i + 1$, we have, for every $s \in [t, t + \delta]$,

$$|X_n(\omega)(t) - X_n(\omega)(s)| \le 2 \max_{0 \le l < n\delta+1} \frac{1}{\sqrt{n}} |S_{i+l+1}(\omega) - S_i(\omega)| .$$

Therefore (T_1) of Corollary 2.107 is proved if : *for every $\varepsilon > 0$ and every $\eta > 0$, there exist $\delta \in (0,1]$ and $n_0 \in \mathbb{N}$ such that, for every $n \ge n_0$ and every $i = 0, \dots, n$,*

$$\frac{1}{\delta} \cdot P\left(\left\{\omega \in \Omega : \max_{1 \le l < n\delta+2} \frac{1}{\sqrt{n}} |S_{i+l}(\omega) - S_i(\omega)| \ge \varepsilon\right\}\right) \le \eta. \tag{1}$$

If $m = [n\delta + 2] \in \mathbb{N}$, then the condition (1) becomes: *for every $\varepsilon > 0$ and every $\eta > 0$, there exist $m \in \mathbb{N}$, $\delta \in (0,1]$ and $n_0 \in \mathbb{N}$ such that, for every $n \ge n_0$ and every $i = 0, \dots, n$,*

$$P\left(\left\{\omega \in \Omega : \max_{1 \le l \le m} |S_{i+l}(\omega) - S_i(\omega)| \ge \sqrt{n} \cdot \varepsilon\right\}\right) \le \eta\delta. \tag{2}$$

As $n \ge \frac{m-2}{\delta}$, if we note $m_0 = [n_0\delta + 2]$, then (2) is satisfied if we have: *for every $m > m_0$ and every $i \in \mathbb{N}$,*

$$P\left(\left\{\omega \in \Omega : \max_{1 \le l \le m} |S_{i+l}(\omega) - S_i(\omega)| \ge \sqrt{\frac{m-2}{\delta}} \cdot \varepsilon\right\}\right) \le \eta\delta. \tag{3}$$

Let now $\lambda = \frac{\varepsilon}{\sqrt{\delta}} \cdot \sqrt{\frac{m-2}{m}}$; since δ be chosen as small, $\lambda > 1$.

So, we can reduce (3) to: *for every $\varepsilon > 0$ and every $\eta > 0$, there exist $m_0 \in \mathbb{N}$ and $\lambda > 1$ such that, for every $m \ge m_0$ and every $i \in \mathbb{N}$,*

$$P\left(\left\{\omega \in \Omega : \max_{1 \le l \le m} |S_{i+l}(\omega) - S_i(\omega)| \ge \sqrt{m} \cdot \lambda\right\}\right) \le \frac{1}{2}\eta\frac{\varepsilon^2}{\lambda^2}, \tag{4}$$

which is equivalent to (T_2). □

Theorem 2.109. *The sequence $(P_n)_{n \in \mathbb{N}}$ defined in 2.105 is tight.*

Proof. Let us mention again that $(\xi_n)_{n \in \mathbb{N}}$ is a independent sequence normally distributed $\mathcal{N}(0,1)$ on the space $(\Omega, \mathcal{A}, P)$, that $(X_n)_{n \in \mathbb{N}}$ is the sequence of mappings $X_n : \Omega \to \mathbf{C}$, $(\mathcal{A} - \mathcal{B})$-measurable such that:

$$X_n(\omega)(t) = \frac{1}{\sqrt{n}} S_{[nt]}(\omega) + \frac{nt - [nt]}{\sqrt{n}} \cdot \xi_{[nt]+1}(\omega),$$

where $S_k(\omega) = \xi_1(\omega) + \cdots + \xi_k(\omega)$ and $S_0(\omega) = 0$ and that, for all $n \in \mathbb{N}$, $P_n : \mathcal{B} \to \mathbb{R}$ is image of P under X_n.

According to Theorem 2.108, to show that the sequence $(P_n)_{n\in\mathbb{N}}$ is tight, it is sufficient to show that, for every $\varepsilon > 0$, there exist $\lambda > 1$ and $n_0 \in \mathbb{N}$ such that, for every $n \geq n_0$ and every $k \in \mathbb{N}$,

$$P\left(\left\{\omega \in \Omega : \max_{i\leq n} |S_{k+i}(\omega) - S_k(\omega)| \geq \lambda\sqrt{n}\right\}\right) \leq \frac{\varepsilon}{\lambda^2}. \qquad (T_2)$$

Let $k \in \mathbb{N}, \lambda > 1$ and $n \in \mathbb{N}$ be fixed. For every $i \leq n-1$, we note

$$E_i = \left\{\omega \in \Omega : \max_{j\leq i-1} \left|S_{k+j}(\omega) - S_k\omega)\right| < \lambda\sqrt{n} \leq |S_{k+i}(\omega) - S_k(\omega)|\right\}.$$

If $i \neq j$, then $E_i \cap E_j = \emptyset$ and

$$\left\{\omega \in \Omega : \max_{i\leq n} S_{k+i}(\omega) - S_k(\omega)| \geq \lambda\sqrt{n}\right\} \subseteq \left\{\omega : |S_{k+n}(\omega) - S_k(\omega)| \geq \frac{\lambda}{2}\sqrt{n}\right\}$$

$$\cup \bigcup_{i=1}^{n-1} E_i \cap \left\{\omega \in \Omega : |S_{k+n}(\omega) - S_{k+i}(\omega)| \geq \frac{\lambda}{2}\sqrt{n}\right\}. \qquad (1)$$

Indeed, let $\omega \in \Omega$ so that

$$\max_{i\leq n} |S_{k+i}(\omega) - S_k(\omega)| \geq \lambda\sqrt{n} \quad \text{and} \quad |S_{k+n}(\omega) - S_k(\omega)| < \frac{\lambda}{2}\sqrt{n}.$$

If we note i_0 the smallest index $i \leq n$ for which $|S_{k+i}(\omega) - S_k(\omega)| \geq \lambda\sqrt{n}$, then $i_0 \leq n-1$ and $\omega \in E_{i_0}$. Also, we have

$$\begin{aligned}\left|S_{k+n}(\omega) - S_{k+i_0}(\omega)\right| &\geq \left|S_{k+i_0}(\omega) - S_k(\omega)\right| - |S_{k+n}(\omega) - S_k(\omega)| \\ &\geq \lambda\sqrt{n} - \frac{\lambda}{2}\sqrt{n} = \frac{\lambda}{2}\sqrt{n}\end{aligned}$$

which proves (1).

Since the random variables $(\xi_n)_{n\in\mathbb{N}}$ are independent, by (ii) of Proposition 2.101, for every $i \leq n-1$, we have

$$\begin{aligned}&P\left(E_i \cap \left\{\omega : |S_{k+n}(\omega) - S_{k+i}(\omega)| \geq \frac{\lambda}{2}\sqrt{n}\right\}\right) \qquad (2)\\ &= P(E_i) \cdot P\left(\left\{\omega : |S_{k+n}(\omega) - S_{k+i}(\omega)| \geq \frac{\lambda}{2}\sqrt{n}\right\}\right).\end{aligned}$$

For every $i = 0, \dots, n-1$, $S_{k+n} - S_{k+i} = \xi_{k+i+1} + \cdots + \xi_{k+n}$ is a random variable normally distributed $\mathcal{N}(0, \sqrt{n-i})$ and, therefore, having the same distribution as $\xi_1 + \cdots + \xi_{n-i} = S_{n-i}$. Then, for all $i = 0, \dots, n-1$,

$$P\left(\left\{\omega : |(S_{k+n} - S_{k+i})(\omega)| \geq \frac{\lambda}{2}\sqrt{n}\right\}\right) = P\left(\left\{\omega : |S_{n-i}(\omega)| \geq \frac{\lambda}{2}\sqrt{n}\right\}\right). \tag{3}$$

Then, by (1), (2) and (3), we obtain:

$$\begin{aligned} &P\left(\left\{\omega : \max_{i \leq n} |(S_{k+i} - S_k)(\omega)| \geq \lambda\sqrt{n}\right\}\right) \qquad (4) \\ &\leq P\left(\left\{\omega : |S_n(\omega)| \geq \frac{\lambda}{2}\sqrt{n}\right\}\right) + \sum_{i=1}^{n-1} P(E_i) \cdot P\left(\left\{\omega : |S_{n-i}(\omega)| \geq \frac{\lambda}{2}\sqrt{n}\right\}\right) \\ &\leq P\left(\left\{\omega : \frac{|S_n|}{\sqrt{n}}(\omega) \geq \frac{\lambda}{2}\right\}\right) + \sum_{i=1}^{n-1} P(E_i) \cdot P\left(\left\{\omega : \frac{|S_{n-i}|}{\sqrt{n-i}}(\omega) \geq \frac{\lambda}{2}\right\}\right). \end{aligned}$$

According to (ii) of Proposition 2.103, the random variable

$$\frac{S_j}{\sqrt{j}} = \frac{1}{\sqrt{j}}(\xi_1 + \cdots + \xi_j)$$

is normally distributed $\mathcal{N}(0, 1)$. Therefore, for every j, we have

$$P\left(\left\{\omega : \frac{|S_j|}{\sqrt{j}}(\omega) \geq \frac{\lambda}{2}\right\}\right) = \sqrt{\frac{2}{\pi}} \int_{\frac{\lambda}{2}}^{\infty} e^{-\frac{x^2}{2}} dx. \tag{5}$$

Because

$$\lim_{\lambda \to \infty} \lambda^3 \int_{\frac{\lambda}{2}}^{\infty} e^{-\frac{x^2}{2}} dx = 0, \tag{6}$$

there exists $\lambda_0 > 1$ such that

$$P\left(\left\{\omega : \frac{|S_j|}{\sqrt{j}}(\omega) \geq \frac{\lambda}{2}\right\}\right) < \frac{1}{\lambda^3}, \quad \text{for every} \quad \lambda \geq \lambda_0 \quad \text{and every} \quad j \in \mathbb{N}^*. \tag{7}$$

Then, by (4) and (7), we obtain

$$\begin{aligned} &P\left(\left\{\omega : \max_{i \leq n} |S_{k+i}(\omega) - S_k(\omega)| \geq \lambda\sqrt{n}\right\}\right) \leq \frac{1}{\lambda^3}\left(1 + \sum_{i=1}^{n-1} P(E_i)\right) \qquad (8) \\ &= \frac{1}{\lambda^3}\left(1 + P\left(\bigcup_{i=1}^{n-1} E_i\right)\right) \leq \frac{2}{\lambda^3}, \qquad \text{for every} \quad \lambda \geq \lambda_0. \end{aligned}$$

For every $\varepsilon > 0$ and $\lambda > \max\{\lambda_0, \frac{2}{\varepsilon}\}$, $\lambda > 1$ and so

$$P\left(\max_{i\le n} |S_{k+i}(\omega) - S_k(\omega)| \ge \lambda\sqrt{n}\right) \le \frac{\varepsilon}{\lambda^2}, \quad \text{for every} \quad k, n \in \mathbb{N}.$$

The sequence $(P_n)_{n\in\mathbb{N}}$ satisfies (T_2) and, therefore, it is tight. □

Now we can introduce the Wiener measure on $\mathbf{C}$.

Definition 2.110. A measure $W : \mathcal{B} \to [0,1]$ is called a *Wiener measure* on $\mathbf{C}$ if it satisfies the two following conditions:

(i) $W(\{x \in \mathbf{C} : x(t) \le \alpha\}) = \frac{1}{\sqrt{2\pi t}} \int_{-\infty}^{\alpha} e^{-\frac{u^2}{2t}} du, \quad \forall t \in]0,1], \forall \alpha \in \mathbb{R}$.

(ii) $\forall n \in \mathbb{N}$, $\forall\, 0 \le t_0 < t_1 < \cdots < t_n \le 1$, $\forall i = 0, \ldots, n-1$, the random variables $\xi_i : \mathbf{C} \to \mathbb{R}$, defined by $\xi_i(x) = x(t_{i+1}) - x(t_i)$, $\forall x \in \mathbf{C}$, are independent with respect to the measure W.

Remark 2.111.

(i) $W(\mathbf{C}) = \lim_{n\to\infty} W(\pi_1^{-1}(-\infty, n]) = \lim_{n\to\infty} \frac{1}{\sqrt{2\pi}} \int_{-\infty}^{n} e^{-\frac{u^2}{2}} du = 1.$

Therefore, W is a probability on $\mathbf{C}$.

(ii) The first condition of Definition 2.105 implies that, for every $t \in]0,1]$, the random variable $\pi_t : (\mathbf{C}, \mathcal{B}, W) \to \mathbb{R}$ (see Definition 2.87) is normally distributed $\mathcal{N}(0, \sqrt{t})$. Indeed, for every $t \in]0,1]$ and every $\alpha \in \mathbb{R}$,

$$W\left(\pi_t^{-1}(-\infty, \alpha]\right) = \frac{1}{\sqrt{2\pi t}} \int_{(-\infty,\alpha]} e^{-\frac{u^2}{2t}} du,$$

so that, for every $t \in]0,1]$ and every $B \in \mathcal{B}_{\mathbb{R}}$,

$$W\left(\pi_t^{-1}(B)\right) = \frac{1}{\sqrt{2\pi t}} \int_B e^{-\frac{u^2}{2t}} du.$$

(iii) $W(\{x \in \mathbf{C} : x(0) = 0\}) = 1$. Indeed, let $t_n \downarrow 0$ and, for every $\alpha \in \mathbb{R}$, let $A_n(\alpha) = \{x \in \mathbf{C} : x(t_n) \le \alpha\}$. Then

$$\limsup_n A_n(\alpha) = \bigcap_n \bigcup_{m\ge n} A_m(\alpha) \subseteq \{x \in \mathbf{C} : x(0) \le \alpha\}$$

so that

$$W(\{x \in \mathbf{C} : x(0) \leq \alpha\}) \geq W(\limsup_n A_n(\alpha)) \geq \limsup_n W(A_n(\alpha))$$
$$= \limsup_n \frac{1}{\sqrt{2\pi t_n}} \int_{-\infty}^{\alpha} e^{-\frac{u^2}{2t_n}} du$$
$$= \lim_n \frac{1}{\sqrt{\pi}} \int_{-\infty}^{\frac{\alpha}{\sqrt{2t_n}}} e^{-v^2} dv = 1.$$

It results that, for every $\alpha > 0$, $W(\{x \in \mathbf{C} : x(0) \leq \alpha\}) \geq 1$ and so,

$$W(\{x \in \mathbf{C} : x(0) \leq \alpha\}) = 1.$$

Particularly, for $\alpha = 0$, we obtain

$$W(\{x \in \mathbf{C} : x(0) \leq 0\}) = 1. \tag{*}$$

On the other hand, $(\{x \in \mathbf{C} : x(0) < 0\}) = \bigcup_{p=1}^{\infty}(\{x \in \mathbf{C} : x(0) < -\frac{1}{p}\})$ from where

$$W(\{x \in \mathbf{C} : x(0) < 0\}) = \lim_p W\left(\left\{x \in \mathbf{C} : x(0) < -\frac{1}{p}\right\}\right).$$

Since $(\{x \in \mathbf{C} : x(0) < -\frac{1}{p}\}) \subseteq \liminf_n A_n(-\frac{1}{p})$, for every $p \in \mathbb{N}^*$,

$$W\left(\left\{x \in \mathbf{C} : x(0) < -\frac{1}{p}\right\}\right) \leq W\left(\liminf_n A_n\left(-\frac{1}{p}\right)\right)$$
$$\leq \liminf_n \frac{1}{\sqrt{\pi}} \int_{-\infty}^{-\frac{1}{p\sqrt{2t_n}}} e^{-v^2} dv = 0.$$

Therefore, for every $p \in \mathbb{N}^*$, $W(\{x \in \mathbf{C} : x(0) < -\frac{1}{p}\}) = 0$ from where we obtain: $W(\{x \in \mathbf{C} : x(0) < 0\}) = 0$ and, by $(*)$,

$$W(\{x \in \mathbf{C} : x(0) = 0\}) = 1. \tag{**}$$

According to the relation $(**)$, the theory of Wiener measure is unchanged if the space $\mathbf{C}$ is replaced by its subspace $\mathbf{C}_0 = \{x \in \mathbf{C} : x(0) = 0\}$.

(iv) From condition (ii) of Definition 2.110 and according to the notations of Definition 2.87, for every $t, s \in [0, 1]$ with $0 \leq s < t \leq 1$, the random variables $\pi_s - \pi_0$ and $\pi_t - \pi_s$ are independent. From the above property $(**)$, it results that $\pi_s - \pi_0 = \pi_s$, W-a.e. hence π_s and $\pi_t - \pi_s$ are independent. According to Proposition 2.103, we have $E(\pi_t) = E(\pi_s) + E(\pi_t - \pi_s)$ and $V(\pi_t) = V(\pi_s) + V(\pi_t - \pi_s)$. Therefore, by (ii), $E(\pi_t - \pi_s) = 0$ and $V(\pi_t - \pi_s) = t - s$, i.e. $\pi_t - \pi_s$ is normally distributed $\mathcal{N}(0, \sqrt{t - s})$.

(v) For the same reason, for all $t_0 = 0 < t_1 < \cdots < t_n \leq 1$, the set of random variables $\{\pi_{t_{i+1}} - \pi_{t_i} : i = 0, \ldots, n-1\}$ is independent. We can therefore write, for every $\alpha_1, \ldots, \alpha_n \in \mathbb{R}$,

$$\begin{aligned} W & \left(\left\{x \in \mathbf{C} : \pi_{t_{i+1}}(x) - \pi_{t_i}(x) \leq \alpha_{i+1} : i = 0, \ldots, n-1\right\}\right) \\ &= \prod_{i=0}^{n-1} W\left(\pi_{t_{i+1}} - \pi_{t_i} \leq \alpha_{i+1}\right) \\ &= \prod_{i=0}^{n-1} \frac{1}{\sqrt{2\pi(t_{i+1}-t_i)}} \int_{-\infty}^{\alpha_{i+1}} e^{-\frac{u^2}{2(t_{i+1}-t_i)}}\,du \\ &= \frac{1}{\sqrt{(2\pi)^n t_1(t_2-t_1)\ldots(t_n-t_{n-1})}} \\ &\quad \times \int_{\prod_{i=1}^n(-\infty,\alpha_i]} e^{-\left(\frac{u_1^2}{2t_1}+\frac{u_2^2}{2(t_2-t_1)}+\cdots+\frac{u_n^2}{2(t_n-t_{n-1})}\right)} du_1 \ldots du_n. \end{aligned}$$

Proposition 2.112.

(i) *Let $W : \mathcal{B} \to [0,1]$ be a Wiener measure on $\mathbf{C}$ and let $\mathcal{C}$ be the algebra of all cylinder sets on $\mathbf{C}$; for every $0 = t_0 < t_1 < \cdots < t_n \leq 1$ and every $B \in \mathcal{B}_n$, let $C = \pi_{t_1,\ldots,t_n}^{-1}(B) \in \mathcal{C}$, then*

$$W(C) = \frac{1}{\sqrt{(2\pi)^n \prod_{k=1}^n (t_k - t_{k-1})}} \int_B e^{-\left(\sum_{k=1}^n \frac{u_k^2}{2(t_k-t_{k-1})}\right)} du_1 \ldots du_n. \tag{*}$$

(ii) *Conversely, if $W : \mathcal{B} \to [0,1]$ is a measure satisfying $(*)$, for every $C \in \mathcal{C}$, then W is a Wiener measure on $\mathbf{C}$.*

Proof. Let $n \in \mathbb{N}^*$ and let $T_n : \mathbb{R}^n \to \mathbb{R}^n$ be defined by $T_n(x) = (x_1, x_2 - x_1, \ldots, x_n - x_{n-1})$, for every $x = (x_1, \ldots, x_n) \in \mathbb{R}^n$. It is clear that T_n is a homeomorphism of $\mathbb{R}^n$ on $\mathbb{R}^n$. The σ-algebra $\mathcal{B}_n$ is generated by the family $\mathcal{S} = \{\prod_{i=1}^n]-\infty, \alpha_i] : \alpha_1, \ldots, \alpha_n \in \mathbb{R}\}$; since T_n is a homeomorphism, $T_n^{-1}(\mathcal{S})$ generates $\mathcal{B}_n$.

(i) Let $0 = t_0 < t_1 < \cdots < t_n \leq 1$ and $B \in T_n^{-1}(\mathcal{S})$. Then, there exists $(\alpha_1, \ldots, \alpha_n) \in \mathbb{R}^n$ such that

$$\begin{aligned} B &= T_n^{-1}\Big(\prod_{i=1}^n]-\infty, \alpha_i]\Big) \\ &= \{(x_1, \ldots, x_n) : x_1 \leq \alpha_1, x_2 - x_1 \leq \alpha_2, \ldots, x_n - x_{n-1} \leq \alpha_n\}. \end{aligned}$$

Let $C = \pi_{t_1,\dots,t_n}^{-1}(B) \in \mathcal{C}$. According to (v) of Remark 2.111, we have

$$\begin{aligned} W(C) &= W\left(\pi_{t_1,\dots,t_n}^{-1}(B)\right) \\ &= \frac{1}{\sqrt{(2\pi)^n \prod_{k=1}^n (t_k - t_{k-1})}} \\ &\quad \times \int_{\prod_{i=1}^n]-\infty,\alpha_i]} e^{-\left(\frac{v_1^2}{2t_1} + \frac{v_2^2}{2(t_2-t_1)} + \dots + \frac{v_n^2}{2(t_n - t_{n-1})}\right)} dv_1, \dots, dv_n. \end{aligned}$$

By using the change of the variable $u = (u_1, \dots, u_n) = T_n(v_1, \dots, v_n) = v$ we obtain $(*)$:

$$W(C) = \frac{1}{\sqrt{(2\pi)^n \prod_{k=1}^n (t_k - t_{k-1})}} \int_B e^{-\left(\sum_{k=1}^n \frac{u_k^2}{2(t_k - t_{k-1})}\right)} du_1, \dots, du_n.$$

Because $T_n^{-1}(\mathcal{S})$ generates $\mathcal{B}_n$ and $(*)$ is satisfied for all sets $C \in \{\pi_{t_1,\dots,t_n}^{-1}(B) : B \in T_n^{-1}(\mathcal{S})\}$, W satisfies $(*)$, for all $C \in \{\pi_{t_1,\dots,t_n}^{-1}(B) : B \in \mathcal{B}_n\}$.

(ii) Let $\mu : \mathcal{B} \to [0,1]$ be a measure that satisfies $(*)$ on $\mathcal{C}$; then, for every $t \in]0,1]$, every $\alpha \in \mathbb{R}$ and $C = \pi_t^{-1}(]-\infty,\alpha]) \in \mathcal{C}$:

$$\mu(C) = \frac{1}{\sqrt{2\pi t}} \int_{-\infty}^{\alpha} e^{-\frac{u^2}{2t}} du,$$

which is condition (i) of Definition 2.110.

Let $n \in \mathbb{N}$, $t_0 = 0 < t_1 < \dots < t_n \le 1$ and $\xi_i : \mathbf{C} \to \mathbb{R}$, $\xi_i(x) = x(t_{i+1}) - x(t_i)$. According to (iii) of Remark 2.111, we can suppose that $x(0) = 0$. Then, $\xi_0(x) = x(t_1)$. In order to show that $\xi_0, \dots, \xi_{n-1}$ are independent it is sufficient, according to (iii) of Proposition 2.100, to show that, for every $\alpha_0, \dots, \alpha_{n-1} \in \mathbb{R}$,

$$\mu\left(\bigcap_{i=0}^{n-1} \xi_i^{-1}(]-\infty,\alpha_i])\right) = \prod_{i=0}^{n-1} \mu\left(\xi_i^{-1}(]-\infty,\alpha_i])\right). \tag{1}$$

On the one hand, we have:

$$\begin{aligned} &\mu\left(\bigcap_{i=0}^{n-1} \xi_i^{-1}(]-\infty,\alpha_i])\right) \\ &\quad = \mu\left(\left\{x \in \mathbf{C} : T_n(x(t_1), \dots, x(t_n)) \in \prod_{i=0}^{n-1}]-\infty,\alpha_i]\right\}\right) \\ &\quad = \mu\left(\left\{x \in \mathbf{C} : (x(t_1), \dots, x(t_n)) \in T_n^{-1}\left(\prod_{i=0}^{n-1}]-\infty,\alpha_i]\right)\right\}\right). \end{aligned}$$

If we note $B = T_n^{-1}(\prod_{i=0}^{n-1}]-\infty,\alpha_i])$, then we can rewrite the previous equality:

$$\mu\left(\bigcap_{i=0}^{n-1}\xi_i^{-1}(]-\infty,\alpha_i])\right) = \mu\left(\pi_{t_1,\dots,t_n}^{-1}(B)\right).$$

Therefore, we have

$$\begin{aligned}
&\mu\left(\bigcap_{i=0}^{n-1}\xi_i^{-1}(]-\infty,\alpha_i])\right) \qquad (2)\\
&= \frac{1}{\sqrt{(2\pi)^n\prod_{k=1}^n(t_k-t_{k-1})}}\int_B e^{-\left(\sum_{k=1}^n\frac{(u_k-u_{k-1})^2}{2(t_k-t_{k-1})}\right)}du_1,\dots,du_n\\
&= \frac{1}{\sqrt{(2\pi)^n\prod_{k=1}^n(t_k-t_{k-1})}}\\
&\quad\times\int_{\prod_{i=0}^{n-1}]-\infty,\alpha_i]} e^{-\left(\frac{v_1^2}{2t_1}+\frac{v_2^2}{2(t_2-t_1)}+\dots+\frac{v_n^2}{2(t_n-t_{n-1})}\right)}dv_1,\dots,dv_n\\
&= \prod_{i=0}^{n-1}\frac{1}{\sqrt{2\pi(t_{i+1}-t_i)}}\int_{-\infty}^{\alpha_i} e^{-\frac{v_{i+1}^2}{2(t_{i+1}-t_i)}}dv_{i+1}\\
&= \mu\left(\xi_0^{-1}(]-\infty,\alpha_0])\right)\cdot\prod_{i=1}^{n-1}\frac{1}{\sqrt{2\pi(t_{i+1}-t_i)}}\int_{-\infty}^{\alpha_i} e^{-\frac{v_{i+1}^2}{2(t_{i+1}-t_i)}}dv_{i+1}.
\end{aligned}$$

On the other hand, for every $i = 1,\dots,n-1$,

$$\begin{aligned}
\mu\left(\xi_i^{-1}(]-\infty,\alpha_i])\right) &= \mu\left(\{x\in\mathbf{C} : x(t_{i+1})-x(t_i)\le\alpha_i\}\right)\\
&= \mu\left(\{x\in\mathbf{C} : T_2(x(t_i),x(t_{i+1}))\in\mathbb{R}\times]-\infty,\alpha_i]\}\right)\\
&= \mu\left(\{x\in\mathbf{C} : (x(t_i),x(t_{i+1}))\in T_2^{-1}(\mathbb{R}\times]-\infty,\alpha_i])\}\right).
\end{aligned}$$

If we note $B_i = T_2^{-1}(\mathbb{R}\times]-\infty,\alpha_i])$, then

$$\begin{aligned}
\mu\left(\xi_i^{-1}(]-\infty,\alpha_i])\right) &= \mu\left(\pi_{t_i,t_{i+1}}^{-1}(B_i)\right)\\
&= \frac{1}{\sqrt{(2\pi)^2 t_i(t_{i+1}-t_i)}}\int_{B_i} e^{-\frac{u_1^2}{2t_i}-\frac{(u_2-u_1)^2}{2(t_{i+1}-t_i)}}du_1du_2.
\end{aligned}$$

Using the change of variable $v = (v_1,v_2) = T_2(u_1,u_2) = T_2(u)$, we obtain

$$\begin{aligned}
\mu\left(\xi_i^{-1}(-\infty,\alpha_i]\right) &= \frac{1}{\sqrt{(2\pi)^2 t_i(t_{i+1}-t_i)}}\int_{\mathbb{R}} e^{-\frac{v_1^2}{2t_i}}dv_1\cdot\int_{-\infty}^{\alpha_i} e^{-\left[\frac{v_2^2}{2(t_{i+1}-t_i)}\right]}dv_2\\
&= \frac{1}{\sqrt{2\pi(t_{i+1}-t_i)}}\int_{-\infty}^{\alpha_i} e^{-\frac{v_{i+1}^2}{2(t_{i+1}-t_i)}}dv_{i+1},
\end{aligned}$$

which transferred back in (2), for $i = 1,\dots,n-1$, leads to relation (1).

Therefore, the random variables $\xi_0, \dots, \xi_{n-1}$ are independent and so μ is a Wiener measure. □

Remark 2.113.

(i) It results that, in order to study if one measure is a Wiener measure, we can replace conditions (i) and (ii) of Definition 2.110 with condition $(*)$ of Proposition 2.112.

(ii) If the measures $\mu, \nu : \mathcal{B} \to [0, 1]$ on $\mathbf{C}$ verify $(*)$, they coincide on $\mathcal{C}$ and, by Remark 2.89, $\mu = \nu$. Therefore, if there is a measure which satisfies conditions (i) and (ii) of Definition 2.110, this measure is unique. In other words, there is at least a Wiener measure on $\mathbf{C}$.

Now, we demonstrate the existence of such a measure.

Theorem 2.114. *There exists only one Wiener measure on* $\mathbf{C}$.

Proof. Let $(\xi_n)_{n\in\mathbb{N}}$ be a independent sequence of random variables on $(\Omega, \mathcal{A}, P)$, every one normally distributed $\mathcal{N}(0, 1)$; let $(X_n)_{n\in\mathbb{N}}$ be the sequence of $(\mathcal{A} - \mathcal{B})$-measurable mappings defined in 2.105. We recall that

$$X_n(\omega)(t) = \frac{1}{\sqrt{n}} S_{[nt]}(\omega) + \frac{nt - [nt]}{\sqrt{n}} \cdot \xi_{[nt]+1}(\omega),$$

where $S_k(\omega) = \xi_1(\omega) + \cdots + \xi_k(\omega)$ and $S_0(\omega) = 0$. For each $n \in \mathbb{N}$, let $P_n : \mathcal{B} \to \mathbb{R}$ be the image of P under the mapping X_n.

We will identify the Wiener measure as the $\mathcal{T}$-limit of $(P_n)_{n\in\mathbb{N}}$.

We have shown in Theorem 2.109 that $(P_n)_{n\in\mathbb{N}}$ is tight. Therefore, since $\mathbf{C}$ is Polish, and so completely regular, we can apply Prohorov's theorem (see 2.71). Then the family $\{P_n : n \in \mathbb{N}\}$ is relatively $\mathcal{T}$-compact. So every subsequence $(P'_n)_{n\in\mathbb{N}}$ of $(P_n)_{n\in\mathbb{N}}$ admits at least a subsequence $(P''_n)_{n\in\mathbb{N}}$ narrowly convergent to a measure $\mu \in Rca^+(\mathcal{B})$. We only have to demonstrate that μ is the Wiener measure under research.

According to Theorem 2.90, for each $p \in \mathbb{N}^*$ and each $0 \le t_1 \le t_2 \le \cdots \le t_p \le 1$,

$$\pi_{t_1,\dots,t_p}(P''_n) \xrightarrow{\mathcal{T}_p} \pi_{t_1,\dots,t_p}(\mu).$$

Particularly, for every $t \in (0, 1]$,

$$\pi_t(P''_n) \xrightarrow{\mathcal{T}_1} \pi_t(\mu).$$

We have therefore, according to (ii) and (iv) of Theorem 2.53, on the one hand, $(\pi_t(P_n''))(\mathbb{R}) = P_n''(\mathbf{C}) = 1 \longrightarrow 1 = (\pi_t(\mu))(\mathbb{R}) = \mu(\mathbf{C})$, from where it results that μ is a probability on $\mathbf{C}$.

On the other hand, for every $\alpha \in \mathbb{R}$ such that $(\pi_t(\mu))(\partial(\,]-\infty,\alpha])) = \mu(\{x \in \mathbf{C} : x(t) = \alpha\}) = 0$,

$$\begin{aligned}\left(\pi_t(P_n'')\right)(\,]-\infty,\alpha]) &= P_n''\left(\{x \in \mathbf{C} \colon x(t) \le \alpha\}\right) \\ &= P\left(\{\omega \in \Omega : X_n''(\omega)(t) \le \alpha\}\right) \to \mu\left(\{x \in \mathbf{C} \colon x(t) \le \alpha\}\right).\end{aligned}$$

According to (ii) of Remark 2.106, for every $n \in \mathbb{N}$, $X_n(.)(t)$ is normally distributed $\mathcal{N}(0, \sigma_n(t))$ where $\sigma_n^2(t) = \frac{[nt]}{n} + \frac{(nt-[nt])^2}{n}$. Then we have $\sigma_n^2(t) \to t$ and

$$P\left(\{\omega \in \Omega : X_n''(\omega)(t) \le \alpha\}\right) = \frac{1}{\sqrt{2\pi\sigma_n''^2(t)}} \int_{-\infty}^{\alpha} e^{-\frac{u^2}{2\sigma_n''^2(t)}}\, du.$$

So, for every $t \in]0,1]$ and every $\alpha \in \mathbb{R}$ with $\mu(\{x \in \mathbf{C} : x(t) = \alpha\}) = 0$,

$$\lim_n \left(\pi_t(P_n'')\right)(\,]-\infty,\alpha]) = \frac{1}{\sqrt{2\pi t}} \int_{-\infty}^{\alpha} e^{-\frac{u^2}{2t}}\, du. \tag{1}$$

Now, we note that there is a countable set $E \subseteq \mathbb{R}$ such that, for all $\alpha \in \mathbb{R} \setminus E$, $\mu(\{x \in \mathbf{C} : x(t) = \alpha\}) = 0$ and so, by (1),

$$\mu\left(\{x \in \mathbf{C} : x(t) \le \alpha\}\right) = \frac{1}{\sqrt{2\pi t}} \int_{-\infty}^{\alpha} e^{-\frac{u^2}{2t}}\, du, \quad \text{for every} \quad \alpha \in \mathbb{R} \setminus E. \tag{2}$$

Since $\mathbb{R} \setminus E$ is dense in $\mathbb{R}$, (2) is satisfied, for every $\alpha \in \mathbb{R}$ and every $t \in]0,1]$. The condition (i) of Definition 2.110 is therefore fulfiled by μ.

We still have to show that μ satisfies (ii) of Definition 2.110; this means that, for every $0 \le t_0 < t_1 < \cdots < t_n \le 1$ and every $i = 0, \ldots, n-1$, the random variables $x \mapsto x(t_{i+1}) - x(t_i)$, are independent.

According to (iii) of Remark 2.111, we can suppose that, for every $x \in \mathbf{C}$, $x(0) = 0$.

Let us start by $n = 2$. Let $0 < t < s \le 1$ and $T_2 : \mathbb{R}^2 \to \mathbb{R}^2$, $T_2(x_1, x_2) = (x_1, x_2 - x_1)$. For every $\alpha, \beta \in \mathbb{R}$, we note $A =]-\infty,\alpha] \times]-\infty,\beta]$.

Using an analogous demonstration to the previous one, we obtain:

$$\begin{aligned}\mu\left(\{x \in \mathbf{C} : x(t) \le \alpha, x(s) - x(t) \le \beta\}\right) &= \mu\left(\{x \in \mathbf{C} : T_2(x(t), x(s)) \in A\}\right) \\ &= \mu\left(\{x \in \mathbf{C} : (x(t), x(s)) \in T_2^{-1}(A)\}\right) \\ &= \left(\pi_{t,s}(\mu)\right)\left(T_2^{-1}(A)\right) = \lim_n \left(\pi_{t,s}(P_n'')\right)\left(T_2^{-1}(A)\right) \\ &= \lim_n P\left(\{\omega \in \Omega : \left(X_n''(\omega)(t), X_n''(\omega)(s)\right) \in T_2^{-1}(A)\}\right) \\ &= \lim_n P\left(\{\omega \in \Omega \colon X_n''(\omega)(t) \le \alpha, X_n''(\omega)(s) - X_n''(\omega)(t) \le \beta\}\right).\end{aligned}$$

Since the random variables $(\xi_n)_{n\in\mathbb{N}}$ are independent,

$$\begin{aligned}P\left(\{\omega\in\Omega : X''_n(\omega)(t) \right. &\left.\le \alpha, X''_n(\omega)(s)-X''_n(\omega)(t)\le\beta\}\right)\\ &= P\left(\{\omega\in\Omega : X''_n(\omega)(t)\le\alpha\}\right)\\ &\quad\times P\left(\{\omega\in\Omega : X''_n(\omega)(s)-X''_n(\omega)(t)\le\beta\}\right)\end{aligned}$$

Therefore, we have:

$$\begin{aligned}\mu\left(\{x\in \mathbf{C} : x(t)\right. &\left.\le \alpha, x(s)-x(t)\le\beta\}\right)\\ &= \lim_n P\left(\{\omega\in\Omega : X''_n(\omega)(t)\le\alpha\}\right)\\ &\quad\times P\left(\{\omega\in\Omega : X''_n(\omega)(s)-X''_n(\omega)(t)\le\beta\}\right)\\ &= \mu\left(\{x\in\mathbf{C} : x(t)\le\alpha\}\right)\cdot\mu\left(\{x\in\mathbf{C} : x(s)-x(t)\le\beta\}\right).\end{aligned}$$

Therefore, we have shown that the random variables $x\mapsto x(t)$ and $x\mapsto x(s)-x(t)$ are independent.

Similarly, we show that, for every $n\in\mathbb{N}$ and every $0\le t_0<t_1<\cdots<t_n\le 1$, the set of random variables $\{x\mapsto x(t_{i+1})-x(t_i) : i=0,\ldots,n-1\}$ is independent with respect to μ.

Hence μ satisfies the condition (ii) of Definition 2.110.

Therefore, $\mu=W$ is the unique Wiener measure on $\mathbf{C}$.

Since every subsequence $(P'_n)_{n\in\mathbb{N}}$ of $(P_n)_{n\in\mathbb{N}}$ has a subsequence $(P''_n)_{n\in\mathbb{N}}$ narrowly convergent to W, then $P_n\xrightarrow{\mathcal{T}} W$. □

Chapter 3

Young Measures

In this last chapter, we will use the compactness results obtained in the two previous chapters in order to study the Young measures.

On the one hand, the Young measures generalize the measurable functions. Even if they were introduced in order to obtain relaxed solutions for the variational problems, the Young measures represent an important tool for other branches of mathematical analysis.

In the first paragraph, we start with the presentation of the preliminaries connected to the disintegration of measures as well as of the various families of integrands.

In the second paragraph, we present the definitions and several characteristics of Young measures and we give some examples.

We will study, in the third and fourth paragraphs, the stable topology on the Young measure spaces and the properties of its trace on the subspace of measurable functions. The two main results of the fourth paragraph point out that the trace of the stable topology on the subspace of measurable functions is the topology of convergence in measure and that this subspace is dense in the space of Young measures.

In the fifth paragraph, we will present Prohorov's theorem of compactness for the Young measures. The main element in this theorem is the condition of tightness; the simplicity of this condition represents the main attraction of the results of compactness in the Young measure spaces.

In the first chapter, we have introduced the scalar version of the biting lemma; the sixth paragraph is dedicated to the study of the vector version of this lemma. By adding to the tightness condition that of boundedness, we can state the general result of compactness of Saadoune–Valadier.

We will extend these two results of compactness (biting lemma and theorem of Saadoune–Valadier) to an unbounded set of measurable functions—the finite-tight sets.

In the seventh paragraph, we will study the two types of products for the Young measures and will give the fiber product lemma.

In the eighth paragraph, we will introduce, in the Euclidean case, the Jordan finite-tight sets. The main result of the paragraph states that a tight subset $H \subseteq W^{1,1}(\Omega, \mathbb{R}^m)$, for which the set of its gradients ∇H is a Jordan finite-tight set, is relatively compact in measure. This result will offer excellent conditions of application of fiber product lemma in order to obtain of lower semicontinuity results for integral functionals.

In the ninth paragraph, we will give several applications of Young measure theory in the study of strong compactness in L^p.

In the last two sections of the book, we consider some aspects of relaxed variational calculus. Are studied the Young measures generated by sequences and particularly the gradient Young measures. We pay special attention to quasiconvexity and its various equivalent definitions. The quasiconvexity is essentially used in the Kinderlehrer–Pedregal's characterization of gradient Young measures, but also in the study of lower semicontinuity of energy functional that appears in variational calculus. Finally, we present some results of existence of solutions in a relaxed variant of variational calculus.

The theory of the Young measures is presented in the general frame of the regular Suslin spaces which allows us to included the special case of a separable Banach spaces equipped with the weak topology (see Example 2.17). To simplify the demonstrations, we will present some results in the particular case of Polish spaces or even of the Euclidean spaces, by mentioning, at the same time, the bibliography for the more general case.

3.1 Preliminaries

In this chapter, we will use the following notations:

- Ω is an arbitrary set,
- $\mathcal{A}$ is a σ-algebra of subsets of Ω,
- $\mu \in ca^+(\mathcal{A})$ is a complete, σ-additive measure,
- (S, τ_S) is a regular Suslin space.

We recall that such a space S is a normal space and, according to Theorem 2.29, S is a Radon space ($ca(\mathcal{B}_S) = Rca(\mathcal{B}_S)$).

- $\mathcal{B}_S$ is the family of Borel sets of S,
- $\mathcal{P}_S$ is the set of all probabilities on S endowed with the narrow topology,
- $\mathcal{C}$ is the σ-algebra of Borel sets of $\mathcal{P}_S$.
- $\mathcal{B}_{\mathbb{R}}$ is the family of Borel sets of $\mathbb{R}$ endowed with the usual topology,
- $\mathcal{B}_{[0,1]}$ is the σ-algebra of Borel sets of subspace $[0, 1] \subseteq \mathbb{R}$.

If X is any topological space and $\mathcal{B}_X$ is the σ-algebra of Borel sets, then a mapping $f : \Omega \to X$ is $(\mathcal{A} - \mathcal{B}_X)$*-measurable* if $f^{-1}(B) \in \mathcal{A}$, for every $B \in \mathcal{B}_X$.

- $\mathcal{M}(X)$ notes the set of all $(\mathcal{A} - \mathcal{B}_X)$-measurable mappings.
- Particularly, if $X = \mathbb{R}$, then we note $\mathcal{M}(\mathbb{R})$ with $\mathcal{M}$.
- For every $f \in \mathcal{M}(X)$, $\bar{f}$ denotes the set of all mappings of $\mathcal{M}(X)$ equal with f μ-almost everywhere (μ-a.e.).

If $(E, \|\cdot\|)$ is a separable Banach space and if $p \geq 1$, then

- $\mathcal{L}^p(\Omega, \mu, E) = \mathcal{L}^p(\Omega, E) = \mathcal{L}^p(\mu, E)$
$= \{f \in \mathcal{M}(E) : \|f\|_p = \int_\Omega \|f\|^p d\mu < +\infty\}$ and
- $L^p(\Omega, \mu, E) = L^p(\Omega, E) = L^p(\mu, E) = \{\bar{f} : f \in \mathcal{L}^p(\mu, E)\}$.

More particular, $\mathcal{L}^p(\mu) = \mathcal{L}^p(\mu, \mathbb{R})$ and $L^p(\mu) = L^p(\mu, \mathbb{R})$.

- $\mathcal{L}^\infty(\Omega, \mu, E) = \mathcal{L}^\infty(\mu, E)$
$= \{f \in \mathcal{M}(E) : \|f\|_\infty = \inf\{c > 0 : \|f\| \leq c, \mu - \text{p.p.}\} < +\infty\}$,
- $L^\infty(\Omega, \mu, E) = L^\infty(\mu, E) = \{\bar{f} : f \in \mathcal{L}^\infty(\mu, E)\}$.

3.1.1 Disintegration

Theorem 3.1. *Let $\tau : \Omega \to \mathcal{P}_S$, $\tau(t) = \tau_t$, for every $t \in \Omega$; the following properties are equivalent:*

(i) *τ is $(\mathcal{A} - \mathcal{C})$-measurable.*

(ii) *For every $B \in \mathcal{B}_S$, the mapping $g_B : \Omega \to [0,1]$, defined by $g_B(t) = \tau_t(B)$, is $(\mathcal{A} - \mathcal{B}_{[0,1]})$-measurable.*

(iii) *For every $C \in \mathcal{A} \otimes \mathcal{B}_S$, the mapping $f_C : \Omega \to [0,1]$, defined by $f_C(t) = \tau_t(C_t)$, is $(\mathcal{A} - \mathcal{B}_{[0,1]})$-measurable, where, for every $t \in \Omega$, $C_t = \{x \in S : (t,x) \in C\}$ denotes the section of C determined by t.*

Proof. (i) $\Rightarrow$ (ii) Let $\mathcal{E} = \{B \in \mathcal{B}_S : g_B \text{ is } (\mathcal{A} - \mathcal{B}_{[0,1]})\text{-measurable}\}$. We show that $\mathcal{E} = \mathcal{B}_S$.

(a) For every open set $D \subseteq S$, χ_D is l.s.c. from where the mapping $F : (\mathcal{P}_S, \mathcal{T}) \to \mathbb{R}_+$, defined by $F(\nu) = \nu(D)$, is l.s.c. (see Corollary 2.41, (i)). Then, for any $\alpha \in \mathbb{R}$, $G = \{\nu \in \mathcal{P}_S : \nu(D) > \alpha\} = F^{-1}(]\alpha, +\infty[)$ is narrowly open and, since τ is $(\mathcal{A} - \mathcal{C})$-measurable, $\tau^{-1}(G) = \{t \in \Omega : g_D(t) > \alpha\} \in \mathcal{A}$ from where it results that g_D is $(\mathcal{A} - \mathcal{B}_{[0,1]})$-measurable and, therefore, $D \in \mathcal{E}$.

(b) For every open set D and every closed set F in S, $g_{D\cap F}(t) = \tau_t(D\cap F) = \tau_t(D) - \tau_t(D\setminus F) = g_D(t) - g_{D\setminus F}(t)$. By (a), $D, D\setminus F \in \mathcal{E}$ and so $g_{D\cap F}$ is $(\mathcal{A} - \mathcal{B}_{[0,1]})$-measurable, hence $D\cap F \in \mathcal{E}$.

(c) We can easily show that $\mathcal{E}$ is a monotone class of sets, closed under countable pairwise disjoint union of sets.

(d) By (b) and (c), it results that $\mathcal{E}$ contains the σ-algebra generated by the open sets of S, therefore it contains $\mathcal{B}_S$, which proved (ii).

(ii) $\Rightarrow$ (iii) Let $\mathcal{F} = \{C \in \mathcal{A}\otimes\mathcal{B}_S : f_C \text{ is } (\mathcal{A} - \mathcal{B}_{[0,1]})\text{-measurable}\}$.

(a) For every $A \in \mathcal{A}$ and every $B \in \mathcal{B}_S$,

$$f_{A\times B}(t) = \tau_t((A\times B)_t)$$
$$= \begin{cases} \tau_t(B), & t\in A \\ 0, & t\notin A \end{cases} = \chi_A(t)\cdot\tau_t(B) = \chi_A(t)\cdot g_B(t),$$

from where it results that $f_{A\times B}$ is $(\mathcal{A} - \mathcal{B}_{[0,1]})$-measurable and $A\times B \in \mathcal{F}$.

(b) For every sequence $(C_n)_{n\in\mathbb{N}}$ of pairwise disjoint subsets of $\mathcal{F}$, let $C = \bigcup_{n=1}^{\infty} C_n$; then, for every $t \in \Omega$, $C_t = \bigcup_{n=1}^{\infty}(C_n)_t$, from where $f_C(t) = \tau_t(\bigcup_{n=1}^{\infty}(C_n)_t) = \sum_{n=1}^{\infty}\tau_t((C_n)_t)$. Then $f_C = \sum_{n=1}^{\infty} f_{C_n}$ is measurable and so $C \in \mathcal{F}$.

(c) Let $\mathcal{R} = \{\bigcup_{k=1}^{n}(A_k\times B_k) : n\in\mathbb{N}, (A_k\times B_k)\cap(A_l\times B_l) = \emptyset$ if $k\neq l, A_k\in\mathcal{A}$ and $B_k\in\mathcal{B}_S\}$; by (a) and (b), $\mathcal{R}$ is a sub-algebra of $\mathcal{F}$.

(d) Let $(C_n)_n \subset \mathcal{F}$ be an increasing sequence and let $C = \cup_{n=1}^{\infty} C_n$; then $f_C(t) = \tau_t(\cup_{n=1}^{\infty}(C_n)_t) = \lim_n \tau_t((C_n)_t) = \lim_n f_{C_n}(t)$. Therefore f_C is $(\mathcal{A} - \mathcal{B}_{[0,1]})$-measurable and $C \in \mathcal{F}$.

In the same way, we will show that $\mathcal{F}$ is closed under the limit of decreasing sequences.

Then $\mathcal{F}$ is a monotone class and, according to (c), the σ-algebra generated by $\mathcal{R}$, $\mathcal{A}\otimes\mathcal{B}_S \subset \mathcal{F}$. It follows that, for every $C\in\mathcal{A}\otimes\mathcal{B}_S$, f_C is $(\mathcal{A} - \mathcal{B}_{[0,1]})$-measurable.

(iii) $\Rightarrow$ (i) For every $B \in \mathcal{B}_S$, let $C = \Omega\times B$; $f_C(t) = \tau_t(B) = \int_S \chi_B(x)d\tau_t(x)$; by hypothesis, the mapping $t\mapsto f_C(t)$ is $(\mathcal{A} - \mathcal{B}_{[0,1]})$-measurable. It follows that, for every simple function f (with regards to $\mathcal{B}_S$), the mapping $t\mapsto \int_S f(x)d\tau_t(x)$ is $(\mathcal{A} - \mathcal{B}_{[0,1]})$-measurable. For every function $f\in C_b(S)$, there exists a sequence of simple functions, $(f_n)_{n\in\mathbb{N}}$, uniformly convergent to f; then the sequence $(f_n)_{n\in\mathbb{N}}$ is uniformly bounded and so, according to Lebesgue theorem,

$$\int_S f_n(x)d\tau_t(x) \longrightarrow \int_S f(x)d\tau_t(x), \quad \forall t\in\Omega.$$

Therefore, the mapping $t \mapsto \int_S f(x) d\tau_t(x)$ is $(\mathcal{A} - \mathcal{B}_{[0,1]})$-measurable. As $\sigma(\mathcal{P}_S, C_b(S))$ is the narrow topology on $\mathcal{P}_S$, we obtain that the mapping $\tau : \Omega \to \mathcal{P}_S$ is measurable. □

The following theorem presents the disintegration result. Its proof follows the one of [167].

Theorem 3.2. *Let $p_\Omega : \Omega \times S \to \Omega$ be the canonical projection on Ω. For every measure $\mathcal{T} \in ca^+(\mathcal{A} \otimes \mathcal{B}_S)$ such that $\mu = p_\Omega(\mathcal{T})$, there exists a mapping $\tau : \Omega \to \mathcal{P}_S$, $(\mathcal{A} - \mathcal{C})$-measurable such that*

$$\mathcal{T}(A \times B) = \int_A \tau_t(B) d\mu(t), \quad \textit{for every} \quad A \in \mathcal{A}, B \in \mathcal{B}_S.$$

Proof. (1) Firstly, let S be a compact Suslin space and let $C(S)$ be the space of all real-valued continuous mapping on S endowed with the norm of uniform convergence $\|\cdot\|_0$.

For every $f \in C(S)$, we can introduce the mapping

$$F_f : L^1(\mu) \to \mathbb{R}, \text{ by } F_f(\bar{g}) = \int_{\Omega \times S} g(t) \cdot f(x) d\mathcal{T}(t,x), \text{ for every } g \in \mathcal{L}^1(\mu).$$

Then F_f is a linear map on $L^1(\mu)$ and, since $|F_f(\bar{g})| \leq \|f\|_0 \cdot \|\bar{g}\|_1$, F_f belongs to the dual of $L^1(\mu)$ which is $L^\infty(\mu)$ (see Theorem 1.49).

Let $h_f \in \mathcal{L}^\infty(\mu)$ such that

$$F_f(\bar{g}) = \int_\Omega g(t) \cdot h_f(t) d\mu(t), \quad \text{for every} \quad g \in \mathcal{L}^1(\mu).$$

Obviously, $\|h_f\|_\infty = \|F_f\| \leq \|f\|_0$.

Let $\varphi : C(S) \to L^\infty(\mu)$, defined by $\varphi(f) = \overline{h_f}$. We can see that φ is a linear map on $C(S)$ and, since $\|\varphi(f)\|_\infty \leq \|f\|_0$, for every $f \in C(S)$, φ is continuous; moreover $\varphi(\underline{1}) = \underline{1}$.

According to Theorem IV.2.3 of [97], there exists a linear lifting $\rho : \mathcal{L}^\infty(\mu) \to \mathcal{L}^\infty(\mu)$, i.e. an application having the following properties:

(i) $\rho(h) = h$, μ-a.e.,

(ii) $\rho(h_1) = \rho(h_2)$, if $h_1 = h_2$, μ-a.e.,

(iii) $\rho(\alpha h_1 + \beta h_2) = \alpha\rho(h_1) + \beta\rho(h_2), \quad \forall\, \alpha, \beta \in \mathbb{R}, h_1, h_2 \in \mathcal{L}^\infty(\mu)$,

(iv) $\rho(h) \geq 0$, if $h \geq 0$,

(v) $\rho(\underline{1}) = \underline{1}$,

(vi) $\rho(h_1 \cdot h_2) = \rho(h_1) \cdot \rho(h_2)$.

Therefore, we can define the map $\Psi : L^\infty(\mu) \to \mathcal{L}^\infty(\mu)$ letting $\Psi(\bar{h}) = \rho(h)$, for every $h \in \mathcal{L}^\infty(\mu)$. Ψ is correctly defined because the definition does not depend on our choice of representatives in the class $\bar{h}$ containing h.

Ψ is a linear map and, since $-\|h\|_\infty \leq h \leq \|h\|_\infty$, μ-a.e.,

$$-\|h\|_\infty \leq \rho(h) \leq \|h\|_\infty.$$

Then $\|\rho(h)\|_\infty \leq \|h\|_\infty$, from where $\|\Psi(\bar{h})\|_\infty \leq \|h\|_\infty = \|\bar{h}\|_\infty$ and therefore Ψ is continuous.

For every $t \in \Omega$ and every $f \in C(S)$, let now

$$\tau_t(f) = \Psi\left(\varphi(f)\right)(t) = \Psi\left(\overline{h_f}\right)(t) = \rho(h_f)(t).$$

We can easily remark that τ_t is linear and that, for every $f \in C(S)$

$$|\tau_t(f)| = \left|\rho(h_f)(t)\right| \leq \|h_f\|_\infty \leq \|f\|_0, \mu\text{-a.e. } t \in \Omega.$$

Therefore τ_t is a Radon measure on S; since $\tau_t(\underline{1}) = \Psi(\varphi(\underline{1}))(t) = \rho(\underline{1})(t) = 1$, τ_t is a probability on S.

Let now the mapping $\tau : \Omega \to \mathcal{P}_S$, defined by $\tau(t) = \tau_t$, for every $t \in \Omega$.

Since S is a compact Suslin space, according to Corollary 2.26 and to Theorem 2.58, $Rca^+(\mathcal{B}_S)$ is a Polish space; as $\mathcal{P}_S$ is closed in $Rca^+(\mathcal{B}_S)$, it is Polish, also. Hence the narrow topology on $\mathcal{P}_S$ has a countable base. $\mathcal{C}$ is the family of all Borel sets on $\mathcal{P}_S$; so, τ is $(\mathcal{A} - \mathcal{C})$-measurable if and only if, $\{t \in \Omega : \tau_t(f) > \alpha\} \in \mathcal{A}$, for every $\alpha \in \mathbb{R}$ and every $f \in C(S)$. But $\tau_t(f) = \Psi(\varphi(f))(t)$ and, since $\Psi(\varphi(f)) \in \mathcal{L}^\infty(\mu)$, the mapping $\Psi(\varphi(f))$ is measurable and the condition stated above is proved. Therefore τ is a mapping $(\mathcal{A} - \mathcal{C})$-measurable of Ω in $\mathcal{P}_S$.

For every $A \in \mathcal{A}$ and every $f \in C(S)$,

$$\begin{aligned}
\int_A \tau_t(f)d\mu(t) &= \int_A \Psi\left(\varphi(f)\right)(t)d\mu(t) = \int_A \rho(h_f)(t)d\mu(t) = \int_A h_f(t)d\mu(t) \\
&= \int_\Omega \chi_A(t) \cdot h_f(t)d\mu(t) = F_f(\overline{\chi_A}) \\
&= \int_{\Omega\times S} \chi_A(t) \cdot f(x)d\tau(t,x) = \int_{A\times S} f(x)d\tau(t,x).
\end{aligned}$$

Since f is arbitrary in $C(S)$, for every open set $D \subseteq S$,

$$\int_A \tau_t(D)d\mu(t) = \int_{A\times S} \chi_D(x)d\tau(t,x) = \tau(A \times D)$$

(we must remark that, according to Theorem 3.1 (ii), the mapping $t \mapsto \tau_t(D)$ is positive and $(\mathcal{A} - \mathcal{B}_{[0,1]})$-measurable).

Therefore, the previous relation is also proved for all $D \in \mathcal{B}_S$.

(2) Let now S be an arbitrary regular Suslin space and let $\nu = p_S(\tau)$, where $p_S : \Omega \times S \to S$ is the canonical projection on S. According to Theorem 2.29, $\nu \in ca^+(\mathcal{B}_S) = Rca^+(\mathcal{B}_S)$. Hence there exists a sequence of pairwise disjoint compact subsets of S, $(K_n)_{n\in\mathbb{N}}$, such that

$$\nu(S \setminus \bigcup_{n=1}^{\infty} K_n) = 0.$$

For every $n \in \mathbb{N}$, let $\tau^n : \mathcal{A} \otimes \mathcal{B}_{K_n} \to \mathbb{R}_+$, $\tau^n(M) = \tau((\Omega \times K_n) \cap M)$ and let $\mu^n = p_\Omega(\tau^n)$ be the projection of τ^n on Ω ($\mu^n(A) = \tau(A \times K_n)$); according to Radon–Nikodym's theorem (Theorem 1.40), since μ^n is μ-continuous, there exists an application $g_n : \Omega \to \mathbb{R}_+$, $g_n \in \mathcal{L}^1(\mu)$ such that

$$\mu^n(A) = \int_A g_n d\mu, \quad \text{for every} \quad A \in \mathcal{A}.$$

For every $A \in \mathcal{A}$, $\tau(A \times S) = \sum_{n=1}^{\infty} \tau(A \times K_n)$ and so we obtain

$$\mu(A) = \sum_{n=1}^{\infty} \mu^n(A) = \int_A \sum_{n=1}^{\infty} g_n(t)d\mu(t), \quad \text{for every} \quad A \in \mathcal{A},$$

from where

$$\sum_{n=1}^{\infty} g_n = \underline{1}, \mu - \text{a.e.}$$

By (1), for every $n \in \mathbb{N}$, there exists an application $\tau^n : \Omega \to \mathcal{P}_{K_n}$, $(\mathcal{A} - \mathcal{C}_n)$-measurable such that

$$\tau^n(A \times B) = \int_A \tau_t^n(B)d\mu^n(t), \quad \text{for every} \quad A \in \mathcal{A}, B \in \mathcal{B}_{K_n}.$$

According to 2.57 (i), for every $B \in \mathcal{B}_S$ and every $n \in \mathbb{N}$, $B \cap K_n \in \mathcal{B}_{K_n}$ and so we can define, for every $t \in \Omega$,

$$\tau_t(B) = \sum_{n=1}^{\infty} g_n(t) \cdot \tau_t^n(B \cap K_n). \tag{*}$$

Then, for every $t \in \Omega$, $\tau_t \in ca^+(\mathcal{B}_S) = Rca^+(\mathcal{B}_S)$ and, as $\tau_t(S) = \sum_{n=1}^{\infty} g_n(t) \cdot \tau_t^n(K_n) = \sum_{n=1}^{\infty} g_n(t) = 1$, $\tau_t \in \mathcal{P}_S$.

According to Proposition 2.57 (v), for every $n \in \mathbb{N}$, $\mathcal{P}_{K_n} \hookrightarrow \mathcal{P}_S$ and then τ is measurable.

Finally, for every $A \in \mathcal{A}$ and every $B \in \mathcal{B}_S$,

$$\begin{aligned}\mathcal{T}(A \times B) &= \sum_{n=1}^{\infty} \mathcal{T}^n(A \times (B \cap K_n)) = \sum_{n=1}^{\infty} \int_A \tau_t^n(B \cap K_n) d\mu^n(t) \\ &= \sum_{n=1}^{\infty} \int_A g_n(t) \cdot \tau_t^n(B \cap K_n) d\mu(t) = \int_A \tau_t(B) d\mu(t). \qquad \square\end{aligned}$$

Remark 3.3.

(i) Let $\mathcal{T} \in ca^+(\mathcal{A} \otimes \mathcal{B}_S)$ and let $\tau : \Omega \to \mathcal{P}_S$ be obtained by using the previous theorem. Let $\tau' : \Omega \to \mathcal{P}_S$ an other application $(\mathcal{A} - \mathcal{C})$-measurable such that, for every $A \in \mathcal{A}$ and $B \in \mathcal{B}_S$,

$$\mathcal{T}(A \times B) = \int_A \tau'_t(B) d\mu(t).$$

Then, for every $B \in \mathcal{B}_S$, $\tau_t(B) = \tau'_t(B)$, μ-almost everywhere on Ω.

The relation $\tau \sim \tau'$ defined by:

$$\tau_{\cdot}(B) = \tau'_{\cdot}(B), \mu - \text{a.e. and for every} \quad B \in \mathcal{B}_S$$

is an equivalence relation. Then, if we accept to identify two equivalent applications, the mapping τ of the previous theorem is the unique application $(\mathcal{A} - \mathcal{C})$-measurable for which

$$\mathcal{T}(A \times B) = \int_A \tau_t(B) d\mu(t), \quad \text{for every} \quad A \in \mathcal{A} \quad \text{and every} \quad B \in \mathcal{B}_S.$$

(ii) Let $\tau, \tau' : \Omega \to \mathcal{P}_S$ two equivalent $(\mathcal{A} - \mathcal{C})$-measurable mappings and, for every $B \in \mathcal{B}_S$, let

$$A_B = \{t \in \Omega : \tau_t(B) \neq \tau'_t(B)\} \quad \text{and} \quad A = \bigcup_{B \in \mathcal{B}_S} A_B = \{t \in \Omega : \tau_t \neq \tau'_t\}.$$

If $\mathcal{B}_S$ is countably generated ($\mathcal{B}_S$ is generated by a countable algebra), then $A \in \mathcal{A}$ and $\mu(A) = 0$.

Indeed, if $\mathcal{B}_0 = \{B_n : n \in \mathbb{N}\} \subseteq \mathcal{B}_S$ is a countable algebra which generates $\mathcal{B}_S$, then $A = \cup_{n \in \mathbb{N}} A_{B_n}$.

In this case $\tau_{\cdot} = \tau'_{\cdot}$, μ-almost everywhere.

Therefore, if $\mathcal{B}_S$ is countably generated (particularly, if (S, τ_S) is second-countable), then $\tau \sim \tau'$ if and only if $\tau = \tau'$, μ-a.e.

Definition 3.4. The mapping $\tau : \Omega \to \mathcal{P}_S$ of Theorem 3.2 is called the *disintegration* of measure $\mathcal{T} \in ca^+(\mathcal{A} \otimes \mathcal{B}_S)$ with respect to μ. According to (i) of Remark 3.3, $\mathcal{T}$ has an unique disintegration $\tau_{\cdot}$.

Proposition 3.5. *Let $\mathcal{T} \in ca^+(\mathcal{A} \otimes \mathcal{B}_S)$ such that $p_\Omega(\mathcal{T}) = \mu$ and let $\tau : \Omega \to \mathcal{P}_S$ the disintegration of $\mathcal{T}$; then:*

(i) $\mathcal{T}(C) = \int_\Omega \tau_t(C_t) d\mu(t)$, *for every $C \in \mathcal{A} \otimes \mathcal{B}_S$; $C_t = \{x \in S : (t, x) \in C\}$ is the section of C determined by t.*

(ii) $\int_{\Omega \times S} f(t, x) d\mathcal{T}(t, x) = \int_\Omega [\int_S f(t, x) d\tau_t(x)] d\mu(t)$, *for every $(\mathcal{A} \otimes \mathcal{B}_S)$-measurable and positive mapping f, or, for every mapping $f \in \mathcal{L}^1(\mathcal{T})$.*

Proof. (i) Let $\mathcal{E} = \{A \times B : A \in \mathcal{A}, B \in \mathcal{B}_S\}$, $\mathcal{F} = \{C \in \mathcal{A} \otimes \mathcal{B}_S : \mathcal{T}(C) = \int_\Omega \tau_t(C_t) d\mu(t)\}$.

We need to remark that, according to Theorem 3.1 (iii), the mapping $t \mapsto \tau_t(C_t)$ is positive and $(\mathcal{A} - \mathcal{B}_{[0,1]})$-measurable.

According to Theorem 3.2, $\mathcal{E} \subseteq \mathcal{F}$. $\mathcal{E}$ is a semiring and then the algebra that it generates is

$$\mathcal{G} = \left\{ \bigcup_{i=1}^{n} C_i : n \in \mathbb{N}^*, C_1, \ldots, C_n \in \mathcal{E}, C_i \cap C_j = \emptyset, \forall i \neq j \right\}.$$

Let $C_i = A_i \times B_i \in \mathcal{E}, i = 1, 2$, with $C_1 \cap C_2 = \emptyset$ and let $C = C_1 \cup C_2$; then $A_1 \cap A_2 = \emptyset$ or $B_1 \cap B_2 = \emptyset$.

If $A_1 \cap A_2 = \emptyset$, then

$$C_t = \begin{cases} B_1, & t \in A_1, \\ B_2, & t \in A_2, \\ \emptyset, & t \in \Omega \setminus (A_1 \cup A_2) \end{cases} \quad \text{and therefore}$$

$$\int_\Omega \tau_t(C_t) d\mu(t) = \int_{A_1} \tau_t(B_1) d\mu(t) + \int_{A_2} \tau_t(B_2) d\mu(t) = \mathcal{T}(C_1) + \mathcal{T}(C_2) = \mathcal{T}(C),$$

from where $C \in \mathcal{F}$.

If $B_1 \cap B_2 = \emptyset$ and $A_1 \cap A_2 \neq \emptyset$, then

$$C_t = \begin{cases} B_1, & t \in A_1 \setminus A_2, \\ B_2, & t \in A_2 \setminus A_1, \\ B_1 \cup B_2, & t \in A_1 \cap A_2, \\ \emptyset, & t \in \Omega \setminus (A_1 \cup A_2) \end{cases} \quad \text{and then}$$

$$\begin{aligned} \int_\Omega \tau_t(C_t)d\mu(t) &= \int_{A_1 \setminus A_2} \tau_t(B_1)d\mu(t) + \int_{A_2 \setminus A_1} \tau_t(B_2)d\mu(t) \\ &\quad + \int_{A_1 \cap A_2} \tau_t(B_1 \cup B_2)d\mu(t) \\ &= \int_{A_1} \tau_t(B_1)d\mu(t) + \int_{A_2} \tau_t(B_2)d\mu(t) \\ &= \mathcal{T}(C_1) + \mathcal{T}(C_2) = \mathcal{T}(C) \end{aligned}$$

and again $C \in \mathcal{F}$.

A similar reasoning applied to any element of the family $\mathcal{G}$ leads to the inclusion $\mathcal{G} \subseteq \mathcal{F}$.

Since $\mathcal{F}$ is a monotone class, $\mathcal{A} \otimes \mathcal{B}_S$—the σ-algebra generated by $\mathcal{G}$—coincides with $\mathcal{F}$.

(ii) According to (i), the condition (ii) is proved for the simple functions; therefore, according to the theorem of monotonous convergence, it is proved for measurable and positive functions and, according to Lebesgue's theorem of dominated convergence, for the functions of $\mathcal{L}^1(\mathcal{T})$. □

Remark 3.6. Let $\tau : \Omega \to \mathcal{P}_S$ be an $(\mathcal{A} - \mathcal{C})$-measurable function; the mapping $\mathcal{T} : \mathcal{A} \otimes \mathcal{B}_S \to \mathbb{R}_+$ defined by

$$\mathcal{T}(C) = \int_\Omega \tau_t(C_t)d\mu(t), \quad \text{for every} \quad C \in \mathcal{A} \otimes \mathcal{B}_S,$$

is a positive and finite measure on $\mathcal{A} \otimes \mathcal{B}_S$ for which $p_\Omega(\mathcal{T}) = \mu$. The disintegration of $\mathcal{T}$ is equivalent with τ (according to Remark 3.3 (i)). By identifying the equivalent disintegrations, we can observe that the mapping $\mathcal{T} \mapsto \tau$ is a bijection between the measures of $ca^+(\mathcal{A} \otimes \mathcal{B}_S)$, for which $p_\Omega(\mathcal{T}) = \mu$, and their disintegrations.

3.1.2 Integrands

Definition 3.7. An *integrand* on S is an $(\mathcal{A}\otimes\mathcal{B}_S)$-measurable map, $\Psi : \Omega\times S \to \overline{\mathbb{R}}$.

An integrand Ψ is called L^1*-bounded* is there is an application $\phi \in L^1(\mu)$ such that $|\Psi(t,x)| \leq \phi(t)$, for every $t \in \Omega$ and every $x \in S$.

Ψ is said *lower semicontinuous-l.s.c.* (*upper semicontinuous-u.s.c.*) if, for every $t \in \Omega$, the application $\Psi(t,\cdot)$ is l.s.c. (u.s.c.) on S.

The integrand Ψ is a *Carathéodory integrand* if, for every $t \in \Omega$, $\Psi(t,\cdot)$ is continuous on S.

The set of all bounded Carathéodory integrands on $\Omega\times S$ is noted by $Cth^b(\Omega\times S)$.

The following proposition shows that we can see the Carathéodory integrands as functions measurable in the first variable and continuous in the second one.

Proposition 3.8 ([42], Lemma III.14 and [45], Lemma 1.2.3).
A mapping $\Psi : \Omega \times S \to \overline{\mathbb{R}}$ is a Carathéodory integrand if and only if the following two conditions are satisfied:

(i) $\Psi(\cdot,x)$ *is $\mathcal{A}$-measurable, for every $x \in S$.*

(ii) $\Psi(t,\cdot)$ *is continuous, for every $t \in \Omega$.*

Proof. It is sufficient to show that the two conditions mentioned above are sufficient so that Ψ to be $(\mathcal{A} \otimes \mathcal{B}_S)$-measurable.

Firstly, let us suppose that S is a metrizable Suslin space; let d be a metric which generates the topology of S ($\tau_S = \tau_d$) and let $A = \{x_n : n \in \mathbb{N}\}$ be a countable dense subset of (S,τ_d) (every Suslin space is separable).

For every $p \in \mathbb{N}^*$ and every $x \in S$, let

$$n(p,x) = \min\left\{n \in \mathbb{N} : x \in B\left(x_n, \frac{1}{p}\right)\right\},$$

where $B(x_n, \frac{1}{p})$ is the open ball of radius $\frac{1}{p}$ centered in x_n in the metric space (S,d).

For any $p \in \mathbb{N}^*$, let $\Psi_p : \Omega \times S \to \mathbb{R}$, $\Psi_p(t,x) = \Psi(t, x_{n(p,x)})$.

Since, for every $x \in S$, $d(x, x_{n(p,x)}) < \frac{1}{p}$, $\lim_{p\to\infty} x_{n(p,x)} = x$; then, according to (ii), for every $(t,x) \in \Omega \times S$,

$$\lim_{p\to\infty} \Psi_p(t,x) = \Psi(t,x). \tag{*}$$

For every $p \in \mathbb{N}^*$ and every Borel set $A \subseteq \overline{\mathbb{R}}$,

$$\begin{aligned}\Psi_p^{-1}(A) &= \bigcup_{n=0}^{\infty}\left\{(t,x)\in\Omega\times\left[B\left(x_n,\frac{1}{p}\right)\setminus\bigcup_{k<n}B\left(x_k,\frac{1}{p}\right)\right] : \Psi_p(t,x)\in A\right\}\\ &= \bigcup_{n=0}^{\infty}\left\{[\Psi(\cdot,x_n)]^{-1}(A)\times\left[B\left(x_n,\frac{1}{p}\right)\setminus\bigcup_{k<n}B\left(x_k,\frac{1}{p}\right)\right]\right\}.\end{aligned}$$

By (i), $\Psi(\cdot,x_n)$ is $\mathcal{A}$-measurable, for each $n \in \mathbb{N}$ and then $[\Psi(\cdot,x_n)]^{-1}(A) \in \mathcal{A}$. Therefore, $\Psi_p^{-1}(A) \in \mathcal{A} \otimes \mathcal{B}_S$, for every Borel set A of $\overline{\mathbb{R}}$, from where Ψ_p is $(\mathcal{A} \otimes \mathcal{B}_S)$-measurable.

According to $(*)$, Ψ is the limit of a sequence of $(\mathcal{A} \otimes \mathcal{B}_S)$-measurable mappings and then Ψ is $(\mathcal{A} \otimes \mathcal{B}_S)$-measurable.

In the general case, where S is only a regular Suslin space, according to Theorem 2.25 (see also [150]), there exists a metric d on S such that: $\tau_S \subseteq \tau_d$, τ_S and τ_d having the same Borel sets $\mathcal{B}_S$ and so that (S,τ_d) is Suslin too. Then $\Psi(\cdot,x)$ is $\mathcal{A}$-measurable, for every $x \in S$ and $\Psi(t,\cdot)$ is τ_d-continuous, for every $t \in \Omega$. According to the first part, Ψ is $(\mathcal{A} \otimes \mathcal{B}_S)$-measurable. □

If the space S is a metrizable space, then we can to approximate the bounded Carathéodory integrands with integrands Lipschitz continuous in the second variable.

Proposition 3.9. *Let d be a metric which generates the topology of S (i.e. $\tau_S = \tau_d$). For every bounded l.s.c. integrand Ψ on $\Omega \times S$, there exists a sequence $(\Psi_n)_{n\in\mathbb{N}}$ of bounded integrands, d-Lipschitz continuous in the second variable such that $\Psi_n \uparrow \Psi$ on $\Omega \times S$.*

Proof. Let $\|\Psi\| = \sup_{(t,x)\in\Omega\times S}|\Psi(t,x)| < +\infty$ and, for every $n \in \mathbb{N}$, let $\Psi_n : \Omega \times S \to \mathbb{R}$,

$$\Psi_n(t,x) = \inf_{y\in S}\{\Psi(t,y) + n\cdot d(x,y)\}.$$

Then $-\|\Psi\| \le \Psi_n(t,x) \le \Psi(t,x) \le \|\Psi\|$, for every $(t,x) \in \Omega \times S$.

Let $p_\Omega : \Omega \times S \to \Omega$ be the canonical projection on Ω; then, for every $n \in \mathbb{N}$, every $x \in S$ and every $\alpha \in \mathbb{R}$,

$$\{t \in \Omega : \Psi_n(t,x) < \alpha\} = p_\Omega\left(\{(t,y) \in \Omega \times S : \Psi(t,y) + n\cdot d(x,y) < \alpha\}\right).$$

Since Ψ is $(\mathcal{A} \otimes \mathcal{B}_S)$-measurable,

$$\{(t,y) \in \Omega \times S : \Psi(t,y) + n\cdot d(x,y) < \alpha\} \in \mathcal{A} \otimes \mathcal{B}_S.$$

Therefore, according to the measurable projection theorem (Theorem III.23 of [42]), since μ is a complete measure on $\mathcal{A}$,

$$\{t \in \Omega : \Psi_n(t, x) < \alpha\} \in \mathcal{A}.$$

Therefore $\Psi_n(\cdot, x)$ is $\mathcal{A}$-measurable, for every $x \in S$ and every $n \in \mathbb{N}$.

Let $x, y \in S$; for all $\varepsilon > 0$, there exists $z_\varepsilon \in S$ such that

$$\Psi(t, z_\varepsilon) + n \cdot d(y, z_\varepsilon) < \Psi_n(t, y) + \varepsilon$$

and then $\Psi_n(t, x) - \Psi_n(t, y) < \Psi(t, z_\varepsilon) + n \cdot d(x, z_\varepsilon) - \Psi(t, z_\varepsilon) - n \cdot d(y, z_\varepsilon) + \varepsilon \leq n \cdot d(x, y) + \varepsilon$. Hence $|\Psi_n(t, x) - \Psi_n(t, y)| < n \cdot d(x, y) + \varepsilon$ and, as ε is arbitrary, $\Psi_n(t, \cdot)$ is d-Lipschitz continuous, for every $t \in \Omega$. Therefore $\Psi_n(t, \cdot)$ is continuous, for every $t \in \Omega$; according to Proposition 3.8, for every $n \in \mathbb{N}$, Ψ_n is an integrand, Lipschitz continuous in the second variable on S.

Let $(t, x) \in \Omega \times S$ fixed; for every $n \in \mathbb{N}^*$, there exists $z_n \in S$ such that

$$\Psi(t, z_n) + n \cdot d(x, z_n) < \Psi_n(t, x) + \frac{1}{n} \leq \Psi(t, x) + \frac{1}{n}. \tag{1}$$

According to (1), $n \cdot d(x, z_n) \leq 2 \cdot \|\Psi\| + \frac{1}{n}$, for every $n \in \mathbb{N}^*$ and so

$$z_n \xrightarrow{d} x. \tag{2}$$

Again, by (1), $n \cdot d(x, z_n) < \Psi(t, x) + \frac{1}{n} - \Psi(t, z_n)$ and, since $\Psi(t, \cdot)$ is l.s.c.,

$$n \cdot d(x, z_n) \to 0. \tag{3}$$

We finally use once again (1) to obtain (4):

$$\Psi_n(t, x) \to \Psi(t, x). \tag{4}$$

It is obvious that the sequence $(\Psi_n(t, x))_{n \in \mathbb{N}}$ is increasing. □

3.2 Definitions and Examples

In this paragraph, we will introduce Young measures and present several of their characterizations. Some subspaces of the space of Young measures will provide important examples.

Definition 3.10. A *Young measure* on $\Omega \times S$ is a positive measure $\tau : \mathcal{A} \otimes \mathcal{B}_S \to \mathbb{R}_+$ for which

$$\tau(A \times S) = \mu(A), \quad \text{for every} \quad A \in \mathcal{A}.$$

According to Theorem 3.2 and to Remark 3.6, we can identify each Young measure τ with its disintegration $\tau_.$ and, in relation to this identification, we can say that *a Young measure is an application* $\tau : \Omega \to \mathcal{P}_S$, $(\mathcal{A} - \mathcal{C})$*-measurable.*

We will indicate by $\mathcal{Y}(\Omega \times S)$, $\mathcal{Y}(S)$ *or simply by* $\mathcal{Y}$ *(if there is not ambiguity on the space* S*) the space of Young measures on* $\Omega \times S$.

Remark 3.11.

(i) $\mathcal{Y}(S) \subseteq ca^+(\mathcal{A} \otimes \mathcal{B}_S)$ and also $\mathcal{Y}(S) \subseteq \mathcal{M}(\mathcal{P}_S)$.

(ii) Let $p_\Omega : \Omega \times S \to \Omega$ be the canonical projection on Ω; the immediate consequence of the definition is that a measure $\tau \in ca^+(\mathcal{A} \otimes \mathcal{B}_S)$ is a Young measure on $\Omega \times S$ if and only if $p_\Omega(\tau) = \mu$ (μ is the image of τ under the mapping p_Ω).

For every positive measurable function $f : \Omega \to \mathbb{R}$ (or $f \in \mathcal{L}^1(\mu)$),

$$\int_\Omega f d\mu = \int_{\Omega \times S} f \circ p_\Omega d\tau.$$

(iii) If $\mathcal{A}$ is the μ-complete σ-algebra generated by an algebra $\mathcal{A}_0$ and if we note by $\mathcal{A}_0 \otimes \mathcal{B}_S$ the algebra generated by the family of sets $\{A \times B : A \in \mathcal{A}_0, B \in \mathcal{B}_S\}$, then we can extend every positive and σ-additive measure $\tau : \mathcal{A}_0 \otimes \mathcal{B}_S \to \mathbb{R}_+$ in a unique manner to a measure on the σ-algebra $\mathcal{A} \otimes \mathcal{B}_S$ (the σ-algebra generated by $\mathcal{A}_0 \otimes \mathcal{B}_S$). If, in addition, $\tau(A \times S) = \mu(A)$, for every $A \in \mathcal{A}_0$, then the extended measure is a Young measure on $\Omega \times S$. Therefore, for every $\tau', \tau'' \in \mathcal{Y}(S)$, $\tau' = \tau''$ if and only if $\tau'(A \times B) = \tau''(A \times B)$, for every $A \in \mathcal{A}_0$ and every $B \in \mathcal{B}_S$.

(iv) Let τ be a Young measure on $\Omega \times S$ and let τ be its disintegration; for any positive integrand Ψ (or for every $\Psi \in \mathcal{L}^1(\tau)$)

$$\int_{\Omega \times S} \Psi(t,x) d\tau(t,x) = \int_\Omega \left(\int_S \Psi(t,x) d\tau_t(x) \right) d\mu(t).$$

(v) Let $\tau \in \mathcal{Y}(S)$; for every $A \in \mathcal{A}$, the mapping $\tau(A \times .) : \mathcal{B}_S \to \mathbb{R}_+$ is a positive measure on S. Therefore, according to 2.29, $\tau(A \times .) \in Rca^+(\mathcal{B}_S)$; so, for every $\varepsilon > 0$, there exists $K \in \mathcal{K}_S$ such that

$$\tau(\Omega \times (S \setminus K)) < \varepsilon.$$

(vi) A Young measure is also a family of probabilities $\{\tau_t : t \in \Omega\}$ on S such that, for every $C \in \mathcal{A} \otimes \mathcal{B}_S$, the mapping $t \mapsto \tau_t(C_t)$ is measurable.

Thus, we can explain the various denominations, such as *parametric measures, transition probabilities*, sometimes used to indicate Young measures.

If $\tau_. : \Omega \to \mathcal{P}_S$ is a Young measure, then almost all values of $\tau_.$ are in a tight subset of $\mathcal{P}_S$. More exactly, we have the following result:

Proposition 3.12. *Let S be a regular Suslin space, let $\tau \in \mathcal{Y}(S)$ and let $\tau_. : \Omega \to \mathcal{P}_S$ be the disintegration of τ; then, for every $\varepsilon > 0$, there exists a tight set $H_\varepsilon \subseteq \mathcal{P}_S$ such that*

$$\tau_.^{-1}(\mathcal{P}_S \setminus H_\varepsilon) \in \mathcal{A} \quad \textit{and} \quad \mu(\{t \in \Omega : \tau_t \in \mathcal{P}_S \setminus H_\varepsilon\}) < \varepsilon.$$

Proof. Since $\nu = \tau(\Omega \times \cdot) \in Rca^+(\mathcal{B}_S)$, $\{\nu\}$ is tight; according to Proposition 2.63, there exists an inf-compact application $\varphi : S \to [0, +\infty]$ such that $M = \int_S \varphi d\nu < +\infty$.

For every $\varepsilon > 0$, let $H_\varepsilon = \{\theta \in \mathcal{P}_S : \theta(\varphi) \le \frac{M}{\varepsilon}\}$. The same Proposition 2.63 says that H_ε is tight. Since the mapping of $\mathcal{P}_S$ in $\mathbb{R}_+$, defined by $\theta \in \mathcal{P}_S \mapsto I_\varphi(\theta) = \theta(\varphi)$, is narrowly l.s.c. (see (i) of Corollary 2.45), H_ε is a narrowly closed subset of $\mathcal{P}_S$ and then, as the mapping $\tau_. : \Omega \to \mathcal{P}_S$ is $(\mathcal{A} \otimes \mathcal{C})$-measurable,

$$\tau_.^{-1}(\mathcal{P}_S \setminus H_\varepsilon) = \left\{t \in \Omega : \tau_t(\varphi) > \frac{M}{\varepsilon}\right\} \in \mathcal{A}, \quad \text{and}$$

$$\begin{aligned} M &= \int_S \varphi d\nu = \int_{\Omega \times S} \varphi(x) d\tau(t,x) = \int_\Omega \tau_t(\varphi) d\mu(t) \\ &\ge \int_{(\tau_t(\varphi) > \frac{M}{\varepsilon})} \tau_t(\varphi) d\mu(t) \ge \frac{M}{\varepsilon} \cdot \mu\left(\tau_.^{-1}(\mathcal{P}_S \setminus H_\varepsilon)\right) \end{aligned}$$

and so $\mu(\{t \in \Omega : \tau_t \notin H_\varepsilon\}) < \varepsilon$. □

Now we will present two particularly important cases of the Young measures.

3.2.1 Young Measure Associated to a Probability

Definition 3.13. For every probability $\nu \in \mathcal{P}_S$, the product measure $\tau^\nu = \mu \otimes \nu$ is a Young measure whose disintegration is the constant mapping $\tau_t^\nu = \nu$, for all $t \in \Omega$. We say that τ^ν is the Young measure *associated to the probability* ν.

The mapping $\nu \mapsto \tau^\nu$ is an injection of $\mathcal{P}_S$ in $\mathcal{Y}(S)$; identifying ν with τ^ν, we allow to write: $\mathcal{P}_S \subseteq \mathcal{Y}(S)$.

More generally, let $n \in \mathbb{N}^*$, let $\{A_1, \dots, A_n\}$ be an $\mathcal{A}$-partition of Ω and let $\nu_1, \dots, \nu_n \in \mathcal{P}_S$; the mapping $\tau : \Omega \to \mathcal{P}_S$, defined by $\tau_t = \sum_{i=1}^n \nu_i \cdot \chi_{A_i}(t)$, is a Young measure. Indeed, τ is an $\mathcal{A}$-simple function with values in $\mathcal{P}_S$; therefore, for every $B \in \mathcal{B}_S$, $t \mapsto \tau_t(B) = \sum_{i=1}^n \nu_i(B)\chi_{A_i}(t)$, is $(\mathcal{A}-\mathcal{B}_{[0,1]})$-measurable. According to Theorem 3.1, τ is a Young measure. τ is the disintegration of measure $\mathcal{T}$ with respect to μ , where

$$\mathcal{T}(A \times B) = \sum_{i=1}^n \nu_i(B) \cdot \mu(A \cap A_i), \quad \text{for every } A \in \mathcal{A} \text{ and every } B \in \mathcal{B}_S.$$

In this case, we say that τ (as well as $\mathcal{T}$), is a *simple Young measure* and we note by $\mathcal{E}(\mathcal{A}, \mathcal{P}_S)$ the subset of all simple Young measures.

We show in first part of the following theorem that, in the case where S is a metrizable Suslin space, we can approximate any Young measure by simple Young measures; the second part is a result of Lusin type.

Theorem 3.14 ([77]). *Let S be a metrizable Suslin space and let $\tau : \Omega \to \mathcal{P}_S$.*

(i) *τ is a Young measure if and only if there exists a sequence of simple Young measures, $(\tau^n)_{n\in\mathbb{N}}$, such that, for all $t \in \Omega$, $(\tau_t^n)_{n\in\mathbb{N}}$ is narrowly convergent to τ_t.*

(ii) *Moreover, let Ω be a topological space, let μ be a Radon measure on Ω and let $\mathcal{A}$ be the μ-completed of σ-algebra $\mathcal{B}_\Omega$ of Borel sets on Ω; then τ is a Young measure if and only if, for every compact K of Ω and every $\varepsilon > 0$, there exists a compact set $L \subseteq K$ such that $\mu(K \setminus L) < \varepsilon$ and $\tau \restriction_L$ is narrowly continuous.*

Proof. (i) According to b) of 2.58, $(Rca^+(\mathcal{B}_S), \mathfrak{T})$ is a second-countable metrizable space and then $(\mathcal{P}_S, \mathfrak{T})$ is a second-countable metrizable subspace. Therefore $\tau : \Omega \to \mathcal{P}_S$ is a Young measure (i.e. a measurable mapping) if and only if there exists a sequence of simple mappings $(\tau^n)_{n\in\mathbb{N}}$ with values in $\mathcal{P}_S$ such that $(\tau_t^n)_{n\in\mathbb{N}}$ converges narrowly to τ_t, for every $t \in \Omega$.

(ii) ($\Rightarrow$) Let $\tau : \Omega \to \mathcal{P}_S$ be a Young measure. $\mathcal{P}_S$ is separable and metrizable space so that there exists an homeomorphic embedding $\phi : \mathcal{P}_S \to I^{\mathbb{N}}$, where $I = [0, 1]$. For every $n \in \mathbb{N}$, let π_n be the nth-projection of $I^{\mathbb{N}}$ in I. Then, for every $n \in \mathbb{N}$, the mapping $\pi_n \circ \phi \circ \tau : \Omega \to I$ is $(\mathcal{A} - \mathcal{B}_{[0,1]})$-measurable.

Since μ is a regular measure, for every compact set $K \subseteq \Omega$, for every $\varepsilon > 0$ and every $n \in \mathbb{N}$, there exists a compact subset $L_n \subset K$ such that $\mu(K \setminus L_n) < \frac{\varepsilon}{2^n}$ and $\pi_n \circ \phi \circ \tau$ is continuous on L_n.

The set $L = \bigcap_{n=1}^{\infty} L_n \subseteq K$ is compact and $\mu(K \backslash L) \leq \sum_{n=1}^{\infty} \mu(K \backslash L_n) < \varepsilon$. For every $n \in \mathbb{N}$, $\pi_n \circ \phi \circ \tau \restriction_L$ is continuous and therefore, $\phi \circ \tau \restriction_L$ is continuous; then $\tau \restriction_L$ is continuous.

(ii) ($\Leftarrow$) μ is tight so that, for every $n \in \mathbb{N}$, there exists a compact set $K_n \subset \Omega$ such that $\mu(\Omega \backslash K_n) < \frac{1}{n}$. We can choose the sequence $(K_n)_{n \in \mathbb{N}}$ to be increasing. Let $L_n \subset K_n$ be a compact set such that $\mu(K_n \setminus L_n) < \frac{1}{n}$ and τ is narrowly continuous on L_n; we can choose the sequence $(L_n)_{n \in \mathbb{N}}$ to be increasing, also. Let

$$\tau^n : \Omega \to \mathcal{P}_S, \tau^n(t) = \begin{cases} \tau(t), & t \in L_n, \\ \delta_0, & t \notin L_n \end{cases},$$

where δ_0 is the Dirac measure concentrated at 0 and let $L = \bigcup_{n=1}^{\infty} L_n$.

For every $n \in \mathbb{N}$, τ^n is $(\mathcal{A} - \mathcal{C})$-measurable and $\tau^n(t) \to \tau(t)$, for every $t \in L$; since $\mu(\Omega \setminus L) \leq \mu(\Omega \setminus L_n) < \frac{2}{n}$, for every $n \in \mathbb{N}^*$, $\tau^n \to \tau$, μ-a.e. so that τ is measurable. Therefore, τ is a Young measure. □

3.2.2 Young Measure Associated to a Measurable Mapping

The second important case of the Young measures is that of the measures which are associated to the measurable applications of Ω in S.

Let $\mathcal{M}(S)$ be the set of all $(\mathcal{A} - \mathcal{B}_S)$-measurable mappings; for every $u \in \mathcal{M}(S)$ let $\tau^u : \Omega \to \mathcal{P}_S, \tau^u(t) \equiv \tau^u_t = \delta_{u(t)}$, for every $t \in \Omega$ ($\delta_.$ is again the Dirac measure). For every $B \in \mathcal{B}_S$, the mapping $g^u_B : \Omega \to [0,1]$, defined by $g^u_B = \tau^u_.(B) = \chi_{u^{-1}(B)}$ is $(\mathcal{A} - \mathcal{B}_{[0,1]})$-measurable and so, according to (ii) of Theorem 3.1, τ^u is $(\mathcal{A} - \mathcal{C})$-measurable and then a Young measure.

Definition 3.15. τ^u is called the Young measure *associated to the measurable application* u.

We will also say that a Young measure τ *comes from an application* if there is a measurable application $u \in \mathcal{M}(S)$ such that $\tau = \tau^u$.

Remark 3.16.

(i) According to Remark 3.6, τ^u is the disintegration of the Young measure $\underline{\tau}^u$, where:

$$\underline{\tau}^u(A \times B) = \mu(A \cap u^{-1}(B)), \quad \text{for every} \quad A \in \mathcal{A} \quad \text{and every} \quad B \in \mathcal{B}_S.$$

According to (iv) of Remark 3.11, which specifies the integration with respect to a Young measure, for every positive integrand Ψ (or, for every

$\Psi \in \mathcal{L}^1(\tau^u)$),

$$\int_{\Omega \times S} \Psi(t,x) d\tau^u(t,x) = \int_{\Omega} \Psi(t,u(t)) d\mu(t).$$

(ii) The mapping $u \mapsto \tau^u$ is an injection of the space $\mathcal{M}(S)$ in $\mathcal{Y}(S)$ (we identify the applications that coincide μ-a.e.); therefore, by identifying u with τ^u, we can consider that $\mathcal{M}(S) \subseteq \mathcal{Y}(S)$.

(iii) The identification of $\mathcal{P}_S$ and $\mathcal{M}(S)$ as subsets of $\mathcal{Y}(S)$ leads to the identification of S as subset of $\mathcal{Y}(S)$ ($x \in S$ being identifies with δ_x). Indeed, for every $x \in S, \delta_x \in \mathcal{P}_S$ and also δ_x is the Young measure associated to constant mapping $\underline{x} \in \mathcal{M}(S)$. Conversely, if $\nu \in \mathcal{P}_S$, then $\tau^\nu \in \mathcal{M}(S) \subseteq \mathcal{Y}(S)$ if we can find a mapping $u \in \mathcal{M}(S)$ such that $\tau^\nu = \tau^u$, from where $\delta_{u(t)} = \nu$, for every $t \in \Omega$. It means that u is a constant function $\underline{x}$ and so $\nu = \delta_x$.

So, according to this identification, we can consider that the Suslin space S is a subset of $\mathcal{Y}(S)$.

(iv) If $u : \Omega \to S$ is a measurable function and if τ^u is the Young measure associated, then $\tau^u(\Omega) = \{\delta_{u(t)} : t \in \Omega\}$ *is a tight subset of* $\mathcal{P}_S$ *if and only if* $u(\Omega)$ *is relatively compact in* S. Indeed, if $\tau^u(\Omega)$ is tight, then there exists a compact subset $K \in \mathcal{K}_S$ such that $\delta_{u(t)}(S \setminus K) < 1$, for every $t \in \Omega$, which means that $u(\Omega) = \{u(t) : t \in \Omega\} \subseteq K$.

Conversely, if $K = \overline{u(\Omega)}$ is compact in S then $\delta_{u(t)}(S \setminus K) = 0 < \varepsilon$, for every $\varepsilon > 0$ and every $t \in \Omega$ which means that $\tau^u(\Omega)$ is tight.

We now establish two conditions assuring that a Young measure comes from a measurable application.

Proposition 3.17. *Let τ be a Young measure and let $u \in \mathcal{M}(S)$. If G_u is the graph of u, then G'_u will denote its complement with respect to $\Omega \times S$. A necessary and sufficient condition so that $\tau = \tau^u$ is that $\tau(G'_u) = 0$ or, equivalent, $\tau(G_u) = \mu(\Omega)$.*

Proof. If $f_u : \Omega \to \Omega \times S$ is defined by $f_u(t) = (t, u(t))$, then $\tau^u(C) = \mu(f_u^{-1}(C))$, for every $C \in \mathcal{A} \otimes \mathcal{B}_S$.
($\Rightarrow$) Let $\tau = \tau^u$; for every $C \in \mathcal{A} \otimes \mathcal{B}_S$, $\tau(C) = \tau^u(C) = \mu(f_u^{-1}(C))$.

Since $u \in \mathcal{M}(S)$, $G'_u = \{(t,x) \in \Omega \times S : u(t) \neq x\} \in \mathcal{A} \otimes \mathcal{B}_S$ and $f_u^{-1}(G'_u) = \{t \in \Omega : (t,u(t)) \in G'_u\} = \emptyset$; therefore $\tau(G'_u) = 0$.
($\Leftarrow$) By hypothesis, $\tau(G'_u) = 0$. We need to show that, for every $C \in \mathcal{A} \otimes \mathcal{B}_S$, $\tau(C) = \tau^u(C) = \mu(f_u^{-1}(C))$.

Let $A = f_u^{-1}(C)$; then $C \setminus G'_u \subseteq A \times S$ and so $\tau(C) = \tau(C \setminus G'_u) \leq \tau(A \times S) = \mu(A)$ and, because $(A \times S) \setminus G'_u \subseteq C$, $\mu(A) = \tau(A \times S) = \tau((A \times S) \setminus G'_u) \leq \tau(C)$. Therefore $\tau(C) = \mu(A) = \mu(f_u^{-1}(C)) = \tau^u(C)$. □

Theorem 3.18 ([77]). *Let τ be a Young measure; τ comes from a measurable application if and only if* $\inf\{\tau(G'_u) : u \in \mathcal{M}(S)\} = 0$ *or, equivalently,* $\sup\{\tau(G_u) : u \in \mathcal{M}(S)\} = \mu(\Omega)$.

Proof. ($\Leftarrow$) If τ comes from a measurable application v, according to the previous proposition, it is evident that $\inf\{\tau(G'_u) : u \in \mathcal{M}(S)\} = 0$.

($\Rightarrow$) Let $(\tau_t)_{t \in \Omega}$ the disintegration of τ. We will show that, for every $t \in \Omega$, there exists $x_t \in S$ such that $\tau_t(\{x_t\}) = \sup_{x \in S} \tau_t(\{x\})$.

Obviously, this is true if $\sup_{x \in S} \tau_t(\{x\}) = 0$.

Let $\sup_{x \in S} \tau_t(\{x\}) > 0$. If the sup is not reached, then, for every $y \in S$, there exists $z \in S$ such that $\tau_t(\{y\}) < \tau_t(\{z\})$. Then, we can define a sequence $(x_n)_{n \in \mathbb{N}} \subseteq S$ such that the sequence $(\tau_t(\{x_n\}))_{n \in \mathbb{N}}$ to be positive, strictly increasing and $\sum_{n=1}^{\infty} \tau_t(\{x_n\}) = \tau_t(\{x_1, \ldots, x_n, \ldots\}) \leq 1$ what is impossible.

By hypothesis, $\inf\{\tau(G'_u) : u \in \mathcal{M}(S)\} = 0$. Then, there exists a sequence $(u_n)_n \subseteq \mathcal{M}(S)$ such that $\tau(G'_{u_n}) \to 0$, when $n \to \infty$.

$$\tau(G'_{u_n}) = \int_{\Omega} \tau_t((G'_{u_n})_t) d\mu(t) = \int_{\Omega} \tau_t(S \setminus \{u_n(t)\}) d\mu(t),$$

from where, $t \mapsto \tau_t(S \setminus \{u_n(t)\})$ defines a sequence strongly convergent to 0 in L^1; let (n_k) be such that $t \mapsto \tau_t(S \setminus u_{n_k}(t))$ to define a subsequence convergent to 0, μ-a.e. Therefore, $t \mapsto \tau_t(u_{n_k}(t))$ converges μ-a.e. to 1.

Since $\tau_t(\{x_t\}) = \sup_{x \in S} \tau_t(\{x\}) \geq \tau_t(\{u_{n_k}(t)\})$, for every $k \in \mathbb{N}$ we have, $\tau_t(\{x_t\}) = 1$, for μ-almost every $t \in \Omega$, which means that $\tau_t = \delta_{x_t}$, μ-a.e. Then let $u : \Omega \to S$, defined by $u(t) = x_t$. $\tau_t(\{u_{n_k}(t)\}) \to 1$, μ-a.e. Hence we have

$$\tau_t(\{x_t\}) = \tau_t(\{u(t)\}) = 1 = \lim_k \tau_t(u_{n_k}(t)).$$

Therefore, $\delta_{u(t)}(\{u_{n_k}(t)\}) \to 1$ when $k \to \infty$, so that there exists k_0 such that $u_{n_k} = u$ a.e., for every $k \geq k_0$.

(u_{n_k}) converges therefore μ-a.e. to u and then u is measurable.

Obviously, $\tau^u = \tau$. □

3.3 The Stable Topology

Let S be a regular Suslin space, let $\mathcal{A}$ be a σ-algebra on the set Ω and let $\mu \in ca^+(\mathcal{A})$ be a positive measure, complete with respect to $\mathcal{A}$. The family of applications

$$Cth_0^b(\Omega \times S) = \{\chi_A \otimes f : A \in \mathcal{A}, f \in C_b(S)\}$$

is a subset of the family $Cth^b(\Omega \times S)$ of all bounded Carathéodory integrands on $\Omega \times S$.

Definition 3.19. Let $\mathcal{Y}(S)$ be the space of Young measures on $\Omega \times S$; for every $\Psi = \chi_A \otimes f \in Cth_0^b(\Omega \times S)$, let $I_\Psi : \mathcal{Y}(S) \to \mathbb{R}$, defined by

$$I_\Psi(\tau) = \int_{\Omega \times S} \Psi(t,x) d\tau(t,x) = \int_A \tau_t(f) d\mu(t),$$

where $\tau_.$ is the disintegration of τ.

The stable topology on $\mathcal{Y}(S)$ is the projective limit topology on $\mathcal{Y}(S)$ generated by the family of mappings $\mathcal{F}_S = \{I_\Psi : \Psi \in Cth_0^b(\Omega \times S)\}$, i.e. the weakest topology on $\mathcal{Y}(S)$ under which every mapping of the family $\mathcal{F}_S$ is continuous; this topology is noted by $\mathcal{S}(\mathcal{Y}(S))$, or simply $\mathcal{S}$ if there is not ambiguity on the Suslin space S.

If the net $(\tau^i)_{i \in I} \subseteq \mathcal{Y}(S)$ is $\mathcal{S}$-convergent to $\tau \in \mathcal{Y}(S)$, then we write $\tau^i \xrightarrow{\mathcal{S}} \tau$.

Remark 3.20.

(i) $\tau^i \xrightarrow{\mathcal{S}} \tau$ if and only if

$$\tau^i(\Psi) \to \tau(\Psi), \quad \text{for every} \quad \Psi \in Cth_0^b(\Omega \times S)$$

either:

$$\tau^i(\chi_A \otimes f) \to \tau(\chi_A \otimes f), \quad \text{for every} \quad A \in \mathcal{A} \quad \text{and every} \quad f \in C_b(S).$$

If we use the disintegrations, then, according to Theorem 1.57, this is equivalent with the weakly convergence of the net $(\tau_.^i(f))_{i \in I} \subseteq \mathcal{L}^1(\mu)$ to $\tau_.(f) \in \mathcal{L}^1(\mu)$, for every $f \in C_b(S)$.

(ii) Let $(\tau^i)_{i \in I} \subseteq \mathcal{Y}(S)$ be a net and let $\tau \in \mathcal{Y}(S)$.

For every $A \in \mathcal{A}$, $(\tau^i(A \times \cdot))_{i \in I} \subseteq Rca^+(\mathcal{B}_S)$ and $\tau(A \times \cdot) \in Rca^+(\mathcal{B}_S)$. Then

$$\tau^i \xrightarrow{\mathcal{S}} \tau \iff \tau^i(A \times \cdot) \xrightarrow{\mathcal{T}} \tau(A \times \cdot), \quad \text{for every} \quad A \in \mathcal{A}.$$

It means that the net $(\tau^i(A \times \cdot))_{i \in I}$ is narrowly convergent to $\tau(A \times \cdot)$.

(iii) The stable topology on $\mathcal{Y}(S)$ is noted in [45] by $\tau^W_{\mathcal{Y}^1}$.

(iv) In [9] this topology is called the narrow topology. We prefer to use the name of stable topology to not confuse with the narrow topology on $ca(\mathcal{B}_{\Omega\times S})$ in the particular case where Ω is a Hausdorff topological space (see also the remark of the Section 2.1 in [45]).

Proposition 3.21. *Let $\mathcal{A}_0$ be an algebra which generated $\mathcal{A}$; for every net $(\tau^i)_{i\in I} \subseteq \mathcal{Y}(S)$ and every $\tau \in \mathcal{Y}(S)$,*

$$\tau^i \xrightarrow{\mathcal{S}} \tau \iff \tau^i(A\times\cdot) \xrightarrow{\mathcal{T}} \tau(A\times\cdot), \quad \textit{for every} \quad A \in \mathcal{A}_0.$$

Proof. The implication "$\Longrightarrow$" results immediately from (ii) of the previous remark.

$\Longleftarrow$: Let us suppose that $\tau^i(A\times\cdot) \xrightarrow{\mathcal{T}} \tau(A\times\cdot)$, for every $A \in \mathcal{A}_0$ and let us denote $\mathcal{F} = \{E \in \mathcal{A} : \tau^i(E\times\cdot) \xrightarrow{\mathcal{T}} \tau(E\times\cdot)\}$.

For every increasing sequence $(E_n)_{n\in\mathbb{N}} \subseteq \mathcal{F}$, let $E = \cup_{n=1}^{\infty} E_n$.

For every open set $D \in \tau_S$ and every $\varepsilon > 0$, there exists $n_0 \in \mathbb{N}$ such that

$$\tau(E_{n_0}\times D) > \tau(E\times D) - \frac{\varepsilon}{2}.$$

Since $E_{n_0} \in \mathcal{F}$, according to (ii) of Theorem 2.53, there exists $i_0 \in I$ such that

$$\tau^i(E\times D) \geq \tau^i(E_{n_0}\times D) > \tau(E_{n_0}\times D) - \frac{\varepsilon}{2} > \tau(E\times D) - \varepsilon, \quad \text{for every} \quad i \geq i_0$$

and then $\liminf_i \tau^i(E\times D) \geq \tau(E\times D)$.

Since $\tau^i(E\times S) = \mu(E) = \tau(E\times S)$, according again to (ii) of Theorem 2.53,

$$\tau^i(E\times\cdot) \xrightarrow{\mathcal{T}} \tau(E\times\cdot).$$

Therefore $E \in \mathcal{F}$.

The same reasoning, using this time (iii) of 2.53, shows us that the family $\mathcal{F}$ also contains the limit of every decreasing sequence of sets of $\mathcal{F}$.

Therefore $\mathcal{F}$ is a monotone class and then $\mathcal{A} = \mathcal{F}$.

Therefore $\tau^i(A\times\cdot) \xrightarrow{\mathcal{T}} \tau(A\times\cdot)$, for every $A \in \mathcal{A}$ and so $(\tau^i)_{i\in I}$ is $\mathcal{S}$-convergent to τ. □

Because of (ii) of Remark 3.20, we can provide several characterizations of the stably convergence.

Proposition 3.22. *Let S be a regular Suslin space, let $(\tau^i)_{i \in I} \subseteq \mathcal{Y}(S)$ be a net and let $\tau \in \mathcal{Y}(S)$; the following conditions are equivalent:*

(1) $\tau^i \overset{\mathrm{S}}{\to} \tau$.

(2) $\tau(\chi_A \otimes f) \leq \liminf_i \tau^i(\chi_A \otimes f)$, *for every $A \in \mathcal{A}$ and every bounded from below l.s.c. function $f : S \to \overline{\mathbb{R}}$.*

(3) $\tau(A \times D) \leq \liminf_i \tau^i(A \times D)$, *for every $A \in \mathcal{A}$ and every open set $D \subseteq S$.*

(4) $\tau(A \times F) \geq \limsup_i \tau^i(A \times F)$, *for every $A \in \mathcal{A}$ and every closed set $F \subseteq S$.*

If, in addition, S is metrizable and d is a metric which generated the topology of S, then the conditions (1)–(4) *from above are also equivalent to each of the following conditions:*

(5) $\tau(\chi_A \otimes f) = \lim_i \tau^i(\chi_A \otimes f)$, *for every $A \in \mathcal{A}$ and every $f \in BL(S,d)$.*

(6) $\tau(\chi_A \otimes f) = \lim_i \tau^i(\chi_A \otimes f)$, *for every $A \in \mathcal{A}$, uniformly w.r.t $f \in BL(S,d)$ with $\|f\|_{BL} \leq 1$.*

If S is a locally compact Suslin space, then the conditions (1)–(6) *from above are equivalent with the following one:*

(7) $\tau(\chi_A \otimes f) = \lim_i \tau^i(\chi_A \otimes f)$, *for every $A \in \mathcal{A}$ and $f \in C_c(S)$.*

Proof. (1)$\Longleftrightarrow$(2) is a consequence of Proposition 2.50.

(1)$\Longleftrightarrow$(3) is a consequence of (ii) of Theorem 2.53.

(1)$\Longleftrightarrow$(4) is a consequence of (iii) of Theorem 2.53.

(1)$\Longleftrightarrow$(5) is a consequence of Proposition 2.51.

(1)$\Longleftrightarrow$(6) is a consequence of Theorem 2.81 (the definition of $BL(S,d)$ is given in the paragraph 2.5).

(1)$\Longleftrightarrow$(7): according to Theorem 6 of Chapter 2 of [152], in this case S is a Polish space; the result comes from (ii) of Theorem 2.36, while observing that, $\frac{1}{\mu(A)} \cdot \sigma(\chi_A \times \cdot) \in \mathcal{P}_S$, for every $\sigma \in \mathcal{Y}(S)$ and every $A \in \mathcal{A}$ with $\mu(A) \neq 0$. □

Remark 3.23.

(i) The condition (2) of the previous proposition characterizes the convergence in the topology $\tau_{\mathcal{Y}^1}^N$ of [45]; according to D) of Theorem 2.1.3 of [45], the conditions (1) and (2) listed above are equivalent in more general spaces than the regular Suslin spaces.

(ii) If $\mathcal{A}_0$ is an algebra that generates $\mathcal{A}$, then, according to Proposition 3.21, we can replace, in all the conditions (2)–(7) of the previous proposition, $A \in \mathcal{A}$ by $A \in \mathcal{A}_0$.

Proposition 3.24. *For every regular Suslin space S, $(\mathcal{Y}(S), \mathcal{S})$ is a completely regular space.*

Proof. Let $\mathcal{F}_S = \{I_\Psi : \Psi \in Cth_0^b(\Omega \times S)\}$ be the family who generates $\mathcal{S}$ and let $\theta : \mathcal{Y}(S) \to \mathbb{R}^{\mathcal{F}_S}$ be defined by

$$\theta(\tau) = (\varphi(\tau))_{\varphi \in \mathcal{F}_S} = (\tau(\Psi))_{\Psi \in Cth_0^b(\Omega \times S)}.$$

Then θ is an injection of $\mathcal{Y}(S)$ in $\mathbb{R}^{\mathcal{F}_S}$; indeed, if we suppose that $\theta(\tau) = \theta(\tau')$, then $\tau(\Psi) = \tau'(\Psi)$, for every $\Psi \in Cth_0^b(\Omega \times S)$ and therefore

$$\tau(\chi_A \otimes f) = \tau'(\chi_A \otimes f), \quad \text{for every} \quad A \in \mathcal{A} \quad \text{and every} \quad f \in C_b(S). \tag{*}$$

We can notice that $\tau(A \times \cdot) \in Rca^+(\mathcal{B}_S)$, for every $A \in \mathcal{A}$ (see 3.11, (v)) and that $\int_S f(x) d\tau(A \times \cdot)(x) = \tau(\chi_A \otimes f)$. Then $(*)$ becomes

$$\int_S f(x) d\tau(A \times \cdot)(x) = \int_S f(x) d\tau'(A \times \cdot)(x), \quad \text{for every} \quad f \in C_b(S),$$

or

$$I_f(\tau(A \times \cdot) = I_f(\tau'(A \times \cdot), \quad \text{for every} \quad f \in C_b(S). \tag{**}$$

The family $\{I_f : f \in C_b(S)\}$ defines the topology on $Rca^+(\mathcal{B}_S)$ which is separated. According to Proposition 2.46, we obtain from $(**)$

$$\tau(A \times B) = \tau'(A \times B), \quad \text{for every} \quad A \in \mathcal{A} \quad \text{and every} \quad B \in \mathcal{B}_S,$$

from where $\tau = \tau'$.

A net $(\tau^i)_{i \in I} \subseteq \mathcal{Y}(S)$ is $\mathcal{S}$-convergent to $\tau \in \mathcal{Y}(S)$ if and only if $(\theta(\tau^i))_{i \in I}$ is convergent to $\theta(\tau)$ in the product space $\mathbb{R}^{\mathcal{F}_S}$ and so θ is a homeomorphism of $(\mathcal{Y}(S), \mathcal{S})$ on a subspace of the product space $\mathbb{R}^{\mathcal{F}_S}$. Therefore $(\mathcal{Y}(S), \mathcal{S})$ is completely regular. □

In the stronger conditions on S, we can obtain metrizability of $\mathcal{Y}(S)$.

Proposition 3.25. *Let S be a metrizable Suslin space; if $\mathcal{A}$ is countably generated, then $(\mathcal{Y}(S), \mathcal{S})$ is metrizable.*

Proof. Let $\mathcal{A}_0 = \{A_1, A_2, \dots, A_n, \dots\}$ be a countable algebra which generated $\mathcal{A}$ and let d be a compatible metric on S.

For every $\tau', \tau'' \in \mathcal{Y}(S)$ and every $n \in \mathbb{N}^*$, $\tau'(A_n \times .), \tau''(A_n \times .) \in Rca^+(\mathcal{B}_S)$. According to Definition 2.80, the Dudley's distance between these two measures is:

$$D(\tau'(A_n \times .), \tau''(A_n \times .)) = \sup_{\|f\|_{BL} \le 1} |\tau'(\chi_{A_n} \otimes f) - \tau''(\chi_{A_n} \otimes f)| \\ \le 2\mu(A_n) \le 2\mu(\Omega).$$

So, we can define $\Delta : \mathcal{Y}(S) \times \mathcal{Y}(S) \to \mathbb{R}_+$ by:

$$\Delta(\tau', \tau'') = \sum_{n=1}^{\infty} \frac{1}{2^n} \cdot D(\tau'(A_n \times .), \tau''(A_n \times .)) \\ = \sum_{n=1}^{\infty} \frac{1}{2^n} \cdot \sup_{\|f\|_{BL} \le 1} |\tau'(\chi_{A_n} \otimes f) - \tau''(\chi_{A_n} \otimes f)|,$$

for every $\tau', \tau'' \in \mathcal{Y}(S)$.

$\Delta(\tau', \tau'') = 0$ if and only if $D(\tau'(A_n \times .), \tau''(A_n \times .)) = 0$, for every $n \in \mathbb{N}^*$, and so if and only if $\tau'(A_n \times .) = \tau''(A_n \times .)$, for every $n \in \mathbb{N}^*$; therefore, according to (iii) of Remark 3.11, $\Delta(\tau', \tau'') = 0$ if and only if $\tau' = \tau''$.

The other axioms of distance being clearly proved, Δ is a metric on $\mathcal{Y}(S)$.

Let $(\tau^i)_{i \in I} \subseteq \mathcal{Y}(S)$ be a net and let $\tau \in \mathcal{Y}(S)$. According to Proposition 3.21,

$$\tau^i \xrightarrow{\mathcal{S}} \tau \iff \tau^i(A_n \times \cdot) \xrightarrow{\mathcal{T}} \tau(A_n \times \cdot), \quad \text{for every} \quad n \in \mathbb{N}^*$$

and, since (S, d) is a second-countable metric space, according to Theorem 2.81 and to Remark 2.82,

$$\tau^i \xrightarrow{\mathcal{S}} \tau \iff D(\tau^i(A_n \times .), \tau(A_n \times .)) \to 0, \quad \text{for every} \quad n \in \mathbb{N}^*$$

which is equivalent to $\Delta(\tau^i, \tau) \to 0$. Therefore, Δ is a metric which generates the topology of the space $(\mathcal{Y}(S), \mathcal{S})$. □

According to Definition 3.13, $\mathcal{P}_S \subseteq \mathcal{Y}(S)$; the following result presents the trace of stable topology $\mathcal{S}$ on $\mathcal{P}_S$.

Proposition 3.26.

(1) *The trace of stable topology $\mathcal{S}$ on $\mathcal{P}_S$ is the narrow topology $\mathcal{T}$.*

(2) *$\mathcal{P}_S$ is a closed subspace of $(\mathcal{Y}(S), \mathcal{S})$.*

Proof. (1) If $(\nu_i)_{i\in I} \subseteq \mathcal{P}_S$ is a net, then

$$\nu_i \xrightarrow{\mathcal{T}} \nu \in \mathcal{P}_S \iff \nu_i(f) \to \nu(f), \quad \text{for every} \quad f \in C_b(S)$$

which means that $\tau^{\nu_i}(\chi_A \otimes f) = \mu(A)\cdot\nu_i(f) \to \mu(A)\cdot\nu(f) = \tau^{\nu}(\chi_A \otimes f)$, for every $A \in \mathcal{A}$ and every $f \in C_b(S)$. Therefore $\nu_i \xrightarrow{\mathcal{T}} \nu \iff (\tau^{\nu_i})_i$ is $\mathcal{S}$-convergent to τ^{ν} what demonstrates (1).

(2) Let $(\nu_i)_{i\in I} \subseteq \mathcal{P}_S$ be such that $\tau^{\nu_i} \xrightarrow{\mathcal{S}} \tau \in \mathcal{Y}(S)$. $\nu : \mathcal{B}_S \to [0,1]$, defined by

$$\nu(B) = \frac{1}{\mu(\Omega)} \int_\Omega \tau_t(B)\,d\mu(t), \quad \text{for every} \quad B \in \mathcal{B}_S,$$

is a probability on S so that $\nu \in \mathcal{P}_S$.

Let $\tau^{\nu} = \mu \otimes \nu$; for every $A \in \mathcal{A}$ and every $f \in C_b(S)$,

$$\begin{aligned}
\tau^{\nu}(\chi_A \otimes f) &= \mu(A)\cdot\nu(f) = \frac{\mu(A)}{\mu(\Omega)} \cdot \int_\Omega \left(\int_S f(x)\,d\tau_t(x) \right) d\mu(t) \\
&= \frac{\mu(A)}{\mu(\Omega)} \cdot \tau(\chi_\Omega \otimes f) = \frac{\mu(A)}{\mu(\Omega)} \cdot \lim_i \tau^{\nu_i}(\chi_\Omega \otimes f) \\
&= \mu(A)\cdot \lim_i \nu_i(f) = \lim_i \tau^{\nu_i}(\chi_A \otimes f).
\end{aligned}$$

Then $(\tau^{\nu_i})_i$ is $\mathcal{S}$-convergent to τ^{ν} and, since the narrow topology is separated, $\tau = \tau^{\nu} \in \mathcal{P}_S$. Therefore $\mathcal{P}_S$ is closed. □

Remark 3.27. Since $\mathcal{S}\restriction_{\mathcal{P}_S}$ is the narrow topology on $\mathcal{P}_S$, the stable topology $\mathcal{S}$ is also called the narrow topology on $\mathcal{Y}(S)$.

Theorem 3.28. *Let T and S be two regular Suslin spaces such that T be homeomorphic to a subspace of S and let $\mathcal{S}_T$ and $\mathcal{S}_S$ be the stable topologies of $\mathcal{Y}(T)$ and $\mathcal{Y}(S)$, respectively; then $(\mathcal{Y}(T), \mathcal{S}_T)$ is homeomorphic to a subspace of $(\mathcal{Y}(S), \mathcal{S}_S)$.*

Proof. Let $i : T \to S$ be a homeomorphism between T and $i(T) \subseteq S$. According to Proposition 2.57, $\mathcal{B}_T = i^{-1}(\mathcal{B}_S)$ and $I : Rca^+(\mathcal{B}_T) \to Rca^+(\mathcal{B}_s)$, defined by $I(\nu)(B) = \nu(i^{-1}(B))$, for every $\nu \in Rca^+(\mathcal{B}_T)$ and every $B \in \mathcal{B}_S$, is a homeomorphism between $Rca^+(\mathcal{B}_T)$ and $I(Rca^+(\mathcal{B}_T)) \subseteq Rca^+(\mathcal{B}_S)$ provided with their narrow topologies.

The restriction of I to the subset of all probabilities, noted again with I, is a homeomorphism between $\mathcal{P}_T$ and the subspace $I(\mathcal{P}_T) \subseteq \mathcal{P}_S$ provided with its narrow topology. Let $\mathcal{C}_T$ and $\mathcal{C}_S$ be the Borel subsets of $\mathcal{P}_T$ and $\mathcal{P}_S$, respectively; then $\mathcal{C}_T = I^{-1}(\mathcal{C}_S)$.

Since the Young measures on $\Omega \times T$ are $(\mathcal{A} - \mathcal{C}_T)$-measurable, for every $\sigma_. \in \mathcal{Y}(T)$, $\tau_. = I \circ \sigma_.$ is $(\mathcal{A} - \mathcal{C}_S)$-measurable and so a Young measure on $\Omega \times S$. For every $t \in \Omega$ and every $B \in \mathcal{B}_S$, $\tau_t(B) = \sigma_t(i^{-1}(B))$.

Let $J : \mathcal{Y}(T) \to \mathcal{Y}(S)$, defined by $J(\sigma_.) = I \circ \sigma_.$, for every $\sigma_. \in \mathcal{Y}(T)$. If $\sigma_.$ is the disintegration of σ and $J(\sigma_.)$ is the disintegration of τ, for every $C \in \mathcal{A} \otimes \mathcal{B}_S$,

$$\begin{aligned}\tau(C) = \int_\Omega (I \circ \sigma_.)_t(C_t)d\mu(t) &= \int_\Omega \sigma_t(i^{-1}(C_t))d\mu(t)\\ &= \int_\Omega \sigma_t(\{x \in T : (t, i(x)) \in C\})d\mu(t)\\ &= \int_\Omega \sigma_t(j^{-1}(C)_t)d\mu(t) = \sigma(j^{-1}(C)),\end{aligned}$$

where $j : \Omega \times T \to \Omega \times S$, $j(t,x) = (t, i(x))$, for every $(t,x) \in \Omega \times S$. Therefore $J(\sigma) = \sigma \circ j^{-1}$.

Obviously, J is an injection.

Let $(\sigma^d)_{d \in D} \subseteq \mathcal{Y}(T)$ be a net and let $\sigma \in \mathcal{Y}(T)$; according to (3) of Proposition 3.22, $(\sigma^d)_{d \in D}$ is stably convergent to σ in $\mathcal{Y}(T)$ if and only if

$$\sigma(A \times G) \le \liminf_d \sigma^d(A \times G), \text{ for every } A \in \mathcal{A} \text{ and every open set } G \in \tau_T,$$

which is equivalent with

$$\begin{aligned}I(\sigma)(A \times U) = \sigma(A \times i^{-1}(U)) &\le \liminf_d \sigma^d(A \times i^{-1}(U))\\ &= \liminf_d I(\sigma^d)(A \times U),\end{aligned}$$

for every $A \in \mathcal{A}$ and every $U \in \tau_S$.

Therefore $(\sigma^d)_{d \in D}$ is stable convergent to σ in $\mathcal{Y}(T)$ if and only if $(J(\sigma^d))_{d \in D}$ is stably convergent to $J(\sigma)$ in $\mathcal{Y}(S)$.

Then J is a homeomorphism between $\mathcal{Y}(T)$ and $J(\mathcal{Y}(T)) \subseteq \mathcal{Y}(S)$. □

Remark 3.29.

(i) A Young measure $\tau \in \mathcal{Y}(S)$ belongs to $J(\mathcal{Y}(T)) \subseteq \mathcal{Y}(S)$ if and only if

$$\tau^*(\Omega \times (S \setminus i(T))) = \inf\{\tau(\Omega \times B) : B \in \mathcal{B}_T, B \supseteq S \setminus i(T)\} = 0.$$

(ii) If S is a regular Suslin space and if T is Suslin subspace of S, then the stable topology on $\mathcal{Y}(T)$ is the trace of stable topology on $\mathcal{Y}(S)$.

In Proposition 3.22, we have mentioned several equivalent properties with the stably convergence. Another important characterization of the stably convergence can be expressed with the help of the Carathéodory integrands.

Theorem 3.30. *Let S be metrizable Suslin space; for every net $(\tau^i)_{i\in I} \subseteq \mathcal{Y}(S)$ and every $\tau \in \mathcal{Y}(S)$, the following two conditions are equivalent.*

(1) $\tau^i \xrightarrow{\mathcal{S}} \tau$.

(2) $\tau^i(\Psi) \to \tau(\Psi)$, *for every* $\Psi \in Cth^b(\Omega \times S)$.

Proof. Obviously, (2) $\Longrightarrow$ (1).

(1) $\Longrightarrow$ (2): Let $(\tau^i)_{i\in I} \subseteq \mathcal{Y}(S)$, $\tau \in \mathcal{Y}(S)$ such that $\tau^i \xrightarrow{\mathcal{S}} \tau$.

(a) Firstly, we suppose that S is a compact Suslin space; then S is Polish (see 2.26). The space $C(S) = C_b(S)$ of all real valued continuous functions on S is equipped with norm of uniform convergence $\|\cdot\|_0$ and $(C(S), \|\cdot\|_0)$ is a separable Banach space (see Theorem I.5.1 of [173]).

For every $\Psi \in Cth^b(\Omega \times S)$, let $\varphi_\Psi : \Omega \to C(S)$ be the mapping defined by $\varphi_\Psi(t) = \Psi(t, .)$, for every $t \in \Omega$.

Let A be a countable set dense in S. For every $f \in C(S)$ and every $r \in \mathbb{Q}_+^*$, let $T(f, r) \subseteq C(S)$ be the closed ball with center f and radius r; then

$$\varphi_\Psi^{-1}(T(f,r)) = \bigcap_{x\in A} \{t \in \Omega : |\Psi(t,x) - f(x)| \le r\} \in \mathcal{A}$$

and, since $C(S)$ is separable, φ_Ψ is $(\mathcal{A} - \mathcal{B}_{C(S)})$-measurable.

On the other hand, since Ψ is bounded,

$$\|\varphi_\Psi\|_1 = \int_\Omega \|\varphi_\Psi(t)\|_0 d\mu(t) = \int_\Omega \sup_{x\in S} |\Psi(t,x)| d\mu(t) \le \|\Psi\|_0 \cdot \mu(\Omega) < +\infty.$$

Then $\varphi_\Psi \in \mathcal{L}^1(\mu, C(S))$.

According to Theorem I.4.30 of [173], the space $\mathcal{E}(\mathcal{A}, C(S))$ of all $C(S)$-valued simple functions is dense in the space $\mathcal{L}^1(\mu, C(S))$. Then, for every $\varepsilon > 0$, there exists $\varphi \in \mathcal{E}(\mathcal{A}, C(S))$ such that

$$\|\varphi - \varphi_\Psi\|_1 = \int_\Omega \|\varphi(t) - \Psi(t,.)\|_0 d\mu(t) < \frac{\varepsilon}{4}. \tag{1}$$

Let $\varphi = \sum_{k=1}^n \chi_{A_k} \otimes f_k$, where $f_k \in C(S)$ and $\{A_k : k = 1, \ldots, n\} \subseteq \mathcal{A}$.

Since $(\tau^i)_{i\in I}$ is $\mathcal{S}$-convergent to τ, $(\tau^i(\varphi))_{i\in I}$ is convergent to $\tau(\varphi)$; let so $i_0 \in I$ such that

$$|\tau^i(\varphi) - \tau(\varphi)| < \frac{\varepsilon}{2}, \quad \text{for every} \quad i \ge i_0. \tag{2}$$

After (1), for every $\sigma \in \mathcal{Y}(S)$,

$$|\sigma(\Psi) - \sigma(\varphi)| = \left|\int_\Omega \left[\int_S (\Psi(t,x) - \varphi(t)(x)) d\sigma_t(x)\right] d\mu(t)\right| \\ \leq \int_\Omega \|\Psi(t,.) - \varphi(t)\|_0 d\mu(t) < \tfrac{\varepsilon}{4} \tag{3}$$

and then, according to (3) and (2), for every $i \geq i_0$,

$$|\tau^i(\Psi) - \tau(\Psi)| \leq |\tau^i(\Psi) - \tau^i(\varphi)| + |\tau^i(\varphi) - \tau(\varphi)| + |\tau(\varphi) - \tau(\Psi)| \\ < \frac{\varepsilon}{4} + \frac{\varepsilon}{2} + \frac{\varepsilon}{4} = \varepsilon.$$

Therefore we have $\tau^i(\Psi) \to \tau(\Psi)$.

(b) Let now S be a metrizable Suslin space; then S is second-countable and so S is homeomorphic to a subset of Hilbert cube $I^{\mathbb{N}} = [0,1]^{\mathbb{N}}$.

Let $\bar{S}$ be a metrizable compactification of S and let d be a metric generating the topology of $\bar{S}$. The embedding $i : S \to \bar{S}$ is a homeomorphism between S and $i(S) \subseteq \bar{S}$; let $j : \Omega \times S \to \Omega \times \bar{S}$, defined by $j(t,x) = (t, i(x))$, for every $(t,x) \in \Omega \times S$.

According to Theorem 3.28, the mapping $J : \mathcal{Y}(S) \to \mathcal{Y}(\bar{S})$, defined by $J(\tau) = \tau \circ j^{-1}$, is a homeomorphism of $(\mathcal{Y}(S), \mathcal{S}_S)$ on the subspace $J(\mathcal{Y}(S)) \subseteq (\mathcal{Y}(\bar{S}), \mathcal{S}_{\bar{S}})$.

Since $\tau^i \xrightarrow{\mathcal{S}_S} \tau$, $\bar{\tau}^i = J(\tau^i) \xrightarrow{\mathcal{S}_{\bar{S}}} J(\tau) = \bar{\tau}$.

For every $\Psi \in Cth^b(\Omega \times S)$ and every $n \in \mathbb{N}$, let $\bar{\Psi}_n : \Omega \times \bar{S} \to \mathbb{R}$, defined by

$$\bar{\Psi}_n(t,\bar{x}) = \inf_{x \in S}\{\Psi(t,x) + n \cdot d(x,\bar{x})\}, \quad \text{for every} \quad (t,\bar{x}) \in \Omega \times \bar{S}.$$

Following the same reasoning as in Proposition 3.9, we obtain that $\bar{\Psi}_n \in Cth^b(\Omega \times \bar{S})$ and that $\bar{\Psi}_n(t,x) \uparrow \Psi(t,x)$, for every $(t,x) \in \Omega \times S$.

Then $\tau(\bar{\Psi}_n \restriction_{\Omega \times S}) \to \tau(\Psi)$.

For every $\varepsilon > 0$, there exists $n_0 \in \mathbb{N}$ such that

$$\tau(\Psi) < \tau\left(\bar{\Psi}_n \restriction_{\Omega \times S}\right) + \frac{\varepsilon}{2} = \bar{\tau}(\bar{\Psi}_{n_0}) + \frac{\varepsilon}{2}. \tag{4}$$

$(\bar{\tau}^i)_{i \in I}$ is $\mathcal{S}_{\bar{S}}$-convergent to $\bar{\tau}$ in the space $(\mathcal{Y}(\bar{S}), \mathcal{S}_{\bar{S}})$ and, according to the part (a) of this proof, $\bar{\tau}^i(\bar{\Psi}_{n_0}) \to \bar{\tau}(\bar{\Psi}_{n_0})$; therefore there exists $i_0 \in I$ such that

$$\bar{\tau}(\bar{\Psi}_{n_0}) < \bar{\tau}^i(\bar{\Psi}_{n_0}) + \frac{\varepsilon}{2} = \tau^i\left(\bar{\Psi}_{n_0} \restriction_{\Omega \times S}\right) + \frac{\varepsilon}{2} \leq \tau^i(\Psi) + \frac{\varepsilon}{2}, \quad \forall i \geq i_0. \tag{5}$$

By (4) and (5),

$$\tau(\Psi) \leq \liminf_i \tau^i(\Psi). \tag{6}$$

If we repeat the same reasoning for $-\Psi$, we obtain

$$\limsup_i \tau^i \leq \tau(\Psi). \tag{7}$$

By (6) and (7), $\tau^i(\Psi) \to \tau(\Psi)$, for every $\Psi \in Cth^b(\Omega \times S)$. □

Remark 3.31. Condition (2) of the previous theorem characterizes the convergence in the topology $\tau_{\mathcal{Y}_1}^M$ of [45]; the point G of Theorem 2.1.3. of [45] contains the previous result.

In the following theorem, we will present other characterizations of the stably convergence in the frame of the metrizable Suslin spaces.

Theorem 3.32. *Let S be a metrizable Suslin space; for every $\tau \in \mathcal{Y}(S)$ and every net $(\tau^i)_{i\in I} \subseteq \mathcal{Y}(S)$, the following conditions are equivalent:*

(1) $\tau^i \xrightarrow{\mathcal{S}} \tau$.

(2) $\tau^i(\Psi) \to \tau(\Psi)$, *for every* $\Psi \in Cth^b(\Omega \times S)$.

(3) $\tau(\Psi) \leq \liminf_i \tau^i(\Psi)$, *for every bounded l.s.c. integrand* Ψ.

(4) $\tau(\Psi) \leq \liminf_i \tau^i(\Psi)$, *for every L^1-bounded l.s.c. integrand* Ψ.

(5) $\tau^i(\Psi) \to \tau(\Psi)$, *for every L^1-bounded Carathéodory integrand* Ψ.

Proof. According to Theorem 3.30, (1)$\Longleftrightarrow$(2).

Since the implications (4)$\Longrightarrow$(5)$\Longrightarrow$(2) are evident, we only have to demonstrate that (2) $\Longrightarrow$ (3) $\Longrightarrow$ (4).

(2)$\Longrightarrow$(3): Let $\Psi : \Omega \times S \to \bar{\mathbb{R}}$ be a bounded l.s.c. integrand. According to Proposition 3.9, there exists a sequence $(\Psi_n)_{n\in\mathbb{N}} \subseteq Cth^b(\Omega \times S)$ such that $\Psi_n \uparrow \Psi$. Then $\tau(\Psi_n) \uparrow \tau(\Psi)$ and so, for every $\varepsilon > 0$, there exists $n_0 \in \mathbb{N}$ such that

$$\tau(\Psi) < \tau(\Psi_{n_0}) + \frac{\varepsilon}{2}. \tag{i}$$

By (2), $\tau^i(\Psi_{n_0}) \to \tau(\Psi_{n_0})$ and then there exists $i_0 \in I$ such that

$$\tau(\Psi_{n_0}) < \tau^i(\Psi_{n_0}) + \frac{\varepsilon}{2}, \quad \text{for every} \quad i \geq i_0. \tag{ii}$$

By (i) and (ii), for every $i \geq i_0$, $\tau(\Psi) < \tau^i(\Psi_{n_0}) + \varepsilon < \tau^i(\Psi) + \varepsilon$, from where $\tau(\Psi) \leq \liminf_i \tau^i(\Psi)$.

(3)$\Longrightarrow$(4): Let $\Psi : \Omega \times S \to \bar{\mathbb{R}}$ be a L^1-bounded l.s.c. integrand and let $\phi \in L^1(\mu)$ be such that $|\Psi(t,x)| \leq \phi(t)$, for every $t \in \Omega$ and every $x \in S$.

For every $\varepsilon > 0$, there exists $M > 0$ such that

$$\int_{(\phi(t)>M)} \phi(t) d\mu(t) < \varepsilon.$$

Let $A = \{t \in \Omega : \phi(t) \leq M\} \in \mathcal{A}$. Then $\Psi_1 = \chi_A \cdot \Psi$ is a bounded l.s.c. integrand ($\|\Psi_1\|_0 \leq M$) and, by (3),

$$\tau(\Psi_1) \leq \liminf_i \tau^i(\Psi_1). \tag{iii}$$

For every $\sigma \in \mathcal{Y}(S)$,

$$\begin{aligned} |\sigma(\Psi) - \sigma(\Psi_1)| = |\sigma(\chi_{\Omega \setminus A} \cdot \Psi)| &\leq \int_{\Omega \setminus A} \int_S |\Psi(t,x)| d\sigma_t(x) d\mu(t) \qquad \text{(iv)} \\ &\leq \int_{\Omega \setminus A} \phi(t) d\mu(t) < \varepsilon. \end{aligned}$$

According to (iii) and (iv), for every $\varepsilon > 0$,

$$\tau(\Psi) < \tau(\Psi_1) + \varepsilon \leq \liminf_i \tau^i(\Psi_1) + \varepsilon \leq \liminf \tau^i(\Psi) + 2\varepsilon. \qquad \square$$

For other particular spaces, we can give supplementary characterizations of the stable convergence.

Theorem 3.33. *Let S be a locally compact Suslin space; for every net $(\tau^i)_{i \in I} \subseteq \mathcal{Y}(S)$ and every $\tau \in \mathcal{Y}(S)$, the following conditions are equivalent:*

(1) $\tau^i \xrightarrow{\mathbb{S}} \tau$.

(2) $\tau^i(\Psi) \to \tau(\Psi)$, *for every bounded integrand Ψ such that $\Psi(t,.) \in C_c(S)$, for every $t \in \Omega$.*

Proof. Since S is a locally compact Suslin space, it is a Polish space (see Corollary 2.26).

According to Theorem 3.30, $\tau^i \xrightarrow{\mathbb{S}} \tau \iff \tau^i(\Psi) \to \tau(\Psi)$, for every $\Psi \in Cth^b(\Omega \times S)$; then (1)$\Longrightarrow$(2).

(2)$\Longrightarrow$(1): Let $\Psi \in Cth^b(\Omega \times S)$; then

$$M = \sup_{(t,x) \in \Omega \times S} |\Psi(t,x)| < +\infty.$$

According to (v) of Remark 3.11, for every $\varepsilon > 0$, there exists a compact set $K \in \mathcal{K}_S$ such that $\tau(\Omega\times(S\setminus K)) < \varepsilon$. Let $g \in C_c(S)$ such that $g(S) \subseteq [0,1]$ and $g\restriction_K = 1$ and let $C = \mathrm{supp} g \in \mathcal{K}_S$. The mapping $\Psi_1 : \Omega \times S \to \mathbb{R}$, $\Psi_1(t,x) = g(x)$, is a bounded integrand, continuous and with compact support with respect to the second variable. By (2), there exists $i_1 \in I$ such that $|\tau^i(\Psi_1)-\tau(\Psi_1)| < \varepsilon$, for every $i \geq i_1$. Since $g \leq \chi_C, \chi_{S\setminus C} \leq 1-g$ and then

$$\chi_{\Omega\times(S\setminus C)} \leq 1 - \Psi_1 \leq \chi_{\Omega\times(S\setminus K)}.$$

Therefore, for every $i \geq i_1$,

$$\begin{aligned}\tau^i(\Omega \times (S \setminus C)) &\leq \tau^i(1-\Psi_1) = \mu(\Omega) - \tau^i(\Psi_1) \leq \mu(\Omega) - \tau(\Psi_1) + \varepsilon \quad \text{(a)}\\ &= \tau(1-\Psi_1) + \varepsilon \leq \tau(\Omega \times (S \setminus K)) + \varepsilon < 2\varepsilon.\end{aligned}$$

Since $C \in \mathcal{K}_S$, there exists $h \in C_c(S)$ such that $h(S) \subseteq [0,1]$ and $h\restriction_C = 1$. Let $\Psi_2 = \Psi \cdot h$; then Ψ_2 is a bounded integrand and $\Psi_2(t,.) \in C_c(S)$, for every $t \in \Omega$. According to (2), $\tau^i(\Psi_2) \to \tau(\Psi_2)$ and then there exists $i_2 \in I, i_2 \geq i_1$ such that

$$|\tau^i(\Psi_2) - \tau(\Psi_2)| < \varepsilon, \quad \text{for every} \quad i \geq i_2. \qquad \text{(b)}$$

On the other hand, for every $(t,x) \in \Omega \times S$,

$$|\Psi(t,x) - \Psi_2(t,x)| = |\Psi(t,x)| \cdot |1-h(x)| \leq M \cdot \chi_{\Omega\times(S\setminus C)}(t). \qquad \text{(c)}$$

By (a), (b) and (c), for every $i \geq i_2$,

$$\begin{aligned}|\tau^i(\Psi) - \tau(\Psi)| &\leq |\tau^i(\Psi) - \tau^i(\Psi_2)| + |\tau^i(\Psi_2) - \tau(\Psi_2)| + |\tau(\Psi_2) - \tau(\Psi)|\\ &\leq \tau^i(|\Psi - \Psi_2|) + \varepsilon + \tau(|\Psi - \Psi_2|)\\ &\leq M \cdot \tau^i(\Omega \times (S \setminus C)) + \varepsilon + M \cdot \tau(\Omega \times (S \setminus C))\\ &\leq 2\varepsilon M + \varepsilon + M \cdot \tau(\Omega \times (S \setminus K)) < \varepsilon(3M+1).\end{aligned}$$

Then $\tau^i(\Psi) \to \tau(\Psi)$ and therefore $\tau^i \overset{\mathcal{S}}{\to} \tau$. □

We complete this paragraph by giving several results of semi-continuity for the unbounded integrands.

Proposition 3.34. *Let S be a metrizable Suslin space; for every net $(\tau^i)_{i\in I} \subseteq \mathcal{Y}(S)$ and every $\tau \in \mathcal{Y}(S)$,*

$$\tau^i \overset{\mathcal{S}}{\to} \tau \quad \textit{if and only if} \quad \tau(\Psi) \leq \liminf_i \tau^i(\Psi),$$

for every l.s.c. integrand $\Psi : \Omega \times S \to \overline{\mathbb{R}}$ bounded from below.

Proof. Let us suppose that $\tau^i \xrightarrow{S} \tau$ and let $\Psi : \Omega \times S \to \overline{\mathbb{R}}$ be a l.s.c. integrand bounded from below; for every $n \in \mathbb{N}$, let $\Psi_n : \Omega \times S \to \mathbb{R}$ defined by $\Psi_n = \min\{\Psi, \underline{n}\}$. Then Ψ_n is a l.s.c. bounded integrand. Since $\Psi_n \uparrow \Psi$ and Ψ is bounded from below, $\tau(\Psi_n) \uparrow \tau(\Psi)$. According to (3) of Theorem 3.32,

$$\tau(\Psi_n) \leq \liminf_i \tau^i(\Psi_n) \leq \liminf_i \tau^i(\Psi), \quad \text{for every} \quad n \in \mathbb{N},$$

from where

$$\tau(\Psi) \leq \liminf_i \tau^i(\Psi).$$

The reciprocal is obvious. □

We can further relax the condition of being bounded from below.

The following results are presented according to [169].

Theorem 3.35. *Let S be a metrizable Suslin space and let $(u^i)_{i\in I} \subseteq \mathcal{M}(S) \subseteq \mathcal{Y}(S)$ be a net such that $(\tau^{u^i})_{i\in I}$ is stably convergent to $\tau \in \mathcal{Y}(S)$.*

Let $\Psi : \Omega \times S \to \mathbb{R}$ be a l.s.c. integrand such that $\{\Psi^-(\cdot, u^i(\cdot)) : i \in I\}$ is an uniformly integrable subset of $\mathcal{L}^1(\mu)$, where $\Psi^- = \sup\{-\Psi, \underline{0}\}$ is the negative part of Ψ; if there exists $\tau(\Psi)$, then

$$\tau(\Psi) = \int_{\Omega\times S} \Psi(t,x)d\tau(t,x) \leq \liminf_{i\in I} \int_\Omega \Psi(t,u^i(t))d\mu(t) = \liminf_{i\in I} \tau^{u^i}(\Psi).$$

Proof. According to Theorem 1.84 (see also Definition 1.85), for every $\varepsilon > 0$, there exists $a_\varepsilon > 0$ such that

$$\sup_{i\in I} \int_{(\Psi^-(\cdot,u^i(\cdot))>a_\varepsilon)} \Psi^-(t,u^i(t))d\mu(t) < \varepsilon. \tag{1}$$

For every $i \in I$, let $A_i = \{t \in \Omega : \Psi(t,u^i(t)) < -a_\varepsilon\}$; then $t \in A_i$ if and only if $\Psi^-(t,u^i(t)) = -\Psi(t,u^i(t)) > a_\varepsilon$ and by (1)

$$\inf_{i\in I} \int_{A_i} \Psi(t,u^i(t))d\mu(t) > -\varepsilon. \tag{2}$$

According to (2), for every $i \in I$,

$$\begin{aligned}\int_\Omega \Psi(t,u^i(t))d\mu(t) &= \int_{A_i} \Psi(t,u^i(t))d\mu(t) + \int_{\Omega\setminus A_i} \Psi(t,u^i(t))d\mu(t)\\ &> -\varepsilon + \int_{\Omega\setminus A_i} \sup\{\Psi(t,u^i(t)), -a_\varepsilon\}d\mu(t)\\ &= -\varepsilon + \int_\Omega \sup\{\Psi(t,u^i(t)), -a_\varepsilon\}d\mu(t)\\ &\quad - \int_{A_i} \sup\{\Psi(t,u^i(t)), -a_\varepsilon\}d\mu(t)\\ &= -\varepsilon + \int_\Omega \sup\{\Psi(t,u^i(t)), -a_\varepsilon\}d\mu(t) + a_\varepsilon \cdot \mu(A_i)\\ &\geq -\varepsilon + \int_\Omega \sup\{\Psi, \underline{-a_\varepsilon}\}(t,u^i(t))d\mu(t).\end{aligned}$$

The mapping $\Psi_\varepsilon : \Omega \times S \to \mathbb{R}$, defined by: $\Psi_\varepsilon(t,x) = \sup\{\Psi(t,x), -a_\varepsilon\}$, for every $(t,x) \in \Omega \times S$, is a l.s.c. integrand bounded from below and

$$\int_\Omega \Psi(t,u^i(t))d\mu(t) > -\varepsilon + \int_\Omega \Psi_\varepsilon(t,u^i(t))d\mu(t), \quad \text{for every} \quad i \in I. \tag{3}$$

Then, according to the previous proposition,

$$\tau(\Psi_\varepsilon) \leq \liminf_{i\in I} \tau^{u^i}(\Psi_\varepsilon). \tag{4}$$

Since $\Psi \leq \Psi_\varepsilon$, by (3) and (4), for every $\varepsilon > 0$,

$$\tau(\Psi) \leq \liminf_{i\in I} \int_\Omega \Psi_\varepsilon(t,u^i(t))d\mu(t) \leq \liminf_{i\in I} \int_\Omega \Psi(t,u^i(t))d\mu(t) + \varepsilon. \qquad \square$$

Corollary 3.36 (see also Corollary 2.3.2 of [44]). *Let $\tau \in \mathcal{Y}(S)$ and let $(u^i)_{i\in I} \subseteq \mathcal{M}(S) \subseteq \mathcal{Y}(S)$ be a net such that $\tau^{u^i} \overset{\mathbb{S}}{\to} \tau$. Then, for every Carathéodory integrand $\Psi : \Omega \times S \to \mathbb{R}$ for which $\{\Psi(\cdot,u^i(\cdot)) : i \in I\}$ is an uniformly integrable subset of $\mathcal{L}^1(\mu)$ and such that there exists $\tau(\Psi)$, we have:*

$$\tau(\Psi) = \int_{\Omega\times S} \Psi(t,x)d\tau(t,x) = \lim_{i\in I} \int_\Omega \Psi(t,u^i(t))d\mu(t) = \lim_{i\in I} \tau^{u^i}(\Psi).$$

Proof. The result follows by application of the previous theorem for Ψ and $-\Psi$. $\square$

Proposition 3.37. *Let* $S = \mathbb{R}^d$ *and let* $(u^i)_{i\in I} \subseteq \mathcal{L}^1(\mu, \mathbb{R}^d)$ *be an uniformly integrable net such that* $\tau^{u^i} \xrightarrow{\mathbb{S}} \tau \in \mathcal{Y}(\mathbb{R}^d)$*; then* $(u^i)_{i\in I}$ *is weakly convergent to the integrable function* $u : \Omega \to \mathbb{R}^d$*, where, for every* $t \in \Omega$*,* $u(t) = \int_{\mathbb{R}^d} x d\tau_t(x) = \mathrm{bar}\tau_t$ *(the barycenter of* τ_t*).*

Proof. Indeed, for each $A \in \mathcal{A}$ and each $j \in \{1, \dots, d\}$, the map $\Psi_j : \Omega \times \mathbb{R}^d \to \mathbb{R}$, $\Psi_j(t, x) = \chi_A(t) \cdot p_j(x)$ (where $p_j : \mathbb{R}^d \to \mathbb{R}$ is the canonical j-projection) is a Carathéodory integrand for which $\{\Psi_j(\cdot, u^i(\cdot)) : i \in I\} = \{\chi_A \cdot p_j(u^i) : i \in I\}$ is uniformly integrable.

$\Psi_1 : \Omega \times \mathbb{R}^d \to \mathbb{R}_+$, $\Psi_1(t, x) = \|x\|$, is a l.s.c. integrand bounded from below; then, according to Proposition 3.34,

$$\tau(\Psi_1) \le \liminf_i \tau^{u^i}(\Psi_1) = \liminf_i \int_\Omega \|u^i(t)\| d\mu(t) < +\infty.$$

Then $\int_\Omega [\int_{\mathbb{R}^d} \|x\| d\tau_t(x)] d\mu(t) < +\infty$ and so there exists $u(t) = \int_{\mathbb{R}^d} x d\tau_t(x)$, almost for every $t \in \Omega$; obviously, $u \in \mathcal{L}^1(\mu, \mathbb{R}^d)$.

For every $j \in \{1, \dots, d\}$,

$$\tau(\Psi_j) = \int_A \left[\int_{\mathbb{R}^d} p_j(x) d\tau_t(x) \right] d\mu(t) \le \int_\Omega \left[\int_{\mathbb{R}^d} \|x\| d\tau_t(x) \right] d\mu(t),$$

so that $\tau(\Psi_j)$ there exists. According to the previous corollary,

$$\lim_{i\in I} \int_A p_j(u^i(t)) d\mu(t) = \tau(\Psi_j) = \int_A p_j(u(t)) d\mu(t), \text{ for every } j \in \{1, \dots, d\}.$$

Therefore $\lim_{i\in I} \int_A u^i(t) d\mu(t) = \int_A u(t) d\mu(t)$, for every $A \in \mathcal{A}$; according to Theorem 1.57, $(u^i)_{i\in I}$ is weakly convergent to u. □

3.4 The Subspace $\mathcal{M}(S) \subseteq \mathcal{Y}(S)$

Let S be a regular Suslin space. In Definition 3.15, we have associated to each application $u \in \mathcal{M}(S)$ a Young measure τ^u and, according to (ii) of Remark 3.16, $\mathcal{M}(S)$ was identified as a subspace of $\mathcal{Y}(S)$. This subspace has interesting specific properties that we will present in this paragraph.

We recall that the disintegrated of τ^u is an application which associates, for each $t \in \Omega$, the Dirac probability $\delta_{u(t)} \in \mathcal{P}_S$ and that, for all positive integrand Ψ (or for all $\Psi \in \mathcal{L}^1(\tau^u)$),

$$\tau^u(\Psi) = \int_{\Omega\times S} \Psi(t, x) d\tau^u(t, x) = \int_\Omega \Psi(t, u(t)) d\mu(t).$$

For every net $(u_i)_{i\in I} \subseteq \mathcal{M}(S) \subseteq \mathcal{Y}(S)$, $\tau^{u_i} \xrightarrow{\mathcal{S}} \tau \in \mathcal{Y}(S)$ will be noted by $u_i \xrightarrow{\mathcal{S}} \tau$ or by $u_i \xrightarrow{\mathcal{S}} u$ if $\tau = \tau^u$.

The main results of this paragraph are:

(1) The trace of stable topology on $\mathcal{M}(S)$ is the topology of convergence in measure.

(2) Under the natural condition that μ is nonatomic, $\mathcal{M}(S)$ is dense in $(\mathcal{Y}(S), \mathcal{S})$.

First of all, we introduce on $\mathcal{Y}(S)$ a finer topology than the stable topology.

Definition 3.38 (see [45], p. 56). Let S be a regular Suslin space; for every $f \in C_b(S)$, let $\Delta_f : \mathcal{Y}(S) \times \mathcal{Y}(S) \to \mathbb{R}_+$, defined by

$$\Delta_f(\tau,\sigma) = \int_\Omega |\tau_t(f) - \sigma_t(f)| d\mu(t), \quad \text{for all} \quad \tau,\sigma \in \mathcal{Y}(S).$$

Then $\{\Delta_f : f \in C_b(S)\}$ is a family of metrics on $\mathcal{Y}(S)$; the topology generated by this family is said *the topology of convergence in measure* on $\mathcal{Y}(S)$. It is noted by $\mathcal{P}$. $\tau^i \xrightarrow{\mathcal{P}} \tau$ notes that the net $(\tau^i)_{i\in I} \subseteq \mathcal{Y}(S)$ is $\mathcal{P}$-convergent to $\tau \in \mathcal{Y}(S)$. If $(u_i)_{i\in I} \subseteq \mathcal{M}(S)$ and $u \in \mathcal{M}(S)$, then we will note $u_i \xrightarrow{\mathcal{P}} u$ instead of $\tau^{u_i} \xrightarrow{\mathcal{P}} \tau^u$.

Remark 3.39.

(i) A net $(\tau^i)_{i\in I} \subseteq \mathcal{Y}(S)$ is $\mathcal{P}$-convergent to $\tau \in \mathcal{Y}(S)$ if and only if

$$\Delta_f(\tau^i, \tau) = \int_\Omega |\tau^i_t(f) - \tau_t(f)| d\mu(t) \to 0, \quad \text{for every} \quad f \in C_b(S);$$

in other words, $\tau^i \xrightarrow{\mathcal{P}} \tau$ if and only if $(\tau^i_\cdot(f))_{i\in I}$ is strongly convergent to $\tau_\cdot(f)$ in $\mathcal{L}^1(\mu)$, for every $f \in C_b(S)$.

(ii) For every $f \in C_b(S)$ and every $\tau,\sigma \in \mathcal{Y}(S)$,

$$\Delta_f(\tau,\sigma) \geq \sup_{A\in\mathcal{A}} \left| \int_A (\tau_t(f) - \sigma_t(f)) d\mu(t) \right| \geq \frac{1}{2}\Delta_f(\tau,\sigma),$$

which can be also written

$$\Delta_f(\tau,\sigma) \geq \sup_{A\in\mathcal{A}} \left| \tau(\chi_A \otimes f) - \sigma(\chi_A \otimes f) \right| \geq \frac{1}{2}\Delta_f(\tau,\sigma).$$

Then

$$\tau^i \xrightarrow{\mathcal{P}} \tau \iff \sup_{A \in \mathcal{A}} \left| \tau^i(\chi_A \otimes f) - \tau(\chi_A \otimes f) \right| \to 0,$$

$$\text{for every } f \in C_b(S).$$

(iii) According to (i) and to Remark 3.20 (i),

$$\tau^i \xrightarrow{\mathcal{P}} \tau \Longrightarrow \tau^i \xrightarrow{\mathcal{S}} \tau$$

and therefore $\mathcal{S} \subseteq \mathcal{P}$.

(iv) For every net $(u_i)_{i \in I} \subseteq \mathcal{M}(S)$ and every $u \in \mathcal{M}(S)$,

$$u_i \xrightarrow{\mathcal{P}} u \iff f(u_i) \xrightarrow[\mathcal{L}^1(\mu)]{\|\cdot\|_1} f(u), \quad \text{for every} \quad f \in C_b(S).$$

We find so the definition given by Hoffmann–Jørgensen in [95] for the convergence in measure which explains the appellation of this convergence on $\mathcal{Y}(S)$. In the case in where S is metrizable and d is a metric that generated the topology of S, the $\mathcal{P}{\restriction}_{\mathcal{M}(S)}$-convergence is the classic convergence in measure.

Proposition 3.40. *Let S be a metrizable Suslin space and let d be a metric generating the topology of S; for every sequence $(u_n)_{n \in \mathbb{N}} \subseteq \mathcal{M}(S)$ and every $u \in \mathcal{M}(S)$,*

$$u_n \xrightarrow{\mathcal{P}} u \Longleftrightarrow u_n \xrightarrow{\mathcal{S}} u \Longleftrightarrow \lim_n \mu(\{t \in \Omega : d(u_n(t), u(t)) \geq \varepsilon\}) = 0, \quad \forall \varepsilon > 0.$$

Therefore, according to the notation already used in paragraph 1.3 in the real case,

$$u_n \xrightarrow{\mathcal{S}} u \iff u_n \xrightarrow{\mu} u.$$

Proof. ($\Longrightarrow$): By (iii) of previous remarks, since $u_n \xrightarrow{\mathcal{P}} u$, $\tau^{u_n} \xrightarrow{\mathcal{S}} \tau^u$; according to Theorem 3.30, $\tau^{u_n}(\Psi) \to \tau^u(\Psi)$, for every $\Psi \in Cth^b(\Omega \times S)$.

Let $\Psi : \Omega \times S \to \mathbb{R}$, defined by $\Psi(t, x) = \min\{d(u(t), x), 1\}$, for every $t \in \Omega$ and every $x \in S$. Then $\Psi \in Cth^b(\Omega \times S)$ and therefore

$$\int_\Omega \Psi(t, u_n(t)) d\mu(t) \to \int_\Omega \Psi(t, u(t)) d\mu(t) = 0.$$

On the other hand, for every $\varepsilon > 0$,

$$\int_\Omega \Psi(t, u_n(t))d\mu(t) \geq \int_{\{t:d(u_n(t),u(t))\geq\varepsilon\}} \min\{1, d(u(t), u_n(t))\}d\mu(t)$$
$$\geq \min\{1, \varepsilon\} \cdot \mu(\{t : d(u_n(t), u(t)) \geq \varepsilon\})$$

from where $\mu(\{t : d(u_n(t), u(t)) \geq \varepsilon\}) \to 0$.

($\Longleftarrow$): Let $u_n \xrightarrow{\mu} u$; since (S, d) is separable, every subsequence $(u'_n)_{n\in\mathbb{N}}$ of $(u_n)_{n\in\mathbb{N}}$ has a subsequence $(u''_n)_{n\in\mathbb{N}}$ convergent μ-almost everywhere to u (Theorem 9.2.1 of [61]). Then, for every $f \in C_b(S)$, $(|f(u''_n) - f(u)|)_{n\in\mathbb{N}}$ is a sequence uniformly bounded of measurable mappings convergent μ-a.e. to 0 and so

$$\int_\Omega |f(u''_n(t)) - f(u(t))|d\mu(t) \to 0.$$

Therefore every subsequence of $(f(u_n))_{n\in\mathbb{N}}$ has a subsequence convergent to $f(u)$ in $(\mathcal{L}^1(\mu), \|\cdot\|_1)$; so $f(u_n) \xrightarrow[\mathcal{L}^1(\mu)]{\|\cdot\|_1} f(u)$, for every $f \in C_b(S)$ and then $u_n \xrightarrow{\mathcal{P}} u$. □

Proposition 3.41. *Let S be a metrizable Suslin space; then $\mathcal{S}\restriction_{\mathcal{M}(S)} = \mathcal{P}\restriction_{\mathcal{M}(S)}$, which means that the trace of the stable topology on $\mathcal{M}(S)$ coincides with the topology of convergence in measure.*

Proof. According to (ii) of Remark 3.39, $\mathcal{S}\restriction_{\mathcal{M}(S)} \subseteq \mathcal{P}\restriction_{\mathcal{M}(S)}$.

Let $(u_i)_{i\in I} \subseteq \mathcal{M}(S)$ be a net and let $u \in \mathcal{M}(S)$ such that $u_i \xrightarrow{\mathcal{S}} u$; we must show that $u_i \xrightarrow{\mathcal{P}} u$.

For every $f \in C_b(S)$ let $\Psi : \Omega \times S \to \mathbb{R}$, $\Psi(t, x) = |f(x) - f(u(t))|$; then $\Psi \in Cth^b(\Omega \times S)$. According to Theorem 3.30, $\tau^{u_i}(\Psi) \to \tau^u(\Psi)$ and therefore

$$\int_\Omega \Psi(t, u_i(t))d\mu(t) = \tau^{u_i}(\Psi) \to \tau^u(\Psi) = \int_\Omega \Psi(t, u(t))d\mu(t) = 0.$$

Then $\int_\Omega |f(u_i(t)) - f(u(t))|d\mu(t) \to 0$, for every $f \in C_b(S)$ and so, by (iv) of Remark 3.39, $u_i \xrightarrow{\mathcal{P}} u$. □

Remark 3.42. Let S be a metrizable Suslin space and let $(u_n)_{n\in\mathbb{N}} \subseteq \mathcal{M}(S)$ and $u \in \mathcal{M}(S)$; then, according to Propositions 3.40 and 3.41,

$$u_n \xrightarrow{\mathcal{S}} u \iff u_n \xrightarrow{\mu} u.$$

In this case, we may say that, almost for all $t \in \Omega$,

$$u(t) \in \mathrm{Ls}(\{u_n(t) : n \in \mathbb{N}\}) = \bigcap_{p=1}^{\infty} \overline{\{u_k(t) : k \geq p\}} = Ls_n\{u_n(t)\}$$

(see [9]).

More generally, if $u_n \xrightarrow{S} \tau \in \mathcal{Y}(S)$, then, almost for all $t \in \Omega$, the support of τ_t is contained in $\mathrm{Ls}(\{u_p(t) : p \in \mathbb{N}\})$.

Proposition 3.43. *Let S be a metrizable Suslin space; if the sequence $(u_n)_n \subseteq \mathcal{M}(S) \subseteq \mathcal{Y}(S)$ is stably convergent to $\tau \in \mathcal{Y}(S)$ then, almost for all $t \in \Omega$,*

$$\operatorname{supp} \tau_t \subseteq \mathrm{Ls}(\{u_p(t) : p \in \mathbb{N}\}).$$

Proof. Let d be a metric generating the topology of S such that $d \leq 1$ and, for every $p \in \mathbb{N}$ and every $t \in \Omega$, let $A_p(t) = \overline{\{u_k(t) : k \geq p\}}$.

The mapping $\Psi_p : \Omega \times S \to \mathbb{R}$, defined by $\Psi_p(t, x) = d(x, A_p(t))$, for every $t \in \Omega$ and every $x \in S$, is continuous with respect to the second variable and measurable with respect to the first one (for every $x \in S$ and every $\alpha \in \mathbb{R}$, $\{t \in \Omega : \Psi_p(t, x) < \alpha\} = \bigcup_{k \geq p}\{t \in \Omega : d(x, u_k(t)) < \alpha\} \in \mathcal{A}$).

According to Proposition 3.8, Ψ is a bounded Carathéodory integrand. Therefore, in accordance with Theorem 3.30, for all $p \in \mathbb{N}$,

$$\tau^{u_n}(\Psi_p) = \int_\Omega \Psi_p(t, u_n(t))d\mu(t) = \int_\Omega d(u_n(t), A_p(t))d\mu(t) \xrightarrow[n\to\infty]{} \tau(\Psi_p).$$

Since, for every $t \in \Omega$ and every $p \in \mathbb{N}$, $\lim_{n\to\infty} d(u_n(t), A_p(t)) = 0$,

$$\tau(\Psi_p) = \int_\Omega \int_S d(x, A_p(t))d\tau_t(x)d\mu(t) = 0$$

and so, for all $p \in \mathbb{N}$,

$$\int_S d(x, A_p(t))d\tau_t(x) = 0, \quad \text{almost for every} \quad t \in \Omega.$$

Therefore, almost for every $t \in \Omega$,

$$\int_S d(x, A_p(t))d\tau_t(x) = 0, \quad \text{for every} \quad p \in \mathbb{N}.$$

Since every $A_p(t)$ is a closed set,

$$\tau_t(S \setminus A_p(t)) = 0, \quad \text{for every} \quad p \in \mathbb{N} \quad \text{and almost for every} \quad t \in \Omega.$$

Then, almost for all $t \in \Omega$,

$$\operatorname{supp} \tau_t \subseteq A_p(t), \quad \text{for every} \quad p \in \mathbb{N} \quad \text{or} \quad \operatorname{supp} \tau_t \subseteq \mathrm{Ls}(\{u_n(t) : n \in \mathbb{N}\}).$$

□

As shown in the following example, the stable topology is strictly weaker than the topology of the convergence in measure on $\mathcal{Y}(S)$.

Example 3.44. Let the Rademacher's sequence $(r_n)_{n\in\mathbb{N}}$ where $r_n : [0,1] \to \mathbb{R}$ is defined by

$$r_n(t) = \begin{cases} +1, & t \in \bigcup_{k=0}^{2^{n-1}-1} \left[\frac{2k}{2^n}, \frac{2k+1}{2^n}\right[, \\ -1, & t \in \bigcup_{k=0}^{2^{n-1}-1} \left[\frac{2k+1}{2^n}, \frac{2k+2}{2^n}\right[. \end{cases}$$

We have shown, in 1.56, b), that $(r_n)_{n\in\mathbb{N}}$ is not convergent in measure on $[0,1]$.

Let $\tau = \mu \otimes (\frac{1}{2}(\delta_1 + \delta_{-1})) \in \mathcal{Y}(\mathbb{R})$. We show that $r_n \overset{\mathcal{S}}{\to} \tau$ in $(\mathcal{Y}(\mathbb{R}), \mathcal{S})$ by establishing that, for every $A \in \mathcal{A}$ and every $f \in C_b(\mathbb{R})$,

$$\begin{aligned} |\tau^{r_n}(\chi_A \otimes f) - \tau(\chi_A \otimes f)| &= \left|\int_A f(r_n(t))d\mu(t) - \int_A \left(\int_{\mathbb{R}} f(x)d\tau_t(x)\right) d\mu(t)\right| \\ &= \left|\int_A f(r_n(t))d\mu(t) - \frac{1}{2}\mu(A)(f(1)+f(-1))\right| \longrightarrow 0. \end{aligned}$$

Let $f \in C_b(\mathbb{R})$ and $A \in \mathcal{A}$ ($\mathcal{A}$ is the σ-algebra of all Lebesgue-measurable sets on $]0,1[$).

(a) For every $\varepsilon > 0$, there exists $D_\varepsilon = D$ an open subset of $]0,1[$, such that $A \subseteq D$ and $\mu(D \setminus A) < \varepsilon$. Then,

$$\begin{aligned} &\left|\int_A f(r_n(t))d\mu(t) - \frac{\mu(A)}{2}[f(1) + f(-1)]\right| \\ &\le \left|\int_D f(r_n(t))d\mu(t) - \frac{\mu(D)}{2}[f(1) + f(-1)]\right| + \left|\int_{D\setminus A} f(r_n(t))d\mu(t)\right| \\ &\quad + \left|\frac{\mu(D \setminus A)}{2}[f(1) + f(-1)]\right| \\ &\le \left|\int_D f(r_n(t))d\mu(t) - \frac{\mu(D)}{2}[f(1) + f(-1)]\right| + \frac{3\varepsilon}{2}\max(|f(1)|, |f(-1)|). \end{aligned}$$

Therefore, it is sufficient to demonstrate the property for the open sets of $]0,1[$.

(b) Any open set D of $]0,1[$ is a countable union of pairwise disjoint open intervals I_k and, therefore

$$\begin{aligned} &\int_D f(r_n(t))d\mu(t) - \frac{\mu(D)}{2} \cdot (f(1) + f(-1)) \\ &= \sum_{k=1}^{\infty} \left(\int_{I_k} f(r_n(t))d\mu(t) - \frac{\mu(I_k)}{2}(f(1) + f(-1))\right). \end{aligned}$$

It is sufficient to demonstrate the property for the open intervals of $[0,1]$.

(c) Let $I =]a,b[=]\frac{k}{2^{n_0}}, \frac{k'}{2^{n_0}}[$; for every $n \geq n_0$, $I = \bigcup_{l=h}^{h'-1}]\frac{l}{2^n}, \frac{l+1}{2^n}[$ where $h = k \cdot 2^{n-n_0}$ and $h' = k' \cdot 2^{n-n_0}$. Then

$$\begin{aligned}\int_I f(r_n(t))d\mu(t) &= \sum_{l=h}^{h'-1} \int_{]\frac{l}{2^n}, \frac{l+1}{2^n}[} f((-1)^l)d\mu(t) = \sum_{l=h}^{h'-1} \frac{1}{2^n} f((-1)^l) \\ &= \frac{h'-h}{2^{n+1}}(f(1)+f(-1)) = \frac{k'-k}{2^{n_0+1}}(f(1)+f(-1)) \\ &= \frac{b-a}{2}(f(1)+f(-1))\end{aligned}$$

if k and k' have the same parity.

The case in which k and k' have different parity we make the demonstration in a similar manner. We obtain:

$$\begin{aligned}&\left| \int_{]a,b[} f(r_n(t))d\mu(t) - \frac{b-a}{2} \cdot (f(1)+f(-1)) \right| \\ &\qquad \leq \frac{1}{2^n} \cdot \max(|f(1)|, |f(-1)|).\end{aligned}$$

Therefore

$$\int_{]a,b[} f(r_n(t))d\mu(t) \longrightarrow \frac{b-a}{2} \cdot (f(1)+f(-1)).$$

(d) If $I =]a,b[\subset [0,1]$, then, for every $n \in \mathbb{N}$,

$$\begin{aligned}\int_{]a,b[} f(r_n(t))d\mu(t) &= \int_{]a, \frac{[2^n a]+1}{2^n}[} f(r_n(t))d\mu(t) \\ &+ \int_{]\frac{[2^n b]-1}{2^n}, b[} f(r_n(t))d\mu(t) + \int_{]\frac{[2^n a]+1}{2^n}, \frac{[2^n b]-1}{2^n}[} f(r_n(t))d\mu(t)\end{aligned}$$

from where it results that

$$\int_{]a,b[} f(r_n(t))d\mu(t) \longrightarrow \frac{b-a}{2} \cdot (f(1)+f(-1))$$

which was to be demonstrated.

The disintegration of τ is the constant mapping $\tau_t = \frac{1}{2}(\delta_1 + \delta_{-1})$ and $\operatorname{supp}\tau_t = \{-1, 1\} = Ls_n\{r_n(t) : n \in \mathbb{N}\}$.

Remark 3.45. According to the previous example, $\mathcal{S} \subsetneq \mathcal{P}$. Moreover, it results that $\mathcal{M}(S)$) is not a closed subset of $\mathcal{Y}(S)$.

The previous example is a special case of a more general construction, which is presented in the following proposition.

Proposition 3.46. *Let S be a regular Suslin space and let $u : \mathbb{R} \to S$ be a periodic measurable application of period T. Let $[0, T[$ provided with the Lebesgue measure μ and let $u_n : [0, T[\to S$, defined by $u_n(t) = u(nt)$, for every $t \in [0, T[$ and every $n \in \mathbb{N}$. Then, the sequence $(u_n)_{n \in \mathbb{N}} \subseteq \mathcal{M}(S) \subseteq \mathcal{Y}(S)$ is stably convergent to $\tau = \mu \otimes \lambda$, where $\lambda \in \mathcal{P}_S$ is defined by $\lambda(B) = \frac{1}{T}\mu(u^{-1}(B) \cap [0, T[)$, for every $B \in \mathcal{B}_S$.*

Proof. Let $\mathcal{A}_0 = \{[\frac{k}{2^p}, \frac{k+1}{2^p}[: p \in \mathbb{N}, k = 0, 1, \ldots, [2^p T] - 1\}$ where $[a]$ is the integer part of $a \in \mathbb{R}$. Then $\mathcal{A}_0$ is an algebra generating the σ-algebra $\mathcal{A}$ of all Lebesgue measurable sets on $[0, T[$.

In accordance to Proposition 3.21, in order to prove that $u_n \overset{\mathcal{S}}{\to} \tau = \mu \otimes \lambda$, it is sufficient to establish that

$$\tau^{u_n}(\chi_A \otimes f) \to \tau(\chi_A \otimes f), \quad \text{for every} \quad A \in \mathcal{A}_0 \quad \text{and every} \quad f \in C_b(S).$$

Let $A = [\frac{k}{2^p}, \frac{k+1}{2^p}[\in \mathcal{A}_0$ and let $f \in C_b(S)$; then

$$\begin{aligned} \tau^{u_n}(\chi_A \otimes f) &= \int_A f(u_n(t))d\mu(t) = \int_A f(u(nt))d\mu(t) \\ &= \frac{1}{n}\int_{nA} f(u(t))d\mu(t), \end{aligned}$$

where $nA = [\frac{nk}{2^p}, \frac{nk+n}{2^p}[$.

Let $q_n = [\frac{n}{2^p T}] \in \mathbb{N}$; then $q_n T \leq \frac{n}{2^p}$ and, since u is periodic,

$$\begin{aligned} \tau^{u_n}(\chi_A \otimes f) &= \frac{1}{n} \cdot \sum_{i=0}^{q_n - 1} \int_{[\frac{nk}{2^p} + iT, \frac{nk}{2^p} + (i+1)T[} f(u(t))d\mu(t) \\ &\quad + \frac{1}{n} \cdot \int_{[\frac{nk}{2^p} + q_n T, \frac{nk}{2^p} + \frac{n}{2^p}[} f(u(t))d\mu(t) \\ &= \frac{1}{n} \cdot q_n \cdot \int_{[0,T[} f(u(t))d\mu(t) + I_n, \quad \text{where} \end{aligned}$$

$$|I_n| \leq \frac{1}{n} \cdot \|f\|_0 \cdot \left(\frac{n}{2^p} - q_n T\right) < \frac{1}{n} \cdot \|f\|_0 \cdot T, \quad \text{so that} \quad I_n \to 0.$$

Then

$$\begin{aligned}\lim_{n\to\infty}\tau^{u_n}(\chi_A\otimes f) &= \int_{[0,T[} f(u(t))d\mu(t)\cdot\lim_{n\to\infty}\frac{q_n}{n}\\ &= \frac{1}{2^pT}\int_{[0,T[} f(u(t))d\mu(t)\\ &= \mu(A)\cdot\frac{1}{T}\cdot\int_{[0,T[} f(u(t))d\mu(t)\\ &= \mu(A)\cdot\int_S f(x)d\lambda(x)\\ &= \int_{[0,T[\times S}(\chi_A\otimes f)(t,x)d\tau(t,x).\end{aligned}$$

□

Remark 3.47.

(i) Let $u:\mathbb{R}\to\mathbb{R}$, defined by $u=\chi_{\cup_{k\in\mathbb{Z}}[2k,2k+\frac{1}{2}[}-\chi_{\cup_{k\in\mathbb{Z}}[2k+\frac{1}{2},2k+1[}$; then u is periodic of period 1.

Let $\lambda=\frac{1}{2}(\delta_1+\delta_{-1})$ and let $u_n:[0,1[\to\mathbb{R}$ defined by

$$u_n(t)=\begin{cases}+1, & t\in\bigcup_{k=0}^{n-1}\left[\frac{2k}{2n},\frac{2k+1}{2n}\right[,\\ -1, & t\in\bigcup_{k=0}^{n-1}\left[\frac{2k+1}{2n},\frac{2k+2}{2n}\right[.\end{cases}$$

Obviously, the Rademacher's sequence $(r_n)_{n\in\mathbb{N}}$ is the subsequence $(u_{2^{n-1}})_{n\in\mathbb{N}}$ of $(u_n)_{n\in\mathbb{N}}$ and therefore it is $\mathcal{S}$-convergent to $\mu\otimes\lambda$.

(ii) If we take $u:\mathbb{R}\to[-1,1]$, $u(t)=\sin t$, then $\lambda:\mathcal{B}_{[-1,1]}\to[0,1]$ is defined by

$$\lambda(B)=\frac{1}{\pi}\int_B\frac{1}{\sqrt{1-x^2}}d\mu(x),\quad\text{for every}\quad B\in\mathcal{B}_{[-1,1]}.$$

The sequence $u_n:[0,2\pi]\to[-1,1]$, defined by $u_n(t)=\sin nt$, is stably convergent to $\mu\otimes\lambda$ on $[0,2\pi[\times[-1,1]$.

In what follows, we will first state a lemma that will be used to show two density results on the space $\mathcal{Y}(S)$.

Lemma 3.48. *Let S be a regular Suslin space; for every $\tau\in\mathcal{Y}(S)$ and every $\varepsilon>0$, there exist a compact set $K\in\mathcal{K}_S$ and a measure $\tau'\in\mathcal{Y}(K)$ such that*

$$|\tau(\Psi)-\tau'(\Psi\restriction_{\Omega\times K})|<\varepsilon\cdot\|\Psi\|_0,\quad\textit{for every}\quad\Psi\in Cth^b(\Omega\times S).$$

Proof. According to (v) of Remark 3.11, for every $\tau \in \mathcal{Y}(S)$ and every $\varepsilon > 0$, there exists a compact set $K \in \mathcal{K}_S$ such that

$$\tau(\Omega \times (S \setminus K)) = \int_\Omega \tau_t(S \setminus K) d\mu(t) < \frac{\varepsilon}{4}. \tag{1}$$

Let $A = \{t \in \Omega : \tau_t(K) = 0\}$; since $K \in \mathcal{B}_S$, the mapping $t \mapsto \tau_t(K)$ is $(\mathcal{A} - \mathcal{B}_{[0,1]})$-measurable (see 3.1, (ii)) and therefore $A \in \mathcal{A}$. Moreover,

$$\begin{aligned} \mu(A) &\leq \int_A \tau_t(S \setminus K) d\mu(t) + \int_{\Omega \setminus A} \tau_t(S \setminus K) d\mu(t) \\ &= \int_\Omega \tau_t(S \setminus K) d\mu(t) < \frac{\varepsilon}{4}. \end{aligned} \tag{2}$$

Let $\nu \in \mathcal{P}_K$ be a fixed probability; then we can define $\tau' : \Omega \to \mathcal{P}_K$ by

$$\tau'_t(B) = \begin{cases} \frac{1}{\tau_t(K)} \cdot \tau_t(B), & t \in \Omega \setminus A \\ \nu(B), & t \in A \end{cases}, \text{ for every } t \in \Omega \text{ and every } B \in \mathcal{B}_K.$$

The mapping $t \mapsto \tau'_t(B)$ is $(\mathcal{A} - \mathcal{B}_{[0,1]})$-measurable and so τ' is $(\mathcal{A} - \mathcal{C}_K)$-measurable. Then, it is the disintegration of a measure $\tau' \in \mathcal{Y}(K)$.

For every integrand $\Psi \in Cth^b(\Omega \times S)$, $\Psi' = \Psi\restriction_{\Omega \times K} \in Cth^b(\Omega \times K)$. Then, according to (1) and (2), we have

$$\begin{aligned} |\tau(\Psi) - \tau'(\Psi')| &\leq \int_{\Omega \setminus A} \left| \int_S \Psi(t,x) d\tau_t(x) - \int_K \Psi(t,x) d\tau'_t(x) \right| d\mu(t) \\ &\quad + \int_A \left[\int_S |\Psi(t,x)| d\tau_t(x) + \int_K |\Psi(t,x)| d\nu(x) \right] d\mu(t) \\ &\leq \int_{\Omega \setminus A} \left| \int_K \Psi(t,x) d\tau_t(x) - \int_K \frac{\Psi(t,x)}{\tau_t(K)} d\tau_t(x) \right| d\mu(t) \\ &\quad + \int_{\Omega \setminus A} \left(\int_{S \setminus K} |\Psi(t,x)| d\tau_t(x) \right) d\mu(t) + 2\|\Psi\|_0 \cdot \mu(A) \\ &\leq \int_{\Omega \setminus A} \frac{\tau_t(S \setminus K)}{\tau_t(K)} \left(\int_K |\Psi(t,x)| d\tau_t(x) \right) d\mu(t) \\ &\quad + \|\Psi\|_0 \cdot \tau\left(\Omega \times (S \setminus K)\right) + 2\frac{\varepsilon}{4} \cdot \|\Psi\|_0 \\ &\leq \|\Psi\|_0 \cdot \int_{\Omega \setminus A} \tau_t(S \setminus K) d\mu(t) + \frac{\varepsilon}{4} \cdot \|\Psi\|_0 + \frac{\varepsilon}{2} \cdot \|\Psi\|_0 \\ &< \frac{\varepsilon}{4} \cdot \|\Psi\|_0 + \frac{3\varepsilon}{4} \cdot \|\Psi\|_0 = \varepsilon \cdot \|\Psi\|_0. \end{aligned}$$

□

Theorem 3.49. *Let S be a regular Suslin space; the space $\mathcal{E}(\mathcal{A}, \mathcal{P}_S)$ of all simple Young measures is dense in $(\mathcal{Y}(S), \mathcal{S})$.*

Proof. According to Definition 3.13, $\tau \in \mathcal{E}(\mathcal{A}, \mathcal{P}_S)$ if $\tau = \sum_{i=1}^n \nu_i \cdot \chi_{A_i}$, where $\{A_1, \dots, A_n\}$ is an $\mathcal{A}$-partition of Ω and $\nu_1, \dots, \nu_n \in \mathcal{P}_S$.

(I) Let S be metrizable. According to (i) of Theorem 3.14, for every $\tau \in \mathcal{Y}(S)$, there exists a sequence $(\tau^n)_{n\in\mathbb{N}} \subseteq \mathcal{E}(\mathcal{A}, \mathcal{P}_S)$ such that $\tau_t^n \xrightarrow{\mathcal{T}} \tau_t$, for every $t \in \Omega$. Then, for every $A \in \mathcal{A}$ and every $f \in C_b(S)$,

$$\tau^n(\chi_A \otimes f) = \int_A \tau_t^n(f) d\mu(t) \to \int_A \tau_t(f) d\mu(t) = \tau(\chi_A \otimes f)$$

therefore $\tau^n \xrightarrow{\mathcal{S}} \tau$ and $\tau \in \overline{\mathcal{E}(\mathcal{A}, \mathcal{P}_S)}^{\mathcal{S}}$.

(II) Let now S be a regular Suslin space and let $\tau \in \mathcal{Y}(S)$; since the stable topology on $\mathcal{Y}(S)$ is generated by the family $\mathcal{F}_S = \{I_\Psi : \Psi \in Cth_0^b(\Omega \times S)\}$, for every $\mathcal{S}$-neighborhood V of τ, there exist $\varepsilon > 0, n \in \mathbb{N}^*$ and $\Psi_1, \dots, \Psi_n \in Cth_0^b(\Omega \times S)$ such that

$$U = \bigcap_{i=1}^n I_{\Psi_i}^{-1}(]\tau(\Psi_i) - \varepsilon, \tau(\Psi_i) + \varepsilon[) \subseteq V,$$

where

$$U = \{\sigma \in \mathcal{Y}(S) : |\sigma(\Psi_i) - \tau(\Psi_i)| < \varepsilon, i = 1, \dots, n\}.$$

For every $i = 1, \dots, n$, let $\Psi_i = \chi_{A_i} \otimes f_i$, where $A_i \in \mathcal{A}$ and $f_i \in C_b(S)$ and let $M = \max\{\| f_i \|_0 : i = 1, \dots, n\}$. According to Lemma 3.48, there exist a compact set $K \in \mathcal{K}_S$ and a measure $\tau' \in \mathcal{Y}(K)$ such that

$$|\tau(\Psi) - \tau'(\Psi\restriction_{\Omega\times K})| < \frac{\varepsilon}{2M} \cdot \|\Psi\|_0, \quad \text{for every} \quad \Psi \in Cth^b(\Omega \times S).$$

For every $i = 1, \dots, n$, let $\Psi_i' = \Psi_i \restriction_{\Omega\times K} = \chi_{A_i} \otimes f_i \restriction_K \in Cth_0^b(\Omega \times K)$; then $U' = \{\sigma' \in \mathcal{Y}(K) : |\sigma'(\Psi_i') - \tau'(\Psi_i')| < \frac{\varepsilon}{2}, i = 1, \dots, n\}$ is a $\mathcal{S}_K$-neighborhood of τ' and then there exists $\sigma' \in \mathcal{E}(\mathcal{A}, \mathcal{P}_K) \bigcap U'$.

Let σ' be the disintegration of σ'; $\sigma' = \sum_{j=1}^p \nu_j' \cdot \chi_{B_j}$, where, for every $j = 1, \dots, p$, $\nu_j' \in \mathcal{P}_K$ and $B_j \in \mathcal{A}$. For every $j = 1, \dots, p$, we define $\nu_j : \mathcal{B}_S \to [0, 1]$ by $\nu_j(B) = \nu_j'(B \cap K)$; then ν_j are probabilities on S and therefore $\sigma = \sum_{j=1}^p \nu_j \cdot \chi_{B_j} \in \mathcal{E}(\mathcal{A}, \mathcal{P}_S)$.

Let us suppose that σ is the disintegration of $\boldsymbol{\sigma}$. We show that $\boldsymbol{\sigma} \in U$. For every $i = 1, \dots, n$,

$$|\boldsymbol{\sigma}(\Psi_i) - \boldsymbol{\tau}(\Psi_i)| \leq |\boldsymbol{\sigma}(\Psi_i) - \boldsymbol{\sigma}'(\Psi_i')| + |\boldsymbol{\sigma}'(\Psi_i') - \boldsymbol{\tau}'(\Psi_i')| + |\boldsymbol{\tau}'(\Psi_i') - \boldsymbol{\tau}(\Psi_i)|$$
$$< 0 + \frac{\varepsilon}{2} + \frac{\varepsilon}{2M} \cdot \|f\|_0 \leq \varepsilon.$$

Therefore $\boldsymbol{\sigma} \in U \subseteq V$.

Then, for every $\mathcal{S}$-neighborhood V of $\boldsymbol{\tau}$, $V \bigcap \mathcal{E}(\mathcal{A}, \mathcal{P}_S) \neq \emptyset$; this means that $\boldsymbol{\tau} \in \overline{\mathcal{E}(\mathcal{A}, \mathcal{P}_S)}^{\mathcal{S}}$ and so $\mathcal{E}(\mathcal{A}, \mathcal{P}_S)$ is dense in $(\mathcal{Y}(S), \mathcal{S})$. □

Theorem 3.50. *Let S be a regular Suslin space; if the measure μ is atomless, then $\mathcal{M}(S)$ is dense in $(\mathcal{Y}(S), \mathcal{S})$.*

Proof. (I) Let us first consider the case in which S is a compact Suslin space. The stable topology on $\mathcal{Y}(S)$ is generated by the family $\mathcal{F}_S = \{I_\Psi : \Psi \in Cth_0^b(\Omega \times S)\}$. Let $\boldsymbol{\tau} \in \mathcal{Y}(S)$; for every $A \in \mathcal{A}$, every $f \in C^b(S) = C(S)$ and every $\varepsilon > 0$, let $I(\boldsymbol{\tau}; A, f, \varepsilon) = I_\Psi^{-1}(]I_\Psi(\boldsymbol{\tau}) - \varepsilon, I_\Psi(\boldsymbol{\tau}) + \varepsilon[)$ where $\Psi = \chi_A \otimes f \in Cth_0^b(\Omega \times S)$.

Then, the family of sets $U = \bigcap_{k=1}^p I(\boldsymbol{\tau}; A_k, f_k, \varepsilon)$, where $p \in \mathbb{N}^*, \{A_1, \dots, A_p\}$ $\subseteq \mathcal{A}, \{f_1, \dots, f_p\} \subseteq C(S)$ and $\varepsilon > 0$, is an open neighborhood basis for $\boldsymbol{\tau}$ in $(\mathcal{Y}(S), \mathcal{S})$. We can choose the pairwise disjoint sets $A_k, k = 1, \dots, p$; indeed, we can always find the pairwise disjoint sets $B_1, \dots, B_q \in \mathcal{A}$ such that, for every $k = 1, \dots, p$, there exist $i_1^k, \dots, i_{r_k}^k \in \{1, \dots, q\}$ with $A_k = \bigcup_{j=1}^{r_k} B_{i_j^k}$. Then, if we note $g_l = f_k$, every time that $B_l \subseteq A_k$, the set $W = \bigcap_{l=1}^q I(\boldsymbol{\tau}; B_l, g_l, \frac{\varepsilon}{q}) \subseteq \bigcap_{k=1}^p I(\boldsymbol{\tau}; A_k, f_k, \varepsilon) = U$.

According to Theorem 3.49, $\mathcal{E}(\mathcal{A}, \mathcal{P}_S)$ is dense in $\mathcal{Y}(S)$.

Therefore, it is sufficient to show that, for every $\boldsymbol{\tau} \in \mathcal{E}(\mathcal{A}, \mathcal{P}_S)$ and every neighborhood V of $\boldsymbol{\tau}$, $V \cap \mathcal{M}(S) \neq \emptyset$.

Let $\boldsymbol{\tau} = \sum_{j=1}^q \nu_j \cdot \chi_{B_j} \in \mathcal{E}(\mathcal{A}, \mathcal{P}_S)$ where $\nu_1, \dots, \nu_q \in \mathcal{P}_S$ and $\{B_1, \dots, B_q\}$ is an $\mathcal{A}$-partition of Ω. According to the remark stated above, for every neighborhood V of $\boldsymbol{\tau}$ in $(\mathcal{Y}(S), \mathcal{S})$, there exist $p \in \mathbb{N}^*, \{A_1, \dots, A_p\} \subseteq \mathcal{A}$ pairwise disjoint, $f_1, \dots, f_p \in C(S)$ and $\varepsilon > 0$ such that

$$U = \bigcap_{k=1}^p I(\boldsymbol{\tau}; A_k, f_k, \varepsilon) \subseteq V. \tag{1}$$

Let $\varepsilon_1 = \frac{\varepsilon}{2\mu(\Omega)} > 0$; then the family

$$\left\{ D_y = \bigcap_{k=1}^{p} f_k^{-1}\left( ]f_k(y) - \varepsilon_1, f_k(y) + \varepsilon_1[\right) : y \in S \right\}$$

is an open cover of compact space S.

Let now $\{D_{y_i} : i = 1, \dots, n\}$ a finite subcover of S; if $C_i = D_{y_i} \setminus \cup_{j<i} D_{y_j}$, for every $i = 2, \dots, n$ and $C_1 = D_{y_1}$, then $\{C_1, \dots, C_n\}$ is a $\mathcal{B}_S$-partition of S.

For every $i = 1, \dots, n$, let $z_i \in C_i$ be a fixed point; then, for every $z \in C_i$ and every $k = 1, \dots, p$,

$$|f_k(z) - f_k(z_i)| \le |f_k(z) - f_k(y_i)| + |f_k(y_i) - f_k(z_i)| < 2\varepsilon_1. \tag{2}$$

For every $j = 1, \dots, q$, every $i = 1, \dots, n$ and every $k = 1, \dots, p$, let

$$\alpha_{ij}^k = \nu_j(C_i) \cdot \mu(B_j \cap A_k). \tag{3}$$

Then, for every $j = 1, \dots, q$ and every $k = 1, \dots, p$,

$$\sum_{i=1}^{n} \alpha_{ij}^k = \left(\sum_{i=1}^{n} \nu_j(C_i) \right) \cdot \mu(B_j \cap A_k) = \nu_j(S) \cdot \mu(B_j \cap A_k) = \mu(B_j \cap A_k).$$

Since μ is atomless, according to Saks' theorem (see [173], Th. I.4.10), for every $j = 1, \dots, q$ and every $k = 1, \dots, p$, there exists an $\mathcal{A}$-partition $\{B_{ij}^k : i = 1, \dots, n\}$ of $B_j \cap A_k$, such that

$$\mu(B_{ij}^k) = \alpha_{ij}^k, \quad \text{for every} \quad i = 1, \dots, n. \tag{4}$$

Then, for every $k = 1, \dots, p$,

$$A_k = \bigcup_{j=1}^{q} (B_j \cap A_k) = \bigcup_{j=1}^{q} \bigcup_{i=1}^{n} B_{ij}^k = \bigcup_{i=1}^{n} \left(\bigcup_{j=1}^{q} B_{ij}^k \right).$$

For every $i = 1, \dots, n$ and every $k = 1, \dots, p$, let

$$D_i^k = \bigcup_{j=1}^{q} B_{ij}^k \tag{5}$$

Then $\{D_i^k : i = 1, \dots, n\}$ is an $\mathcal{A}$-partition of A_k, for every $k = 1, \dots, p$.

Let $u : \Omega \to S, u = \sum_{i=1}^{n} z_i \cdot \chi_{\cup_{l=1}^{p} D_i^l}$; then $u \in \mathcal{M}(S)$ and, for every $k = 1, \ldots, p$,

$$
\begin{aligned}
&\left|\tau(\chi_{A_k} \otimes f_k) - \tau^u(\chi_{A_k} \otimes f_k)\right| \\
&\quad = \left|\sum_{j=1}^{q} \tau(\chi_{A_k \cap B_j} \otimes f_k) - \sum_{i=1}^{n} \tau^u(\chi_{D_i^k} \otimes f_k)\right| \\
&\quad = \left|\sum_{j=1}^{q} \nu_j(f_k) \cdot \mu(A_k \cap B_j) - \sum_{i=1}^{n} f_k(z_i)\mu(D_i^k)\right| \\
&\quad \overset{(5)}{=} \left|\sum_{j=1}^{q}\sum_{i=1}^{n} \nu_j(\chi_{C_i} \otimes f_k) \cdot \mu(A_k \cap B_j) - \sum_{i=1}^{n}\sum_{j=1}^{q} f_k(z_i)\mu(B_{ij}^k)\right| \\
&\quad \overset{(4)}{\le} \sum_{j=1}^{q}\sum_{i=1}^{n} \left|\nu_j(\chi_{C_i} \otimes f_k) \cdot \mu(A_k \cap B_j) - f_k(z_i)\alpha_{ij}^k\right| \\
&\quad \overset{(3)}{=} \sum_{j=1}^{q}\sum_{i=1}^{n} \left|\nu_j(\chi_{C_i} \otimes f_k) \cdot \mu(A_k \cap B_j) - f_k(z_i) \cdot \nu_j(C_i) \cdot \mu(A_k \cap B_j)\right| \\
&\quad = \sum_{j=1}^{q}\sum_{i=1}^{n} \left|\nu_j(\chi_{C_i} \otimes (f_k - f_k(z_i)))\right| \cdot \mu(A_k \cap B_j) \\
&\quad \le \sum_{j=1}^{q}\sum_{i=1}^{n} \int_{C_i} |f_k(z) - f_k(z_i)| d\nu_j(z) \cdot \mu(A_k \cap B_j) \\
&\quad \overset{(2)}{\le} 2\varepsilon_1 \sum_{j=1}^{q}\sum_{i=1}^{n} \nu_j(C_i)\mu(A_k \cap B_j) \\
&\quad = 2\varepsilon_1 \sum_{j=1}^{q} \mu(A_k \cap B_j) = 2\varepsilon_1 \mu(A_k) \le \varepsilon.
\end{aligned}
$$

Then, according to (1), we obtain $\tau^u \in \bigcap_{k=1}^{p} I(\tau; A_k, f_k, \varepsilon) = U \subseteq V$ and $V \cap \mathcal{M}(S) \neq \emptyset$.

(II) Let now S be a regular Suslin space and let $\tau \in \mathcal{Y}(S)$.

For every $\mathcal{S}$-neighborhood V of τ there exist $p \in \mathbb{N}^*, \Psi_1, \ldots, \Psi_p \in Cth_0^b(\Omega \times S)$ and $\varepsilon > 0$ such that

$$U = \{\sigma \in \mathcal{Y}(S) : |\sigma(\Psi_k) - \tau(\Psi_k)| < \varepsilon, k = 1, \ldots, p\} \subseteq V. \tag{6}$$

Let $M = \max\{\|\Psi_k\|_0 : k = 1, \dots, p\}$; according to Lemma 3.48, there exist a compact set $K \in \mathcal{K}_S$ and a measure $\tau' \in \mathcal{Y}(K)$ such that

$$|\tau(\Psi) - \tau'(\Psi \restriction_{\Omega \times K})| < \frac{\varepsilon}{2M} \|\Psi\|_0, \quad \text{for every} \quad \Psi \in Cth^b(\Omega \times S). \tag{7}$$

For every $k = 1, \dots, p$, let $\Psi'_k = \Psi_k \restriction_{\Omega \times K}$. Then $\Psi'_k \in Cth^b_0(\Omega \times K)$; indeed, if $\Psi_k = \chi_{A_k} \otimes f_k$, then $\Psi'_k = \chi_{A_k} \otimes (f_k \restriction_K)$.

Then

$$\begin{aligned} U' &= \bigcap_{k=1}^{p} I(\tau'; A_k, f_k \restriction_K, \varepsilon) \\ &= \{\sigma' \in \mathcal{Y}(K) : |\sigma'(\Psi'_k) - \tau'(\Psi'_k)| < \varepsilon, k = 1, \dots, p\} \end{aligned}$$

is a $\mathcal{S}_K$-neighborhood of $\tau' \in \mathcal{Y}(K)$.

According to (I), there exists an application $(\mathcal{A}-\mathcal{B}_K)$-measurable $u' : \Omega \to K$, such that $\tau^{u'} \in U'$; then

$$|\tau^{u'}(\Psi'_k) - \tau'(\Psi'_k)| < \frac{\varepsilon}{2}, \quad \text{for every} \quad k = 1, \dots, p. \tag{8}$$

Therefore the application $u : \Omega \to S$, defined by $u(t) = u'(t)$, for every $t \in \Omega$, is $(\mathcal{A} - \mathcal{B}_S)$-measurable. According to (7) and (8), for every $k = 1, \dots, p$,

$$\begin{aligned} |\tau^u(\Psi_k) - \tau(\Psi_k)| &= |\tau^{u'}(\Psi'_k) - \tau(\Psi_k)| \\ &\le |\tau^{u'}(\Psi'_k) - \tau'(\Psi'_k)| + |\tau'(\Psi'_k) - \tau(\Psi_k)| \\ &< \frac{\varepsilon}{2} + \frac{\varepsilon}{2M} \cdot \|\Psi_k\|_0 \le \varepsilon. \end{aligned}$$

Then $\tau^u \in U \subseteq V$ and therefore $V \cap \mathcal{M}(S) \neq \emptyset$. □

3.5 Compactness

Let S be a regular Suslin space and let $\mathcal{K}_S$ be the family of all compact subsets of S. Let us remind that a mapping $\varphi : S \to [0, +\infty]$ is inf-compact if, for every $a \in \mathbb{R}_+, \varphi^{-1}([0, a]) \in \mathcal{K}_S$. Every inf-compact mapping is l.s.c. and therefore it is a Borel mapping.

An integrand $\Psi : \Omega \times S \to \bar{\mathbb{R}}_+$ is said *inf-compact* if, for every $t \in \Omega$, $\Psi(t, .)$ is an inf-compact mapping. Obviously, every inf-compact integrand is a lower semicontinuous integrand.

Theorem 3.51. *Let S be a regular Suslin space, let $\mathcal{Y}(S)$ be the space of all Young measures on $\Omega \times S$ and let $\mathcal{H} \subseteq \mathcal{Y}(S)$. Let us consider the following conditions:*

(T_1) *For every $\varepsilon > 0$, there exists $K \in \mathcal{K}_S$ such that $\tau(\Omega \times (S \setminus K)) < \varepsilon$, for every $\tau \in \mathcal{H}$.*

(T_2) *There exists an inf-compact application $\varphi : S \to [0, +\infty]$ such that $\sup_{\tau \in \mathcal{H}} \tau(\chi_\Omega \otimes \varphi) < +\infty$.*

(T_3) *For every $\varepsilon > 0$, there is a tight set $H \subseteq \mathcal{P}_S$ such that, for $\tau_. \in \mathcal{H}$, $\tau_.^{-1}(\mathcal{P}_S \setminus H) \in \mathcal{A}$ and $\mu(\{t \in \Omega : \tau_t \in \mathcal{P}_S \setminus H\}) \leq \varepsilon$.*

(T_4) *For every $\varepsilon > 0$ there exists $C \in \mathcal{A} \otimes \mathcal{B}_S$ such that, for every t, $C_t = \{x \in S : (t, x) \in C\} \in \mathcal{K}_S$ and $\sup_{\tau \in \mathcal{H}} \tau((\Omega \times S) \setminus C) < \varepsilon$.*

(T_5) *There exists an inf-compact integrand $\Psi : \Omega \times S \to [0, +\infty]$ such that $\sup_{\tau \in \mathcal{H}} \tau(\Psi) < +\infty$.*

(T_6) *For every $\varepsilon > 0$ there exists a measurable set-valued function $\Gamma : \Omega \to \mathcal{K}_S$ such that $\sup_{\tau \in \mathcal{H}} \int_\Omega \tau_t(S \setminus \Gamma(t)) d\mu(t) < \varepsilon$.*

Then:

$$\begin{array}{ccccc} (T_1) & \Longleftrightarrow & (T_2) & \Longleftrightarrow & (T_3) \\ \Downarrow & & & & \\ (T_4) & \Longleftrightarrow & (T_5) & \Longleftrightarrow & (T_6) \end{array}$$

Proof. Condition (T_1) states that the family $\{\tau(\Omega \times \cdot) : \tau \in \mathcal{H}\} \subseteq Rca^+(\mathcal{B}_S)$ is tight; indeed the condition a) of Definition 2.61 says that $\sup_{\tau \in \mathcal{H}} \tau(\Omega \times S) = \mu(\Omega) < +\infty$. Therefore, the equivalence (T_1) $\Longleftrightarrow$ (T_2) is a consequence of Proposition 2.63.

(T_2) $\Longrightarrow$ (T_3) Let $\varphi : S \to [0, +\infty]$ be an inf-compact mapping and let $M = \sup_{\tau \in \mathcal{H}} \tau(\chi_\Omega \otimes \varphi) < +\infty$.

For every $\varepsilon > 0$, let $H = \{\theta \in \mathcal{P}_S : \theta(\varphi) \leq \frac{M}{\varepsilon}\}$. According to Proposition 2.63, H is tight and it is a narrowly closed subset of $\mathcal{P}_S$ (see the proof of Proposition 3.12).

Therefore, for every $\tau \in \mathcal{H}$, $\tau^{-1}(\mathcal{P}_S \setminus H) \in \mathcal{A}$ and $\mu(\tau^{-1}(\mathcal{P}_S \setminus H)) < \varepsilon$.

(T_3) $\Longrightarrow$ (T_1) According to (T_3), for every $\varepsilon > 0$ there exists a tight $H \subseteq \mathcal{P}_S$ such that, for every $\tau \in \mathcal{H}$, $\tau^{-1}(\mathcal{P}_S \setminus H) \in \mathcal{A}$ and $\mu(\tau^{-1}(\mathcal{P}_S \setminus H)) < \frac{\varepsilon}{2}$. Let

$K \in \mathcal{K}_S$ such that $\nu(S \setminus K) < \frac{\varepsilon}{2\mu(\Omega)}$, for every $\nu \in H$. Then, for every $\tau \in \mathcal{H}$,

$$\begin{aligned}\tau(\Omega \times (S \setminus K)) &= \int_\Omega \tau_t(S \setminus K) d\mu(t) \\ &= \int_{\tau^{-1}(\mathcal{P}_S \setminus H)} \tau_t(S \setminus K) d\mu(t) + \int_{\tau^{-1}(H)} \tau_t(S \setminus K) d\mu(t) \\ &\leq \mu(\tau^{-1}(\mathcal{P}_S \setminus H)) + \frac{\varepsilon}{2\mu(\Omega)} \cdot \mu(\tau^{-1}(H)) < \varepsilon.\end{aligned}$$

$(T_1) \Longrightarrow (T_4)$ For every $\varepsilon > 0$, let $K \in \mathcal{K}_S$ such that $\tau(\Omega \times (S \setminus K)) < \varepsilon$, for every $\tau \in \mathcal{H}$. Then $C = \Omega \times K$ satisfies (T_4).

$(T_4) \Longleftrightarrow (T_6)$ For every $C \in \mathcal{A} \otimes \mathcal{B}_S$ with compact sections, $\Gamma : \Omega \to \mathcal{K}_S$, defined by $\Gamma(t) = C_t$, for every $t \in \Omega$, is a measurable set-valued function (i.e. $G_\Gamma = \{(t, x) \in \Omega \times S : x \in \Gamma(t)\} \in \mathcal{A} \otimes \mathcal{B}_S$).

Conversely, for every measurable set-valued function $\Gamma : \Omega \to \mathcal{K}_S$, if $C = G_\Gamma$, $C \in \mathcal{A} \otimes \mathcal{B}_S$ and $C_t \in \mathcal{K}_S$, for every $t \in \Omega$. The equivalence of (T_4) and (T_6) results if we notice that, for every $\tau \in \mathcal{H}$,

$$\tau((\Omega \times S) \setminus C) = \int_\Omega \tau_t(S \setminus C_t) d\mu(t) = \int_\Omega \tau_t(S \setminus \Gamma(t)) d\mu(t).$$

$(T_5) \Longrightarrow (T_6)$ Let $\Psi : \Omega \times S \to [0, +\infty]$ an inf-compact integrand such that $M = \sup_{\tau \in \mathcal{H}} \tau(\Psi) < +\infty$. For every $\varepsilon > 0$, we define the set-valued function $\Gamma : \Omega \to \mathcal{K}_S$ letting $\Gamma(t) = \{x \in S : \Psi(t, x) \leq \frac{M}{\varepsilon}\}$, for every $t \in \Omega$ (since Ψ is inf-compact, $\Gamma(t) \in \mathcal{K}_S$, for every $t \in \Omega$).

Ψ is $(\mathcal{A} \otimes \mathcal{B}_S)$-measurable; then the graph of Γ, $G_\Gamma = \{(t, x) : x \in \Gamma(t)\} = \{(t, x) : \Psi(t, x) \leq \frac{M}{\varepsilon}\} \in \mathcal{A} \otimes \mathcal{B}_S$ and therefore Γ is measurable.

Moreover, we have, for every $\tau \in \mathcal{H}$,

$$\begin{aligned}\int_\Omega \tau_t(S \setminus \Gamma(t)) d\mu(t) &= \int_\Omega \tau_t\left(\left\{x \in S : \Psi(t, x) > \frac{M}{\varepsilon}\right\}\right) d\mu(t) \\ &\leq \frac{\varepsilon}{M} \cdot \int_\Omega \tau_t(\Psi) d\mu(t) = \frac{\varepsilon}{M} \cdot \tau(\Psi) \leq \varepsilon.\end{aligned}$$

$(T_6) \Longrightarrow (T_5)$ For every $n \in \mathbb{N}^*$, let $\Gamma_n : \Omega \to \mathcal{K}_S$ be a measurable set-valued function such that

$$\sup_{\tau \in \mathcal{H}} \int_\Omega \tau_t(S \setminus \Gamma_n(t)) d\mu(t) < \frac{1}{3^n}$$

and let $\Gamma_0(t) = \emptyset$, for every $t \in \Omega$. It is clear that we can choose the set-valued functions $(\Gamma_n)_{n \in \mathbb{N}}$ in a such manner that, for every $t \in \Omega$ and every $n \in \mathbb{N}, \Gamma_n(t) \subseteq \Gamma_{n+1}(t)$. Let now $\Psi : \Omega \times S \to [0, +\infty]$ defined by

$$\Psi(t, x) = \begin{cases} 2^n, & x \in \Gamma_{n+1}(t) \setminus \Gamma_n(t) \\ +\infty, & x \in S \setminus \cup_{n=0}^\infty \Gamma_n(t) \end{cases}, \quad \text{for every} \quad (t, x) \in \Omega \times S.$$

For every $a > 0$,

$$\{(t,x) : \Psi(t,x) < a\} = \bigcup_{2^n < a} \{(t,x) : x \in \Gamma_{n+1}(t) \setminus \Gamma_n(t)\}$$
$$= \bigcup_{2^n < a} [\{(t,x) : x \in \Gamma_{n+1}(t)\} \setminus \{(t,x) : x \in \Gamma_n(t)\}] \in \mathcal{A}.$$

Therefore Ψ is an integrand.

For every $t \in \Omega$ and every $a \in \mathbb{R}_+$, $\Psi(t,\cdot)^{-1}([0,a]) = \Gamma_{n+1}(t)$, where n is the integer part of $\frac{\ln a}{\ln 2}$. Then Γ is an inf-compact integrand.

Let $\tau \in \mathcal{H}$; then $\int_\Omega \tau_t(\cap_{n=0}^\infty (S \setminus \Gamma_n(t)) d\mu(t) = \lim_n \int_\Omega \tau_t(S \setminus \Gamma_n(t)) d\mu(t) = 0$ and so, for almost every $t \in \Omega$, $\tau_t(\cap_{n=0}^\infty (S \setminus \Gamma_n(t)) = 0$.

Then, for almost every $t \in \Omega, \tau_t(\Psi) = \int_{\cap_{n=0}^\infty (S \setminus \Gamma_n(t))} \Psi(t,x) d\tau_t(x) + \sum_{n=0}^\infty \int_{\Gamma_{n+1}(t) \setminus \Gamma_n(t)} 2^n d\tau_t(x) = \sum_{n=0}^\infty 2^n \cdot \tau_t(\Gamma_{n+1}(t) \setminus \Gamma_n(t))$ and therefore

$$\tau(\Psi) = \int_\Omega \tau_t(\Psi) d\mu(t) = \sum_{n=0}^\infty 2^n \cdot \int_\Omega \tau_t\left(\Gamma_{n+1}(t) \setminus \Gamma_n(t)\right) d\mu(t)$$
$$\leq \sum_{n=0}^\infty 2^n \cdot \int_\Omega \tau_t(S \setminus \Gamma_n(t)) d\mu(t) < \mu(\Omega) + \sum_{n=1}^\infty \left(\frac{2}{3}\right)^n = \mu(\Omega) + 2.$$

□

Definition 3.52. Let S be a regular Suslin space; a set $\mathcal{H} \subseteq \mathcal{Y}(S)$ is said *tight* if, for every $\varepsilon > 0$, there exists a compact set $K \in \mathcal{K}_S$ such that $\tau(\Omega \times (S \setminus K)) < \varepsilon$, for every $\tau \in \mathcal{H}$.

A net $(\tau^i)_{i \in I} \subseteq \mathcal{Y}(S)$ is *tight* if the set $\{\tau^i : i \in I\}$ is tight.

Remark 3.53.

(i) The condition of tightness is equivalent to conditions (T_2) and (T_3) of previous theorem.

(ii) According to (v) of Remark 3.11, for every $\tau \in \mathcal{Y}(S), \{\tau\}$ is tight.

(iii) Let $H \subseteq \mathcal{P}_S$ be a family of probabilities on S; by using Definition 3.13, we can note that $\mathcal{H} = \{\tau^\nu : \nu \in H\} \subseteq \mathcal{Y}(S)$. Then $\mathcal{H}$ is tight if and only if H is tight in $\mathcal{P}_S$ (see Definition 2.61).

This remark validates the use of the same term "tight" for sets of Young measures as well as for sets of probabilities.

(iv) In [45] is used the term "strictly tight" for a set that satisfies one of the equivalent conditions (T_1) to (T_3) of Theorem 3.51 and the term "flexibly tight" if one the equivalent conditions (T4) to (T6) is satisfied.

(v) The condition (T_5) of Theorem 3.51 is introduced in [9] and the condition (T_6) in [98]; we can find in [99] a proof for the sequential case of equivalence $(T_5) \iff (T_6)$.

(vi) In Corollary 3.61, we will show that, if S is a Polish space, then all conditions (T_1) to (T_6) are equivalent. More generally, in Theorem 4.3.14 of [45], it is shown that, if S is a regular Suslin space and a Prohorov space (see (i) of Remark 2.75), then the conditions (T_1) to (T_6) are equivalent.

In the following proposition, we will formulate a new condition equivalent to that of tightness.

Proposition 3.54. *Let S be a regular Suslin space; a set $\mathcal{H} \subseteq \mathcal{Y}(S)$ is tight if and only if, for every $\varepsilon > 0$ there exists a compact $K \in \mathcal{K}_S$ such that*

$$\mu\left(\{t \in \Omega : \tau_t(S \setminus K) \geq \varepsilon\}\right) < \varepsilon, \quad \textit{for every} \quad \tau_{.} \in \mathcal{H}.$$

Proof. ($\Longrightarrow$) If $\mathcal{H}$ is tight, for every $\varepsilon > 0$, there exists $K \in \mathcal{K}_S$ such that $\tau(\Omega \times (S \setminus K)) < \varepsilon^2$, for every $\tau \in \mathcal{H}$. Then, for every $\tau \in \mathcal{H}$,

$$\begin{aligned} \mu\left(\{t \in \Omega : \tau_t(S \setminus K) \geq \varepsilon\}\right) &\leq \frac{1}{\varepsilon} \int_{(\tau_t(S \setminus K) \geq \varepsilon)} \tau_t(S \setminus K) d\mu(t) \\ &\leq \frac{1}{\varepsilon} \int_{\Omega} \tau_t(S \setminus K) d\mu(t) = \frac{1}{\varepsilon} \cdot \tau(\Omega \times (S \setminus K)) < \varepsilon. \end{aligned}$$

($\Longleftarrow$) Let us now suppose that, for every $\varepsilon > 0$, there exists $K \in \mathcal{K}_S$ such that

$$\mu\left(\{t \in \Omega : \tau_t(S \setminus K) \geq \varepsilon\}\right) < \varepsilon, \quad \text{for every} \quad \tau \in \mathcal{H}.$$

Then, for every $\tau \in \mathcal{H}$,

$$\begin{aligned} \tau(\Omega \times (S \setminus K)) &= \int_{\Omega} \tau_t(S \setminus K) d\mu(t) \\ &= \int_{(\tau_t(S \setminus K) \geq \varepsilon)} \tau_t(S \setminus K) d\mu(t) + \int_{(\tau_t(S \setminus K) < \varepsilon)} \tau_t(S \setminus K) d\mu(t) \\ &\leq \mu\left(\{t \in \Omega : \tau_t(S \setminus K) \geq \varepsilon\}\right) + \varepsilon \cdot \mu(\Omega) < \varepsilon \cdot (1 + \mu(\Omega)). \end{aligned}$$

□

Definition 3.55. Let $H \subseteq \mathcal{M}(S)$ be a family of measurable mappings and let $\mathcal{H} = \{\tau^u : u \in H\} \subseteq \mathcal{Y}(S)$ (see Definition 3.15 and Remark 3.16); $\mathcal{H}$ is tight if and only if

$$\forall \varepsilon > 0, \exists K \in \mathcal{K}_S \quad \text{s.t.} \quad \mu\left(u^{-1}(S \setminus K)\right) < \varepsilon, \quad \forall u \in H. \tag{T}$$

In this case, we will say that the family $H \subseteq \mathcal{M}(S)$ is *tight*.

A net $(u^i)_{i \in I} \subseteq \mathcal{M}(S)$ will be said *tight* if the set $\{u^i : i \in I\} \subseteq \mathcal{M}(S)$ is tight.

Particularly, if $S = \mathbb{R}^d$, then H is tight if and only if

$$\forall \varepsilon > 0, \exists k > 0 \quad \text{t.q.} \quad \mu(\|u\| \geq k) = \mu(\{t : \|u(t)\| \geq k\}) < \varepsilon, \quad \forall u \in H, \tag{T_e}$$

where $\|\cdot\|$ is the usual norm on $\mathbb{R}^d$.

In this particular case, we have the following property:

Proposition 3.56.

(i) *Every bounded set* $H \subseteq L^1(\mu, \mathbb{R}^d) \subseteq \mathcal{M}(\mathbb{R}^d)$ *is tight.*

(ii) *More generally, let* $(E, \|\cdot\|)$ *be a reflexive separable Banach space; every bounded set* $H \subseteq L^1(\mu, E)$ *is tight with respect to weak topology on E (w-tight).*

Proof. (i) Let H be a bounded set in $L^1(\mu, \mathbb{R}^d)$ and

$$M = \sup_{u \in H} \int_\Omega \|u(t)\| d\mu(t) < +\infty.$$

Then, for every $k > 0$ and every $u \in H$, $M \geq \int_{(\|u\| \geq k)} \|u\| d\mu \geq k \cdot \mu(\|u\| \geq k)$. Then, for every $\varepsilon > 0$, there exists $k > \frac{M}{\varepsilon}$ such that $\mu(\|u\| \geq k) < \varepsilon$, for every $u \in H$. Therefore, according to Definition 3.55, H is tight.

(ii) E endowed with the weak topology is a regular Suslin space (see the Example 2.17); the proof comes from that of (i) by remarking that the set $K = \{x \in E : \|x\| \leq k\}$ is weakly compact and that $\tau^u(\Omega \times (E \setminus K)) = \mu(\|u\| > k)$. □

Remark 3.57. If E is a separable Banach space, then, generally, a bounded set of $L^1(\mu, E)$ is not a tight with respect to strong topology of E. Indeed, let $E = C_{[0,1]}$ be the space of all continuous real-valued mappings on $[0, 1]$ provided with the norm of uniform convergence and, for every $n \in \mathbb{N}$, let $f_n : [0, 1] \to \mathbb{R}$, $f_n(x) = x^n$. Then $t \in [0, 1] \mapsto u_n(t) = f_n$ defines a E-valued function on $[0, 1]$. The sequence $(u_n)_n \subseteq L^1(\mu, E)$ is bounded. If we suppose that $(u_n)_n$ is tight, then there exists a compact set $K \subseteq C_{[0,1]}$ such that $\mu(u_n^{-1}(E \setminus K)) < \frac{1}{2}$, for every $n \in \mathbb{N}$ and so $(f_n)_n \subseteq K$ which is impossible since $(f_n)_n$ is not equicontinuous.

Theorem 3.58. *Let S be a compact Suslin space; then $(\mathcal{Y}(S), \mathcal{S})$ is a compact space.*

Proof. Let $C(S) = C_b(S)$ be the space of all real-valued continuous functions on S provided with the norm of uniform convergence, $\|\cdot\|_0$; let $\mathcal{L}^1(\mu, C(S)) = \{f \in \mathcal{M}(C(S)) : \|f\|_1 = \int_\Omega \|f\|_0 d\mu < +\infty\}$.

If $L^1(\mu, C(S)) = \{\bar{f} : f \in \mathcal{L}^1(\mu, C(S))\}$, then $L^1(\mu, C(S))$ is a Banach space; let $[L^1(\mu, C(S))]^*$ its dual space.

In the proof of Theorem 3.30, we have associated to each $\Psi \in Cth^b(\Omega \times S)$ the mapping $\varphi_\Psi : \Omega \to C(S)$, defined by $\varphi_\Psi(t) = \Psi(t, \cdot)$, for every $t \in \Omega$ and we showed that $\varphi_\Psi \in L^1(\mu, C(S))$.

Since $\sup_{t\in\Omega} \|\varphi_\Psi(t)\|_0 = \sup_{t\in\Omega} \sup_{x\in S} |\Psi(t,x)| < +\infty$, φ_Ψ is a bounded mapping.

The mapping $\Psi \mapsto \varphi_\Psi$ is a bijection of $Cth^b(\Omega \times S)$ on the subset E of all bounded functions of $L^1(\mu, C(S))$. Indeed, for every bounded function $f \in L^1(\mu, C(S))$, the mapping $\Psi : \Omega \times S \to \overline{\mathbb{R}}$, $\Psi(t,x) = f(t)(x)$, is measurable in t and continuous in x (if, for every $x \in S$, $\phi_x : C(S) \to \mathbb{R}$, $\phi_x(g) = g(x)$, for every $g \in C(S)$, then $\Psi(\cdot, x) = \phi_x \circ f$ is a composition of the continuous function ϕ_x with the measurable function f). Then, according to Proposition 3.8, Ψ is a Carathéodory integrand. Moreover, for every $(t,x) \in \Omega \times S$,

$$|\Psi(t,x)| \le \|f(t)\|_0 \le \sup_{t\in\Omega} \|f(t)\|_0 < +\infty.$$

Then $\Psi \in Cth^b(\Omega \times S)$. Obviously, $\varphi_\Psi = f$.

For every $\tau \in \mathcal{Y}(S)$, let $f_\tau^* : L^1(\mu, C(S)) \to \mathbb{R}$ defined by

$$f_\tau^*(f) = \int_\Omega \left(\int_S f(t)(x) d\tau_t(x) \right) d\mu(t), \quad \text{for every} \quad f \in L^1(\mu, C(S)).$$

We remark that, for every $f \in L^1(\mu, C(S))$,

$$\int_\Omega \sup_{x\in S} |f(t)(x)|\, d\mu(t) = \int_\Omega \|f(t)\|_0 d\mu(t) < +\infty \qquad \text{and therefore}$$

$$|f_\tau^*(f)| \le \int_\Omega \|f(t)\|_0 d\mu(t) = \|f\|_1.$$

Then the application $\tau \mapsto f_\tau^*$ is an application of $\mathcal{Y}(S)$ in $[L^1(\mu, C(S))]^*$. Moreover, $\|f_\tau^*\| = \sup_{\|f\|_1 \le 1} |f_\tau^*(f)| \le 1$; which means that the range of this application, $F = \{f_\tau^* : \tau \in \mathcal{Y}(S)\}$, is a subset of the unit closed ball of$[L^1(\mu, C(S))]^*$.

According to Theorem 3.30, a net $(\tau^i)_{i\in I} \subseteq \mathcal{Y}(S)$ is stably convergent to τ if and only if $\tau^i(\Psi) \to \tau(\Psi)$, for every $\Psi \in Cth^b(\Omega \times S)$, which is equivalent to say that $f_{\tau^i}^*(\varphi_\Psi) \to f_\tau^*(\varphi_\Psi)$, or that $f_{\tau^i}^*(f) \to f_\tau^*(f)$, for every $f \in E$.

Since the set E is dense in $L^1(\mu, C(S))$, it results that $\tau^i \xrightarrow{\mathcal{S}} \tau$ if and only if $(f^*_{\tau^i})_{i\in I}$ is weak*-convergent to f^*_τ in $[L^1(\mu, C(S))]^*$. Then $(\mathcal{Y}(S), \mathcal{S})$ is homeomorphic to the subset F of the unit closed ball of $[L^1(\mu, C(S))]^*$ endowed with the weak* topology.

In order to complete the proof, it is sufficient to show that F is a weak*-closed set.

Let f^* be in the weak* closure of F; there exists a net $(\tau^i)_{i\in I} \subseteq \mathcal{Y}(S)$ such that $(f^*_{\tau^i})_{i\in I}$ to be weak*-convergent to f^*. Then, for every $\Psi \in Cth^b(\Omega \times S)$,

$$f^*_{\tau^i}(\varphi_\Psi) \to f^*(\varphi_\Psi) \quad \text{or} \quad \tau^i(\Psi) \to f^*(\varphi_\Psi).$$

For every $A \in \mathcal{A}$, the mapping $\tau(A \times \cdot) : C(S) \to \mathbb{R}$, defined by: $\tau(A \times f) = \lim_i \tau^i(\chi_A \otimes f)$ is a positive linear functional; therefore, according to Theorem 4-1 of Ch. 3.n° 2 from [32], a Radon measure on S (i.e. $\tau(A \times \cdot) \in Rca^+(\mathcal{B}_S)$). Moreover $\tau(A \times S) = \lim_i \tau^i(A \times S) = \mu(A)$, for every $A \in \mathcal{A}$ and then $\tau \in \mathcal{Y}(S)$.

Since, for each $\Psi \in Cth^b(\Omega \times S), \tau(\Psi) = \lim_i \tau^i(\Psi) = f^*(\varphi_\Psi)$, $f^* = f^*_\tau \in F$. □

Theorem 3.59 (Prohorov's theorem).

(a) *Let S be a regular Suslin space and let $\mathcal{H} \subseteq \mathcal{Y}(S)$ be a tight; then $\mathcal{H}$ is relatively $\mathcal{S}$-compact.*

(b) *Let S be a regular Suslin space and a Prohorov space; if $\mathcal{H}$ is a relatively stably compact subset of $\mathcal{Y}(S)$, then $\mathcal{H}$ is tight.*

Proof. (a) Let us recall that, according to (i) of Remark 3.53, $\mathcal{H}$ is tight if and only if there exists an inf-compact application $\varphi : S \to [0, +\infty]$ such that $M = \sup_{\tau\in\mathcal{H}} \tau(\chi_\Omega \otimes \varphi) < +\infty$.

(a1) Firstly, let us consider that S is metrizable; in this case, it can be embedded in a compact metrizable space $\bar{S}$. According to Theorem 3.58, $(\mathcal{Y}(\bar{S}), \mathcal{S}_{\bar{S}})$ is a compact space and according to Theorem 3.28, $(\mathcal{Y}(S), \mathcal{S})$ is homeomorphic to a subspace of $(\mathcal{Y}(\bar{S}), \mathcal{S}_{\bar{S}})$.

For every net $(\tau^i)_{i\in I} \subseteq \mathcal{H} \subseteq \mathcal{Y}(S) \subseteq \mathcal{Y}(\bar{S})$, there exist a subnet $(\tau^{i_j})_{j\in J}$ and a Young measure $\tau \in \mathcal{Y}(\bar{S})$ such that $\tau^{i_j} \xrightarrow{\mathcal{S}_{\bar{S}}} \tau$.

Let $\Phi : \bar{S} \to [0, +\infty]$ be the mapping defined by

$$\Phi(x) = \begin{cases} \varphi(x) & \text{if} \quad x \in S, \\ +\infty & \text{if} \quad x \in \bar{S} \setminus S. \end{cases}$$

Then, for every $a \in \mathbb{R}_+$, $\Phi^{-1}([0,a]) = \varphi^{-1}([0,a]) \in \mathcal{K}_S \subseteq \mathcal{K}_{\bar{S}}$; therefore Φ is an inf-compact application on $\bar{S}$ and then Φ is l.s.c. and bounded from below. According to (2) of Proposition 3.22,

$$\tau(\chi_\Omega \otimes \Phi) \le \liminf_j \tau^{i_j}(\chi_\Omega \otimes \Phi) = \liminf_{j \in J} \tau^{i_j}(\chi_\Omega \otimes \varphi) \le M$$

and then

$$\begin{aligned} +\infty > M &\ge \int_\Omega \left(\int_{\bar{S}} \Phi(x) d\tau_t(x) \right) d\mu(t) \\ &= \int_\Omega \left(\int_S \varphi(x) d\tau_t(x) \right) d\mu(t) + \int_\Omega \left(\int_{\bar{S} \setminus S} (+\infty) d\tau_t(x) \right) d\mu(t). \end{aligned}$$

We can see that, almost for every $t \in \Omega$, $\tau_t(\bar{S} \setminus S) = 0$ and therefore $\tau \in \mathcal{Y}(S)$. Then $\tau^{i_j} \xrightarrow{\mathcal{S}} \tau$; hence $\mathcal{H}$ is relatively $\mathcal{S}$-compact.

(a2) Let now (S, σ) be any regular Suslin space and let $\mathcal{H}$ be a subset σ-tight of $\mathcal{Y}(S)$ (i.e. tight with respect to compacts of (S, σ)). According to (ii) of Theorem 2.25, there exists a metric d on S such that the generated topology σ_d is coarser than σ, second-countable and so that $\mathcal{B}_S(\sigma_d) = \mathcal{B}_S(\sigma)$. Let $\mathcal{S}_d$ be the stable topology on $\mathcal{Y}(S)$ with respect to (S, σ_d). Since $\mathcal{K}_{(S,\sigma)} \subseteq \mathcal{K}_{(S,\sigma_d)}$, $\mathcal{H}$ is σ_d-tight. According to (a1), since (S, σ_d) is a metrizable Suslin space, $\mathcal{H}$ is relatively $\mathcal{S}_d$-compact.

A net $(\tau^i)_{i \in I} \subseteq \mathcal{Y}(S)$ is $\mathcal{S}_d$-convergent to $\tau \in \mathcal{Y}(S)$ if and only if, for every $A \in \mathcal{A}$ and every $f \in C_b(S, \sigma_d)$, $\tau^i(\chi_A \otimes f) \to \tau(\chi_A \otimes f)$. Since $C_b(S, \sigma_d) \subseteq C_b(S, \sigma)$, $\mathcal{S}_d \subseteq \mathcal{S}$.

Since $\mathcal{H}$ is σ-tight, there exists a σ-inf-compact application $\varphi : (S, \sigma) \to [0, +\infty]$ such that $M = \sup_{\tau \in \mathcal{H}} \tau(\chi_\Omega \otimes \varphi) < +\infty$.

Let $(\tau^i)_{i \in I} \subseteq \mathcal{H}$ be a net; since $\mathcal{H}$ is relatively $\mathcal{S}_d$-compact there exist a subnet $(\tau^{i_j})_{j \in J}$ and a Young measure $\tau \in \mathcal{Y}(S)$ such that $\tau^{i_j} \xrightarrow{\mathcal{S}_d} \tau$.

In order to show that $\tau^{i_j} \xrightarrow{\mathcal{S}} \tau$, we will go back to an argument used in the proof of Theorem 2.6 (ii) of [18].

Let $A \in \mathcal{A}$ and let $f : (S, \sigma) \to \overline{\mathbb{R}}_+$ be a σ-l.s.c. application; for every $\varepsilon > 0$, let $\varphi_\varepsilon = f + \varepsilon \cdot \varphi : (S, \sigma) \to \overline{\mathbb{R}}_+$. Then φ_ε is σ-inf-compact; indeed, for every $a \in \mathbb{R}_+$, $\varphi_\varepsilon^{-1}([0,a]) = \{x \in S : f(x) + \varepsilon \cdot \varphi(x) \le a\} \subseteq \{x \in S : \varphi(x) \le \frac{a}{\varepsilon}\} = \varphi^{-1}([0, \frac{a}{\varepsilon}])$. Since φ_ε is σ-l.s.c., $\varphi_\varepsilon^{-1}([0,a])$ is a σ-closed subset of the σ-compact set $\varphi^{-1}([0, \frac{a}{\varepsilon}])$. Then $\varphi_\varepsilon^{-1}([0,a]) \in \mathcal{K}_{(S,\sigma)} \subseteq \mathcal{K}_{(S,\sigma_d)}$, from where $\varphi_\varepsilon^{-1}([0,a])$ is a closed subset of (S, σ_d) and so φ_ε is an application τ_d-l.s.c. and bounded from below. Therefore

$$\tau(\chi_A \otimes \varphi_\varepsilon) \le \liminf_j \tau^{i_j}(\chi_A \otimes \varphi_\varepsilon)$$

and then

$$\liminf_j \tau^{i_j}(\chi_A \otimes f) + \varepsilon \cdot M \geq \liminf_j \tau^{i_j}(\chi_A \otimes \varphi_\varepsilon) \geq \tau(\chi_A \otimes \varphi_\varepsilon)$$
$$\geq \tau(\chi_A \otimes f), \quad \text{for every} \quad \varepsilon > 0;$$

therefore, for every positive σ-l.s.c. application,

$$\liminf_j \tau^{i_j}(\chi_A \otimes f) \geq \tau(\chi_A \otimes f).$$

It is evident that the inequality is also true for any application bounded from below and σ-l.s.c. and then, according to (2) of Proposition 3.22, $\tau^{i_j} \xrightarrow{\mathcal{S}} \tau$.

(b) Let S be a regular Suslin space and a Prohorov space (according to the definition given in 2.75(i), this means that every relatively narrowly compact subset of $Rca^+(\mathcal{B}_S)$ is tight). The mapping $\Phi : \mathcal{Y}(S) \to Rca^+(\mathcal{B}_S)$, defined by $\Phi(\tau) = \tau(\Omega \times \cdot)$, for every $\tau \in \mathcal{Y}(S)$, is $(\mathcal{S} - \mathcal{T}_S)$-continuous (where $\mathcal{T}_S$ is the narrow topology on $Rca^+(\mathcal{B}_S)$). If $\mathcal{H} \subseteq \mathcal{Y}(S)$ is relatively $\mathcal{S}$-compact, then $\Phi(\mathcal{H})$ is relatively $\mathcal{T}_S$-compact in $Rca^+(\mathcal{B}_S)$ and so $\Phi(\mathcal{H})$ is tight. Therefore, for every $\varepsilon > 0$, there exists $K \in \mathcal{K}_S$ such that, for every $\tau \in \mathcal{H}$, $\Phi(\tau)(S \setminus K) = \tau(\Omega \times (S \setminus K)) < \varepsilon$. □

In the special case in which S is a Polish space, we can show, by using a similar method to that used in (a) of Theorem 3.59, that condition (T_5) of Theorem 3.51 is a sufficient condition of stable compactness.

Theorem 3.60. *Let S be a Polish space and let $\mathcal{H} \subseteq \mathcal{Y}(S)$ be a set satisfying the condition (T_5) of Theorem 3.51; then $\mathcal{H}$ is relatively stably compact.*

Proof. According to condition (T_5) of Theorem 3.51, there exists an integrand inf-compact $\Psi : \Omega \times S \to [0, +\infty]$ such that

$$M = \sup_{\tau \in \mathcal{H}} \tau(\Psi) < +\infty.$$

Since S is Polish, it is homeomorphic to a G_δ subset of compact space $\bar{S} = [0, 1]^{\mathbb{N}}$ (see Theorem 2.19) and therefore $\mathcal{B}_S \subseteq \mathcal{B}_{\bar{S}}$. According to Theorem 3.58, $(\mathcal{Y}(\bar{S}), \mathcal{S}_{\bar{S}})$ is a compact space and by Theorem 3.28, $(\mathcal{Y}(S), \mathcal{S})$ is homeomorphic to a subspace of $(\mathcal{Y}(\bar{S}), \mathcal{S}_{\bar{S}})$.

For every net $(\tau^i)_{i \in I} \subseteq \mathcal{H} \subseteq \mathcal{Y}(S) \subseteq \mathcal{Y}(\bar{S})$, there exist a subnet $(\tau^{i_j})_{j \in J}$ and a Young measure $\tau \in \mathcal{Y}(\bar{S})$ such that $\tau^{i_j} \xrightarrow{\mathcal{S}_{\bar{S}}} \tau$.

Let $\bar{\Psi} : \Omega \times \bar{S} \to [0, +\infty]$ be the mapping defined by

$$\bar{\Psi}(t, x) = \begin{cases} \Psi(t, x), & \text{if } \; t \in \Omega, x \in S, \\ +\infty, & \text{if } \; t \in \Omega, x \in \bar{S} \setminus S. \end{cases}$$

Then $\bar{\Psi}$ is an integrand on $\bar{S}$.

For every $t \in \Omega$ and every $a \in \mathbb{R}_+$, $(\bar{\Psi}(t, \cdot))^{-1}([0, a]) = \Psi(t, \cdot)^{-1}([0, a]) \in \mathcal{K}_S \subseteq \mathcal{K}_{\bar{S}}$, therefore $\bar{\Psi}$ is an integrand inf-compact on $\bar{S}$ and so l.s.c. and bounded from below. According to Proposition 3.34,

$$\tau(\bar{\Psi}) \leq \liminf_j \tau^{i_j}(\Psi) \leq M < \infty$$

and then

$$\begin{aligned} +\infty > M \geq \tau(\bar{\Psi}) &= \int_\Omega \left(\int_{\bar{S}} \bar{\Psi}(t, x) d\tau_t(x) \right) d\mu(t) \\ &= \int_\Omega \left(\int_S \Psi(t, x) d\tau_t(x) \right) d\mu(t) + \int_\Omega \left(\int_{\bar{S} \setminus S} (+\infty) d\tau_t(x) \right) d\mu(t). \end{aligned}$$

Then $\tau_t(\bar{S} \setminus S) = 0$, almost for every $t \in \Omega$, and so $\tau \in \mathcal{Y}(S)$. Therefore $\tau^{i_j} \xrightarrow{\mathcal{S}} \tau$ and then $\mathcal{H}$ is relatively stably compact. □

Corollary 3.61. *Let S be a Polish space; for every $\mathcal{H} \subseteq \mathcal{Y}(S)$, the conditions (T_1) to (T_6) of Theorem* 3.51 *are equivalent.*

Proof. In accordance to Theorem 3.51, we have to prove that $(T_5) \implies (T_1)$. Let $\mathcal{H} \subseteq \mathcal{Y}(S)$ be a set satisfying (T_5); according to the previous theorem, $\mathcal{H}$ is relatively $\mathcal{S}$-compact. Since every Polish space is a Prohorov one (see (i) of Remark 2.75), $\mathcal{H}$ is tight (see (b) of Theorem 3.59). □

In the following corollary, we will rewrite the equivalent conditions (T_1) to (T_6) of Theorem 3.51 for the particular case in which $H \subseteq \mathcal{M}(S) \subseteq \mathcal{Y}(S)$.

Corollary 3.62. *Let S be a Polish space and let $H \subseteq \mathcal{M}(S) \subseteq \mathcal{Y}(S)$; The following conditions are equivalent:*

(T_1) *For every $\varepsilon > 0$, there exists $K \in \mathcal{K}_S$ such that* $\sup_{u \in H} \mu(u^{-1}(S \setminus K)) < \varepsilon$.

(T_2) *There exists an inf-compact mapping $\varphi : S \to [0, +\infty]$ such that* $\sup_{u \in H} \int_\Omega \varphi(u(t)) d\mu(t) < +\infty$.

(T_3) *For every* $\varepsilon > 0$, *there exists a tight set* $L \subseteq \mathcal{P}_S$ *such that* $\sup_{u\in H} \mu(\{t \in \Omega : \delta_{u(t)} \notin L\}) \leq \varepsilon$.

(T_4) *For every* $\varepsilon > 0$, *there exists* $C \in \mathcal{A} \otimes \mathcal{B}_S$ *such that*, $\forall t \in \Omega$, $C_t = \{x \in S : (t,x) \in C\} \in \mathcal{K}_S$ *and* $\sup_{u\in H} \mu(\Omega \setminus \bar{u}^{-1}(C)) < \varepsilon$ (*where* $\bar{u} : \Omega \to \Omega \times S, \bar{u}(t) = (t, u(t))$).

(T_5) *There exists an integrand inf-compact* $\Psi : \Omega \times S \to [0, +\infty]$ *such that* $\sup_{u\in H} \int_\Omega \Psi(t, u(t))d\mu(t) < +\infty$.

(T_6) *For every* $\varepsilon > 0$, *there exists a measurable set-valued mapping* $\Gamma : \Omega \to \mathcal{K}_S$ *such that* $\sup_{u\in H} \mu(\{t \in \Omega : u(t) \notin \Gamma(t)\}) < \varepsilon$.

Remark 3.63.

(i) Just as in the special case of the Prohorov theorem, we have: *If* S *is a Polish space, then* $\mathcal{H} \subseteq \mathcal{Y}(S)$ *is relatively* $\mathcal{S}$*-compact if and only if* $\mathcal{H}$ *satisfies to one of equivalent conditions* (T_1)–(T_6).

(ii) The fact that any uniformly integrable subset of $L^1(\mu, \mathbb{R}^d)$ is relatively weakly compact in $(L^1(\mu, \mathbb{R}^d), \|\cdot\|_1)$ (see Dunford–Pettis' Theorem 1.65 and Theorem 1.84) is now a consequence of the Prohorov theorem. Indeed, if H is an uniformly integrable subset of $L^1(\mu, \mathbb{R}^d)$, then H is bounded and, according to Proposition 3.56, H is tight. The Prohorov's theorem 3.59 assures us that H is relatively stably compact. Then every net $(u^i)_{i\in I} \subseteq H$ admits a subnet $(u^{i_j})_{j\in J}$ – $\mathcal{S}$-convergent to a Young measure $\tau \in \mathcal{Y}(\mathbb{R}^d)$. Since $(u^{i_j})_{j\in J}$ is uniformly integrable and $\tau^{u_{i_j}} \xrightarrow{\mathcal{S}} \tau$, $(u^{i_j})_{j\in J}$ is weakly convergent to $u \in L^1(\mu, \mathbb{R}^d)$, where $u(t) = \int_{\mathbb{R}^d} x d\tau_t(x)$, for every $t \in \Omega$ (see 3.37); therefore H is relatively weakly compact in $(L^1(\mu, \mathbb{R}^d), \|\cdot\|_1)$.

We will now give two sequentially compactness results on $(\mathcal{Y}(S), \mathcal{S})$. Their demonstration is inspired by those of Lemma 4.4.1 and of Proposition 4.5.2 of [45].

Theorem 3.64. *Let* S *be a regular Suslin space. If* $\mathcal{H} \subseteq \mathcal{Y}(S)$ *is tight, then it is sequentially* $\mathcal{S}$*-compact.*

Proof. (1) Firstly, let us assume that S is a metrizable Suslin space. According to Theorem 2.58, $Rca^+(\mathcal{B}_S)$ and $\mathcal{P}_S$ are metrizable and second-countable. Then the Borel σ-algebra $\mathcal{C}$ of $\mathcal{P}_S$ is countably generated.

Let $(\tau^n)_{n\in\mathbb{N}}$ be a sequence of Young measures in $\mathcal{H}$; for every $n \in \mathbb{N}, \tau^n$: $(\Omega, \mathcal{A}) \to (\mathcal{P}_S, \mathcal{C})$ is then a measurable mapping. The smallest σ-algebra on

Ω for which all applications $\tau^n_{\cdot}$ are measurable, $\mathcal{A}_0$, is then countably generated, also; obviously, $\mathcal{A}_0 \subseteq \mathcal{A}$. Let μ_0 be the restriction of μ to $\mathcal{A}_0$ and let $\mathcal{Y}_0(S)$ be the space of Young measures on $(\Omega, \mathcal{A}_0, \mu_0)$, therefore the space of all $(\mathcal{A}_0 - \mathcal{C})$-measurable applications $\tau_{\cdot} : \Omega \to \mathcal{P}_S$ or the space of measures $\tau : \mathcal{A}_0 \otimes \mathcal{B}_S \to \mathbb{R}_+$ for which $\tau(A \times S) = \mu_0(A)$, for every $A \in \mathcal{A}_0$. We can notice that μ_0 is not, generally, a complete measure on $\mathcal{A}_0$.

Let $\mathcal{S}_0$ be the stable topology on $\mathcal{Y}_0(S)$; $\mathcal{Y}_0(S) \subseteq \mathcal{Y}(S)$ and $\mathcal{S}_0 \subseteq \mathcal{S}\restriction_{\mathcal{Y}_0(S)}$.

Let $\Phi : \mathcal{Y}(S) \to \mathcal{Y}_0(S)$ be the mapping defined by $\Phi(\tau) = \tau\restriction_{\mathcal{A}_0 \otimes \mathcal{B}_S}$, for every $\tau \in \mathcal{Y}(S)$; then Φ is a surjection $(\mathcal{S} - \mathcal{S}_0)$-continuous. Since $\mathcal{H} \subseteq \mathcal{Y}(S)$ is tight, it is relatively $\mathcal{S}$-compact (Theorem 3.59) and then $\Phi(\mathcal{H})$ is relatively $\mathcal{S}_0$-compact. According to Proposition 3.25, $(\mathcal{Y}_0(S), \mathcal{S}_0)$ is metrizable, therefore $\Phi(\mathcal{H})$ is sequentially compact.

Let now $(\tau^{k_n})_{n\in\mathbb{N}}$ be a subsequence of $(\tau^n)_{n\in\mathbb{N}}$ and let $\tau \in \mathcal{Y}_0(S) \subseteq \mathcal{Y}(S)$ such that $\Phi(\tau^{k_n}) \overset{\mathcal{S}_0}{\longrightarrow} \tau$. We will show that $\tau^{k_n} \overset{\mathcal{S}}{\to} \tau$.

For every $A \in \mathcal{A}$ let $\nu : \mathcal{A}_0 \to \mathbb{R}_+$ defined by $\nu(B) = \mu(A \cap B)$, for every $B \in \mathcal{A}_0$; ν is a finite, σ-additive measure, absolutely continuous with respect to $\mu_0 : \mathcal{A}_0 \to \mathbb{R}_+$. According to Radon–Nikodym theorem (Theorem 1.40), there exists a mapping $g \in \mathcal{L}^1(\Omega, \mathcal{A}_0, \mu_0)$ such that $\nu(B) = \mu(A \cap B) = \int_B g(t)d\mu_0(t)$, for every $B \in \mathcal{A}_0$; particularly, $\mu(A) = \int_\Omega g(t)d\mu_0(t)$. Then, for every $f \in C_b(S)$,

$$\begin{aligned}\tau^{k_n}(\chi_A \otimes f) &= \int_\Omega \chi_A(t)\cdot\tau^{k_n}_t(f)d\mu(t) = \int_\Omega g(t)\cdot\Phi(\tau^{k_n})_t(f)d\mu_0(t)\\ &\to \int_\Omega g(t)\cdot\tau_t(f)d\mu_0(t) = \int_\Omega \chi_A(t)\cdot\tau_t(f)d\mu(t)\\ &= \tau(\chi_A \otimes f).\end{aligned}$$

(2) Let us now suppose that (S, σ) is a regular Suslin space. According to (ii) of Theorem 2.25, there exists a metric d on S such that the generated topology, σ_d, is second-countable, less fine than σ and such that $\mathcal{B}_S(\sigma_d) = \mathcal{B}_S(\sigma)$. Then (S, σ_d) is a metrizable Suslin space and, as we proved in Theorem 3.59, the stable topology $\mathcal{S}_d$ on $\mathcal{Y}(S, \sigma_d)$ is less fine than the stable topology $\mathcal{S}$ on $\mathcal{Y}(S, \sigma)$.

Let $\mathcal{H} \subseteq \mathcal{Y}(S)$ be σ-tight. The closure $\overline{\mathcal{H}}$ of $\mathcal{H}$ in the space $(\mathcal{Y}(S), \mathcal{S})$ is $\mathcal{S}$-compact and then $\mathcal{S}\restriction_{\overline{\mathcal{H}}} = \mathcal{S}_d\restriction_{\overline{\mathcal{H}}}$. But, according to the first part of proof, $\mathcal{H}$ is sequentially $\mathcal{S}_d$-compact and then it is sequentially $\mathcal{S}$-compact. □

The following proposition gives a reciprocal of this property.

Proposition 3.65. *Let S be a Polish space. If $\mathcal{H} \subseteq \mathcal{Y}(S)$ is a sequentially $\mathcal{S}$-compact set, then it is tight.*

Proof. The mapping $\Phi : \mathcal{Y}(S) \to Rca^+(\mathcal{B}_S)$, defined by $\Phi(\tau) = \tau(\Omega \times \cdot)$, for every $\tau \in \mathcal{Y}(S)$, is $(\mathcal{S} - \mathcal{T}_S)$-continuous. Then $\Phi(\mathcal{H})$ is sequentially $\mathcal{T}_S$-compact in $Rca^+(\mathcal{B}_S)$. Since $(Rca^+(\mathcal{B}_S), \mathcal{T}_S)$ is metrizable (Theorem 2.58), $\Phi(\mathcal{H})$ is relatively compact and, according to Theorem 2.74, $\Phi(\mathcal{H})$ is tight; then $\mathcal{H}$ is tight. □

We will conclude this paragraph by giving two convergence results on $\mathcal{M}(S)$.

Theorem 3.66. *Let S be locally compact Suslin space and let $(u^n)_n \subseteq \mathcal{M}(S)$; then $(u^n)_n$ is $\mathcal{S}$-convergent in $\mathcal{Y}(S)$ if and only if*

$$\begin{cases} \text{(i)} \quad (u^n)_n \quad \textit{is tight and} \\ \text{(ii)} \quad (f(u^n))_n \quad \textit{is weakly convergent in} \quad L^1(\mu), \forall f \in C_c(S). \end{cases}$$

Proof. $(\Rightarrow)$ Let $\tau \in \mathcal{Y}(S)$ such that $u^n \xrightarrow{\mathcal{S}} \tau$; the set $H = \{u^n : n \in \mathbb{N}\}$ is sequentially $\mathcal{S}$-compact. According to Corollary 2.26, every locally compact Suslin space is a Polish one; by the previous proposition H is tight. According to (7) of Proposition 3.22, for every $f \in C_c(S)$, $(f(u^n))_{n\in\mathbb{N}}$ is weakly convergent in $L^1(\mu)$ to a function $u_f \in L^1(\mu)$, defined by $u_f(t) = \int_S f(x)d\tau_t(x)$, for every $t \in \Omega$.

$(\Leftarrow)$ Let $(u^n)_n \subseteq \mathcal{M}(S)$ be a tight such that, for every $f \in C_c(S)$, $(f(u^n))$ is weakly convergent in $L^1(\mu)$. Then, for every $f \in C_c(S)$, there exists $u_f \in L^1(\mu)$ such that $f(u^n) \xrightarrow[L^1(\mu)]{w} u_f$. If $f \geq 0$, then, for every $A \in \mathcal{A}$, $\int_A u_f d\mu \geq 0$; therefore $u_f \geq 0$, μ-almost everywhere. By replacing u_f with $\max(u_f, 0)$, we obtain: $f \geq 0 \Rightarrow u_f \geq 0$. For every $t \in \Omega$, we define $\tau_t : C_c(S) \to \mathbb{R}$, by $\tau_t(f) = u_f(t)$; then τ_t is positive and linear and therefore a Radon measure on S (Theorem 4-1, Ch. 3, no. 2 of [32]). Moreover, for every $A \in \mathcal{A}$ and every $f \in C_c(S)$,

$$\int_A f(u^n(t))d\mu(t) \to \int_A \tau_t(f)d\mu(t).$$

Since $(u^n)_n$ is tight, according to Theorem 3.64, it has a subsequence $(u^{k_n})_n$ convergent to a Young measure $\tau' \in \mathcal{Y}(S)$; then, for every $f \in C_c(S)$, $\tau_t(f) = \tau'_t(f)$, for μ-almost every $t \in \Omega$.

Let $(f_p)_{p\in\mathbb{N}} \subseteq C_c(S)$ be a sequence pointwise convergent to $\underline{1}$; then, almost for every $t \in \Omega$, $\tau_t(\underline{1}) = \lim_p \tau_t(f_p) = \lim_p \tau'_t(f_p) = \tau'_t(\underline{1}) = 1$. Therefore $\tau \in \mathcal{Y}(S)$ and, obviously, $u^n \xrightarrow{\mathcal{S}} \tau$. □

Theorem 3.67. *Let $S = \mathbb{R}^d$ and let $(u^n)_{n\in\mathbb{N}} \subseteq \mathcal{M}(\mathbb{R}^d)$; $(u^n)_n$ is convergent in measure in $\mathcal{M}(\mathbb{R}^d)$ if and only if*

$$\begin{cases} \text{(i)} & (u^n)_n \quad \textit{is tight and} \quad , \\ \text{(ii)} & (f(u^n))_n \quad \textit{is strongly convergent in} \quad L^1(\mu), \forall f \in C_b(S). \end{cases}$$

Proof. ($\Rightarrow$) The implication is a consequence of (iii) and (iv) of Remark 3.39 and of Proposition 3.65.

($\Leftarrow$) Let $(u^n)_n \subseteq \mathcal{M}(\mathbb{R}^d)$ be a tight such that, for every $f \in C_b(\mathbb{R}^d)$, $(f(u_n))_n$ converges strongly in $L^1(\mu)$. According to Definition 3.55, for every $\varepsilon > 0$, there exists $k > 0$ such that, for every $n \in \mathbb{N}$, $\mu(\|u_n\| \geq k) < \frac{\varepsilon}{4}$. For every $i \leq m$, let $f_i \in C_b(\mathbb{R}^d)$ such that, $f_i \restriction_{\{\|y\|\leq k\}} (y) = y_i$ (y_i is the canonical i-projection of y). For every $i \leq m$, $(f_i(u^n))_n$ is strongly convergent so that there exists n_0 such that, for every $i \leq m$, every n and every $p \geq n_0$, $\|f_i(u^n) - f_i(u^p)\|_1 < \frac{\varepsilon}{4}$. Therefore, for every $\eta > 0$, every $i \leq m$, and every $n, p \geq n_0$,

$$\begin{aligned} \frac{\varepsilon}{4} &> \int_\Omega |f_i(u^n) - f_i(u^p)| d\mu \\ &\geq \int_{(|f_i(u^n)-f_i(u^p)|>\eta)\cap(||u^n||\leq k)\cap(||u^p||\leq k)} |f_i(u^n) - f_i(u^p)| d\mu \\ &> \eta \cdot \mu\left[(|f_i(u^n) - f_i(u^p)| > \eta) \cap (||u^n|| \leq k) \cap (||u^p|| \leq k)\right] \end{aligned}$$

and so, for every $i \leq m$,

$$\mu\left[(|f_i(u^n) - f_i(u^p)| > \eta) \cap (|\|u^n\| \leq k) \cap (\|u^p\| \leq k)\right] < \frac{\varepsilon}{4\eta}.$$

By considering $||y|| = \sup_{i\leq m} |y_i|$, we obtain:

$$\mu(||u^n - u^p|| > \eta, ||u^n|| \leq k, ||u^p|| \leq k) < \frac{m\varepsilon}{4\eta}.$$

Therefore, for every $\varepsilon > 0$,

$$\mu(||u^n - u^p|| > \eta) < \frac{m \cdot \varepsilon}{4\eta} + \frac{\varepsilon}{2}.$$

Then the sequence $(u^n)_n$ is convergent in measure. □

3.6 Biting Lemma

In Chapter 1, Theorem 1.103, we have presented a scalar version of the biting lemma; in the first part of this section, we give its vector version. By adding to the condition of "boundedness" that of "tightness", we obtain a vector version of the Saadoune and Valadier compactness result [149].

In the second part of this section, we extend the results to a family of unbounded sets of measurable functions—the finite-tight sets.

Let $(E, \|\cdot\|)$ be a separable Banach space; then E is a Polish space and so a regular Suslin space.

Since E is separable, every $(\mathcal{A} - \mathcal{B}_E)$-measurable function $u : \Omega \to E$ is strongly measurable. For every $u \in \mathcal{M}(E), \bar{u}$ denotes the set of all applications of $\mathcal{M}(E)$ μ-almost everywhere equal with u. We recall the usual notations:

$$\mathcal{L}^1(\mu, E) = \{u \in \mathcal{M}(E) : \|u\|_1 = \int_\Omega \|u(t)\| d\mu(t) < +\infty\} \text{ and}$$
$$L^1(\mu, E) = \{\bar{u} : u \in \mathcal{L}^1(\mu, E)\}.$$

Commonly, and if there is no ambiguity, $\bar{u} \in L^1(\mu, E)$ is replaced by $u \in L^1(\mu, E)$.

We recall that a set $H \subseteq L^1(\mu, E)$ is absolutely continuous if

$$\lim_{\mu(A)\to 0} \sup_{u \in H} \int_A \|u\| d\mu = 0.$$

H is uniformly integrable if

$$\lim_{a\to+\infty} \sup_{u\in H} \int_{(\|u\|\geq a)} \|u\| d\mu = 0.$$

H is uniformly integrable if and only if it is absolutely continuous and bounded in $L^1(\mu, E)$.

For every bounded set $H \subseteq L^1(\mu, E)$, the modulus of uniform integrability of H is defined by:

$$\begin{aligned}\eta(H) &= \inf_{\varepsilon>0} \left[\sup \left\{ \int_A \|u(t)\| d\mu(t) : A \in \mathcal{A}, \mu(A) < \varepsilon, u \in H \right\}\right] \\ &= \lim_{a\to\infty} \left[\sup_{u\in H} \int_{(\|u\|>a)} \|u(t)\| d\mu(t)\right]\end{aligned}$$

(see Definition 1.82 and Proposition 1.83).

H is uniformly integrable if and only if $\eta(H) = 0$.

Theorem 3.68 (Biting Lemma). *Let $(E, \|\cdot\|)$ be a separable Banach space, let $(u_n)_n$ be a bounded sequence of $L^1(\mu, E)$ and let $\eta = \eta((u_n)_n)$ be its modulus of uniform integrability. There exist a subsequence $(u_{k_n})_n$ of $(u_n)_n$ and a decreasing sequence of "bits" $(B_p)_{p\in\mathbb{N}} \subseteq \mathcal{A}$ with $\mu(B_p) \to 0$ such that:*

(1) $(\chi_{\Omega\setminus B_n} \cdot u_{k_n})_{n\in\mathbb{N}}$ *is uniformly integrable.*

(2) $\eta((u_{l_{k_n}})_n) = \eta((u_n)_n)$, *for every subsequence* $(u_{l_{k_n}})_n$ *of* $(u_{k_n})_n$.

(3) $\lim_n \int_{B_n} \|u_{k_n}(t)\| d\mu(t) = \eta((u_n)_n)$.

If E is reflexive, then we can choose the sequences $(u_{k_n})_n$ and $(B_p)_p$ such that $(\chi_{\Omega\setminus B_n} \cdot u_{k_n})_{n\in\mathbb{N}}$ to be weakly convergent. For every $p \in \mathbb{N}$, $(u_{k_n}\restriction_{\Omega\setminus B_p})_{n\in\mathbb{N}}$ is then weakly convergent in $L^1(\Omega \setminus B_p, E)$.

Proof. For every $n \in \mathbb{N}$, let $f_n : \Omega \to \mathbb{R}$, $f_n = \|u_n\|$. Then $(f_n)_n \subseteq L^1(\mu)$ is a bounded sequence; according to Corollary 1.106 ("Subsequence Splitting Lemma"), there exist a subsequence $(f_{k_n})_n$ of $(f_n)_n$ and a sequence of pairwise disjoint sets $(A_n)_n \subseteq \mathcal{A}$ such that $\lim_n \int_{A_n} |f_{k_n}(t)| d\mu(t) = \eta((f_n)_n) = \eta$ and so that the sequence $(\chi_{\Omega\setminus A_n} \cdot f_{k_n})_n$ to be weakly convergent. Then

$$\lim_n \int_{A_n} \|u_{k_n}(t)\| d\mu(t) = \eta \tag{1}$$

and, according to Theorem 1.84 and Proposition 1.83,

$$\eta\left((\chi_{\Omega\setminus A_n} \cdot f_{k_n})_n\right) = \eta\left((\chi_{\Omega\setminus A_n} \cdot u_{k_n})_n\right) = 0. \tag{2}$$

For every $p \in \mathbb{N}$, we note $B_p = \cup_{n=p}^{\infty} A_n \in \mathcal{A}$; it is clear that $(B_p)_p$ is a decreasing sequence and that $\mu(B_p) = \sum_{n=p}^{\infty} \mu(A_n) \to 0$.

Since $\chi_{\Omega\setminus B_n} \le \chi_{\Omega\setminus A_n}$, by (2), $\eta((\chi_{\Omega\setminus B_n} \cdot u_{k_n})_n) = 0$ and then the sequence $(\chi_{\Omega\setminus B_n} \cdot u_{k_n})_n$ is uniformly integrable.

For every subsequence $(u_{l_{k_n}})_n$ of $(u_{k_n})_n$,

$$\eta((u_{l_{k_n}})_n) \le \eta((u_n)_n) = \eta. \tag{3}$$

$$\eta((u_{l_{k_n}})_n) = \inf_{\varepsilon>0} \sup\left\{\int_A \|u_{l_{k_n}}(t)\| d\mu(t) : n \in \mathbb{N}, \mu(A) < \varepsilon\right\}.$$

Therefore, for every $\theta > 0$, there exists $\varepsilon > 0$ such that, for every $n \in \mathbb{N}$ and every $A \in \mathcal{A}$ with $\mu(A) < \varepsilon$,

$$\int_A \|u_{l_{k_n}}(t)\| d\mu(t) < \eta((u_{l_{k_n}})_n) + \theta. \tag{4}$$

By (1), $\int_{A_n} \|u_{k_n}(t)\| d\mu(t) \to \eta$ and $\mu(A_n) \to 0$; then there exists $n \in \mathbb{N}$ such that $\mu(A_n) < \varepsilon$ and

$$\int_{A_n} \|u_{l_{k_n}}(t)\| d\mu(t) > \eta - \theta. \tag{5}$$

According to (4) and (5), $\eta((u_{l_{k_n}})) + \theta > \eta - \theta$, for every $\theta > 0$ and, using (3), we obtain $\eta((u_{l_{k_n}})) = \eta$.

By (2), for every $\varepsilon > 0$, there exists $\delta > 0$ such that, for every $A \in \mathcal{A}$ with $\mu(A) < \delta$ and every $n \in \mathbb{N}$, $\int_{A \setminus A_n} \|u_{k_n}(t)\| d\mu(t) < \varepsilon$. Since $\mu(B_n) \to 0$, there exists $n_0 \in \mathbb{N}$ such that $\mu(B_n) < \delta$, for every $n \geq n_0$ and therefore

$$\int_{B_n \setminus A_n} \|u_{k_n}(t)\| d\mu(t) < \varepsilon, \quad \text{for every} \quad n \geq n_0. \tag{6}$$

According to (6), for every $n \geq n_0$,

$$\begin{aligned}\int_{A_n} \|u_{k_n}(t)\| d\mu(t) \leq \int_{B_n} \|u_{k_n}(t)\| d\mu(t) &= \int_{A_n} \|u_{k_n}(t)\| d\mu(t) \\ &+ \int_{B_n \setminus A_n} \|u_{k_n}(t)\| d\mu(t) < \int_{A_n} \|u_{k_n}(t)\| d\mu(t) + \varepsilon.\end{aligned}$$

Therefore, from (1) we obtain 3 of the theorem.

We recall that, according to Dunford's theorem (Theorem 1, p. 101 of [56]), if E is reflexive, then the uniformly integrable sets are relatively weakly compact. □

Remark 3.69. Generally, the bounded sets of $L^1(\mu, E)$ are not tight (see 3.57). If we add to the condition of "boundedness" that of "tightness", then we can give a general result of convergence (for the Euclidean case, see Theorem 9 of [169]).

Theorem 3.70. *Let E be a separable and reflexive normed space and let $(u_n)_n$ be a bounded tight sequence in $L^1(\mu, E)$; then we can find a subsequence $(u_{k_n})_n$, a Young measure $\tau \in \mathcal{Y}(E)$ and a decreasing sequence of sets $(B_p)_p \subseteq \mathcal{A}$ with $\mu(B_p) \downarrow 0$ such that:*

(1) $u_{k_n} \xrightarrow{\mathcal{S}_E} \tau$.

(2) *Almost for every $t \in \Omega$, τ_t has a barycenter,* $\mathrm{bar}(\tau_t) = \int_E x \, d\tau_t(x)$ *and the mapping $t \mapsto u(t) =$* $\mathrm{bar}(\tau_t)$ *is integrable.*

(3) $\chi_{\Omega \setminus B_n} \cdot u_{k_n} \xrightarrow[L^1(\mu,E)]{w} u$.

(4) *If $M = \{t \in \Omega : \int_E \|x - u(t)\| d\tau_t(x) = 0\} = \{t \in \Omega : \tau_t = \delta_{u(t)}\}$, then $(u_{k_n} \upharpoonright_M)_n$ is convergent in measure to $u \upharpoonright_M$.*

(5) $\lim_{n\to\infty} \|u_{k_n} - u\|_1 = \eta((u_n)) + \int_{\Omega \times E} \|x - u(t)\| d\tau(t, x)$.

Proof. According to Theorem 3.68, there exist a subsequence $(u'_n)_n$ of $(u_n)_n$ and a decreasing sequence $(B_p)_p \subseteq \mathcal{A}$ with $\mu(B_p) \downarrow 0$, such that the sequence $(\chi_{\Omega\setminus B_n} \cdot u'_n)_n$ is weakly convergent to a mapping $v \in L^1(\mu, E)$, $\eta((u_{k_n})) = \eta((u_n))$, for every subsequence $(u_{k_n})_n$ of $(u'_n)_n$ and $\eta((u_n)) = \lim_n \int_{B_n} \|u'_n(t)\| d\mu(t)$.

Since $(u'_n)_n$ is tight, there exists a subsequence $(u_{k_n})_n$ of $(u'_n)_n$ stably convergent to a Young measure $\tau \in \mathcal{Y}(E)$ (see Theorem 3.64).

Let $f : E \to \mathbb{R}$, defined by $f(x) = \|x\|$, for every $x \in E$. Then f is continuous and bounded from below; according to (2) of Proposition 3.22,

$$\tau(\chi_\Omega \otimes f) \leq \liminf_n \tau^{u_{k_n}}(\chi_\Omega \otimes f), \quad \text{or, since} \quad (u_{k_n})_n \quad \text{is bounded,}$$

$$\int_\Omega \left(\int_E \|x\| d\tau_t(x) \right) d\mu(t) \leq \liminf_n \int_\Omega \|u_{k_n}(t)\| d\mu(t) \leq m < +\infty.$$

Therefore, for almost every $t \in \Omega$, $\int_E \|x\| d\tau_t(x) < +\infty$. Subsequently, for almost all $t \in \Omega$,

$$u(t) = \int_E x d\tau_t(x) = \mathrm{bar}(\tau_t) \in E$$

and $u \in L^1(\mu, E)$.

Let us show that $u = v$. Since E is reflexive, the dual space of $L^1(\mu, E)$ is $L^\infty(\Omega, E^*) = \{g : \Omega \to E^* : g \text{ is measurable and } \|g\|_\infty = \inf\{c > 0 : \|g(t)\|_{E^*} \leq c, \mu - \text{a.e.}\ \} < +\infty\}$ (see Theorem 1, p. 98 of [56]).

Let g be arbitrary in $L^\infty(\Omega, E^*)$; since $\chi_{\Omega\setminus B_n} \cdot u_{k_n} \xrightarrow[L^1(\mu,E)]{w} v$,

$$\int_\Omega g(t) \left(\chi_{\Omega\setminus B_n}(t) \cdot u_{k_n}(t) \right) d\mu(t) \to \int_\Omega g(t) (v(t))\, d\mu(t). \tag{1}$$

Then $\Psi : \Omega \times E \to \mathbb{R}, \Psi(t, x) = g(t)(x)$, for every $x \in E$ and every $t \in \Omega$, is measurable in t and continuous in x; according to Proposition 3.8, Ψ is a Carathéodory integrand. Moreover, for all $t \in \Omega$ and all $n \in \mathbb{N}$,

$$\begin{aligned} \left| \Psi(t, \chi_{\Omega\setminus B_n}(t) \cdot u_{k_n}(t)) \right| &= \left| g(t) \left(\chi_{\Omega\setminus B_n}(t) \cdot u_{k_n}(t) \right) \right| \\ &\leq \|g(t)\|_{E^*} \cdot \|\chi_{\Omega\setminus B_n}(t) \cdot u_{k_n}(t)\|_E \\ &\leq \|g\|_\infty \cdot \|\chi_{\Omega\setminus B_n}(t) \cdot u_{k_n}(t)\|_E \end{aligned}$$

and, since $(\chi_{\Omega\setminus B_n} \cdot u_{k_n})_n$ is uniformly integrable, $(\Psi(., \chi_{\Omega\setminus B_n}(.) \cdot u_{k_n}(.)))_n$ is uniformly integrable too.

According to Corollary 3.36,

$$\int_{\Omega} \Psi(t, \chi_{\Omega\setminus B_n}(t) \cdot u_{k_n}(t)) d\mu(t) \to \int_{\Omega} \left(\int_E \Psi(t,x) d\tau_t(x) \right) d\mu(t)$$

or

$$\int_{\Omega} g(t) \left(\chi_{\Omega\setminus B_n}(t) \cdot u_{k_n}(t) \right) d\mu(t) \to \int_{\Omega} \left(\int_E g(t)(x) d\tau_t(x) \right) d\mu(t). \quad (2)$$

According to (1) and (2),

$$\begin{aligned} \int_{\Omega} g(t)(v(t)) d\mu(t) &= \int_{\Omega} \left(\int_E g(t)(x) d\tau_t(x) \right) d\mu(t) \\ &= \int_{\Omega} g(t) \left(\int_E x d\tau_t(x) \right) d\mu(t) = \int_{\Omega} g(t)(u(t)) d\mu(t) \end{aligned}$$

and so, as g is arbitrary in the dual space, $u = v$.

For every $t \in M$, $\tau_t(\{u(t)\}) = 1$, which means that $\tau_t = \delta_{u(t)}$. According to Proposition 3.41, $(u_{k_n} \upharpoonright_M)_n$ is convergent in measure to $u \upharpoonright_M$.

In order to show (5), let us first remark that

$$\eta = \eta((u_n)) = \eta((u_{k_n})) = \inf_{\delta>0} \sup_{n\in\mathbb{N}} \sup_{\mu(A)<\delta} \int_A \|u_{k_n}(t)\| d\mu(t).$$

For every $\varepsilon > 0$, let $\varepsilon_1 = \frac{\varepsilon}{2}$ and $\varepsilon_2 = \frac{\varepsilon}{6}$; there exists $\delta > 0$ such that, for every $n \in \mathbb{N}$,

$$\int_A \|u_{k_n}(t)\| d\mu(t) < \eta + \varepsilon_1, \quad \text{for every} \quad A \in \mathcal{A} \quad \text{with} \quad \mu(A) < \delta. \quad (3)$$

Since $\lim_n \int_{B_{k_n}} \|u_{k_n}(t)\| d\mu(t) = \eta$, there exists $p \in \mathbb{N}$ such that

$$\int_{B_{k_n}} \|u_{k_n}(t)\| d\mu(t) > \eta - \varepsilon_1, \quad \text{for every} \quad n \geq p. \quad (4)$$

Since $\mu(B_p) \downarrow 0$ and $\int_{\Omega\times E} \|x\| d\tau(t,x) \leq m$, one can choose $p \in \mathbb{N}$ so that $\mu(B_p) < \delta$ and that $\int_{B_p \times E} \|x\| d\tau(t,x) < \varepsilon_2$.

Then, by (3) and (4), for every $n \geq p$,

$$\eta - \varepsilon_1 < \int_{B_{k_n}} \|u_{k_n}(t)\| d\mu(t) \leq \int_{B_p} \|u_{k_n}(t)\| d\mu(t) < \eta + \varepsilon_1.$$

Therefore, for every $\varepsilon > 0$, there exists $p \in \mathbb{N}$ such that

$$\left| \int_{B_p} \|u_{k_n}(t)\| d\mu(t) - \eta \right| < \varepsilon_1, \quad \text{for all} \quad n \geq p. \quad (5)$$

Then, for every $n \geq p$,

$$\begin{aligned}\|u_{k_n} - u\|_1 &= \int_\Omega \|u_{k_n}(t) - u(t)\| d\mu(t) \\ &= \int_{B_p} \|u_{k_n}(t) - u(t)\| d\mu(t) + \int_{\Omega\setminus B_p} \|u_{k_n}(t) - u(t)\| d\mu(t).\end{aligned}$$

By (5), for every $n \geq p$,

$$\begin{aligned}\eta - \varepsilon_1 - \varepsilon_2 &< \int_{B_p} \|u_{k_n}(t)\| d\mu(t) - \int_{B_p} \|u(t)\| d\mu(t) \leq \int_{B_p} \|u_{k_n}(t) - u(t)\| d\mu(t) \\ &\leq \int_{B_p} \|u_{k_n}(t)\| d\mu(t) + \int_{B_p} \|u(t)\| d\mu(t) < \eta + \varepsilon_1 + \varepsilon_2\end{aligned}$$

and therefore

$$\left| \int_{B_p} \|u_{k_n}(t) - u(t)\| d\mu(t) - \eta \right| < \varepsilon_1 + \varepsilon_2, \quad \text{for all} \quad n \geq p. \tag{6}$$

Let $\Psi_p : \Omega \times E \to \mathbb{R}$, $\Psi_p(t,x) = \chi_{\Omega\setminus B_p}(t) \cdot \|x - u(t)\|$; then Ψ_p is a Carathéodory integrand and, for every $n \geq p$ and every $t \in \Omega$,

$$\left|\Psi_p(t, u_{k_n}(t))\right| \leq \chi_{\Omega\setminus B_p}(t) \cdot \|u_{k_n}(t) - u(t)\|.$$

Since $(\chi_{\Omega\setminus B_p} \cdot u_{k_n})_n$ is uniformly integrable, $(\chi_{\Omega\setminus B_p} \cdot (u_{k_n} - u))_n$ is uniformly integrable too and then $(\Psi_p(.,u_{k_n}(.)))_n$ is uniformly integrable in $L^1(\mu)$. According to Corollary 3.36,

$$\begin{aligned}\lim_n \int_{\Omega\setminus B_p} \|u_{k_n}(t) - u(t)\| d\mu(t) &= \tau(\Psi_p) \\ &= \int_{\Omega\setminus B_p} \left(\int_E \|x - u(t)\| d\tau_t(x) \right) d\mu(t) \\ &= \int_{\Omega\times E} \|x - u(t)\| d\tau(t,x) + \int_{B_p\times E} \|x - u(t)\| d\tau(t,x).\end{aligned}$$

Then there exists $n_0 \geq p$ such that, for every $n \geq n_0$,

$$\begin{aligned}&\left| \int_{\Omega\setminus B_p} \|u_{k_n}(t) - u(t)\| d\mu(t) - \int_{\Omega\times E} \|x - u(t)\| d\tau(t,x) \right| \\ &\qquad < \int_{B_p\times E} \|x - u(t)\| d\tau(t,x) \\ &\qquad \leq \int_{B_p\times E} \|x\| d\tau(t,x) + \int_{B_p} \|u(t)\| d\mu(t) \\ &\qquad < \varepsilon_2 + \int_{B_p} \left(\int_E \|x\| d\tau_t(x) \right) d\mu(t) < 2\varepsilon_2.\end{aligned}$$

Therefore, for every $n \geq n_0$,

$$\left| \int_{\Omega \setminus B_p} \|u_{k_n}(t) - u(t)\| d\mu(t) - \int_{\Omega \times E} \|x - u(t)\| d\tau(t,x) \right| < 2\varepsilon_2. \quad (7)$$

Then, by (6) and (7), for all $n \geq n_0 \geq p$,

$$\begin{aligned} \Big| \|u_{k_n} - u\|_1 - \eta - \int_{\Omega \times E} \|x - u(t)\| d\tau(t,x) \Big| & \\ \leq \left| \int_{B_p} \|u_{k_n}(t) - u(t)\| d\mu(t) - \eta \right| & \\ + \left| \int_{\Omega \setminus B_p} \|u_{k_n}(t) - u(t)\| d\mu(t) - \int_{\Omega \times E} \|x - u(t)\| d\tau(t,x) \right| & \\ < \varepsilon_1 + \varepsilon_2 + 2\varepsilon_2 = \varepsilon, & \end{aligned}$$

from where

$$\lim_n \|u_{k_n} - u\|_1 = \eta + \int_{\Omega \times E} \|x - u(t)\| d\tau(t,x). \qquad \square$$

Remark 3.71.

(i) In the conditions of the previous theorem:

(1) If $(u_n)_n$ is uniformly integrable, then

$$\|u_{k_n} - u\|_1 \to \int_{\Omega \times E} \|x - u(t)\| d\tau(t,x).$$

(2) If $(u_n)_n$ is convergent in measure, then

$$\|u_{k_n} - u\|_1 \to \eta((u_n)).$$

Indeed, in this case, according to Proposition 3.40, $u_n \xrightarrow{\mu} u$ and then $\mu(\Omega \setminus M) = 0$.

(3) If $(u_n)_n$ is uniformly integrable and convergent in measure, then

$$u_{k_n} \xrightarrow[L^1(\mu,E)]{\|\cdot\|_1} u$$

and we obtain again the Vitali's Theorem 1.54, this time in the vector case.

(ii) Even if the sequence $(u_n)_n$ is $\mathcal{S}_E$-convergent to a Young measure $\tau \in \mathcal{Y}(E)$, it is necessary to extract a subsequence. Indeed, if $\Omega = [0,1]$, μ is the Lebesgue measure on Ω and if $u_n : \Omega \to \mathbb{R}$ is defined by $u_{2n} = 2n\chi_{[0,\frac{1}{2n}]}$ and $u_{2n+1} = 0$, $\forall n \in \mathbb{N}$, then $u_n \xrightarrow{\mu} 0$ and $u_n \xrightarrow{\mathcal{S}_\mathbb{R}} \mu \otimes \delta_0 \in \mathcal{Y}(\mathbb{R})$, $M = \Omega$ and $\eta((u_n)) = 1$ but $\lim_n \|u_n\|_1$ does not exist.

(iii) $\lim_n \|u_{k_n} - u\|_1 = \eta((u_n)) + \int_{(\Omega\times E)\setminus G_u} \|x - u(t)\| d\tau(t,x)$, where G_u is the graph of u (see Proposition 3.17 and Theorem 3.18). We can show that $\tau(G_u') = \sup_{a>0} \limsup_n \mu(\|u_n - u\| \geq a)$. Therefore, the previous formula highlights a decomposition of $\lim_n \|u_{k_n} - u\|_1$ in a part that can be called "weak" and one that can be called "in measure".

(iv) $u_n^1 \xrightarrow{\mu} u \iff \int_{\Omega\times E} \|x - u(t)\| d\tau(t,x) = 0$.

Example 3.72. Let $u_n : [-2\pi, 2\pi] \to \mathbb{R}$ be defined by

$$u_n(t) = n\chi_{[-\frac{1}{n},0]}(t) + \sin nt \cdot \chi_{]0,2\pi]}(t).$$

In this case $\mathcal{A}$ is the family of all Lebesgue measurable subsets and μ is the Lebesgue measure on $\Omega = [-2\pi, 2\pi]$.

According to Proposition 3.40 and to (ii) of Remark 3.47, $u_n \xrightarrow{\mathcal{S}_\mathbb{R}} \tau$ where, $\forall A \in \mathcal{A}, B \in \mathcal{B}_\mathbb{R}$,

$$\tau(A \times B) = \mu(A \cap [-2\pi, 0]) \cdot \delta_0(B) + \frac{1}{\pi} \cdot \mu(A \cap]0, 2\pi]) \int_{B\cap[-1,1]} \frac{1}{\sqrt{1-x^2}} d\mu(x).$$

The disintegration of τ, $(\tau_t)_{t\in\Omega}$, is defined by

$$\tau_t(B) = \begin{cases} \delta_0(B), & t \in [-2\pi, 0], \\ \frac{1}{\pi} \int_{B\cap[-1,1]} \frac{1}{\sqrt{1-x^2}} d\mu(x), & t \in]0, 2\pi[. \end{cases}$$

$u(t) = \text{bar}(\tau_t) = 0$, for all $t \in [-2\pi, 2\pi]$. $B_p = [-\frac{1}{p}, 0]$, $\forall p \in \mathbb{N}$. $\chi_{\Omega\setminus B_n} \cdot u_n \xrightarrow[L^1(\mu,\mathbb{R})]{w} \underline{0}$ and $\eta((u_n)) = 1$. $M = [-2\pi, 0]$ and

$$\lim_n \|u_n\|_1 = 1 + \int_{\Omega\times\mathbb{R}} |x| d\tau(t,x) = 1 + 2\int_0^{2\pi} \left[\int_0^1 x \, d\tau_t(x)\right] d\mu(t) = 5.$$

In the Euclidean case, we obtain Theorem 9 of [169] as a corollary of Theorem 3.70.

Corollary 3.73. *Let $(u_n)_n \subseteq L^1(\mu, \mathbb{R}^d)$ be a sequence weakly convergent to $u \in L^1(\mu, \mathbb{R}^d)$; then, we can find a subsequence $(u_{k_n})_n$ and a Young measure $\tau \in \mathcal{Y}(\mathbb{R}^d)$ such that:*

(a) $u_{k_n} \xrightarrow{\mathcal{S}_{\mathbb{R}^d}} \tau$;

(b) *almost for every $t \in \Omega$, τ_t has a barycenter,* $\mathrm{bar}(\tau_t) = \int_{\mathbb{R}^d} x\, d\tau_t(x)$ *and* $\mathrm{bar}(\tau_t) = u(t)$;

(c) $\lim_{n\to\infty} \|u_{k_n} - u\|_1 = \int_{\Omega\times\mathbb{R}^d} \|x - u(t)\| d\tau(t,x)$.

The following result is a consequence of (5) of Theorem 3.70.

Theorem 3.74 ([105]-T.2.7.). *Let $(u_n) \subseteq L^1(\Omega, \mu; \mathbb{R})$ be a bounded sequence and let $u \in L^1(\Omega, \mu, \mathbb{R})$ such that* $\limsup_n \|u_n - u\|_1 \le \eta((u_n))$. *Then there exists a subsequence $(u_{k_n})_n$ such that $u_{k_n} \xrightarrow{\mu} u$ and $\eta((u_n)) = \lim_n \|u_{k_n} - u\|_1$.*

Now we extend the compactness results presented for some non-bounded sets of measurable functions—the finite-tight sets. A finite-tight set localizes the great growths of a set of measurable functions on a finite family of sets of small measure. More precisely, we will give the following definition:

Definition 3.75 ([81]). A subset $H \subseteq \mathcal{M}(E)$ is said *finite-tight set* if, for every $\varepsilon > 0$, there exist a compact $K \in \mathcal{K}_E$ and a finite subfamily $\mathcal{A}_f \subseteq \mathcal{A}$ with $\mu(A) < \varepsilon$, for every $A \in \mathcal{A}_f$, such that, for every $u \in H$, there exists $A_u \in \mathcal{A}_f$ with $u^{-1}(E \setminus K) \subseteq A_u$. A sequence $(u_n)_{n\in\mathbb{N}} \subseteq \mathcal{M}(E)$ is a *finite-tight sequence* if the set $H = \{u_n : n \in \mathbb{N}\}$ is a finite-tight set.

Proposition 3.76. *Let $\mu^* : 2^\Omega \to \mathbb{R}_+$ be the outer measure generated by μ (for every $A \subseteq \Omega$, $\mu^*(A) = \inf\{\mu(C) : C \in \mathcal{A}, A \subseteq C\}$).*

$H \subseteq \mathcal{M}(E)$ is a finite-tight set if and only if, for every $\varepsilon > 0$, there exist a compact $K \in \mathcal{K}_E$ and a finite cover of H, $\{H_1, \dots, H_p\}$, such that, for every $i \in \{1, \dots, p\}$, $\mu^*(\bigcup_{u\in H_i} u^{-1}(E \setminus K)) < \varepsilon$.

In the case where H is countable, we can use the measure μ instead of μ^*.

Proof. Let $H \subseteq \mathcal{M}(E)$ be a finite-tight set; for every $\varepsilon > 0$ there exist $K \in \mathcal{K}_E$ and a finite subfamily $\mathcal{A}_f = \{A_1, \dots, A_p\} \subseteq \mathcal{A}$ with $\mu(A_i) < \varepsilon$, for every $i = 1, \dots, p$, such that, for any $u \in H$, there exists $A_u \in \mathcal{A}_f$ with $u^{-1}(E \setminus K) \subseteq A_u$. For every $i = 1, \dots, p$, let $H_i = \{u \in H : u^{-1}(E \setminus K) \subseteq A_i\}$; then $\{H_1, \dots, H_p\}$ is the required cover of H.

Conversely, let $K \in \mathcal{K}_E$ and let $\{H_1, \dots, H_p\}$ be a finite cover of H such that, for every $i = 1, \dots, p$, $\mu^*(\cup_{u\in H_i} u^{-1}(E \setminus K)) < \varepsilon$. Then, for every i, there exists $A_i \in \mathcal{A}$ such that $\cup_{u \in H_i} u^{-1}(E \setminus K) \subseteq A_i$ and $\mu(A_i) < \varepsilon$; therefore $\mathcal{A}_f = \{A_1, \dots, A_p\}$ is the finite subfamily of $\mathcal{A}$ required by Definition 3.75. □

Remark 3.77.

(i) If E is a Euclidean space then $H \subseteq \mathcal{M}(E)$ is a finite-tight set if and only if, for every $\varepsilon > 0$, there exist $k > 0$ and a finite subfamily $\mathcal{A}_f \subseteq \mathcal{A}$ with $\mu(A) < \varepsilon$, for every $A \in \mathcal{A}_f$, such that, for every $u \in H$, there exists $A_u \in \mathcal{A}_f$ with $\{t \in \Omega : \|u(t)\| > k\} \subseteq A_u$.

Obviously, a bounded sequence in $(L^\infty(\Omega, E), \|\cdot\|_\infty)$ is a finite-tight sequence.

(ii) According to Definition 3.75, for every $u \in H, \mu(u^{-1}(E \setminus K)) < \varepsilon$; thus any finite-tight set is a tight set.

The following example shows that the reciprocal is not valid.

Example 3.78. Let $n = \sum_{p=1}^{\infty} a_{n,p} \cdot 2^{p-1}$ be the binary form of the natural number n; thus, for every $p \in \mathbb{N}^*, a_{n,p} \in \{0, 1\}$ and there exists $l_n \in \mathbb{N}$ with $a_{n,l_n} = 1$ and $a_{n,p} = 0$, for all $p > l_n$.

For all $n, p \in \mathbb{N}^*$, we note

$$k_{n,p} = a_{n,p} + a_{n,p+1} \cdot 2 + \cdots + a_{n,2p-1} \cdot 2^{p-1};$$

then $0 \leq k_{n,p} \leq 2^p - 1, k_{n,l_n} = a_{n,l_n} = 1$ and, for every $p > l_n, k_{n,p} = 0$.

For any $n \in \mathbb{N}$, let $u_n :]0, 1[\to \mathbb{R}, u_n = \sum_{p=1}^{\infty} p \cdot \chi_{]\frac{k_{n,p}}{2^p}, \frac{k_{n,p}+1}{2^p}[}$.

Therefore, for every $t \in]0, 1[$,

$$u_n(t) = \sum_{p=1}^{l_n} p \cdot \chi_{\left]\frac{k_{n,p}}{2^p}, \frac{k_{n,p}+1}{2^p}\right[}(t) + \sum_{p=l_n+1}^{\infty} p \cdot \chi_{]0, \frac{1}{2^p}[}(t) \in \mathbb{R};$$

thus u_n is well defined, for every $n \in \mathbb{N}$.

Proposition 3.79. *$(u_n)_n \subseteq \mathcal{M}(\mathbb{R})$ is a tight sequence but is not a finite-tight one.*

Proof. For every $n \in \mathbb{N}, \int_0^1 |u_n(t)| dt = \sum_{p=1}^{\infty} p \cdot \frac{1}{2^p} = 2$; so $(u_n)_n$ is a bounded sequence in $L^1(]0, 1[, \mathbb{R})$ and then it is tight (see Proposition 3.56).

Let $\varepsilon = \frac{1}{2}$; to show that $(u_n)_n$ is not finite-tight it is enough to notice that, for every $k > 0$, for every $q \in \mathbb{N}$ and all $A_1, \dots, A_q \in \mathcal{A}$ with $\mu(A_i) < \varepsilon, \quad \forall i = 1, \dots, q$, there exists $n \in \mathbb{N}$ such that $(|u_n| > k) \nsubseteq A_i, \quad \forall i = 1, \dots, q$.

Let $k > 0, q \in \mathbb{N}$ and $A_1, \dots, A_q \in \mathcal{A}$ arbitrarily such that, for any $i = 1, \dots, q$, $\mu(A_i) < \frac{1}{2}$; fix $p \in \mathbb{N}, p > k$. $]\frac{k_{n,p}}{2^p}, \frac{k_{n,p}+1}{2^p}[\subseteq (|u_n| > k)$, for all $n \in \mathbb{N}$.

Since $\mu(A_1) < \frac{1}{2}$, there exists $i_1 \in \{0, \ldots, 2^p - 1\}$ such that $]\frac{i_1}{2^p}\frac{i_1+1}{2^p}[\not\subseteq A_1$. Let $n_1 = 2^{p-1} \cdot i_1$; then $k_{n_1,p} = i_1$ and, according to

$$(|u_{n_1}| > k) \supseteq \left]\frac{k_{n_1,p}}{2^p}, \frac{k_{n_1,p}+1}{2^p}\right[= \left]\frac{i_1}{2^p}, \frac{i_1+1}{2^p}\right[,$$

we can deduce that $(|u_{n_1}| > k) \not\subseteq A_1$.

Since $\mu(A_2) < \frac{1}{2}$, there exists $i_2 \in \{0, \ldots, 2^{2p} - 1\}$ such that $]\frac{i_2}{2^{2p}}\frac{i_2+1}{2^{2p}}[\not\subseteq A_2$. Let $n_2 = n_1 + 2^{2p-1} \cdot i_2$; then $k_{n_2,p} = i_1$ and $k_{n_2,2p} = i_2$. According to

$$(|u_{n_2}| > k) \supseteq \left]\frac{i_1}{2^p}, \frac{i_1+1}{2^p}\right[\cup \left]\frac{i_2}{2^{2p}}, \frac{i_2+1}{2^{2p}}\right[,$$

we deduce that $(|u_{n_2}| > k) \not\subseteq A_1$ and $(|u_{n_2}| > k) \not\subseteq A_2$.

Let us continue this reasoning; since $\mu(A_q) < \frac{1}{2}$, there exists $i_q \in \{0, \ldots, 2^{2^{q-1}\cdot p} - 1\}$ such that $]\frac{i_q}{2^{2^{q-1}\cdot p}}, \frac{i_q+1}{2^{2^{q-1}\cdot p}}[\not\subseteq A_q$. Let $n_q = n_1 + \ldots + n_{q-1} + 2^{2^{q-1}\cdot p-1} \cdot i_q$; then $k_{n_q,p} = i_1, k_{n_q,2p} = i_2, \ldots, k_{n_q,2^{q-1}\cdot p} = i_q$ and

$$(|u_{n_q}| > k) \supseteq \left]\frac{i_1}{2^p}, \frac{i_1+1}{2^p}\right[\cup \left]\frac{i_2}{2^{2p}}, \frac{i_2+1}{2^{2p}}\right[\cup \cdots \cup \left]\frac{i_q}{2^{2^{q-1}\cdot p}}, \frac{i_q+1}{2^{2^{q-1}\cdot p}}\right[.$$

Therefore, for every $i = 1, \ldots, q$, $(|u_{n_q}| > k) \not\subseteq A_i$. □

In the following, we will present some sufficient conditions for a set to be finite-tight.

Proposition 3.80. *Let E be a separable Banach space.*

(i) *Every finite set $H \subseteq \mathcal{M}(E)$ is a finite-tight set.*

(ii) *Let $H = \{u_n : n \in \mathbb{N}\} \subseteq \mathcal{M}(E)$ be a tight set; if H satisfies one of the following two conditions:*

$$\exists K_0 \in \mathcal{K}_E \quad \textit{such that} \quad \sum_{n=0}^{\infty} \mu(u_n^{-1}(E \setminus K_0)) < +\infty, \tag{S}$$

$$\forall K \in \mathcal{K}_E, \exists n_K \in \mathbb{N} \textit{ such that } u_{n+1}^{-1}(E \setminus K) \subseteq u_n^{-1}(E \setminus K), \ \forall n \geq n_K, \tag{M}$$

then H is finite-tight.

Proof. (i) For every $u \in \mathcal{M}(E)$, $\mu \circ u^{-1} : \mathcal{B}_E \to \mathbb{R}_+$, defined by $(\mu \circ u^{-1})(C) = \mu(u^{-1}(C))$, for every $C \in \mathcal{B}_E$, is a σ-additive measure; since E is a Polish space, $\mu \circ u^{-1}$ is a Radon measure and then, for every $\varepsilon > 0$, there exists $K \in \mathcal{K}_E$ such that $\mu(u^{-1}(E \setminus K)) < \varepsilon$. Therefore, every finite set $H \subseteq \mathcal{M}(E)$ is finite-tight.

(ii) Let H be a tight set satisfying condition (S); for every $\varepsilon > 0$, there exists $K \in \mathcal{K}_E$ such that $\mu(u_n^{-1}(E \setminus K)) < \varepsilon$, for all $n \in \mathbb{N}$; obviously, we can suppose that $K \supseteq K_0$. Let us note, for every $n \in \mathbb{N}$, $A_n = u_n^{-1}(E \setminus K)$.

According to condition (S), there is $p \in \mathbb{N}$ such that $\sum_{n=p}^{\infty} \mu(A_n) < \varepsilon$.

Therefore, the family $\mathcal{A}_f = \{A_0, A_1, \dots, A_{p-1}, \cup_{n=p}^{\infty} A_n\}$ satisfies the conditions of Definition 3.75.

If H satisfies the condition (M), then the family $\mathcal{A}_f = \{A_0, A_1, \dots, A_{n_K}\}$ satisfies also the conditions of 3.75. □

In the following examples, we will show that conditions (S) and (M) of the previous proposition are not necessary.

Example 3.81.

(1) Let $\Omega = [0,1]$ and let $A_1 =]\frac{1}{3}, \frac{2}{3}[$, $A_2 =]\frac{1}{3^2}, \frac{2}{3^2}[\cup]\frac{7}{3^2}, \frac{8}{3^2}[$, $A_3 =]\frac{1}{3^3}, \frac{2}{3^3}[\cup]\frac{7}{3^3}, \frac{8}{3^3}[\cup]\frac{19}{3^3}, \frac{20}{3^3}[\cup]\frac{25}{3^3}, \frac{26}{3^3}[\cdots$ be the sets extracted from the interval $[0,1]$ to obtain the Cantor set; we recall that A_n is an union of 2^{n-1} intervals of length $\frac{1}{3^n}$ every one. Since $\mu(A_n) \to 0$, the set $H = \{u_n = 3^n \cdot \chi_{A_n} : n \in \mathbb{N}^*\} \subseteq \mathcal{M}(\mathbb{R})$ is a tight set. We note that H is not bounded in $L^1([0,1], \mathbb{R})$.

Let $K_0 = [-1, 1] \in \mathcal{K}_{\mathbb{R}}$; then $\sum_{n=1}^{\infty} \mu(u_n^{-1}(\mathbb{R} \setminus K_0)) = \sum_{n=1}^{\infty} \mu(|3^n \cdot \chi_{A_n}| > 1) = \sum_{n=1}^{\infty} \mu(A_n) = 1 < +\infty$ and, according to condition (S) of previous proposition, H is finite-tight.

However, H does not satisfy condition (M) of (ii). In fact, for $K = [-1, 1]$ and for all $n \in \mathbb{N}^*$, $u_n^{-1}(\mathbb{R} \setminus K) = A_n$, even though the sets A_n are pairwise disjoint.

(2) For every $n \in \mathbb{N}^*$, let $u_n : [0,1] \to \mathbb{R}$, $u_n = n^2 \cdot \chi_{[0,\frac{1}{n}]}$; then $(u_n)_n \subseteq \mathcal{M}(\mathbb{R})$. The set $H = \{u_n : n \in \mathbb{N}^*\}$ is tight but it is not bounded in $L^1([0,1], \mathbb{R})$.

For every $K \in \mathcal{K}_{\mathbb{R}}$, there exists $n_K \in \mathbb{N}$ such that $n^2 \notin K$, for all $n \geq n_K$. Then, for every $n \geq n_K$,

$$u_n^{-1}(\mathbb{R} \setminus K) = \begin{cases} [0, \frac{1}{n}], & 0 \in K \\ [0, 1], & 0 \notin K \end{cases};$$

therefore $(u_n^{-1}(\mathbb{R} \setminus K))_{n \geq n_K}$ is a decreasing sequence of sets, so that H satisfies the condition (M) and then H is finite-tight.

However, condition (S) is not satisfied; in fact

$$\sum_{n=1}^{\infty} \mu\left(u_n^{-1}(\mathbb{R} \setminus \{0\})\right) = \sum_{n=1}^{\infty} \frac{1}{n} = +\infty.$$

Definition 3.82. Let E be a separable Banach space; a sequence $(u_n)_{n\in\mathbb{N}} \subseteq \mathcal{M}(E)$ is w^2-*convergent* to a measurable function $u : \Omega \to E$ if the following two conditions are satisfied:

(a) There exists a decreasing sequence $(B_p)_{p\in\mathbb{N}} \subseteq \mathcal{A}$ with $\mu(B_p) \to 0$ and, for all $p \in \mathbb{N}$, $(u_n)_n \subseteq L^1(\Omega \setminus B_p, E), u \in L^1(\Omega \setminus B_p, E)$.

(b) For every $p \in \mathbb{N}$, $(\chi_{\Omega\setminus B_p} \cdot u_n)_{n\in\mathbb{N}}$ is weakly convergent to $\chi_{\Omega\setminus B_p} \cdot u$.

In this case we note $u_n \overset{w^2}{\longrightarrow} u$.

If $(u_n)_{n\in\mathbb{N}}$ is w^2-convergent to u, then every subsequence of $(u_n)_{n\in\mathbb{N}}$ is w^2-convergent to u.

Remark 3.83. This definition slightly extends Definition 1.95 in which $u \in L^1(\mu, E)$ and $u_n \in L^1(\mu, E)$, for every $n \in \mathbb{N}$.

The following result is an alternative of the biting lemma (Theorem 3.68) for the finite-tight sets.

Theorem 3.84 (Theorem 2.10 of [81]). *Let E be a separable Banach space such that E and its dual space E^* have the Radon–Nikodym property (particularly, let E be a reflexive normed space) and let $(u_n)_{n\in\mathbb{N}} \subseteq \mathcal{M}(E)$ be a finite-tight sequence; then there exist a subsequence $(u_{k_n})_{n\in\mathbb{N}}$ and a measurable mapping $u \in \mathcal{M}(E)$ such that $u_{k_n} \overset{w^2}{\longrightarrow} u$.*

Proof. (I) Since $(u_n)_n$ is finite-tight, for every $\varepsilon > 0$, there exist $K \in \mathcal{K}_E$ and a finite subfamily $\mathcal{A}_f$ of $\mathcal{A}$ with $\mu(A) < \varepsilon$, for every $A \in \mathcal{A}_f$, such that, for every $n \in \mathbb{N}$, there exists $A_n \in \mathcal{A}_f$ with $u_n^{-1}(E \setminus K) \subseteq A_n$. Then, there exist a set $A \in \mathcal{A}_f$ and an infinite subset $N \subseteq \mathbb{N}$ such that $u_n^{-1}(E \setminus K) \subseteq A$, for any $n \in N$.

Therefore, for every $\varepsilon > 0$, there exist $K \in \mathcal{K}_E$, $A \in \mathcal{A}$ with $\mu(A) < \varepsilon$ and an infinite subset $N \subseteq \mathbb{N}$, such that $u_n^{-1}(E \setminus K) \subseteq A$, for every $n \in N$.

For $\varepsilon = 1$ let $K_1 \in \mathcal{K}_E$, $B_1 \in \mathcal{A}$ with $\mu(B_1) < 1$ and let N_1 be an infinite subset of $\mathbb{N}$ such that $u_n^{-1}(E \setminus K_1) \subseteq B_1$, for every $n \in N_1$.

For $\varepsilon = \frac{1}{2}$ let $K_2 \in \mathcal{K}_E$, $B_2 \in \mathcal{A}$ with $\mu(B_2) \leq \frac{1}{2}$ and let N_2 be an infinite subset of N_1 such that $u_n^{-1}(E \setminus K_2) \subseteq B_2$, for every $n \in N_2$; obviously, we can choose $K_2 \supseteq K_1$ and $B_2 \subseteq B_1$.

Generally, for $p \in \mathbb{N}^*$ and $\varepsilon = \frac{1}{p}$, let $K_p \in \mathcal{K}_E$ with $K_p \supseteq K_{p-1}$, let $B_p \in \mathcal{A}$ with $B_p \subseteq B_{p-1}$ and $\mu(B_p) < \frac{1}{p}$ and let N_p be an infinite subset of N_{p-1} such that $u_n^{-1}(E \setminus K_p) \subseteq B_p$, for every $n \in N_p$.

Let now $n_p \in N_p$ with $n_p > n_{p-1}$, for every $p \in \mathbb{N}^*$; $(u_{n_i})_{i\in\mathbb{N}}$ is a subsequence of $(u_n)_{n\in\mathbb{N}}$ and, for every $p \in \mathbb{N}$ and $i \geq p$, let $n_i \in N_p$ such that $u_{n_i}^{-1}(E \setminus K_p) \subseteq B_p$. Then, for every $p \in \mathbb{N}^*$ and every $i \geq p$,

$$u_{n_i}(\Omega \setminus B_p) \subseteq K_p \subseteq E.$$

Thus $(u_{n_i}|_{\Omega\setminus B_p})_{i\geq p}$ is uniformly bounded and then $(u_{n_i}|_{\Omega\setminus B_p})_{i\in\mathbb{N}}$ is bounded in $L^1(\Omega \setminus B_p, E)$ and absolutely continuous.

For every $A \in \mathcal{A}$, with $A \subseteq \Omega \setminus B_p$ and $\mu(A) > 0$ and for every $i \geq p$,

$$\int_A u_{n_i}\, d\mu \in \mu(A) \cdot \overline{\text{co}}(u_{n_i}(A)) \subseteq \mu(A) \cdot \overline{\text{co}}(K_p).$$

Therefore $\{\int_A u_{n_i}\, d\mu \ : \ i \ \in \mathbb{N}\}$ is relatively weakly compact. According to Dunford's theorem (see Theorem 1 of [56], pp. 101), $(u_{n_i}|_{\Omega\setminus B_p})_{i\in\mathbb{N}}$ is relatively weakly compact in $L^1(\Omega \setminus B_p, E)$, for every $p \in \mathbb{N}$. Using Eberlein–Šmulyan's theorem, $(u_{n_i}|_{\Omega\setminus B_p})_{i\in\mathbb{N}}$ is relatively sequentially weakly compact in $L^1(\Omega \setminus B_p, E)$, for every $p \in \mathbb{N}$.

(II) Let M_1 be an infinite subset of $\mathbb{N}$ and let $v_1 \in L^1(\Omega \setminus B_1, E)$ such that $(u_{n_i})_{i\in M_1}$ is weakly convergent to v_1. Let M_2 be an infinite subset of M_1 and let $v_2 \in L^1(\Omega \setminus B_2, E)$ such that $(u_{n_i})_{i\in M_2}$ is weakly convergent to v_2 in $L^1(\Omega \setminus B_2, E)$. Since $\Omega \setminus B_1 \subseteq \Omega \setminus B_2$, $v_2 = v_1$ almost everywhere on $\Omega \setminus B_1$.

Generally, let M_p be an infinite subset of M_{p-1} and let $v_p \in L^1(\Omega \setminus B_p)$, such that $(u_{n_i})_{i\in M_p}$ is weakly convergent to v_p in $L^1(\Omega \setminus B_p, E)$; then $v_p = v_{p-1}$ almost everywhere on $\Omega \setminus B_{p-1}$.

For every $p \in \mathbb{N}^*$, we can choose $i_p \in M_p$ with $i_p > i_{p-1}$; then $(u_{n_{i_q}})_{q\in\mathbb{N}}$ is still a subsequence of $(u_n)_n$ and, for every $p \in \mathbb{N}$, $(u_{n_{i_q}})_{q\in\mathbb{N}}$ is weakly convergent to v_p in $L^1(\Omega \setminus B_p, E)$.

We define $u : \Omega \to E$ letting $u = v_p$ on $\Omega \setminus B_p$ and $u = 0$ on $\cap_{p=1}^{\infty} B_p$. Then $u \in \mathcal{M}(E)$ and $u_{n_{i_q}} \xrightarrow{w^2} u$. □

Now, let us give an alternative for the Saadoune–Valadier theorem (Theorem 3.70).

Theorem 3.85 (Theorem 2.11 of [81]). *Let E be a separable Banach space such that E and E^* have the Radon–Nikodym property (particularly, this happens if E is a separable reflexive normed space) and let $(u_n)_{n\in\mathbb{N}} \subseteq \mathcal{M}(E)$ be a finite-tight* *sequence; there exist a subsequence* $(u_{k_n})_{n\in\mathbb{N}}$ *and a Young measure* $\mathcal{T} \in \mathcal{Y}(E)$ *such that* $u_{k_n} \xrightarrow{\mathcal{S}} \mathcal{T}$.

If $\tau : \Omega \to \mathcal{P}_E$ *is the disintegration of* $\mathcal{T}$*, then, for almost every* $t \in \Omega$*,* τ_t *has a barycenter*

$$u(t) = \mathrm{bar}(\tau_t) = \int_E x d\tau_t(x) \in E.$$

The mapping $u : \Omega \to E$ *is measurable* ($u \in \mathcal{M}(E)$), $u_{k_n} \xrightarrow{w^2} u$ *and* $u(t) \in \overline{\mathrm{co}}(\mathrm{Ls}_n(u_{k_n}(t)))$*, for almost every* $t \in \Omega$ *(here,* $\overline{\mathrm{co}}(A)$ *is the closed convex hull of* A *and* $\mathrm{Ls}_n(u_{k_n}(t)) = \bigcap_{p=1}^{\infty} \overline{\{u_{k_i}(t) : i \geq p\}}$*).*

Proof. According to Theorem 3.84, there exist a subsequence $(u'_n)_{n\in\mathbb{N}}$ of $(u_n)_{n\in\mathbb{N}}$ and a mapping $v \in \mathcal{M}(E)$ such that $u'_n \xrightarrow{w^2} v$.

Let $(B_p)_{p\in\mathbb{N}} \subseteq \mathcal{A}$ be a decreasing sequence with $\mu(B_p) \downarrow 0$ such that, for every $p \in \mathbb{N}$, $(u'_n)_n \subseteq L^1(\Omega \setminus B_p, E)$, $v \in L^1(\Omega \setminus B_p, E)$ and $(\chi_{\Omega\setminus B_p} \cdot u'_n)_{n\in\mathbb{N}}$ is weakly convergent to $\chi_{\Omega\setminus B_p} \cdot v$. From Prohorov's theorem (Theorem 3.59), there exist a subsequence $(u_{k_n})_{n\in\mathbb{N}}$ of $(u'_n)_n$ and a Young measure $\mathcal{T}$ such that $u_{k_n} \xrightarrow{\mathcal{S}} \mathcal{T}$; obviously, $u_{k_n} \xrightarrow{w^2} v$.

For every $p \in \mathbb{N}$, the mapping $\Psi_p : \Omega \times E \to \mathbb{R}$, defined by $\Psi_p(t,x) = \chi_{\Omega\setminus B_p}(t) \cdot \|x\|$, for every $(t,x) \in \Omega \times E$, is a positive l.s.c. integrand. Thus, according to Proposition 3.34,

$$\int_{\Omega\times E} \Psi(t,x) d\mathcal{T}(t,x) \leq \liminf_n \int_{\Omega\setminus B_p} \|u_{k_n}(t)\| d\mu(t) < +\infty$$

($(u_{k_n})_n$ is bounded in $L^1(\Omega \setminus B_p), E)$).

Therefore

$$\int_{\Omega\setminus B_p} \left(\int_E \|x\| d\tau_t(x) \right) d\mu(t) < +\infty.$$

So, for almost every $t \in \Omega \setminus B_p$, there exists $u^p(t) = \mathrm{bar}(\tau_t) = \int_E x d\tau_t(x)$. $u^p : \Omega \setminus B_p \to E$ is a measurable mapping.

Since $\Omega \setminus B_p \subseteq \Omega \setminus B_{p+1}$, $u^{p+1}|_{\Omega\setminus B_p} = u^p$ almost everywhere.

Thus, we can define $u : \Omega \to E$ letting

$$u(t) = \begin{cases} \mathrm{bar}(\tau_t), & \text{if } \ t \in \cup_{p=1}^{\infty}(\Omega \setminus B_p) \ \text{ and if there exists bar } (\tau_t), \\ 0_E, & \text{otherwise.} \end{cases}$$

Then $u \in \mathcal{M}(E)$ and $u \in L^1(\mu \setminus B_p, E)$, for every $p \in \mathbb{N}$.

Let us show that $u = v$. Since E^* has the Radon–Nikodym property, $L^\infty(\Omega, E^*)$ is dual space of $L^1(\mu, E)$ (Theorem 1, pp. 98, of [56]).

Let g be an arbitrary function of $L^\infty(\Omega, E^*)$ and let $p \in \mathbb{N}$.

Since $\chi_{\Omega \setminus B_p} \cdot u_{k_n} \xrightarrow[L^1(\mu,E)]{w} \chi_{\Omega \setminus B_p} \cdot v$,

$$\int_{\Omega \setminus B_p} \langle g(t), u_{k_n}(t) \rangle \, d\mu(t) \xrightarrow[n\to\infty]{} \int_{\Omega \setminus B_p} \langle g(t), v(t) \rangle \, d\mu(t). \tag{1}$$

Let $\Phi_p : \Omega \times E \to \mathbb{R}$, $\Phi(t, x) = \chi_{\Omega \setminus B_p} \cdot < g(t), x >$, for every $t \in \Omega$ and every $x \in E$; then Φ_p is measurable in t and continuous in x. By Proposition 3.8, it is a Carathéodory integrand.

Furthermore, for every $t \in \Omega$ and every $n \in \mathbb{N}$,

$$\left| \Phi_p \left(t, \chi_{\Omega \setminus B_p}(t) \cdot u_{k_n}(t) \right) \right| \leq \|g\|_\infty \cdot \chi_{\Omega \setminus B_p}(t) \cdot \|u_{k_n}(t)\|_E$$

and since $(\chi_{\Omega \setminus B_p} \cdot u_{k_n})_n$ is uniformly integrable, $(\Phi_p(\cdot, \chi_{\Omega \setminus B_p}(\cdot) u_{k_n}(\cdot)))_n$ is also uniformly integrable.

According to Corollary 3.36,

$$\int_\Omega \Phi_p \left(t, \chi_{\Omega \setminus B_p}(t) \cdot u_{k_n}(t) \right) d\mu(t) \to \int_{\Omega \times E} \Phi_p(t, x) d\tau(t, x) \quad \text{or}$$

$$\int_{\Omega \setminus B_p} \langle g(t), u_{k_n}(t) \rangle \, d\mu(t) \to \int_{\Omega \setminus B_p} \left(\int_E \langle g(t), x \rangle \, d\tau_t(x) \right) d\mu(t). \tag{2}$$

By (1) and (2), $\int_{\Omega \setminus B_p} < g(t), v(t) > d\mu(t) =$

$$= \int_{\Omega \setminus B_p} \left\langle g(t), \int_E x \, d\tau_t(x) \right\rangle d\mu(t) = \int_{\Omega \setminus B_p} \langle g(t), u(t) \rangle \, d\mu(t).$$

Since g is arbitrary in the dual space, $\chi_{\Omega \setminus B_p}(t) \cdot v(t) = \chi_{\Omega \setminus B_p}(t) \cdot u(t)$, for almost every $t \in \Omega$ and for every $p \in \mathbb{N}$; hence $v = u$ almost everywhere and consequently $u_{k_n} \xrightarrow{w^2} u$.

Since $u_{k_n} \xrightarrow{\mathcal{S}} \tau$, $\operatorname{supp}\tau_t \subseteq \operatorname{Ls}(u_{k_n}(t)) = \cap_{n=1}^\infty \overline{\{u_{k_p}(t) : p \geq n\}}$ (see Proposition 3.43). Therefore $u(t) = \int_E x \, d\tau_t(x) = \int_{\operatorname{supp}\tau_t} x \, d\tau_t(x) \in \overline{\operatorname{co}}(\operatorname{supp}\tau_t) \subseteq \overline{\operatorname{co}}\operatorname{Ls}(u_{k_n}(t))$, for almost every $t \in \Omega$. □

3.7 Product of Young Measures

In this section, we will introduce two products for Young measures.

3.7.1 Fiber Product

Let S and T be two regular Suslin spaces. According to (iv) of Theorem 2.23, $S \times T$ is a regular Suslin space; since S and T are hereditarily Lindelöf ((ii) of Theorem 2.21), $S \times T$ is hereditarily Lindelöf and then $\mathcal{B}_{S\times T} = \mathcal{B}_S \otimes \mathcal{B}_T$.

For every $\sigma \in \mathcal{Y}(S)$ and every $\tau \in \mathcal{Y}(T)$ we define $\sigma \otimes \tau : \Omega \to \mathcal{P}_{S\times T}$ letting $(\sigma \otimes \tau)_t = \sigma_t \otimes \tau_t$, for every $t \in \Omega$, where $\sigma_.$ and $\tau_.$ are the disintegrations of σ and τ.

For every $A \in \mathcal{B}_S$, every $B \in \mathcal{B}_T$ and every $t \in \Omega$, $(\sigma \otimes \tau)_t(A \times B) = \sigma_t(A) \cdot \tau_t(B)$. Therefore the mapping $t \mapsto (\sigma \otimes \tau)_t(A \times B)$ is measurable and so the mapping $t \mapsto (\sigma \otimes \tau)_t(C)$ is measurable, for every $C \in \mathcal{B}_{S\times T}$.

Therefore $\sigma \otimes \tau$ is a Young measure on $S \times T$; this mapping is the disintegration of a Young measure noted $\sigma \otimes \tau \in \mathcal{Y}(S \times T)$; $\sigma \otimes \tau : \mathcal{A} \otimes \mathcal{B}_{S\times T} \to \mathbb{R}_+$ and, for every positive integrand $\Psi : \Omega \times (S \times T) \to \bar{\mathbb{R}}_+$,

$$(\sigma \otimes \tau)(\Psi) = \int_\Omega \left(\int_{S\times T} \Psi(t, x, y) d(\sigma_t \otimes \tau_t)(x, y) \right) d\mu(t).$$

If we suppose that $\sigma = \tau^u$ and that $\tau = \tau^v$ are the Young measures associated to the measurable applications $u \in \mathcal{M}(S)$ and respectively $v \in \mathcal{M}(T)$, then, for every $t \in \Omega$, $(\sigma \otimes \tau)_t = \delta_{(u(t),v(t))}$.

Definition 3.86. The measure $\sigma \otimes \tau$ is called *the fiber product* of σ and τ.

Let us note by $\mathcal{S}_S, \mathcal{S}_T$ and $\mathcal{S}_{S\times T}$ the stable topologies on $\mathcal{Y}(S), \mathcal{Y}(T)$ and respectively on $\mathcal{Y}(S \times T)$.

The following theorem has a special importance due to its multiple applications. The proof is that of [169].

Theorem 3.87 (fiber product lemma). *Let $\mathcal{H} \subseteq \mathcal{Y}(S)$ and $\mathcal{L} \subseteq \mathcal{Y}(T)$ be two tight sets; then:*

(1) *$\mathcal{H} \otimes \mathcal{L} = \{\sigma \otimes \tau : \sigma \in \mathcal{H}, \tau \in \mathcal{L}\} \subseteq \mathcal{Y}(S \times T)$ is tight.*

(2) *Let $(\sigma^i)_{i\in I} \subseteq \mathcal{H}$ and $(\tau^i)_{i\in I} \subseteq \mathcal{L}$ be two nets such that*

$$\sigma^i \xrightarrow{\mathcal{S}_S} \sigma \in \mathcal{Y}(S) \quad \textit{and} \quad \tau^i \xrightarrow{\mathcal{S}_T} \tau \in \mathcal{Y}(T).$$

If $\sigma = \tau^u$, where $u \in \mathcal{M}(S)$, then $\sigma^i \otimes \tau^i \xrightarrow{\mathcal{S}_{S\times T}} \sigma \otimes \tau$.

Proof. (1) Since $\mathcal{H}$ and $\mathcal{L}$ are tight, for every $\varepsilon > 0$, there exists $K \in \mathcal{K}_S, C \in \mathcal{K}_T$ such that

$$\sigma(\Omega \times (S \setminus K)) < \frac{\varepsilon}{2}, \quad \text{for every} \quad \sigma \in \mathcal{H} \quad \text{and}$$
$$\tau(\Omega \times (T \setminus C)) < \frac{\varepsilon}{2}, \quad \text{for every} \quad \tau \in \mathcal{L}.$$

Then $K \times C \in \mathcal{K}_{S\times T}$ and, since $(S\times T)\setminus(K\times C) \subseteq [S\times(T\setminus C)]\cup[(S\setminus K)\times T]$, for every $\sigma \in \mathcal{H}$ and every $\tau \in \mathcal{L}$,

$$\begin{aligned}
&(\sigma \otimes \tau)(\Omega \times [(S \times T) \setminus (K \times C)]) \\
&= \int_\Omega (\sigma \otimes \tau)_t[(S \times T) \setminus (K \times C)]d\mu(t) \\
&\leq \int_\Omega (\sigma_t \otimes \tau_t)(S \times (T \setminus C))d\mu(t) + \int_\Omega (\sigma_t \otimes \tau_t)((S \setminus K) \times T)d\mu(t) \\
&= \int_\Omega \tau_t(T \setminus C)d\mu(t) + \int_\Omega \sigma_t(S \setminus K)d\mu(t) \\
&= \tau(\Omega \times (T \setminus C)) + \sigma(\Omega \times (S \setminus K)) < \varepsilon.
\end{aligned}$$

Therefore $\mathcal{H} \otimes \mathcal{L}$ is tight.

(2) Let now $(\sigma^i)_{i\in I} \subseteq \mathcal{H}$ and $(\tau^i)_{i\in I} \subseteq \mathcal{L}$ such that

$$\sigma^i \xrightarrow{\mathcal{S}_S} \sigma \in \mathcal{Y}(S), \tau^i \xrightarrow{\mathcal{S}_T} \tau \in \mathcal{Y}(T)$$

where $\sigma = \tau^u$ and $u \in \mathcal{M}(S)$.

(a) Firstly, let us suppose that S and T are metrizable; then $S \times T$ is also metrizable; $\sigma^i \otimes \tau^i \xrightarrow{\mathcal{S}_{S\times T}} \sigma \otimes \tau$ if and only if every subnet of $(\sigma^i \otimes \tau^i)_{i\in I}$ has a subnet convergent to $\sigma \otimes \tau$.

Then let $(\sigma^j \otimes \tau^j)_{j\in J} \subseteq \mathcal{H} \otimes \mathcal{L}$ be a subnet of $(\sigma^i \otimes \tau^i)_{i\in I}$; since $\mathcal{H} \otimes \mathcal{L}$ is tight, according to Prohorov's theorem 3.59, there exists a subnet $(\sigma^l \otimes \tau^l)_{l\in L}$ of $(\sigma^j \otimes \tau^j)_{j\in J}$, $\mathcal{S}_{S\times T}$-convergent to a Young measure $\Lambda \in \mathcal{Y}(S \times T)$.

Let d be a bounded metric on S which generates the topology of S; then $\Psi : \Omega \times S \to \mathbb{R}$, defined by $\Psi(t,x) = d(u(t), x)$, is a bounded Carathéodory integrand. Since $Cth^b(\Omega\times S) \subseteq Cth^b(\Omega\times(S\times T))$, according to Theorem 3.30,

$$\left(\sigma^l \otimes \tau^l\right)(\Psi) \to \Lambda(\Psi). \tag{1}$$

On the other hand

$$\left(\sigma^l \otimes \tau^l\right)(\Psi) = \int_\Omega \int_{S\times T} d(u(t), x) d(\sigma_t^l \otimes \tau_t^l)(x, y) d\mu(t) \tag{2}$$
$$= \int_\Omega \int_S d(u(t), x) d\sigma_t^l(x) d\mu(t)$$
$$= \sigma^l(\Psi) \to \sigma(\Psi) = \tau^u(\Psi) = 0.$$

By (1) and (2)

$$\Lambda(\Psi) = 0. \tag{3}$$

But $\Lambda(\Psi) = \int_\Omega \lambda_t(\Psi) d\mu(t)$. Then $\lambda_t(\Psi) = 0$, for almost every $t \in \Omega$ ($\lambda_.$ is the disintegration of Λ). Therefore, for almost every $t \in \Omega$,

$$0 = \int_{S\times T} \Psi(t, x) d\lambda_t(x, y) = \int_{(S\setminus\{u(t)\})\times T} d(u(t), x) d\lambda_t(x, y).$$

Since, on the open set $S \setminus \{u(t)\}$, $d(u(t), x) > 0$, $\lambda_t((S \setminus \{u(t)\}) \times T) = 0$ and then $\lambda_t(\{u(t)\} \times T) = 1$, almost for every $t \in \Omega$.

Let $\nu_t : \mathcal{B}_T \to \mathbb{R}_+$ defined by $\nu_t(B) = \lambda_t(\{u(t)\} \times B)$, for every $B \in \mathcal{B}_T$; $\nu_t \in \mathcal{P}_T$ and $\lambda_t(A\times B) = \delta_{u(t)}(A)\cdot\nu_t(B)$, for every $A \in \mathcal{B}_S$ and every $B \in \mathcal{B}_T$. Therefore $\lambda_t = \delta_{u(t)} \otimes \nu_t$, almost for every $t \in \Omega$.

Let $A \in \mathcal{A}$ and $f \in C_b(T) \subseteq C_b(S \times T)$. Then

$$\left(\sigma^l \otimes \tau^l\right)(\chi_A \otimes f) \to \Lambda(\chi_A \otimes f) = \int_A \nu_t(f) d\mu(t).$$

On the other hand

$$(\sigma^l \otimes \tau^l)(\chi_A \otimes f) = \int_A \tau_t^l(f) d\mu(t)$$
$$= \tau^l(\chi_A \otimes f) \to \tau(\chi_A \otimes f)$$
$$= \int_A \tau_t(f) d\mu(t).$$

Therefore $\nu_t(f) = \tau_t(f)$, for every $f \in C_b(T)$ and then $\lambda_t = \delta_{u(t)} \otimes \tau_t$, almost for every $t \in \Omega$; therefore, $\Lambda = \sigma \otimes \tau$.

(b) Let now S and T be two regular Suslin spaces; according to (ii) of Theorem 2.25, there exist a metric d on S and a metric g on T such that the generated topologies are coarser than the initial topologies, but still having the same Borel sets. Just as in the proof of (a2) of Theorem 3.59, we will note by $\mathcal{S}_S^d$, $\mathcal{S}_T^g$, $\mathcal{S}_{S\times T}^{d\times g}$ the stable topologies on $\mathcal{Y}(S)$, $\mathcal{Y}(T)$ and respectively on $\mathcal{Y}(S \times T)$.

Then $\mathcal{S}_S^d \subseteq \mathcal{S}_S, \mathcal{S}_T^g \subseteq \mathcal{S}_T, \mathcal{S}_{S\times T}^{d\times g} \subseteq \mathcal{S}_{S\times T}$. Since $\sigma^i \xrightarrow{\mathcal{S}_S^d} \sigma$ and $\tau^i \xrightarrow{\mathcal{S}_T^g} \tau$, we can use (a) to obtain $\sigma^i \otimes \tau^i \xrightarrow{\mathcal{S}_{S\times T}^{d\times g}} \sigma \otimes \tau$. The same argument already used in demonstration of 3.59 shows that $\sigma^i \otimes \tau^i \xrightarrow{\mathcal{S}_{S\times T}} \sigma \otimes \tau$. □

Remark 3.88.

(i) It is clear that we can change the condition $\sigma \in \mathcal{M}(S)$ by $\tau \in \mathcal{M}(T)$.

If $\sigma \notin \mathcal{M}(S)$ and $\tau \notin \mathcal{M}(T)$, then, generally, $\sigma^i \otimes \tau^i \overset{\mathcal{S}_{S\times T}}{\nrightarrow} \sigma \otimes \tau$, as highlighted in the following example.

We have shown in Example 3.44 that, if $(r_n)_n \subseteq \mathcal{M}(\mathbb{R})$ is the Rademacher's sequence, then

$$\tau^{r_n} \xrightarrow{\mathcal{S}_{\mathbb{R}}} \tau = \mu \otimes \left(\frac{1}{2} \cdot (\delta_1 + \delta_{-1})\right).$$

If we make the fiber product between this sequence with itself, we obtain, through a simple reasoning similar to that of 3.44:

$$\tau^{r_n} \otimes \tau^{r_n} \xrightarrow{\mathcal{S}_{\mathbb{R}\times\mathbb{R}}} \mu \otimes \left(\frac{1}{2} \cdot \left(\delta_{(1,1)} + \delta_{(-1,-1)}\right)\right)$$

and this limit is different from the Young measure

$$\tau \otimes \tau = \mu \otimes \left(\frac{1}{4} \cdot \left(\delta_{(1,1)} + \delta_{(1,-1)} + \delta_{(-1,1)} + \delta_{(-1,-1)}\right)\right).$$

(ii) In Theorem 3.3.1 of [45], the fiber product lemma is stated in a more general manner:

$$\left.\begin{array}{l} \sigma^i \xrightarrow{\mathcal{P}_S} \sigma \in \mathcal{Y}(S) \\ \tau^i \xrightarrow{\mathcal{S}_T} \tau \in \mathcal{Y}(T) \end{array}\right\} \Longrightarrow \sigma^i \otimes \tau^i \xrightarrow{\mathcal{S}_{S\times T}} \sigma \otimes \tau.$$

In fact, the condition $\sigma \in \mathcal{M}(S)$ of the previous theorem assures, according to Corollary 3.2.2 of [45], that $(\sigma^i)_{i\in I}$ is convergent in measure towards σ.

The most frequently, we will use the fiber product lemma in the following manner:

Corollary 3.89. *Let S and T be two regular Suslin spaces. For every sequence $(u^n)_{n\in\mathbb{N}} \subseteq \mathcal{M}(S)$ which is tight and convergent in measure to $u \in \mathcal{M}(S)$ and every sequence $(v^n)_{n\in\mathbb{N}} \subseteq \mathcal{M}(T)$ which is tight and stably convergent to $\tau \in \mathcal{Y}(T)$,*

$$\tau^{u^n} \otimes \tau^{v^n} \xrightarrow{\mathcal{S}_{S\times T}} \tau^u \otimes \tau \quad or \quad (u^n, v^n) \xrightarrow{\mathcal{S}_{S\times T}} \tau^u \otimes \tau.$$

If S and T are two Polish spaces, then the hypothesis of being tight for the two sequences is no longer necessary.

Proof. The result comes from (iii) of Remark 3.39.

In Polish spaces, any sequence of Young measures, stably convergent, is tight (Proposition 3.65). □

3.7.2 Tensor Product

We will present another product of the Young measures-tensor product. Tensor product was introduced by E. J. Balder in [12] (see also [168] and [18]).

Let $(\Omega_1, \mathcal{A}_1, \mu_1), (\Omega_2, \mathcal{A}_2, \mu_2)$ be two measure spaces, where $\mu_1 \in ca^+(\mathcal{A}_1)$ and $\mu_2 \in ca^+(\mathcal{A}_2)$ and let $(\Omega, \mathcal{A}, \mu)$ be the product space: $\Omega = \Omega_1 \times \Omega_2, \mathcal{A} = \mathcal{A}_1 \otimes \mathcal{A}_2, \mu = \mu_1 \otimes \mu_2$.

Let P_1 and P_2 be two Polish spaces; according to (ii) of Theorem 2.18, the space $P = P_1 \times P_2$ is Polish and, according to (iv) of the same theorem, it is hereditarily Lindelöf; then $\mathcal{B}_P = \mathcal{B}_{P_1} \otimes \mathcal{B}_{P_2}$.

Finally, let $\mathcal{Y}(\Omega_1 \times P_1)$ and $\mathcal{Y}(\Omega_2 \times P_2)$ be the Young measure spaces on $\Omega_1 \times P_1$ and $\Omega_2 \times P_2$, respectively.

The mapping $\varphi : (\Omega_1 \times P_1) \times (\Omega_2 \times P_2) \to \Omega \times P, \varphi((t_1, x_1), (t_2, x_2)) = ((t_1, t_2), (x_1, x_2))$, is obviously measurable and then, for every $\tau^1 \in \mathcal{Y}(\Omega_1 \times P_1)$, $\tau^2 \in \mathcal{Y}(\Omega_2 \times P_2)$, we can define $\tau : \mathcal{A} \otimes \mathcal{B}_P \to \mathbb{R}_+$ letting $\tau(E) = (\tau^1 \otimes \tau^2)(\varphi^{-1}(E))$, for every $E \in \mathcal{A} \otimes \mathcal{B}_P$; then $\tau \in ca^+(\mathcal{A} \otimes \mathcal{B}_P)$.

For every $A \in \mathcal{A}$, $F = \varphi^{-1}(A \times P) = \{((t_1, x_1), (t_2, x_2)) : (t_1, t_2) \in A\} \in (\mathcal{A}_1 \otimes \mathcal{B}_{P_1}) \otimes (\mathcal{A}_2 \otimes \mathcal{B}_{P_2})$ and then

$$\begin{aligned}
\tau(A \times P) = (\tau^1 \otimes \tau^2)(F) &= \int_{\Omega_1 \times P_1} \tau^2(F_{(t_1,x_1)}) d\tau^1(t_1, x_1) \\
&= \int_{\Omega_1 \times P_1} \tau^2(A_{t_1} \times P_2) d\tau^1(t_1, x_1) = \int_{\Omega_1 \times P_1} \mu_2(A_{t_1}) d\tau^1(t_1, x_1) \\
&= \int_{\Omega_1} \left(\int_{P_1} \mu_2(A_{t_1}) d\tau^1_{t_1}(x_1) \right) d\mu_1(t_1) = \int_{\Omega_1} \mu_2(A_{t_1}) d\mu_1(t_1) \\
&= (\mu_1 \otimes \mu_2)(A) = \mu(A).
\end{aligned}$$

Therefore $\mathcal{T}$ is a Young measure on $\Omega \times P$; in order to differentiate this product from the fiber product, we note $\mathcal{T} = \mathcal{T}^1 \dot{\otimes} \mathcal{T}^2$ and we say that $\mathcal{T}$ is *the tensor product* of $\mathcal{T}^1$ and $\mathcal{T}^2$. Let $\tau_.$ be the disintegration of $\mathcal{T}$ and let $\tau_.^1, \tau_.^2$ be the disintegrations of $\mathcal{T}^1$ and $\mathcal{T}^2$, respectively.

For every $E = A \times B$, where $A \in \mathcal{A}$ and $B \in \mathcal{B}_P$,

$$\begin{aligned}
(\mathcal{T}^1 \dot{\otimes} \mathcal{T}^2)(E) &= (\mathcal{T}^1 \otimes \mathcal{T}^2)(\varphi^{-1}(E)) \\
&= \int_{\Omega_1 \times P_1} \mathcal{T}^2 \left((\varphi^{-1}(E))_{(t_1,x_1)}\right) d\mathcal{T}^1(t_1, x_1) \\
&= \int_{\Omega_1} \left(\int_{P_1} \mathcal{T}^2(A_{t_1} \times B_{x_1}) d\tau_{t_1}^1(x_1) \right) d\mu_1(t_1) \\
&= \int_{\Omega_1} \left[\int_{P_1} \left(\int_{A_{t_1}} \tau_{t_2}^2(B_{x_1}) d\mu_2(t_2) \right) d\tau_{t_1}^1(x_1) \right] d\mu_1(t_1) \\
&= \int_{\Omega_1} \int_{A_{t_1}} \underbrace{\left(\int_{P_1} \tau_{t_2}^2(B_{x_1}) d\tau_{t_1}^1(x_1) \right)}_{(\tau_{t_1}^1 \otimes \tau_{t_2}^2)(B)} d\mu_2(t_2) d\mu_1(t_1) \\
&= \int_{\Omega_1} \int_{\Omega_2} \chi_{A_{t_1}}(t_2) \cdot (\tau_{t_1}^1 \otimes \tau_{t_2}^2)(B) d\mu_2(t_2) d\mu_1(t_1) \\
&= \int_A (\tau_{t_1}^1 \otimes \tau_{t_2}^2)(B) d\mu(t).
\end{aligned}$$

On the other hand, $\mathcal{T}(E) = \int_A \tau_t(B) d\mu(t)$ from where $\tau_t = \tau_{t_1}^1 \otimes \tau_{t_2}^2$, for almost every $t = (t_1, t_2) \in \Omega$.

The tensor product $\dot{\otimes}$ has a simpler behavior in connection to the stable convergence (see Theorem 2.5 of [12]).

Theorem 3.90.

(i) *Let* $\mathcal{H}^1 \subseteq \mathcal{Y}(\Omega_1 \times P_1)$ *and* $\mathcal{H}^2 \subseteq \mathcal{Y}(\Omega_2 \times P_2)$ *be two tight sets; then*

$$\mathcal{H}^1 \dot{\otimes} \mathcal{H}^2 = \{\mathcal{T}^1 \dot{\otimes} \mathcal{T}^2 : \mathcal{T}^1 \in \mathcal{H}^1, \mathcal{T}^2 \in \mathcal{H}^2\} \quad \textit{is tight.}$$

(ii) *Let* $(\mathcal{T}^{1,n})_n \subseteq \mathcal{Y}(\Omega_1 \times P_1)$, $(\mathcal{T}^{2,n})_n \subseteq \mathcal{Y}(\Omega_2 \times P_2)$ *be two sequences such that*

$$\mathcal{T}^{1,n} \xrightarrow{\mathcal{S}_{P_1}} \mathcal{T}^1 \in \mathcal{Y}(\Omega \times P_1) \quad \textit{and} \quad \mathcal{T}^{2,n} \xrightarrow{\mathcal{S}_{P_2}} \mathcal{T}^2 \in \mathcal{Y}(\Omega_2 \times P_2);$$

then

$$\mathcal{T}^{1,n} \dot{\otimes} \mathcal{T}^{2,n} \xrightarrow{\mathcal{S}_P} \mathcal{T}^1 \dot{\otimes} \mathcal{T}^2.$$

Proof. (i) Since $\mathcal{H}_1$ and $\mathcal{H}_2$ are tight, for every $\varepsilon > 0$, there exist $K_1 \in \mathcal{K}_{P_1}$, $K_2 \in \mathcal{K}_{P_2}$ such that

$$\tau^1(\Omega_1 \times (P_1 \setminus K_1)) < \varepsilon_1, \quad \text{for every} \quad \tau^1 \in \mathcal{H}_1 \quad \text{and}$$

$$\tau^2(\Omega_2 \times (P_2 \setminus K_2)) < \varepsilon_2, \quad \text{for every} \quad \tau^2 \in \mathcal{H}_2$$

where $\varepsilon_1 = \frac{1}{2T_2} \cdot \varepsilon, \varepsilon_2 = \frac{1}{2T_1} \cdot \varepsilon$ and $T_1 = \mu_1(\Omega_1), T_2 = \mu_2(\Omega_2)$.

Then $K = K_1 \times K_2 \in \mathcal{K}_P$ and, for every $\tau = \tau^1 \dot{\otimes} \tau^2 \in \mathcal{H}_1 \dot{\otimes} \mathcal{H}_2$,

$$\begin{aligned}
\tau(\Omega \times (P \setminus K)) &= (\tau^1 \otimes \tau^2)(\varphi^{-1}(\Omega \times (P \setminus K)) \\
&= (\tau^1 \otimes \tau^2)(\varphi^{-1}(\Omega \times [(P_1 \setminus K_1) \times K_2 \cup P_1 \times (P_2 \setminus K_2)])) \\
&= (\tau^1 \otimes \tau^2)(\varphi^{-1}(\Omega \times (P_1 \setminus K_1) \times K_2) \\
&\quad \cup \varphi^{-1}(\Omega \times (P_1 \times (P_2 \setminus K_2)))) \\
&= (\tau^1 \otimes \tau^2)(\Omega_1 \times (P_1 \setminus K_1) \times \Omega_2 \times K_2) \\
&\quad + (\tau^1 \otimes \tau^2)(\Omega_1 \times P_1 \times \Omega_2 \times (P_2 \setminus K_2)) \\
&= \tau^1(\Omega_1 \times (P_1 \setminus K_1)) \cdot \tau^2(\Omega_2 \times K_2) \\
&\quad + \tau^1(\Omega_1 \times P_1) \cdot \tau^2(\Omega_2 \times (P_2 \setminus K_2)) \\
&< \varepsilon_1 \cdot T_2 + T_1 \cdot \varepsilon_2 = \varepsilon.
\end{aligned}$$

Therefore $\mathcal{H}_1 \dot{\otimes} \mathcal{H}_2$ is tight.

(ii) Let us suppose that $(\tau^{1,n} \dot{\otimes} \tau^{2,n})_n$ is not $\mathcal{S}_P$-convergent to $\tau^1 \dot{\otimes} \tau^2$; let then $\varepsilon_0 > 0$, $A_0 \in \mathcal{A}$, $f_0 \in C_b(P)$ and let a subsequence of $(\tau^{1,n} \dot{\otimes} \tau^{2,n})_n$—still noted $(\tau^{1,n} \dot{\otimes} \tau^{2,n})_n$—such that

$$\left| (\tau^{1,n} \dot{\otimes} \tau^{2,n})(\chi_{A_0} \otimes f_0) - (\tau^1 \dot{\otimes} \tau^2)(\chi_{A_0} \otimes f_0) \right| \geq \varepsilon_0, \quad \forall n \in \mathbb{N}.$$

Since the sets $\{\tau^{1,n} : n \in \mathbb{N}\}$ and $\{\tau^{2,n} : n \in \mathbb{N}\}$ are sequentially compact, they are tight (see Proposition 3.65); then, according to (i), $\{\tau^{1,n} \dot{\otimes} \tau^{2,n} : n \in \mathbb{N}\}$ is tight and so sequentially $\mathcal{S}_P$-compact (Theorem 3.64). Then there exists a subsequence of $(\tau^{1,n} \dot{\otimes} \tau^{2,n})_n$—still noted $(\tau^{1,n} \dot{\otimes} \tau^{2,n})_n$—$\mathcal{S}_P$-convergent to a measure $\sigma \in \mathcal{Y}(\Omega \times P)$. This subsequence satisfies simultaneously the two following conditions:

$$\left| (\tau^{1,n} \dot{\otimes} \tau^{2,n})(\chi_{A_0} \otimes f_0) - (\tau^1 \dot{\otimes} \tau^2)(\chi_{A_0} \otimes f_0) \right| \geq \varepsilon_0, \quad \forall n \in \mathbb{N}. \tag{1}$$

$$\tau^{1,n} \dot{\otimes} \tau^{2,n} \xrightarrow{\mathcal{S}_P} \sigma. \tag{2}$$

For every $A_1 \in \mathcal{A}_1, A_2 \in \mathcal{A}_2$ and every $f_1 \in C_b(P_1), f_2 \in C_b(P_2), A = A_1 \times A_2 \in \mathcal{A}$ and $f = f_1 \otimes f_2 \in C_b(P)$ and, according to (2),

$$(\tau^{1,n}\dot{\otimes}\tau^{2,n})(\chi_A \otimes f) \to \sigma(\chi_A \otimes f).$$

On the other hand

$$\begin{aligned}(\tau^{1,n}\dot{\otimes}\tau^{2,n})(\chi_A \otimes f) &= (\tau^{1,n} \otimes \tau^{2,n})\left((\chi_A \otimes f) \circ \varphi\right)\\ &= (\tau^{1,n} \otimes \tau^{2,n})(\chi_{A_1} \otimes f_1, \chi_{A_2} \otimes f_2)\\ &= \tau^{1,n}(\chi_{A_1} \otimes f_1) \cdot \tau^{2,n}(\chi_{A_2} \otimes f_2)\\ &\to \tau^1(\chi_{A_1} \otimes f_1) \cdot \tau^2(\chi_{A_2} \otimes f_2).\end{aligned}$$

It results that

$$\sigma(\chi_A \otimes f) = \tau^1(\chi_{A_1} \otimes f_1) \cdot \tau^2(\chi_{A_2} \otimes f_2) = (\tau^1\dot{\otimes}\tau^2)(\chi_A \otimes f).$$

The family $\mathcal{A}_0 = \{A_1 \times A_2 : A_1 \in \mathcal{A}_1, A_2 \in \mathcal{A}_2\}$ is a semi-algebra related to which $\sigma(\cdot \otimes f) = (\tau^1\dot{\otimes}\tau^2)(\cdot \otimes f)$; therefore, the two measures coincide on the generated σ-algebra $\mathcal{A}$ and therefore

$$\sigma(\chi_A \otimes f) = (\tau^1\dot{\otimes}\tau^2)(\chi_A \otimes f), \quad \forall A \in \mathcal{A}, \quad \forall f = f_1 \otimes f_2 \in C_b(P).$$

Let now $\bar{P}_1$ and $\bar{P}_2$ be two metrizable compactifications of P_1 and P_2, respectively; for every $f \in C_b(P)$, let $(f_n)_n$ be the sequence of Yosida's transforms of f defined by:

$$f_n(x) = \inf\{f(y) + n \cdot d(x, y) : y \in P\}, \text{ for every } x \in P \text{ and every } n \in \mathbb{N},$$

where d is a metric compatible with topology of P.

For every $n \in \mathbb{N}$, f_n are Lipschitz mappings and $f_n \uparrow f$ (see Lemma 2.35).

For every $n \in \mathbb{N}$, let $\bar{f_n}$ be the Lipschitz extension of f_n on $\bar{P} = \bar{P}_1 \times \bar{P}_2$.

A consequence of the Stone–Weierstrass theorem assures us that the set $\{\sum_{i=1}^n f_1^i \cdot f_2^i : n \in \mathbb{N}, f_1^i \in C(\bar{P}_1), f_2^i \in C(\bar{P}_2), \quad \forall i = 1, \dots, n\}$ is dense in $C(\bar{P})$. Then, for every $n \in \mathbb{N}$ and every $\varepsilon > 0$, there exist $k_n \in \mathbb{N}$ and $f_1^1, \dots, f_1^{k_n} \in C(\bar{P}_1), f_2^1, \dots, f_2^{k_n} \in C(\bar{P}_2)$ such that

$$\sup_{x=(x_1,x_2)\in\bar{P}} \left|\bar{f_n}(x) - \sum_{i=1}^{k_n} f_1^i(x_1) \cdot f_2^i(x_2)\right| < \varepsilon.$$

Therefore we can deduce that

$$\sigma(\chi_A \otimes f_n) = (\tau^2 \dot{\otimes} \tau^2)(\chi_A \otimes f_n), \quad \forall n \in \mathbb{N}.$$

Since $f_n \uparrow f$, the monotone convergence theorem leads us to

$$\sigma(\chi_A \otimes f) = (\tau^1 \dot{\otimes} \tau^2)(\chi_A \otimes f), \quad \forall A \in \mathcal{A}, \quad \forall f \in C_b(P).$$

It results that $\tau^1 \dot{\otimes} \tau^2 = \sigma$ and, in this case, conditions (1) and (2) become contradictory. □

3.8 Jordan Finite Tight Sets

The notions and the results of this section come from [80] and [81].

Firstly, we will explain the terms used in this section.

A d-dimensional interval $I \subseteq \mathbb{R}^d$ is a product of bounded closed intervals of $\mathbb{R}$: $I = \prod_{i=1}^d [a_i, b_i]$. An elementary set A is the union of a finite family of non-overlapping d-dimensional intervals, i.e. $A = \cup_{k=1}^p I_k$ and $\mathring{I}_k \cap \mathring{I}_l = \emptyset$, for every $k, l \in \{1, \ldots, p\}$ with $k \neq l$.

Let us denote by $\mathcal{E}$ the family of all elementary sets; the σ-algebra generated by $\mathcal{E}$ is the family of all Borel sets of $\mathbb{R}^d$.

Let $\Omega \subseteq \mathbb{R}^d$ be a bounded Lebesgue measurable set, let $\mathcal{A}$ be the family of all Lebesgue measurable subsets of Ω and let μ be the Lebesgue measure on Ω.

Definition 3.91. Let $(E, \|\cdot\|)$ be a separable Banach space; a set $H \subseteq \mathcal{M}(E)$ is called *Jordan finite-tight* if, for every $\varepsilon > 0$, there exist $K \in \mathcal{K}_E$ and a finite subfamily $\mathcal{E}_f \subseteq \mathcal{E}$ with $\mu(A) < \varepsilon$, for every $A \in \mathcal{E}_f$, such that, for every $u \in H$, there exists $A_u \in \mathcal{E}_f$ with $u^{-1}(A \setminus K) \subseteq A_u$.

A sequence $(u_n)_n \subseteq \mathcal{M}(E)$ is *Jordan finite-tight* if the set $H = \{u_n : n \in \mathbb{N}\}$ is Jordan finite-tight.

Remark 3.92.

(i) Since $\mathcal{E} \subseteq \mathcal{A}$, every Jordan finite-tight set is finite-tight and therefore a tight set.

(ii) If E is a Euclidean space, then $H \subseteq \mathcal{M}(E)$ is Jordan finite-tight if and only if, for every $\varepsilon > 0$, there exist $k > 0$ and a finite subfamily $\mathcal{E}_f \subseteq \mathcal{E}$ with $\mu(A) < \varepsilon$, for every $A \in \mathcal{E}_f$, such that, for any $u \in H$, there exists $A_u \in \mathcal{E}_f$ with $\{t \in \Omega : \|u(t)\| > k\} \subseteq A_u$.

The following proposition gives a justification for the denomination *Jordan finite-tight set.*

For every $B \subseteq \Omega$ let $\mu_J^*(B) = \inf\{\mu(A) : A \in \mathcal{E}, B \subseteq A\}$ be the Jordan outer measure of B; obviously, $\mu_J^*(B) = 0$ if and only if B is a Jordan-negligible set.

Proposition 3.93. *$H \subseteq \mathcal{M}(E)$ is a Jordan finite-tight set if and only if, for every $\varepsilon > 0$, there exist a compact set $K \in \mathcal{K}_E$ and a finite cover of H, $\{H_1, \dots, H_p\}$, such that*

$$\mu_J^*\left(\cup_{u \in H_i} u^{-1}(E \setminus K)\right) < \varepsilon, \quad \textit{for every} \quad i = 1, \dots, p.$$

The demonstration is the same as that of Proposition 3.76.

Theorem 3.94. *Let E be a Euclidean space; for every $H \subseteq \mathcal{M}(E)$ let*

$$I_H(\infty) = \left\{t \in \bar{\Omega} : \limsup_{s \to t} \left(\sup_{u \in H} \|u(s)\| \right) = +\infty \right\}.$$

A set $H \subseteq \mathcal{M}(E)$ is Jordan finite-tight if and only if, for every $\varepsilon > 0$, there exists a finite cover of H, $\{H_1, \dots, H_p\}$, such that, for any $i = 1, \dots, p$,

$$\mu_J^*\left(I_{H_i}(\infty)\right) < \varepsilon.$$

Proof. ($\Longrightarrow$): Firstly, let us note that $\limsup_{s \to t}(\sup_{u \in H} \|u(s)\|) = +\infty$ is equivalent to the existence of two sequences $(u_n)_{n \in \mathbb{N}} \subseteq H$ and $(s_n)_{n \in \mathbb{N}} \subseteq \Omega$ with $s_n \to t$ and $\|u_n(s_n)\| \to +\infty$.

Let us suppose that $H \subseteq \mathcal{M}(E)$ is a Jordan finite-tight set; according to (ii) of Remark 3.92, for every $\varepsilon > 0$, there exist $k > 0$ and a finite subfamily $\mathcal{E}_f \subseteq \mathcal{E}$ with $\mu(A) < \varepsilon$, for every $A \in \mathcal{E}_f$, such that, for all $u \in H$, there exists $A_u \in \mathcal{E}_f$ with $(\|u\| > k) \subseteq A_u$. For every $A \in \mathcal{E}_f$, let $H_A = \{u \in H : (\|u\| > k) \subseteq A\}$; then $\{H_A : A \in \mathcal{E}_f\}$ is a finite cover of H.

Let us show that, for every $A \in \mathcal{E}_f$, $I_{H_A}(\infty) \subseteq A$.

Indeed, for every $t \in I_{H_A}(\infty)$, $\limsup_{s \to t}(\sup_{u \in H_A} \|u(s)\|) = +\infty$ and thus there exist two sequences $(s_n)_{n \in \mathbb{N}} \subseteq \Omega$ with $s_n \to t$ and $(u_n)_{n \in \mathbb{N}} \subseteq H_A$ such that $\|u_n(s_n)\| \to +\infty$; we can suppose that, for any $n \in \mathbb{N}$, $\|u_n(s_n)\| > k$ and therefore $(s_n)_{n \in \mathbb{N}} \subseteq (\|u_n\| > k) \subseteq A$.

Since A is closed, $t \in A$. Hence $\mu_J^*(I_{H_A}(\infty)) < \varepsilon$, for every $A \in \mathcal{E}_f$.

($\Longleftarrow$): For every $\varepsilon > 0$, let $\{H_1, \dots, H_p\}$ be a finite cover of H such that, for any $i = 1, \dots, p$, $\mu_J^*(I_{H_i}(\infty)) < \varepsilon$ and let $A_i \in \mathcal{E}$ such that $I_{H_i}(\infty) \subseteq A_i$ and $\mu(A_i) < \varepsilon$. Obviously, we can suppose that every $I_{H_i}(\infty)$ is contained in the interior $\mathring{A}_i$ of A_i. Then, for any $i = 1, \dots, p$, there exists $k_i > 0$ such that $\|u(t)\| \leq k_i$, for every $t \in \Omega \setminus \mathring{A}_i$ and every $u \in H_i$ (if, on the contrary, we suppose that there is a sequence $(t_n)_{n\in\mathbb{N}} \subseteq \Omega \setminus \overset{\circ}{A}_i$ and a sequence $(u_n)_{n\in\mathbb{N}} \subseteq H_i$ with $\|u_n(t_n)\| > n$, for all $n \in \mathbb{N}$, then $(t_n)_{n\in\mathbb{N}}$ has a subsequence $(t_{k_n})_{n\in\mathbb{N}}$ convergent to $t \in \bar{\Omega} \setminus \overset{\circ}{A}_i$ from where $t \in I_{H_i}(\infty)$; but $I_{H_i}(\infty) \subseteq \overset{\circ}{A}_i$).

Let $k = \max\{k_1, \dots, k_p\}$. For every $u \in H$, there exists $i \in \{1, \dots, p\}$ such that $u \in H_i$; thus $(\|u\| > k) \subseteq (\|u\| > k_i) \subseteq A_i$. □

Remark 3.95.

(i) Let E be a Euclidean space and let $u \in \mathcal{M}(E)$; then, according to the previous theorem, $\{u\}$ is a Jordan finite-tight set if and only if $I_u(\infty) = \{t \in \bar{\Omega} : \limsup_{s\to t} \|u(s)\| = +\infty\}$ is a Jordan-negligible set. We can notice that $I_u(\infty)$ is the set of points in the vicinity of which u is not bounded.

(ii) Let $\mathbb{Q}\cap]0,1[= \{q_0, q_1, \dots, q_n, \dots\}$ be the set of all rational numbers of $]0,1[$ and let $u :]0,1[\to \mathbb{R}, u = \sum_{n=0}^{\infty} n \cdot \chi_{\{q_n\}}$. Then, according to (i) of Proposition 3.80, $H = \{u\}$ is a finite-tight set.

We remark that $I_u(\infty) = [0,1]$. Indeed, for each $t \in [0,1]$ and each $n \in \mathbb{N}^*$, there exists $k_n \geq n$ such that $|q_{k_n} - t| < \frac{1}{n}$; then, $q_{k_n} \to t$ and $u(q_{k_n}) = k_n \to +\infty$.

Therefore, according to the previous remark, H is not a Jordan finite-tight set.

We can notice that $u = \underline{0}$, almost everywhere and that $H_1 = \{\underline{0}\}$ is a Jordan finite-tight set.

This example is a special case of the following remark.

(iii) Let E be a Euclidean space; if the sequence $(u_n)_n \subseteq \mathcal{L}^{\infty}(\Omega, E)$ is bounded, then there exists a Jordan finite-tight sequence $(v_n)_n$ such that, for every $n \in \mathbb{N}$, $u_n = v_n$ almost everywhere.

Indeed, let $k > 0$ such that, for every $n \in \mathbb{N}$, $\|u_n\|_{\infty} \leq k$.

If we define

$$v_n(t) = \begin{cases} u_n(t), & \|u_n\|_E \leq k \\ 0_E, & \|u_n\|_E > k \end{cases},$$

then $(v_n)_n$ is an uniformly bounded sequence and therefore $(v_n)_n$ is a Jordan finite-tight sequence.

Obviously, for every $n \in \mathbb{N}$, $v_n = u_n$ almost everywhere.

Particularly, any application of $\mathcal{L}^{\infty}(\Omega, E)$ coincides almost everywhere with a Jordan finite-tight application.

In conclusion, the property of being Jordan finite-tight can be seen as a property of the sets of measurable functions and not as a property of the sets of classes of functions equal almost everywhere.

We will now present a sufficient condition that a set may be Jordan finite-tight.

Proposition 3.96. *Let $\Omega \subseteq \mathbb{R}^d$ be a measurable bounded set; for every $A \subseteq \Omega$, let $\delta(A) = \sup\{\|t-s\|_{\mathbb{R}^d} : t,s \in A\}$ be the diameter of A.*

Every $H \subseteq \mathcal{M}(E)$ satisfying the condition

$$\forall \varepsilon > 0, \exists K \in \mathcal{K}_E \quad \textit{tel que} \quad \delta\left(u^{-1}(E \setminus K)\right) < \varepsilon, \quad \forall u \in H, \tag{δ}$$

is a Jordan finite-tight set.

Proof. Let $a,b \in \mathbb{R}^d, a = (a_1,\dots,a_d), b = (b_1,\dots,b_d), a < b$ such that $\Omega \subseteq [a,b] = \prod_{i=1}^d [a_i,b_i]$; $\mu([a,b]) = (b_1-a_1)\cdots(b_d-a_d)$.

For every $\varepsilon > 0$, let $n \in \mathbb{N}$ such that

$$\frac{3^d \cdot \mu([a,b])}{n^d} < \varepsilon. \tag{1}$$

Let $\Gamma = \{0,1,\dots,n-1\}^d$. For every $i \in \{1,\dots,d\}$ and every $j \in \{0,1,\dots,n\}$ let us denote $a_i^j = a_i + j \cdot \frac{b_i-a_i}{n}$ and, if $\gamma = (j_1,\dots,j_d) \in \Gamma$, let $I_\gamma = \prod_{i=1}^d [a_i^{j_i}, a_i^{j_i+1}]$.

$\{I_\gamma : \gamma \in \Gamma\}$ is a partition of $[a,b]$; thus $\Omega \subseteq \cup_{\gamma\in\Gamma} I_\gamma$. Furthermore $\mu(I_{\gamma'} \cap I_{\gamma''}) = 0$ if $\gamma' \neq \gamma''$ and $\mu(I_\gamma) = \frac{b_1-a_1}{n} \cdots \frac{b_d-a_d}{n} = \frac{\mu([a,b])}{n^d}$.

According to condition (δ), there exists $K \in \mathcal{K}_E$ such that

$$\delta\left(u^{-1}(E \setminus K)\right) < \varepsilon_1 = \min\left\{\frac{b_i-a_i}{n} : i = 1,\dots,d\right\}. \tag{2}$$

Let $\mathcal{E}_f = \{\prod_{i=1}^d [a_i^{j_i-1}, a_i^{j_i+2}] : (j_1,\dots,j_d) \in \Gamma\}$; then $\mathcal{E}_f$ is a finite subfamily of $\mathcal{E}$ and, according to (1), $\mu(\prod_{i=1}^d [a_i^{j_i-1}, a_i^{j_i+2}]) = \frac{3^d \cdot \mu([a,b])}{n^d} < \varepsilon$, for every interval $\prod_{i=1}^d [a_i^{j_i-1}, a_i^{j_i+2}] \in \mathcal{E}_f$.

For every function $u \in H$ for which $u^{-1}(E \setminus K) \neq \emptyset$, there exists $\gamma = (j_1,\dots,j_d) \in \Gamma$ such that $u^{-1}(E \setminus K) \cap I_\gamma \neq \emptyset$.

Then, according to (2), $u^{-1}(E \setminus K) \subseteq \prod_{i=1}^d [a_i^{j_i-1}, a_i^{j_i+2}] \in \mathcal{E}_f$.

Indeed, if $t^0 = (t_1^0,\dots,t_d^0) \in u^{-1}(E \setminus K) \cap I_\gamma$, then, for every $t = (t_1,\dots,t_d) \in u^{-1}(E \setminus K)$ and every $i = 1,\dots,d$,

$$t_i < t_i^0 + \varepsilon_1 \leq a_i^{j_i+1} + \varepsilon_1 \leq a_i^{j_i+1} + \frac{b_i-a_i}{n} = a_i^{j_i+2} \quad \text{and}$$

$$t_i > t_i^0 - \varepsilon_1 \geq a_i^{j_i} - \varepsilon_1 \geq a_i^{j_i} - \frac{b_i-a_i}{n} = a_i^{j_i-1}. \qquad \square$$

Example 3.97.

(1) Let $\mathbb{Q} \cap]0,1[= \{q_0, q_1, \ldots, q_n, \ldots\}$ and, for every $n \in \mathbb{N}^*$, let $u_n = n^2 \cdot \chi_{]q_n, q_n + \frac{1}{n}[}$. Obviously, $H = \{u_n : n \in \mathbb{N}^*\}$ is a tight set but it is not bounded in $\mathcal{L}^1(]0,1[, \mathbb{R})$. For every $k > 0$ and every $n \in \mathbb{N}^*$,

$$\{t \in]0,1[: |u_n(t)| > k\} = \begin{cases} \emptyset, & n^2 \leq k, \\ ]q_n, q_n + \frac{1}{n}[, & n^2 > k. \end{cases}$$

Then $\delta(|u_n| > k) \leq \frac{1}{\sqrt{k}}$, for every $n \in \mathbb{N}^*$; thus $H = \{u_n : n \in \mathbb{N}^*\}$ satisfies condition (δ) of the previous proposition and then H is a Jordan finite-tight set, so that a finite-tight set.

However, $\sum_{n=1}^{\infty} \mu(|u_n| > k) = \sum_{n \geq \sqrt{k}} \frac{1}{n} = +\infty$; thus H does not satisfy condition (S) of Proposition 3.80; evidently, H does not satisfy condition (M) of the same proposition.

(2) For every $n \in \mathbb{N}^*$, let $u_n : [0,1] \to \mathbb{R}, u_n = n \cdot \chi_{[0,\frac{1}{n}] \cup [1-\frac{1}{n}, 1]}$; then, for every $\varepsilon > 0$, there exists $n_0 \in \mathbb{N}$ such that $\frac{2}{n_0} < \varepsilon$. Let $K = [0, n_0] \in \mathcal{K}_{\mathbb{R}}$ and $A = [0, \frac{1}{n_0}] \cup [1 - \frac{1}{n_0}, 1] \in \mathcal{E}$; then $\mu(A) < \varepsilon$.

For every $n \in \mathbb{N}^*$,

$$u_n^{-1}(\mathbb{R} \setminus K) = u_n^{-1}(]\, n_0, +\infty[= \begin{cases} \emptyset, & n \leq n_0, \\ [0, \frac{1}{n}] \cup [1 - \frac{1}{n}, 1], & n > n_0 \end{cases} \subseteq A.$$

Thus $H = \{u_n : n \in \mathbb{N}^*\}$ is a Jordan finite-tight set. However, for every $K \in \mathcal{K}_{\mathbb{R}}$ and $n_1 \in \mathbb{N}$ with $K \subseteq [-n_1, n_1]$,

$$\delta\left(u_n^{-1}(\mathbb{R} \setminus K)\right) = 1, \quad \text{for every} \quad n > n_1.$$

The main result of this section states that a tight set $H \subseteq W^{1,1}(\Omega, \mathbb{R}^m)$, for which there exists a Jordan finite-tight set of gradients ∇H, is necessarily relatively compact in measure.

At first, let us recall and clarify some notions and results.

For every $k \in \mathbb{N}$, let us note by $C^k(\Omega)$ the set of all functions f, k-times differentiable and with $f^{(k)}$ continuous on Ω; let $C^\infty(\Omega) = \bigcap_{k=0}^{\infty} C^k(\Omega)$ and let $C_c^\infty(\Omega)$ be the subset of all functions $\phi \in C^\infty(\Omega)$ with compact support.

Definition 3.98. Let $\Omega \subseteq \mathbb{R}^d$ be a bounded open set. $v = (v_j^i)_{\substack{1 \leq i \leq m \\ 1 \leq j \leq d}} \in \mathcal{L}^1(\mu, \mathbb{R}^{md})$ is the gradient of $u = (u_i)_{1 \leq i \leq m} \in \mathcal{L}^1(\mu, \mathbb{R}^m)$ if, for every

$i \in \{1, \dots, m\}$, $j \in \{1, \dots, d\}$ and every application $\phi \in C_c^\infty(\Omega)$,

$$\int_\Omega u_i(t) \cdot \frac{\partial \phi}{\partial t_j}(t) d\mu(t) = -\int_\Omega v_j^i(t) \cdot \phi(t) d\mu(t).$$

If v' and v'' are two gradients of u, then $v' = v''$ almost everywhere.

We will note the gradient of u by ∇u and then

$$\nabla u = (\nabla u_i)_{1 \le i \le m} = (\nabla_j u_i)_{\substack{1 \le i \le m \\ 1 \le j \le d}} = (v_j^i)_{\substack{1 \le i \le m \\ 1 \le j \le d}}.$$

For every $p \geq 1$, the Sobolev space $W^{1,p}(\Omega, \mathbb{R}^m)$ consists of all mappings $u \in \mathcal{L}^p(\Omega, \mathbb{R}^m)$ with $\nabla u \in \mathcal{L}^p(\Omega, \mathbb{R}^{md})$.

For every $H \subseteq W^{1,p}(\Omega, \mathbb{R}^m)$ we write $\nabla H = \{\nabla u : u \in H\}$ and we say that ∇H is a gradient of H.

Definition 3.99. For every mapping $u : \Omega \to \mathbb{R}^m$, let us denote

$$L(u, \Omega) = \sup\left\{\frac{\|u(t) - u(s)\|}{\|t - s\|} : t, s \in \Omega, t \neq s\right\};$$

u is a Lipschitz mapping on Ω if and only if $L(u, \Omega) < +\infty$.

We recall that, if $\Omega' \subseteq \mathbb{R}^d$ is a bounded open convex set and if $u \in \mathcal{L}^1(\Omega', \mathbb{R}^m)$ is a continuous mapping with $\nabla u = (\nabla_j u_i)_{\substack{1 \le i \le m \\ 1 \le j \le d}} \in \mathcal{L}^\infty(\Omega', R^{md})$, then u is a Lipschitz application on Ω' and

$$L(u, \Omega') \le \left(\sum_{i=1}^m \sum_{j=1}^d \|\nabla_j u_i\|^2_{\mathcal{L}^\infty(\Omega', \mathbb{R})}\right)^{\frac{1}{2}} < +\infty.$$

Proposition 3.100 (Proposition 3.10 of [81]). *Let $\Omega \subseteq \mathbb{R}^d$ be a bounded open convex set and let $H \subseteq W^{1,1}(\Omega, \mathbb{R}^m) \cap C(\Omega, \mathbb{R}^m)$ be a tight set. If, for every $u \in H$, there exists a gradient ∇u such that $\nabla H = \{\nabla u : u \in H\} \subseteq \mathcal{L}^1(\mu, \mathbb{R}^{md})$ is a Jordan finite-tight set, then H is also a Jordan finite-tight set.*

Proof. Let us suppose that, for every $u \in H$, there exists a gradient ∇u such that $\nabla H = \{\nabla u : u \in H\}$ is a Jordan finite-tight set; we notice that, for any other gradient v_u of u, $v_u = \nabla u$ almost everywhere, but, according to (ii) of Remark 3.95, we cannot assert that the set $\{v_u : u \in H\}$ is also a Jordan finite-tight set.

For every $\varepsilon > 0$, let $k > 0$ and let $\mathcal{E}_f = \{A_1, \dots, A_p\} \subseteq \mathcal{E}$ be a finite family of elementary sets, with $\mu(A_i) < \varepsilon$, for every $i \in \{1, \dots, p\}$, such that, for all $u \in H$, there exists $i \in \{1, \dots, p\}$ with $\{t \in \Omega : \|\nabla u(t)\| > k\} \subseteq A_i$.

For any $i \in \{1, \dots, p\}$, we can find a finite family of open convex sets $\{\Omega^i_j : j = 1, \dots, p_i\}$ such that $\Omega \setminus A_i = \bigcup_{j=1}^{p_i} \Omega^i_j$; let $\delta = \min\{\mu(\Omega^i_j) : 1 \le i \le p, 1 \le j \le p_i\} > 0$.

Since H is tight, there exists $k_1 > 0$ such that

$$\mu(\|u\| > k_1) < \delta, \quad \text{for every} \quad u \in H. \tag{1}$$

Fix an arbitrary $u \in H$; then $u \in \mathcal{L}^1(\mu, \mathbb{R}^m) \cap C(\Omega, \mathbb{R}^m)$.

Let $i \in \{1, \dots, p\}$ such that $(\|\nabla u\| > k) \subseteq A_i$; then, for any $j \in \{1, \dots, p_i\}$ and any $t \in \Omega^i_j$, $\|\nabla u(t)\| \le k$ from where $\nabla u \in \mathcal{L}^\infty(\Omega^i_j, \mathbb{R}^{md})$. Therefore u is a Lipschitz function on Ω^i_j and

$$L(u, \Omega^i_j) \le \left(\sum_{a=1}^{m} \sum_{b=1}^{d} \|\nabla_b u_a\|^2_{\mathcal{L}^\infty(\Omega^i_j, \mathbb{R})} \right)^{\frac{1}{2}} \le \sqrt{md} \cdot k. \tag{2}$$

Since $\mu(\Omega^i_j) \ge \delta$, by (1), $\Omega^i_j \not\subseteq (\|u\| > k_1)$; then there exists $t_0 \in \Omega^i_j$ with

$$\|u(t_0)\| \le k_1. \tag{3}$$

According to (2) and (3), for every $t \in \Omega^i_j$,

$$\begin{aligned} \|u(t)\| &\le \|u(t) - u(t_0)\| + \|u(t_0)\| \le \sqrt{md} \cdot k \cdot \|t - t_0\| + k_1 \\ &\le \sqrt{md} \cdot k \cdot \operatorname{diam}(\Omega) + k_1 \equiv k_2. \end{aligned}$$

Since j is arbitrary in $\{1, \dots, p_i\}$,

$$\|u(t)\| \le k_2, \quad \text{for every} \quad t \in \bigcup_{j=1}^{p_i} \Omega^i_j. \tag{4}$$

By (4), $(\|u\| > k_2) \subseteq A_i$ and therefore H is a Jordan finite-tight set. □

Remark 3.101.

(i) According to the demonstration of previous proposition, if H is a tight subset of $W^{1,1}(\Omega, \mathbb{R}^m)$, for which ∇H Jordan finite-tight, then, for every $\varepsilon > 0$, there exist $k > 0$ and a finite family of elementary sets $\mathcal{E}_f = \{A_1, \dots, A_p\}$ with $\mu(A_i) < \varepsilon$ for any $i \in \{1, \dots, p\}$, such that, for every $u \in H$, there exists $i \in \{1, \dots, p\}$ with

$$\{t \in \Omega : \|u(t)\| > k\} \subseteq A_i \quad \text{and} \quad \{t \in \Omega : \|\nabla u(t)\| > k\} \subseteq A_i.$$

(ii) We cannot remove the condition of continuity for the applications of H in the previous proposition. Indeed, if $H = \{u\} \subseteq W^{1,1}(]0,1[,\mathbb{R})$, where $u = \sum_{n=0}^{\infty} n \cdot \chi_{\{q_n\}}$ is the function of Remark 3.95(ii), then H is tight and $\nabla H = \{\underline{0}\}$ is Jordan finite-tight but H is not a Jordan finite-tight set.

(iii) In the previous proposition, the condition of being tight for H is a necessary condition. Indeed, if, for every $n \in \mathbb{N}$, $u_n = \underline{n}$ is the constant function n on $]0,1[$, then $H = \{u_n : n \in \mathbb{N}\} \subseteq W^{1,1}(]0,1[,\mathbb{R}) \cap C(]0,1[,\mathbb{R})$ and $\nabla H = \{\underline{0}\}$ is a Jordan finite-tight set but H is not even a tight set.

Theorem 3.102 (Theorem 3.12 of [81]). *Let $\Omega \subseteq \mathbb{R}^d$ be a bounded open convex set and let $H \subseteq W^{1,1}(\Omega,\mathbb{R}^m) \cap C(\Omega,\mathbb{R}^m)$ be a tight set such that there exists a Jordan finite-tight gradient ∇H; then H is relatively compact in the topolgy of convergence in measure on $\mathcal{M}(\mathbb{R}^m)$.*

Proof. According to the previous proposition and to (i) of Remark 3.101, for every $\varepsilon > 0$, there exist $k > 0$ and a finite family of d-dimensional intervals $\mathcal{I} = \{I_1,\dots,I_p\}$ such that, for every $u \in H$, we can find a subfamily $\mathcal{I}_u \subseteq \mathcal{I}$ with $\mu(\cup\mathcal{I}_u) < \varepsilon$ and

$$(\|u\| > k) \subseteq \cup\mathcal{I}_u, \tag{1}$$

$$(\|\nabla u\| > k) \subseteq \cup\mathcal{I}_u, \tag{2}$$

where $\cup\mathcal{I}_u$ is the union of the intervals of the subfamily $\mathcal{I}_u \subseteq \mathcal{I}$.

For every $i \in \{1,\dots,p\}$, let $I_i = \prod_{j=1}^d [a_j^i, b_j^i]$ and let $q \in \mathbb{N}$ be such that

$$d \cdot k \cdot \sqrt{m} < q \cdot \varepsilon. \tag{3}$$

Let us show that H satisfies the conditions of Fréchet theorem of compactness in measure (see Theorem IV.11.1 of [62]).

Let $a_1,\dots,a_d,b_1,\dots,b_d \in \mathbb{Z}$ be such that $\Omega \subseteq \prod_{j=1}^d [\frac{a_j}{q}, \frac{b_j}{q}]$. For all $j \in \{1,\dots,d\}$, we note $t_j^1,\dots,t_j^{r_j}$ the elements of the set

$$\left\{\frac{a_j}{q}, \frac{a_j+1}{q}, \dots, \frac{b_j-1}{q}, \frac{b_j}{q}, a_j^1,\dots,a_j^d, b_j^1,\dots,b_j^d\right\}$$

so that $t_j^1 < t_j^2 < \dots < t_j^{r_j}$.

Let $N = \{1,\dots,r_1-1\} \times \dots \times \{1,\dots,r_d-1\} \subseteq \mathbb{N}^d$ and, for every $n = (n_1,\dots,n_d) \in N$, let $A_n = [t_1^{n_1}, t_1^{n_1+1}[\times \dots \times [t_d^{n_d}, t_d^{n_d+1}[$; then

$$\{\Omega \cap A_n : n \in N\} \quad \text{is a partition of } \Omega. \tag{4}$$

For every $u \in H$, let $F_u = (\cup \mathcal{I}_u) \cup (\cup_{n \in N} \partial(A_n))$, where $\partial(A_n)$ is the boundary of A_n; then

$$\mu(F_u) = \mu(\cup \mathcal{I}_u) < \varepsilon, \quad \text{for every} \quad u \in H \tag{5}$$

and, according to (1)

$$\sup_{t \in \Omega \setminus F_u} \|u(t)\| \leq k, \quad \text{for every} \quad u \in H. \tag{6}$$

We can notice that, for every $n \in N$ and every $u \in H$, $(\Omega \cap A_n) \setminus F_u$ is either empty or equal to the open convex set $\Omega \cap \mathring{A}_n$.

Therefore, according to (2), $\nabla u \in \mathcal{L}^\infty((\Omega \cap A_n) \setminus F_u, \mathbb{R}^{md})$, u is a Lipschitz function on $(\Omega \cap A_n) \setminus F_u$ and

$$L(u, (\Omega \cap A_n) \setminus F_u) \leq \sqrt{md} \cdot k, \quad \text{for every } n \in N \quad \text{and every} \quad u \in H. \tag{7}$$

Thus, for every $n \in N$, every $u \in H$ and every $t, s \in (\Omega \cap A_n) \setminus F_u$,

$$\begin{aligned} \|u(t) - u(s)\| &\leq \sqrt{md} \cdot k \cdot \|t - s\| \leq \sqrt{md} \cdot k \cdot \operatorname{diam}(A_n) \\ &\leq \sqrt{md} \cdot k \cdot \frac{\sqrt{d}}{q} = \frac{\sqrt{m} \cdot dk}{q} \end{aligned}$$

and, according to (3),

$$\sup_{t,s \in (\Omega \cap A_n) \setminus F_u} \|u(t) - u(s)\| < \varepsilon. \tag{8}$$

The result comes from (4), (5), (6) and (8) by applying Theorem IV.11.1 of [62]. □

Under certain conditions, we can avoid the continuity hypothesis in the previous theorem.

Corollary 3.103. *Let $\Omega \subseteq \mathbb{R}^d$ be a bounded open convex set, let $p \geq 1$ and let H be a tight subset of $W^{1,p}(\Omega, \mathbb{R}^m)$ such that ∇H is a Jordan finite-tight set; if either $p > d$ or $d = 1$ is fulfilled, then H is relatively compact in measure.*

Proof. In both cases, $p > d$ or $d = 1$, for every $u \in H$, there exists $\bar{u} \in W^{1,p}(\Omega, \mathbb{R}^m) \cap C(\Omega, \mathbb{R}^m)$ such that $u = \bar{u}$ almost everywhere. Then $\bar{H} = \{\bar{u} : u \in H\} \subseteq W^{1,1}(\Omega, \mathbb{R}^m) \cap C(\Omega, \mathbb{R}^m)$ is a tight set for which there exists a Jordan finite-tight gradient $\nabla \bar{H} = \nabla H$. Consequently, $\bar{H}$, and then H, is relatively compact in measure. □

Remark 3.104. Let $\Omega \subseteq \mathbb{R}^d$ be a bounded open convex set and let $H \subseteq W^{1,1}(\Omega,\mathbb{R}^m) \cap C(\Omega,\mathbb{R}^m)$ be tight such that $\nabla H \subseteq (\mathcal{L}^\infty(\Omega,\mathbb{R}^m), \|\cdot\|_\infty)$ is bounded. According to (iii) of Remark 3.95, for every sequence $(u_n)_n \subseteq H$, there exists $(v_n)_n \subseteq \mathcal{L}^\infty(\Omega,\mathbb{R}^m)$ such that, for every $n \in \mathbb{N}$, $\nabla u_n = v_n$ almost everywhere and $\{v_n : n \in \mathbb{N}\}$ is a Jordan finite-tight set. Then, according to previous theorem, H is relatively compact in measure. But, in this case, we can obtain a stronger result. Following the Arzelà–Ascoli theorem, H is relatively compact in the topology of uniform convergence on $C(\Omega,\mathbb{R}^m)$. In fact, H is uniformly Lipschitz and therefore it is uniformly bounded and uniformly equicontinuous.

Under some relaxed conditions, we can obtain an alternative of the Rellich–Kondrachov theorem (see Theorem 12.12 in [50]).

Corollary 3.105. *Let $\Omega \subseteq \mathbb{R}^d$ be a bounded open convex set and let $(u_n)_{n\in\mathbb{N}} \subseteq W^{1,1}(\Omega,\mathbb{R}^m) \cap C(\Omega,\mathbb{R}^m)$ be a uniformly integrable sequence with the sequence of gradients, $(\nabla u_n)_{n\in\mathbb{N}}$, Jordan finite-tight; then, up to a subsequence, $(u_n)_{n\in\mathbb{N}}$ converges in $\mathcal{L}^1(\mu,\mathbb{R}^m)$.*

Proof. $(u_n)_n$ being uniformly integrable it is bounded in $\mathcal{L}^1(\mu,\mathbb{R}^m)$ and so $(u_n)_{n\in\mathbb{N}}$ is a tight sequence for which $(\nabla u_n)_{n\in\mathbb{N}}$ is a Jordan finite-tight sequence. Therefore $(u_n)_{n\in\mathbb{N}}$ has a subsequence $(u_{k_n})_{n\in\mathbb{N}}$ convergent in measure to an application u. Since $(u_n)_{n\in\mathbb{N}}$ is uniformly integrable, $u \in \mathcal{L}^1(\mu,\mathbb{R}^m)$ and $u_{k_n} \xrightarrow{\mathcal{L}^1} u$. □

The condition of being Jordan finite-tight offers a very good background for the application of the fiber product lemma; we can demonstrate it in the following corollary.

Corollary 3.106. *Let $\Omega \subseteq \mathbb{R}^d$ be a bounded open convex set and let $H \subseteq W^{1,1}(\Omega,\mathbb{R}^m) \cap C(\Omega,\mathbb{R}^m)$ be a tight set such that ∇H is a Jordan finite-tight set; then, for every sequence $(u_n)_n \subseteq \mathcal{M}(\mathbb{R}^m)$, there exist a subsequence $(u_{k_n})_n$ of $(u_n)_n$, a mapping $u \in \mathcal{M}(\mathbb{R}^m)$ and a Young measure $\tau \equiv \tau_. \in \mathcal{Y}(\mathbb{R}^{md})$ such that:*

(i) *$(u_{k_n})_n$ is convergent in measure to u,*

(ii) $\nabla u_{k_n} \xrightarrow{\mathcal{S}} \tau$,

(iii) $\nabla u_{k_n} \xrightarrow{w^2} \text{bar}\, \tau_.$.

(iv) *For every bounded l.s.c. integrand, $\Psi : \Omega \times \mathbb{R}^m \times \mathbb{R}^{md} \to \mathbb{R}$,*

$$\int_{\Omega\times\mathbb{R}^{md}} \Psi(t,u(t),y)d\tau(t,y) \leq \liminf_n \int_\Omega \Psi\left(t,u_{k_n}(t),\nabla u_{k_n}(t)\right) d\mu(t).$$

Proof. According to Theorem 3.102, H is relatively compact in measure in $\mathcal{M}(\mathbb{R}^m)$ and, by 3.85, ∇H is sequentially $\mathcal{S}$-compact in $\mathcal{Y}(\mathbb{R}^{md})$. Therefore, for every $(u_n)_n \subseteq \mathcal{M}(\mathbb{R}^m)$, there exist a subsequence $(u_{k_n})_n$ of $(u_n)_n$, a mapping $u \in \mathcal{M}(\mathbb{R}^m)$ and a Young measure $\tau \equiv \tau_. \in \mathcal{Y}(\mathbb{R}^{md})$ such that (i), (ii) and (iii) are accomplished. According to fiber product lemma (Theorem 3.87),

$$\left(u_{k_n}, \nabla u_{k_n}\right) \xrightarrow{\mathcal{S}} \delta_{u(\cdot)} \otimes \tau.$$

which proves (iv). □

Such a result can be used in order to find relaxed solution for non-convex problems in the calculus of variations for which $(u_n)_n$ is a minimizing sequence.

3.9 Strong Compactness in $L^p(\mu, E)$

In this paragraph, we will use the tool which is offered by the Young measures in order to study the strongly compact sets in L^1.

Two main results are the Visintin–Balder theorem and the Rossi–Savaré theorem which may have, as corollaries, the Lions–Aubin and Gutman theorems.

3.9.1 Visintin–Balder's Theorem

This theorem presents the sufficient conditions in which weak convergence leads to strong convergence. A first condition was given by Visintin in [171] without using Young measures. In [11], E. Balder demonstrated the same result in a more general hypothesis.

In [169], we can find a detailed presentation of these results.

Firstly, let us recall some definitions.

Let $A \subseteq \mathbb{R}^d$; the closed convex hull of A—noted $\overline{\text{co}}\ A$—is the intersection of all closed convex sets containing A.

Let $C \subseteq \mathbb{R}^d$ be a convex set. We say that a point $x \in C$ is an extremal point of C if $(x = (1-\lambda)y + \lambda z \quad \text{with} \quad \lambda \in]0,1[\quad \text{and} \quad y,z \in C) \Longrightarrow (y = z = x)$.

Finally, let us recall that, for every sequence $(u_n)_n \subseteq \mathcal{M}(\mathbb{R}^d)$ and every $t \in \Omega$, $\text{Ls}_n\{u_n(t)\} = \bigcap_{p=1}^{\infty} \overline{\{u_k(t) : k \geq p\}}$ (the Kuratowski upper limit).

Proposition 3.107 (Proposition C of [5]). *If the sequence $(u_n)_n$ is weakly convergent in $L^1(\mu,\mathbb{R}^d)$ to $u \in L^1(\mu,\mathbb{R}^d)$, then*

$$u(t) \in \overline{\text{co}}\ (Ls_n\{u_n(t)\}), \qquad \textit{for almost every} \quad t \in \Omega.$$

Proof. Since $(u_n)_n$ is weakly convergent, according to Dunford–Pettis theorem (Theorem 1.65), the set $\{u_n : n \in \mathbb{N}\}$ is bounded in $L^1(\mu, \mathbb{R}^d)$ and, according to Proposition 3.56, it is tight. By Theorem 3.64, there exists a subsequence $(u_{k_n})_n$ of $(u_n)_n$ $\mathcal{S}$-convergent to a Young measure $\tau \in \mathcal{Y}(\mathbb{R}^d)$; let $\{\tau_t : t \in \Omega\}$ be the disintegration of τ. According to Proposition 3.37, for every $t \in \Omega, u(t) = \int_{\mathbb{R}^d} x d\tau_t(x)$.

Let now the mapping $\mathrm{id}_{\mathbb{R}^d} : \mathbb{R}^d \to \mathbb{R}^d, \mathrm{id}_{\mathbb{R}^d}(x) = x$, for every $x \in \mathbb{R}^d$; then, for every $t \in \Omega$, $u(t) \in \overline{\mathrm{co}}\,(\mathrm{id}_{\mathbb{R}^d}(\mathrm{supp}\tau_t)) = \overline{\mathrm{co}}\,(\mathrm{supp}\tau_t)$.

The result follows from Proposition 3.43:

$$\mathrm{supp}\tau_t \subseteq \mathrm{Ls}_n\{u_{k_n}(t)\} \subseteq \mathrm{Ls}_n\{u_n(t)\}, \quad \text{almost for every} \quad t \in \Omega. \qquad \square$$

Remark 3.108. Let $(u_n)_n \subseteq L^1(\mu, \mathbb{R}^d)$ and $u \in L^1(\mu, \mathbb{R}^d)$. According to Proposition 3.43, if $(u_n)_n$ is stably convergent to $\tau \in \mathcal{Y}(\mathbb{R}^d)$, then, almost for every $t \in \Omega$, $\mathrm{supp}\tau_t \subseteq \mathrm{Ls}\{u_n(t)\}$ and, according to previous theorem, if $(u_n)_n$ is weakly convergent to u, $u(t) \in \overline{\mathrm{co}}\,\mathrm{Ls}_n\{u_n(t)\}$, almost for every $t \in \Omega$.

Generally, $u(t) \notin \mathrm{Ls}_n\{u_n(t)\}$. Indeed, in Example 3.44, we have shown that the Rademacher's sequence $(r_n)_n \subseteq L^1([0,1], \mathbb{R})$ is stably convergent to $\tau = \mu \otimes (\frac{1}{2}(\delta_1 + \delta_{-1}))$ and weakly convergent to $\underline{0}$. For every $t \in [0,1], 0 \notin \mathrm{Ls}_n\{r_n(t)\} = \{-1, 1\}$, but $0 \in \overline{\mathrm{co}}(\mathrm{Ls}_n\{r_n(t)\}) = [-1, 1]$.

Lemma 3.109 (Lemma 6.20 of [15]). *Let E be a separable Banach space and let $\nu \in \mathcal{P}_E$ (ν is a probability on E) such that there exists $x_0 = \mathrm{bar}\,\nu = \int_E x d\nu(x) \in \overline{\mathrm{co}}\,(\mathrm{supp}\nu) \subseteq E$.*

If x_0 is an extremal point of $\overline{\mathrm{co}}\,(\mathrm{supp}\nu)$, then $\nu = \delta_{x_0}$ (ν is the Dirac measure—the unit mass concentrated in x_0).

Proof. Let B be a closed convex subset of E such that $x_0 \notin B$; then $\mathrm{supp}\,\nu \nsubseteq B$ and therefore $\nu(B) < 1$.

If we suppose that $\nu(B) > 0$, we can define $\nu_1, \nu_2 \in \mathcal{P}_E$ letting

$$\nu_1(A) = \frac{1}{\nu(B)} \cdot \nu(A \cap B) \text{ and } \nu_2(A) = \frac{1}{1 - \nu(B)} \cdot \nu(A \setminus B), \text{ for every } A \in \mathcal{B}_E.$$

Then $\nu = \nu(B) \cdot \nu_1 + [1 - \nu(B)] \cdot \nu_2$ and therefore

$$x_0 = \int_E x d\nu = \nu(B) \cdot \mathrm{bar}\,\nu_1 + [1 - \nu(B)] \cdot \mathrm{bar}\,\nu_2.$$

Since $\mathrm{bar}\,\nu_1 \in \overline{\mathrm{co}}\,(\mathrm{supp}\,\nu_1) \subseteq \overline{\mathrm{co}}\,(\mathrm{supp}\,\nu)$, $\mathrm{bar}\,\nu_2 \in \overline{\mathrm{co}}\,(\mathrm{supp}\,\nu_2) \subseteq \overline{\mathrm{co}}\,(\mathrm{supp}\,\nu)$ and x_0 is an extremal point of $\overline{\mathrm{co}}\,(\mathrm{supp}\,\nu)$ it results that $x_0 = \mathrm{bar}\,\nu_1 = \mathrm{bar}\,\nu_2$.

Then $x_0 \in \overline{\mathrm{co}}(\mathrm{supp}\nu_1) \subseteq \overline{\mathrm{co}}(B) = B$, who contradicts the hypothesis.

Therefore, for every closed convex set $B \subseteq E$ for which $x_0 \notin B$, $\nu(B) = 0$.

Let $S \subseteq E$ be an open ball of radius r centered at x; then $S = \bigcup_{n>\frac{1}{r}} B_n$, where, for all $n \in \mathbb{N}^*$, B_n is the closed ball of radius $r - \frac{1}{n}$ centered at x. If $x_0 \notin S, x_0 \notin B_n$, for every $n > \frac{1}{r}$, therefore $\nu(B_n) = 0$ and then $\nu(S) = 0$.

Since every open subset D of E is a countable union of open balls, we have $\nu(D) = 0$, for every open set D for which $x_0 \notin D$. It follows that $\nu(E \backslash \{x_0\}) = 0$ and then $\nu = \delta_{x_0}$. □

We can now give the Visintin–Balder theorem.

Theorem 3.110 (Theorem 10 of [169]). *Let $(u_n)_n \subseteq L^1(\mu, \mathbb{R}^d)$ be a sequence weakly convergent to $u \in L^1(\mu, \mathbb{R}^d)$ and let $\mathcal{E}_{cc}$ be the family of all closed convex subsets of $\mathbb{R}^d$. Let us consider the following conditions:*

(V) *There exists a set-valued mapping $\Gamma : \Omega \to \mathcal{E}_{cc}$ such that, for every $n \in \mathbb{N}$ and almost for every $t \in \Omega$, $u_n(t) \in \Gamma(t)$ and $u(t) \in \Gamma(t)$ is an extremal point of $\Gamma(t)$.*

(B) *$u(t)$ is an extremal point of $\overline{\text{co}}(\text{Ls}_n\{u_n(t)\})$, almost for every $t \in \Omega$.*

(C) *$(u_n)_n$ is strongly convergent to u in $L^1(\mu, \mathbb{R}^d)$.*

Then (V) $\Longrightarrow$ (B) $\Longrightarrow$ (C).

Proof. (V) $\Longrightarrow$ (B): Since $\text{Ls}_n\{u_n(t)\} = \bigcap_{p=1}^{\infty} \overline{\{u_n(t) : n \geq p\}} \subseteq \Gamma(t)$, for almost every $t \in \Omega$ and since $\Gamma(t)$ is a closed convex set,

$$\overline{\text{co}}\,(\text{Ls}_n\{u_n(t)\} \subseteq \Gamma(t), \quad \text{almost for every} \quad t \in \Omega. \tag{1}$$

Since $u_n \xrightarrow[L^1]{w} u$, according to Proposition 3.107,

$$u(t) \in \overline{\text{co}}\,(\text{Ls}_n\{u_n(t)\}), \quad \text{almost for every} \quad t \in \Omega. \tag{2}$$

By (1) and (2), because $u(t)$ is an extremal point of $\Gamma(t)$ it is an extremal point of $\overline{\text{co}}\,(\text{Ls}_n\{u_n(t)\})$, all the more.

(B) $\Longrightarrow$ (C): Since $(u_n)_n$ is weakly convergent it is tight and, according to Theorem 3.64, every subsequence $(u'_n)_n$ of $(u_n)_n$ has a subsequence $(u''_n)_n$ $\mathcal{S}$-convergent to a Young measure $\mathcal{T} \in \mathcal{Y}(\mathbb{R}^d)$; let $\tau_.$ be the disintegration of $\mathcal{T}$. According to Proposition 3.37, $u(t) = \text{bar}\ \tau_t = \int_{\mathbb{R}^d} x d\tau_t(x)$, for every $t \in \Omega$. According to Proposition 3.43, $\text{supp}\ \tau_t \subseteq \text{Ls}_n\{u''_n(t)\}$.

In accordance with hypothesis (B), almost for every $t \in \Omega$, $u(t) = \text{bar}\ \tau_t$ is an extremal point of $\overline{\text{co}}\,(\text{Ls}_n\{u_n(t)\}$ and, since

$$\text{bar}\ \tau_t \in \overline{\text{co}}\,(\text{supp}\ \tau_t) \subseteq \overline{\text{co}}\,(\text{Ls}_n\{u''_n(t)\}) \subseteq \overline{\text{co}}\,(\text{Ls}_n\{u_n(t)\}),$$

bar τ_t is an extremal point of $\overline{\text{co}}$ (supp τ_t), almost for every $t \in \Omega$. From Lemma 3.109 it follows that $\tau_t = \delta_{u(t)}$, almost for every $t \in \Omega$; therefore $\tau = \tau^u$ is the Young measure associated to the mapping $u \in L^1(\mu, \mathbb{R}^d)$ and, according to Proposition 3.40, $(u_n'')_n$ is convergent in measure to u.

According to Vitali theorem (Theorem 1.54), $(u_n'')_n$ is strongly convergent to u.

Therefore, every subsequence $(u_n')_n$ of $(u_n)_n$ has a subsequence $(u_n'')_n$ strongly convergent to u; then $u_n \xrightarrow{\|\cdot\|_1} u$. □

Remark 3.111.

(i) Condition (V) is introduced by Visintin in [171] where it is presented in the equivalent form: (V′) $u(t)$ is an extremal point of $\overline{\text{co}}\ \{\{u(t)\} \cup \{u_n(t) : n \in \mathbb{N}\}\}$, almost for every $t \in \Omega$.

(ii) The following example (Example 1 of Section 3 of [169]) shows that the Balder's condition (B) (and also Visintin's condition (V) much less) is not necessary for the strong convergence.

For every $m \in \mathbb{N}^*$ and every $k = 1, \ldots, m$, let $v_n : [0, 1] \to \mathbb{R}$,

$$v_n = \chi_{\left[\frac{k-1}{m}, \frac{k}{m}\right]}, \quad \text{if} \quad n = \frac{m(m-1)}{2} + k.$$

Let $u_n : [0, 1] \to \mathbb{R}, u_{2n} = v_n$, and $u_{2n+1} = -v_n$. Then $(u_n)_n$ is strongly convergent to $\underline{0}$ but $\underline{0}$ is not an extremal point of $\overline{\text{co}}\ (\text{Ls}_n\{u_n(t)\}) = [-1, 1]$.

(iii) In [171], one shows that the previous theorem cannot be generalized by replacing $\mathbb{R}^d$ by a infinite-dimensional normed space.

3.9.2 Rossi–Savaré's Theorem

In this subsection, $\Omega =]0, T[\subseteq \mathbb{R}$ is provided with the σ-algebra $\mathcal{A}$ of Lebesgue measurable sets and with the Lebesgue measure $\mu = dt$.

Let $(E, \|\cdot\|)$ be a separable Banach space and let $1 \leq p < +\infty$. We recall that $L^p(\mu, E) = \{u \in \mathcal{M}(E) : \int_0^T \|u\|_E^p dt < +\infty\}$ is a Banach space with respect to the norm

$$\|\cdot\|_p : L^p(\mu, E) \to \mathbb{R}_+, \|u\|_p = \left(\int_0^T \|u\|_E^p dt\right)^{\frac{1}{p}}.$$

A sequence $(u_n)_n \subseteq L^p(\mu, E)$ is strongly convergent to $u \in L^p(\mu, E)$ if $\|u_n - u\|_p \to 0$; the strong convergence is noted by $u_n \xrightarrow{\|\cdot\|_p} u$.

A sequence $(u_n)_n \subseteq \mathcal{M}(E)$ is convergent in measure to $u \in \mathcal{M}(E)$ if $\mu(\{t \in]0, T[: \|u_n(t) - u(t)\|_E \geq \varepsilon\}) \to 0$, for every $\varepsilon > 0$; the convergence in measure of the sequence $(u_n)_n$ to u with respect to μ is noted by $u_n \overset{\mu}{\to} u$.

We introduce an important condition in the study of the strong convergence in $L^p(\mu, E)$—the strong concentration condition.

Definition 3.112. A set $U \subseteq L^p(\mu, E)$ satisfies to *the strong concentration condition* if

$$\limsup_{h \downarrow 0} \sup_{u \in U} \int_0^{T-h} \|u(t+h) - u(t)\|_E^p \, dt = 0. \tag{CF}$$

In the Euclidean case, this condition is used in order to characterize the strong compactness in $L^p(\mu, E)$ (see Theorem IV.26 of [36]):

Theorem 3.113 (Riesz–Fréchet–Kolmogorov theorem). *Let $1 \leq p < +\infty$ and let $U \subseteq L^p(\mu, \mathbb{R}^m)$ be a bounded set; U is relatively strongly compact in $L^p(\mu, \mathbb{R}^m)$ if and only if U satisfies to the strong concentration condition.*

If $U \subseteq W^{1,p}(\Omega, \mathbb{R}^m)$, then a sufficient condition for (CF) is that ∇U to be bounded in $L^\infty(\mu, \mathbb{R}^m)$; we obtain the following corollary:

Corollary 3.114. *Let $U \subseteq W^{1,p}(\Omega, \mathbb{R}^m)$ be a bounded set in $W^{1,p}$; if $\nabla U = \{\nabla u : u \in U\}$ is bounded in $L^\infty(\mu, \mathbb{R}^m)$, then U is relatively strongly compact in $L^p(\mu, \mathbb{R}^m)$.*

Remark 3.115. In the infinite dimensional case, the condition (CF) is not sufficient for the strong compactness. Indeed, let $E = C_{[0,1]}$ be the space of all real-valued continuous functions on [0, 1] provided with the norm of uniform convergence and let, for every $n \in \mathbb{N}$, $f_n :]0,1[\to \mathbb{R}$, $f_n(x) = x^n$. For every $n \in \mathbb{N}$, let $u_n :]0, T[\to C_{[0,1]}$ be the constant mapping $u_n(t) = f_n$, for every $t \in]0, T[$ and let $U = \{u_n : n \in \mathbb{N}\}$.

For every $n \in \mathbb{N}$,

$$\|u_n\|_\infty = \inf\{c > 0 : \|u_n(t)\|_E \leq c, \quad \text{for almost every} \quad t \in]0, T[\,\} = 1$$

and then $U \subseteq L^\infty(\mu, C_{[0,1]}) \subseteq L^1(\mu, C_{[0,1]})$.

Obviously, U satisfies to condition (CF) but U is not relatively strongly compact in $L^1(\mu, C_{[0,1]})$ because $\|u_n - u_m\|_1 = \int_0^T \|u_n(t) - u_m(t)\|_{C_{[0,1]}} dt = \int_0^T \|f_n - f_m\|_{C_{[0,1]}} dt = T \cdot (\frac{m}{n})^{\frac{m}{n-m}} \cdot \frac{n-m}{n} \xrightarrow[n \to \infty]{} 1$ and then $(u_n)_n$ does not have strongly convergent subsequences.

In the infinite dimensional case, in order to obtain the strong compacity, we need to add to the strong concentration condition the condition of being tight.

We have already noticed in 3.57 that the set U defined above is not tight.

Definition 3.116. A set $U \subseteq L^p(\mu, E)$ is called *p-uniformly integrable* if

$$\lim_{a\to+\infty} \sup_{u\in U} \int_{(\|u\|_E \geq a)} \|u\|_E^p \, dt = 0. \tag{p-UI}$$

U is called *p-equicontinuous* if

$$\lim_{\mu(A)\to 0} \sup_{u\in U} \int_A \|u\|_E^p \, dt = 0. \tag{p-E}$$

A sequence $(u_n)_n \subseteq L^p(\mu, E)$ is p-uniformly integrable (p-equicontinuous) if the set $\{u_n : n \in \mathbb{N}\}$ is p-uniformly integrable (p-equicontinuous).

Remark 3.117.

(i) U is p-uniformly integrable if and only if $U_p = \{\|u\|_E^p : u \in U\} \subseteq L^1(\mu)$ is uniformly integrable (see Definition 1.85 and (iii) of Theorem 1.84).

U is p-equicontinuous if and only if U_p is equicontinuous in $L^1(\mu)$ (see (ii) of Remark 1.53 and Definition 1.52).

(ii) Since μ is atomless, then, according to Saks' theorem (Th. I.4.10 of [173]), every p-equicontinuous set is bounded in $L^p(\mu, E)$ and then U_p is bounded in $L^1(\mu)$. According to Theorem 1.84, $U \subseteq L^p(\mu, E)$ is p-uniformly integrable if and only if it is p-equicontinuous.

(iii) Every p-uniformly integrable set $U \subseteq L^p(\mu, E)$ is bounded in L^p. Indeed, if U is p-uniformly integrable, there exists $a > 0$ such that

$$\sup_{u\in U} \int_{(\|u\|_E \geq a)} \|u\|_E^p \, dt < 1; \quad \text{then } \sup_{u\in U} \|u\|_p < (1 + a^p \cdot T)^{\frac{1}{p}}.$$

(iv) If $1 \leq p < q < +\infty$ and if U is bounded in $L^q(\mu, E) \subseteq L^p(\mu, E)$, then U is p-uniformly integrable.

Indeed, if $\sup_{u\in U} \|u\|_q \leq M$, then, for every $A \in \mathcal{A}$,

$$\sup_{u\in U} \int_A \|u\|_E^p \, dt \leq (\mu(A))^{\frac{q-p}{q}} \cdot \left(\int_A \|u\|_E^q \, dt\right)^{\frac{p}{q}} \leq (\mu(A))^{\frac{q-p}{q}} \cdot M^p.$$

Therefore U is p-equicontinuous and, according to (ii), it is p-uniformly integrable.

We now give the vector version of the Vitali theorem (Theorem 1.54).

Theorem 3.118 (Theorem III.3.6 of [62]). *Let* $(u_n)_n \subseteq L^p(\mu, E)$ *and* $u \in \mathcal{M}(E)$. *Then* $u \in L^p(\mu, E)$ *and* $u_n \xrightarrow{\|\cdot\|_p} u$ *if and only if:*

(i) $u_n \xrightarrow{\mu} u$ *and*

(ii) $(u_n)_n$ *is p-uniformly integrable.*

Proof. Firstly, let us suppose that $u \in L^p(\mu, E)$ and $u_n \xrightarrow{\|\cdot\|_p} u$.

Since, for every $\varepsilon > 0$, $\mu(\|u_n - u\|_E \geq \varepsilon) \leq \frac{1}{\varepsilon^p} \cdot \|u_n - u\|_p^p$, $(u_n)_n$ is convergent in measure to u.

Let us show that the set $\{u_n : n \in \mathbb{N}\}$ is a p-equicontinuous set. For every $\varepsilon > 0$, there exist $n_0 \in \mathbb{N}$ and $\delta > 0$ such that, for all $A \in \mathcal{A}$ such that $\mu(A) < \delta$,

(1) $\|u_n - u\|_p < \varepsilon$, for every $n \geq n_0$,

(2) $\int_A \|u\|_E^p dt < \varepsilon^p$ and

(3) $\int_A \|u_n\|_E^p dt < \varepsilon$, for every $n < n_0$. According to (1) and (2), for every $n \geq n_0$ and $A \in \mathcal{A}$ with $\mu(A) < \delta$,

(4) $\int_A \|u_n\|_E^p dt \leq [(\int_A \|u_n - u\|_E^p dt)^{\frac{1}{p}} + (\int_A \|u\|_E^p dt)^{\frac{1}{p}}]^p < (2\varepsilon)^p$ and then, by (3) and (4), $(u_n)_n$ is p-equicontinuous.

According to (ii) of previous remark, $(u_n)_n$ is p-uniformly integrable.

Reciprocally, let us suppose that conditions (i) and (ii) are satisfied.

For every $n \in \mathbb{N}$, let $v_n = \|u_n\|_E^p \in L^1(\mu)$ and let $v = \|u\|_E^p \in \mathcal{M}(\mathbb{R})$. According to (i) of Remark 3.117, $(v_n)_n$ is uniformly integrable.

Every subsequence $(u_{k_n})_n$ of $(u_n)_n$ has a subsequence $(u_{l_{k_n}})_n$ convergent to u, almost everywhere. Then $(\|u_{l_{k_n}}\|_E^p)_n$ is convergent to $\|u\|_E^p$, almost everywhere and, since $\mu(\Omega) = T < +\infty$, $\|u_{l_{k_n}}\|_E^p \xrightarrow{\mu} \|u\|_E^p$.

Therefore $v_n \xrightarrow{\mu} v$.

According to Vitali's theorem (Theorem 1.54), $v \in L^1(\mu)$ and $v_n \xrightarrow{\|\cdot\|_1} v$.

Then $u \in L^p(\mu, E)$ and

$$\int_0^T \|u_n\|_E^p dt \to \int_0^T \|u\|_E^p dt. \tag{*}$$

For every $n \in \mathbb{N}$, let $w_n = 2^{p-1}(\|u\|_E^p + \|u_n\|_E^p) - \|u - u_n\|_E^p \geq 0$; then $(w_n)_n \subseteq L^1(\mu)$.

Every subsequence $(u_{k_n})_n$ of $(u_n)_n$ has a subsequence $(u_{l_{k_n}})_n$ convergent to u, almost everywhere and therefore $\liminf_{n \to +\infty} w_{l_{k_n}} = 2^p \cdot \|u\|_E^p$.

By applying the Fatou's lemma to the sequence $(w_{l_{k_n}})_n$ and by using $(*)$, we obtain

$$\begin{aligned}2^p \cdot \int_0^T \|u\|_E^p dt &\leq \liminf_{n\to+\infty} \int_0^T w_{l_{k_n}} dt \\ &= \liminf_{n\to+\infty} \Bigg[2^{p-1} \int_0^T \|u\|_E^p dt + 2^{p-1} \cdot \int_0^T \|u_{l_{k_n}}\|_E^p dt \\ &\qquad - \int_0^T \|u - u_{l_{k_n}}\|_E^p dt \Bigg] \\ &= 2^p \cdot \int_0^T \|u\|_E^p dt - \limsup_{n\to+\infty} \int_0^T \|u - u_{l_{k_n}}\|_E^p dt,\end{aligned}$$

from where $u_{l_{k_n}} \xrightarrow{\|\cdot\|_p} u$.

Therefore, every subsequence $(u_{k_n})_n$ of $(u_n)_n$ has a subsequence $(u_{l_{k_n}})_n$ strongly convergent to u in $L^p(\mu, E)$ and then $u_n \xrightarrow{\|\cdot\|_p} u$. □

Corollary 3.119. *A set $U \subseteq L^p(\mu, E)$ is relatively strongly compact if and only if it is p-uniformly integrable and relatively compact in measure.*

We have noted in (iii) of Remark 3.117 that each p-uniformly integrable set is bounded in $L^p(\mu, E)$. Reciprocally, a bounded set that satisfies to the strong concentration condition is p-uniformly integrable.

Proposition 3.120. *Let U be a bounded subset of $L^p(\mu, E)$ which satisfies to the strong concentration condition; then U is p-uniformly integrable.*

Proof. Since, for every $h > 0$ and every $t \in T - h$, $| \|u(t+h)\|_E - \|u(t)\|_E| \leq \|u(t+h) - u(t)\|_E$,

$$\limsup_{h\downarrow 0}\sup_{u\in U} \int_0^{T-h} | \|u(t+h)\|_E - \|u(t)\|_E|^p dt = 0,$$

then $V = \{\|u\|_E : u \in U\}$ satisfies to the strong concentration condition in $L^p(\mu)$. Since V is bounded in $L^p(\mu)$, according to Theorem 3.113, V is relatively strongly compact in $L^p(\mu)$. According to Corollary 3.119, V and then U are p-uniformly integrable. □

Since the strong concentration condition assures the condition $(p - UI)$, in order to obtain the strongly compactness of a set, it is sufficient, according to Corollary 3.119, to characterize the compactness in measure.

For the following results, we will follow the line of reasoning of [144].

Firstly, we need two preliminary lemmas.

Lemma 3.121. *Let* $0 < S < T, \Omega_1 =]0, S[\subseteq]0, T[= \Omega$ *and* $\mathcal{A}_1 = \mathcal{A}\restriction_{\Omega_1}$. *For every Polish space* P *and every Young measure* $\tau \equiv (\tau_t)_{t\in\Omega}$ *on* $\Omega \times P$, *let* $\tau^1 \equiv (\tau_t)_{t\in\Omega_1} \in \mathcal{Y}(\Omega_1 \times P)$.

For every $h \in]0, T-S[$, *the mapping* $\tau^h : \Omega_1 \to \mathcal{P}_P, \tau_t^h = \tau_{t+h}$, *is* $(\mathcal{A}_1 - \mathcal{C})$-*measurable; it is the disintegration of a measure* $\tau^h \in \mathcal{Y}(\Omega_1 \times P)$.

In addition, $\tau^h \xrightarrow[h\to 0]{\mathcal{S}_1} \tau^1$, *where* $\mathcal{S}_1$ *is the stable topology on* $\mathcal{Y}(\Omega_1 \times P)$.

Proof. Firstly, let us remark that, for every $C \in \mathcal{B}_P$, the mapping $t \mapsto \tau_t^h(C)$ is $(\mathcal{A}_1 - \mathcal{B}_P)$-measurable. Indeed, for all $B \in \mathcal{B}_{\mathbb{R}}$, $[\tau_\cdot^h(C)]^{-1}(B) = -h + [\tau_\cdot(C)]^{-1}(B) \in \mathcal{A}_1$ and then, according to Theorem 3.1, $\tau_\cdot^h$ is $(\mathcal{A}_1 - \mathcal{C})$-measurable. Therefore, for every $h \in]0, T - S[$, $\tau_\cdot^h$ is the disintegration of a Young measure $\tau^h \in \mathcal{Y}(\Omega_1 \times P)$.

The family of all finite disjoint unions of open intervals is an algebra which generated the σ-algebra $\mathcal{A}_1$ and then, according to Proposition 3.21,

$$\tau^h \xrightarrow[h\to 0]{\mathcal{S}_1} \tau \Leftrightarrow \tau^h(\chi_A \otimes f) \xrightarrow[h\to 0]{} \tau(\chi_A \otimes f),$$
$$\forall A =]a, b[\subseteq]0, S[, \quad \forall f \in C_b(P).$$

Let now $A =]a, b[\subseteq \Omega_1$ and let $f \in C_b(P)$; the function $t \mapsto \varphi(t) = \int_P f(x) d\tau_t(x)$ belongs to $L^1(\mu, P)$ and then

$$\begin{aligned}
\tau^h(\chi_A \otimes f) &= \int_0^S \chi_A(t) \cdot \varphi(t+h)dt = \int_h^{S+h} \chi_{h+A}(t) \cdot \varphi(t)dt \\
&= \int_h^{S+h} \chi_{]a+h,b+h[}(t) \cdot \varphi(t)dt = \int_{a+h}^{b+h} \varphi(t)dt \\
&= \int_a^b \varphi(t)dt + \int_b^{b+h} \varphi(t)dt - \int_a^{a+h} \varphi(t)dt.
\end{aligned}$$

Because

$$\left| \int_b^{b+h} \varphi(t)dt \right| \le \|f\|_\infty \cdot h \quad \text{and} \quad \left| \int_a^{a+h} \varphi(t)dt \right| \le \|f\|_\infty \cdot h,$$

it results immediately that $\tau^h(\chi_A \otimes f) \to \int_a^b \varphi(t)dt = \tau(\chi_A \otimes f)$. □

Lemma 3.122. *Let P be a Polish space, let $\tau \in \mathcal{Y}(\Omega \times P)$ and let $e : P \times P \to [0, +\infty]$ be a l.s.c. application which vanishes only on the diagonal of P, $\Delta_P = \{(x, x) : x \in P\}$.*

The mapping $\Psi :]0, T[\times P \to [0, +\infty]$ defined by

$$\Psi(t, x) = \int_P e(x, y) d\tau_t(y), \quad \textit{for every} \quad (t, x) \in]0, T[\times P,$$

is a positive l.s.c. integrand.

Proof. Firstly, let us suppose that e is continuous and bounded on $P \times P$; then, since $\tau_.$ is a Young measure, Ψ is measurable with respect to t and continuous in the second variable x ($x_n \to x$ implies $e(x_n, y) \to e(x, y)$ and then $\int_P e(x_n, y) d\tau_t(y) \to \int_P e(x, y) d\tau_t(y)$).

According to Proposition 3.8, Ψ is a Carathéodory integrand.

In the general case, since $P \times P$ is metrizable, let, for every $n \in \mathbb{N}$, $e_n : P \times P \to [0, +\infty[$ be the Yosida's transform of e; $(e_n)_n$ is an increasing sequence of Lipschitz mappings such that $e_n \uparrow e$ (see Lemma 2.35). According to the Lebesgue monotone convergence theorem

$$\Psi_n(t, x) = \iint_{P \times P} e_n(x, y) d\tau_t(y) \uparrow \Psi(t, x).$$

$(\Psi_n)_n$ is a sequence of Carathéodory integrands and so $\Psi = \sup_n \Psi_n$ is measurable and l.s.c. in the second variable. □

The following result characterizes the compactness in measure by adding a weak concentration condition to that of being tight.

Theorem 3.123 (Theorem 2 of [144]).

(I) *Let P be a Polish space and let $e : P \times P \to [0, +\infty]$ be a l.s.c. application with the property: $e(x, y) = 0$ if and only if $x = y$.*

A tight set $U \subseteq \mathcal{M}(P)$ which satisfies to the weak concentration condition with respect to e:

$$\limsup_{h \downarrow 0} \sup_{u \in U} \int_0^{T-h} e(u(t+h), u(t)) dt = 0 \qquad \text{(CF')}$$

is relatively compact in measure.

(II) *Conversely, if $U \subseteq \mathcal{M}(P)$ is relatively compact in measure, then U is tight and satisfies to the weak concentration condition (CF') with respect to any bounded continuous metric e on P.*

Proof. (I) Let $U \subseteq \mathcal{M}(P)$ be a tight set which satisfies to the weak concentration condition with respect to a l.s.c. application $e : P \times P \to [0, +\infty]$ which vanishes only on the diagonal Δ_P of P.

Since the topology of the convergence in measure is metrizable, it is sufficient to show that U is sequentially compact in measure.

Let so $(u_n)_n \subseteq U$; since U is tight, it results from Theorem 3.64 that $(u_n)_n$ has a subsequence (still noted $(u_n)_n$) $\mathcal{S}$-convergent to a Young measure $\tau \in \mathcal{Y}(P)$. If τ is associated to a measurable mapping u, then, according to Proposition 3.40, $u_n \xrightarrow{\mu} u$.

We still need to proof that there exists $u \in \mathcal{M}(P)$ such that $\tau = \tau^u$.

Firstly, let us show that, for almost every $t \in \Omega$, supp τ_t contains only one point $u(t)$; the demonstration has three stages:

$$\lim_{h \downarrow 0} \frac{1}{h} \int_0^h \int_0^{T-s} \left[\iint_{P \times P} e(u, v) d(\tau_{t+s} \otimes \tau_t)(u, v) \right] dt ds = 0. \tag{A}$$

$$\int_0^T \left[\iint_{P \times P} e(u, v) d\tau_t(u) d\tau_t(v) \right] dt = 0. \tag{B}$$

$$\text{supp}(\tau_t) \quad \text{contains only one point, almost for every} \quad t \in (0, T). \tag{C}$$

(A) Let $r(h) = \sup_{s \in]0,h]} \sup_n \int_0^{T-s} e(u_n(t+s), u_n(t)) dt$. According to condition (CF'), $\forall \varepsilon > 0, \exists \delta > 0$ such that, $\forall h \in]0, \delta[, \quad \forall n \in \mathbb{N}, \int_0^{T-h} e(u_n(t + h), u_n(t)) dt < \varepsilon$.

$\forall h \in]0, \delta[, \forall s \in]0, h], 0 < s < \delta$; therefore

$$\sup_n \int_0^{T-s} e(u_n(t+s), u_n(t)) dt \leq \varepsilon.$$

Hence

$$\sup_{s \in]0,h]} \sup_n \int_0^{T-s} e(u_n(t+s), u_n(t)) dt \leq \varepsilon$$

and then

$$\lim_{h \downarrow 0} r(h) = 0. \tag{1}$$

$\int_0^{T-s} e(u_n(t+s), u_n(t)) dt \leq r(h), \quad \forall h > 0, \quad \forall s \in]0, h], \quad \forall n \in \mathbb{N}.$

By integrating the last inequality with respect to $s \in]0,h]$, we obtain

$$\int_0^h \left[\int_0^{T-s} e(u_n(t+s),u_n(t))dt\right] ds \le h \cdot r(h), \quad \text{or}$$

$$\iint_{A(h)} e(u_n(t+s),u_n(t))dtds \le h \cdot r(h), \quad \forall h > 0, \quad \forall n \in \mathbb{N}, \tag{2}$$

in which we noted $A(h) = \{(s,t) \in]0,T[\times]0,T[: s \in]0,h], 0 < t \le T-s\}$.

After the change of variables $(s,t) \mapsto (s-t,t)$, relation (2) becomes

$$\iint_{B(h)} e(u_n(s),u_n(t))dsdt \le h \cdot r(h), \quad \forall h > 0, \quad \forall n \in \mathbb{N}, \tag{3}$$

where $B(h) = \{(s,t) \in]0,T[\times]0,T[: 0 < t < T, t < s \le t+h\}$.

According to Theorem 3.90, $\tau^{u_n} \dot{\otimes} \tau^{u_n} \xrightarrow{\mathcal{S}_{]0,T[\times]0,T[}} \tau \dot{\otimes} \tau$. Since $\Lambda(s,t,x,y) = \chi_{B(h)}(s,t) \cdot e(x,y)$ defines a positive l.s.c. integrand, according to Proposition 3.34,

$$(\tau \dot{\otimes} \tau)(\Lambda) \le \liminf_n (\tau^{u_n} \dot{\otimes} \tau^{u_n})(\Lambda).$$

Then, using (3), we obtain

$$\begin{aligned} &\iint_{B(h)} \left[\iint_{P\times P} e(x,y)d(\tau_t \otimes \tau_s)(x,y)\right] dtds \\ &\le \liminf_n \iint_{B(h)} e(u_n(t),u_n(s))dtds \le h \cdot r(h). \end{aligned} \tag{4}$$

After a reversed change of variables $(s,t) \mapsto (t+s,t)$, we obtain

$$\begin{aligned} &\iint_{B(h)} \left[\iint_{P\times P} e(x,y)d(\tau_t \otimes \tau_s)(x,y)\right] dtds \\ &= \iint_{A(h)} \left[\iint_{P\times P} e(x,y)d(\tau_{t+s} \otimes \tau_t)(x,y)\right] dtds \end{aligned}$$

and, according to (4),

$$\int_0^h \int_0^{T-s} \left[\iint_{P\times P} e(x,y)d(\tau_{t+s} \otimes \tau_t)(x,y)\right] dtds \le h \cdot r(h).$$

The last inequality and (1) lead us to (A).

(B) According to Lemma 3.122, the mapping $\Psi :]0,T[\times P \to [0,+\infty]$ defined by $\Psi(t,x) = \int_P e(x,y)d\tau_t(y)$, for every $(t,x) \in]0,T[\times P$, is a positive l.s.c. integrand. We can therefore write the condition of stage (A) in the following form:

$$\lim_{h\downarrow 0}\frac{1}{h}\cdot\int_0^h\int_0^{T-s}\int_P \Psi(t,x)d\tau_{t+s}(x)dt\,ds = 0.$$

By making the change of variables $(s,t)\mapsto(hs,t)$ and by applying the Fatou lemma, we obtain, for every $\varepsilon > 0$,

$$0 = \liminf_{h\downarrow 0}\frac{1}{h}\cdot\int_0^h\int_0^{T-s}\int_P \Psi(t,x)d\tau_{t+s}(x)dt\,ds \tag{5}$$

$$= \liminf_{h\downarrow 0}\int_0^1\int_0^{T-hs}\int_P \Psi(t,x)d\tau_{t+hs}(x)dt\,ds$$

$$\geq \liminf_{h\downarrow 0}\int_0^1\int_0^{T-\varepsilon}\int_P \Psi(t,x)d\tau_{t+hs}(x)dt\,ds$$

$$= \liminf_{h\downarrow 0}\int_0^1 \tau^{hs}(\chi_{]0,T-\varepsilon[}\cdot\Psi)ds$$

$$\geq \int_0^1 \liminf_{h\downarrow 0}\tau^{hs}(\chi_{]0,T-\varepsilon[}\cdot\Psi)ds.$$

Lemma 3.121 assures us that $\tau^{hs}\xrightarrow[h\to 0]{\mathcal{S}}\tau$ and, since Ψ is positive and l.s.c., we can apply again Proposition 3.34 to obtain

$$\liminf_{h\downarrow 0}\tau^{hs}(\chi_{]0,T-\varepsilon[}\cdot\Psi)\geq\tau(\chi_{]0,T-\varepsilon[}\cdot\Psi).$$

By going back to condition (5), we can write

$$0\geq\int_0^1\tau(\chi_{]0,T-\varepsilon[}\cdot\Psi)ds=\tau(\chi_{]0,T-\varepsilon[}\cdot\Psi)=\int_0^{T-\varepsilon}\int_P\Psi(t,x)\tau_t(x)dt$$
$$=\int_0^{T-\varepsilon}\int_P\int_P e(x,y)d\tau_t(x)d\tau_t(y)dt.$$

Since $e\geq 0$ is arbitrary, we obtain condition (B)

$$\int_0^T\left[\iint_{P\times P} e(x,y)d\tau_t(x)d\tau_t(y)\right]dt = 0.$$

(C) Condition (B) leads us to

$$\iint_{P\times P} e(x,y)d(\tau_t\otimes\tau_t)(x,y)=0, \quad \text{for a.e.} \quad t\in\,]0,T[$$

or

$$(\tau_t\otimes\tau_t)\left(\{(x,y)\in P\times P : e(x,y)>0\}\right)=0, \quad \text{for a.e.} \quad t\in\,]0,T[,$$

which, according to the properties of e, comes back to

$$\operatorname{supp}(\tau_t\otimes\tau_t)\subseteq\triangle_P, \quad \text{for a.e.} \quad t\in]0,T[.$$

If we assume that supp τ_t contains two distinct points, $x,y\in P$, then, there are two open sets $U,V\subseteq P$ such that $x\in U, y\in V$ and $U\cap V=\emptyset$. Since $x,y\in\operatorname{supp}\tau_t$, $\tau_t(U)>0$ and $\tau_t(V)>0$; therefore $(\tau_t\otimes\tau_t)(U\times V)=\tau_t(U)\cdot\tau_t(V)>0$. On the other hand $(\tau_t\otimes\tau_t)(U\times V)=(\tau_t\otimes\tau_t)((U\times V)\cap\triangle_P)=(\tau_t\otimes\tau_t)(\emptyset)=0$ which is a contradiction.

So, supp $\tau_t=\{u(t)\}$, almost for every $t\in\Omega$.

The application $u:\,]0,T[\to\mathbb{R}$ is therefore well defined. For every $A\in\mathcal{B}_P$, the mapping $g_A:\,]0,T[\,\to\mathbb{R}$, defined by $g_A(t)=\tau_t(A)$, is measurable; so $u^{-1}(A)=\{t\in\,]0,T[: u(t)\in A\}=\{t\in\,]0,T[: g_A(t)=1\}=g_A^{-1}(1)\in\mathcal{A}$. Then $u\in\mathcal{M}(P)$ and $\tau=\tau^u$.

(II) Let $U\subseteq\mathcal{M}(P)$ be a set relatively compact in measure; therefore its closure $\bar{U}$ is compact in measure in $\mathcal{M}(P)$ and, according to Proposition 3.41, $\bar{U}$ is $\mathcal{S}$-compact. So $\bar{U}$ is tight (Proposition 3.65).

Let now $e: P\times P\to\mathbb{R}_+$ be a bounded continuous metric and let $M=\sup_{x,y\in P}e(x,y)<+\infty$. Let us show that U satisfies to the weak concentration condition with respect to e.

For every $u,v\in\mathcal{M}(P)$, let $\delta_e(u,v)=\int_0^T e(u(t),v(t))dt$. Then δ_e is a pseudo-metric on $\mathcal{M}(P)$; let us show that δ_e is continuous with respect to the topology of convergence in measure on $\mathcal{M}(P)$.

Let $(u_n)_n\subseteq\mathcal{M}(P)$ be a sequence convergent in measure to a function $u\in\mathcal{M}(P)$; we can suppose that $(u_n(t))_n$ is convergent to $u(t)$ and then $e(u_n(t),u(t))\to 0$, almost for every $t\in\,]0,T[$. Since e is bounded, according to the bounded convergence theorem, $\delta_e(u_n,u)=\int_0^T e(u_n(t),u(t))dt\to 0$.

If $(v_n)_n$ is an other sequence convergent in measure to v, then

$$|\delta_e(u_n,v_n)-\delta_e(u,v)|\le\delta_e(u_n,u)+\delta_e(v_n,v)\to 0$$

therefore δ_e is continuous.

Then $(\bar{U},\delta_e)$ is a compact space.

For every $h\in\,]0,T[$, let $f_h:\bar{U}\to\mathbb{R}$, $f_h(u)=\int_0^{T-h}e(u(t+h),u(t))dt$.

For every $u, v \in \bar{U}$ and every $h \in]0, T[$,

$$\begin{aligned} |f_h(u) - f_h(v)| &\le \int_0^{T-h} |e(u(t+h), u(t)) - e(v(t+h), v(t))|dt \\ &\le \int_0^{T-h} [e(u(t+h), v(t+h)) + e(u(t), v(t))]\, dt \le 2\delta_e(u, v). \end{aligned}$$

Therefore, the mappings f_h are uniformly Lipschitz on $\bar{U}$ and then $A = \{f_h : h \in]0, T[\,\} \subseteq C(\bar{U}, \mathbb{R})$ is an uniformly equicontinuous set.

Moreover, for every $h \in]0, T[$,

$$|f_h(u)| \le M(T-h) \le M \cdot T, \quad \forall u \in \bar{U}$$

and then A is uniformly bounded.

According to Arzelà–Ascoli theorem, A is relatively compact in $C(\bar{U}, \mathbb{R})$ with respect to the topology of uniform convergence.

Let us show that, for every $u \in \bar{U}$, $\lim_{h\to 0} f_h(u) = 0$.

(a) If $u :]0, T[\to P$ is continuous, then $\lim_{h\to 0} u(t+h) = u(t)$, for every $t \in]0, T-h[$ and, since e is bounded,

$$f_h(u) = \int_0^{T-h} e(u(t+h), u(t))dt \xrightarrow[h\to 0]{} 0.$$

(b) Let now $u \in \bar{U}$ be a measurable mapping; there exists a sequence of continuous maps $(u_p)_p$ such that $u_p \to u$, almost everywhere and then $\delta_e(u_p, u) \to 0$.

For every $\varepsilon > 0$, let $p_0 \in \mathbb{N}$ such that $\delta_e(u_{p_0}, u) < \frac{\varepsilon}{4}$. By (a), $f_h(u_{p_0}) \xrightarrow[h\to 0]{} 0$ and then there exists $\delta > 0$ such that $f_h(u_{p_0}) < \frac{\varepsilon}{2}$, for every $h < \delta$. Therefore, for every h with $0 < h < \delta$,

$$\begin{aligned} f_h(u) &= f_h(u) - f_h(u_{p_0}) + f_h(u_{p_0}) \\ &\le \int_0^{T-h} |e(u(t+h), u(t)) - e(u_{p_0}(t+h), u_{p_0}(t))|dt + \frac{\varepsilon}{2} \\ &\le \int_0^{T-h} e(u(t+h), u_{p_0}(t+h))dt + \int_0^{T-h} e(u(t), u_{p_0}(t))dt + \frac{\varepsilon}{2} \\ &= \int_h^{T} e(u(t), u_{p_0}(t))dt + \int_0^{T-h} e(u(t), u_{p_0}(t))dt + \frac{\varepsilon}{2} \\ &< 2\delta_e(u_{p_0}, u) + \frac{\varepsilon}{2} < \varepsilon. \end{aligned}$$

Since $A = \{f_h : h \in]0, T[\,\}$ is relatively compact with respect to the topology of uniform convergence and $\lim_{h\to 0} f_h(u) = 0$, for every $u \in \bar{U}$, the limit is 0 uniformly with respect to $u \in \bar{U}$ and so

$$\lim_{h \to 0} \sup_{u \in U} \int_0^{T-h} e(u(t+h), u(t))dt = 0$$

which is exactly the condition of weak concentration with respect to e. □

This last theorem allows us to demonstrate the result of strong convergence due to Rossi and Savaré.

Theorem 3.124 (Theorem 1 of [144]). *Let E be a separable Banach space, let $1 \leq p < +\infty$ and let $U \subseteq L^p(\mu, E)$ be a L^p-bounded set.*

If U is tight and if it satisfies to strong concentration condition

$$\lim_{h \downarrow 0} \sup_{u \in U} \int_0^{T-h} \|u(t+h) - u(t)\|_E^p dt = 0, \tag{CF}$$

then U is relatively compact in $L^p(\mu, E)$.

Proof. Let $e : E \times E \to \mathbb{R}_+$ defined by $e(x, y) = \|x - y\|_E^p$, for every $x, y \in E$. Then e is l.s.c. on $E \times E$ and $e(x, y) = 0$ if and only if $x = y$.

Since U is tight and, according to (CF), it satisfies to weak concentration condition with respect to e, the previous theorem assures us that U is relatively compact in measure. According to Proposition 3.120, U is p-uniformly integrable and therefore, according to Corollary 3.119, U is relatively strongly compact in $L^p(\mu, E)$. □

We obtain some known compactness results as corollaries of Rossi–Savaré theorem.

The following theorem is presented in §4 of Ch. IV of [114] for the Hilbert spaces and in [7] for the Banach spaces.

Theorem 3.125 (Lions–Aubin theorem). *Let E_0 be a Banach space such that $E_0 \Subset E$ (every bounded subset of E_0 is relatively compact in E) and let U be a bounded subset of $L^p(\mu, E_0)$ which satisfies to the strong concentration condition in $L^p(\mu, E)$:*

$$\lim_{h \downarrow 0} \sup_{u \in U} \int_0^{T-h} \|u(t+h) - u(t)\|_E^p dt = 0. \tag{CF}$$

Then U is relatively strongly compact in $L^p(\mu, E)$.

Proof. According to Theorem 3.124, we only need to demonstrate that U is tight. Let $\varphi : E \to [0, +\infty]$, defined by

$$\varphi(x) = \begin{cases} \|x\|_{E_0}^p, & x \in E_0, \\ +\infty, & x \in E \setminus E_0. \end{cases}$$

Then, for every $a \geq 0, \varphi^{-1}([0, a]) = \{x \in E_0 : \|x\|_{E_0} \leq a^{\frac{1}{p}}\}$ is bounded in E_0 and therefore it is relatively compact in E ($\varphi^{-1}([0, a])$ is a closed set). Then φ is an inf-compact application. Since U is bounded in $L^p(\mu, E_0)$,

$$\sup_{u \in U} \int_0^T \varphi(u(t))dt = \sup_{u \in U} \int_0^T \|u(t)\|_{E_0}^p dt < +\infty$$

and then U is tight (U satisfies to the condition (T_2) of Corollary 3.62). □

Theorem 3.126. *Let E_0 be a Banach space such that $E_0 \Subset E$ and let $U \subseteq L^p(\mu, E_0) \subseteq L^p(\mu, E)$ be a bounded set in $L^p(\mu, E)$ satisfying to the strong concentration condition (CF) in $L^p(\mu, E)$. If*

$$\forall \varepsilon > 0, \exists k > 0 \quad \textit{such that} \quad \sup_{u \in U} \mu\left(\{t \in]0, T[: \|u(t)\|_{E_0} > k\}\right) < \varepsilon,$$

then U is relatively strongly compact in $L^p(\mu, E)$.

Proof. If we remark that the set $K = \{x \in E_0 : \|x\|_{E_0} \leq k\}$ is compact in E, then

$$\sup_{u \in U} \mu(u^{-1}(E \setminus K)) = \sup_{u \in U} \mu(\{t \in]0, T[: \|u(t)\|_{E_0} > k\}) < \varepsilon$$

and therefore, according to condition (T_1) of Corollary 3.62, U is tight. □

Theorem 3.127 (Gutman's Theorem [92]). *Let $U \subseteq L^p(\mu, E)$ be a set satisfying to strong concentration condition (CF) such that*

$$\forall \varepsilon > 0, \exists K \in \mathcal{K}_E \quad \textit{with} \quad \sup_{u \in U} \mu(u^{-1}(E \setminus K)) < \varepsilon.$$

Then U is relatively strongly compact in $L^p(\mu, E)$.

Proof. According to condition (T_1) of Corollary 3.62, U is tight. □

3.10 Gradient Young Measures

Young measures were introduced in two different but equivalent ways. Initially, they were presented as being generated by sequences of measurable functions. This direction is based on the fundamental theorem of Young; it is followed in the works of L. Tartar [163, 164, 165], J. M. Ball [22], L. C. Evans [69], D. Kinderlehrer and P. Pedregal [102, 103], T. Roubiček [145], I. Fonseca and S. Müller [83]. Among more recent works using this approach, we quote [158, 181, 161, 85, 8, 72, 109].

Here, we have used a recent definition, used by E. Balder [10, 12] and M. Valadier [168] (see also [45]). In this section, we compare the two definitions.

In the first part of section, we present the fundamental theorem of Young.

We define and characterize p-Young measures as limits of sequences of L^p. In the case where the Young measure τ is generated by a sequence of gradients, we introduce the p-gradient Young measures.

The p-gradient Young measures and lower semicontinuity are the main ingredients in the study of solutions existence in relaxed variational calculus. Both of these ingredients have characterizations with quasiconvex functions. The study of this class of functions was started by Ch. B. Morrey, Jr. (see [121, 122] and was continued and developed by many others of that we quote: I. Fonseca and S. Müller [83], J. M. Ball, B. Kirchheim and J. Kristensen [24], A. Braides, I. Fonseca and G. Leoni [35], P. Pedregal [131]; see also [84, 71, 117, 160, 66, 34].

In the second part of this section, we define and study quasiconvexity. We find in literature quasiconvexity defined in different ways; we present a comparative study of these different definitions and, finally, the theorem of Kinderlehrer–Pedregal for characterization of p-gradient Young measures (see [102]).

The third part of section is devoted to the study of sequentially weakly lower semicontinuous (wlsc) functionals in $W^{1,p}(\Omega, \mathbb{R}^m)$ and their characterization by quasiconvexity (Morrey–Acerbi–Fusco theorem). Studies on wlsc can also be found in [9, 118, 100, 83, 108, 159, 53, 54, 170, 113] and [125].

In this section, we will use the following framework:

- Ω is a bounded open subset of the Euclidean space $\mathbb{R}^d$ $(d \geq 1)$.
- $\mathcal{A}$ is the σ-algebra of Lebesgue measurable subsets of Ω.
- $\mu = dt$ is Lebesgue measure on Ω.
- $\mathcal{B}(\mathbb{R}^m)$ is the σ-algebra of Borel subsets of the Euclidean space $\mathbb{R}^m$ $(m \geq 1)$.
- $\mathcal{M}(\mathbb{R}^m)$ notes the set of all $(\mathcal{A} - \mathcal{B}(\mathbb{R}^m))$-measurable mappings.

- $\mathcal{P}(\mathbb{R}^m)$ is the set of probabilities on $\mathbb{R}^m$ endowed with the narrow topology.
- $C_0(\mathbb{R}^m)$ is the set of all continuous functions $f : \mathbb{R}^m \to \mathbb{R}$ such that, for every $\varepsilon > 0$, there is $k > 0$, with $|f(x)| < \varepsilon$, for every $x \in \mathbb{R}^m$, $\|x\| > k\}$— the set of all continuous functions vanishing at infinity.
- $\mathrm{Lip}(\Omega, \mathbb{R}^m)$ is the set of all Lipschitz functions $f : \Omega \to \mathbb{R}^m$.
- $C_c^\infty(\Omega, \mathbb{R}^m)$ is the set of all $\mathbb{R}^m$-valued functions with compact support in Ω, differentiable of all orders.

3.10.1 Young Measures Generated by Sequences

The density of the $\mathcal{M}(\mathbb{R}^m)$ in $\mathcal{Y}(\mathbb{R}^m)$ (see Theorem 3.50 for $S = \mathbb{R}^m$) is one of the most important properties of the stable topology; historically, the Young measures have been constructed as limits of the measurable functions. Thus, some mathematicians use the notion of *Young measure associated to a sequence* instead of that of *Young measure*. The existence of these measures is furnished by the fundamental theorem of Young; in [22], J. M. Ball stated the following version of this theorem (see also Theorem 8.2 from [85]):

Theorem 3.128. *Let $\Omega \subseteq \mathbb{R}^d$ be a measurable bounded set and let $(u_n)_n \subseteq \mathcal{M}(\mathbb{R}^m)$ a sequence satisfying the following condition:*

$$\sup_{n\in\mathbb{N}} \mu(\|u_n\| \geq k) \xrightarrow[k\to+\infty]{} 0. \tag{T}$$

Then there exist a subsequence $(u_{k_n})_n$ of $(u_n)_n$ and a measurable mapping $t \mapsto \tau_t \in \mathcal{P}(\mathbb{R}^m)$ such that:

(1) *For every closed set $K \subseteq \mathbb{R}^m$ such that $u_{k_n}(t) \in K$, for every $n \in \mathbb{N}$ and for almost every $t \in \Omega$,*

$$\operatorname{supp}\tau_t \subseteq K, \quad \textit{for almost every} \quad t \in \Omega.$$

(2) *For every Carathéodory integrand $\Psi : \Omega \times \mathbb{R}^m \to \mathbb{R}$ bounded from below for which $(\Psi(\cdot, u_{k_n}))_n$ is uniformly integrable in $L^1(\Omega, \mathbb{R}^m)$, the mapping $u_\Psi : \Omega \to \mathbb{R}$, $u_\Psi(t) = \int_{\mathbb{R}^m} \Psi(t,x)d\tau_t(x)$, is integrable and*

$$\Psi(\cdot, u_{k_n}) \xrightarrow[L^1(\Omega)]{w} u_\Psi.$$

(3) *For every $f \in C_0(\mathbb{R}^m)$, the mapping $u_f : \Omega \to \mathbb{R}$, defined by $u_f(t) = \int_{\mathbb{R}^m} f(x)d\tau_t(x)$, is in $L^\infty(\Omega)$ and*

$$f(u_{k_n}) \xrightarrow[L^\infty(\Omega)]{w^*} u_f.$$

Proof. We observe that, in the hypotheses of the theorem, $(u_n)_n$ is tight in $\mathcal{Y}(\mathbb{R}^m)$ and then, according to Prohorov's theorem (see Theorem 3.64), it is sequentially $\mathcal{S}$-compact. Let now $(u_{k_n})_n$ be a subsequence of $(u_n)_n$ $\mathcal{S}$-convergent to $\tau \in \mathcal{Y}(\mathbb{R}^m)$.

(1) By Proposition 3.43:

$$\operatorname{supp}\tau_t \subseteq \operatorname{Ls}\left(\{u_{k_n}(t) : n \in \mathbb{N}\}\right) = \bigcap_{p=1}^{\infty} \overline{\{u_{k_n}(t) : k \geq p\}} \subseteq K, \text{ for a.e. } t \in \Omega.$$

(2) $(\Psi(\cdot, u_{k_n}))_n$ is uniformly integrable in $L^1(\Omega, \mathbb{R}^m)$ so that it is bounded (see Definition 1.85 and Theorem 1.84; let $M > 0$ such that

$$\sup_n \int_\Omega |\Psi(t, u_{k_n}(t))|dt \leq M < +\infty.$$

According to Proposition 3.34,

$$-\infty < \tau(\Psi) \leq \liminf_n \tau^{u_{k_n}}(\Psi) \leq M.$$

Therefore $\tau(\Psi) = \int_\Omega u_\Psi(t)dt \in \mathbb{R}$ and then $u_\Psi \in L^1(\Omega)$.

Since $u_{k_n} \xrightarrow[\mathbb{R}^m]{\mathcal{S}} \tau$ and $(\Psi(\cdot, u_{k_n}))_n$ is uniformly integrable, Corollary 3.36 assure us that, for every $A \in \mathcal{A}$,

$$\int_A \Psi(t, u_{k_n}(t))d\mu(t) \to \int_A \left(\int_{\mathbb{R}^m} \Psi(t,x)d\tau_t(x)\right) d\mu(t) = \int_A u_\Psi(t)d\mu(t),$$

from where

$$\Psi(\cdot, u_{k_n}) \xrightarrow[L^1(\Omega)]{w} u_\Psi.$$

(3) Let $f \in C_0(\mathbb{R}^m)$; then f is bounded and so $(f(u_{k_n}))_n \subseteq L^\infty(\Omega)$.

Since, for every $t \in \Omega$, $|u_f(t)| \leq \|f\|_\infty$, $u_f \in L^\infty(\Omega)$.

Let $g \in L^1(\Omega)$ and let $\Psi : \Omega \times \mathbb{R}^m \to \mathbb{R}$, $\Psi(t,x) = g(t) \cdot f(x)$. Then Ψ is a Carathéodory integrand L^1-bounded (for every $(t,x) \in \Omega \times \mathbb{R}^m$, $|\Psi(t,x)| \leq \|f\|_\infty \cdot |g(t)|$).

According to (5) of Theorem 3.32, $\tau^{u_{k_n}}(\Psi) \to \tau(\Psi)$ or

$$\int_\Omega g(t) \cdot f(u_{k_n}(t))dt \to \int_\Omega \int_{\mathbb{R}^m} g(t) \cdot f(x)d\tau_t(x) = \int_\Omega g(t) \cdot u_f(t)dt.$$

Therefore

$$f(u_{k_n}) \xrightarrow[L^\infty(\Omega)]{w^*} u_f. \qquad \square$$

Remark 3.129. From the previous theorem, every sequence $(u_n)_n \subseteq \mathcal{M}(\mathbb{R}^m)$ which satisfies the tightness condition (T), generates a Young measure $\mathcal{T}$ on $\mathbb{R}^m$ with the properties (1)–(3). Conversely, if we suppose that μ is atomless on Ω, then, according to Theorem 3.50, $\mathcal{M}(\mathbb{R}^m)$ is dense in $(\mathcal{Y}(\mathbb{R}^m), \mathcal{S})$. $\mathbb{R}^m$ is a metrizable Suslin space and $\mathcal{A}$ is countable generated; then, according to Proposition 3.25, $(\mathcal{Y}(\mathbb{R}^m), \mathcal{S})$ is metrizable. Therefore, every Young measure on $\mathbb{R}^m$ is the limit of a sequence of measurable functions $(u_n)_n$. Because of Proposition 3.65, $(u_n)_n$ satisfies the condition (T) and, therefore, the properties (1)–(3) are satisfied.

Now we give the definition of a p-Young measure generated by a sequence.

Definition 3.130. Let $1 \leq p < +\infty$; a Young measure $\mathcal{T} \in \mathcal{Y}(\mathbb{R}^m)$ is a *p-Young measure* if there exists a sequence $(u_n)_n \subseteq L^p(\Omega, \mathbb{R}^m)$ such that:

$$\begin{cases} \text{(a)} \quad u_n \xrightarrow{\mathcal{S}} \mathcal{T} \quad \text{and} \\ \text{(b)} \quad (\|u_n\|^p)_{n\in\mathbb{N}} \quad \text{is uniformly integrable.} \end{cases}$$

In this definition, $\|\cdot\|$ notes the usual norm on $\mathbb{R}^m$; according to Definition 3.116, (b) says that $(u_n)_n$ is p-uniformly integrable.

With the above notations, we say that $\mathcal{T}$ is *generated by the sequence* $(u_n)_n \subseteq L^p(\Omega, \mathbb{R}^m)$.

Remark 3.131. Let $(u_n)_n \subseteq L^p(\Omega, \mathbb{R}^m)$ be a sequence which generates the p-Young measure $\mathcal{T}$.

(i) $(u_n)_n$ is uniformly integrable in $L^1(\Omega, \mathbb{R}^m)$. Indeed, if $p > 1$, then, according to (iv) of Remark 3.117, $(\|u_n\|)_n$ is uniformly integrable in $L^1(\Omega, \mathbb{R}^m)$ and then $(u_n)_n$ it is.

(ii) By previous remark and by Proposition 3.37, $u_n \xrightarrow[L^1(\Omega,\mathbb{R}^m)]{w} u \in L^1(\Omega, \mathbb{R}^m)$, where $u : \Omega \to \mathbb{R}^m$ is defined by $u(t) = \int_{\mathbb{R}^m} x \, d\tau_t(x) = \operatorname{bar}\tau_t$ (the barycenter of τ_t).

We use Theorem 3.128 to give in the following theorem two characterizations for p-Young measures (see [102] and [103]).

Theorem 3.132. *Let $1 \leq p < +\infty$ and let $\mathcal{T} \in \mathcal{Y}(\mathbb{R}^m)$; the following are equivalent:*

(1) *$\mathcal{T}$ is a p-Young measure.*

(2) *There is a sequence* $(u_n)_n \subseteq L^p(\Omega, \mathbb{R}^m)$ *and a* $g \in L^1(\Omega)$ *such that:*

$$\begin{cases} \text{(a)} \quad \|u_n\|^p \xrightarrow[L^1(\Omega)]{w} g, \\ \text{(b)} \quad \varphi(u_n) \xrightarrow[L^1(\Omega)]{w} u_\varphi, \quad \textit{for every} \quad \varphi \in E^p, \end{cases}$$

where

$$E^p = \left\{\varphi : \mathbb{R}^m \to \mathbb{R} | \varphi \textit{ continuous on } \mathbb{R}^m, \lim_{\|x\|\to\infty} \frac{\varphi(x)}{1 + \|x\|^p} \in \mathbb{R}\right\}$$

$$\textit{and} \quad u_\varphi : \Omega \to \mathbb{R}^m, u_\varphi(t) = \int_{\mathbb{R}^m} \varphi(x) d\tau_t(x).$$

(3) *There exists a sequence* $(u_n)_n \subseteq L^p(\Omega, \mathbb{R}^m)$ *such that:*

$$\begin{cases} \text{(a)} \quad (\|u_n\|^p)_{n\in\mathbb{N}} \quad \textit{is uniformly integrable and} \\ \text{(b)} \quad f(u_n) \xrightarrow[L^\infty(\Omega)]{w*} u_f, \quad \textit{for every} \quad f \in C_0(\mathbb{R}^m), \end{cases}$$

where $u_f : \Omega \to \mathbb{R}, u_f(t) = \int_{\mathbb{R}^m} f(x) d\tau_t(x)$.

Proof. (1)$\Longrightarrow$ (2): Let $(u_n)_n \subseteq L^p(\Omega, \mathbb{R}^m)$, a sequence which generates τ; therefore $(\|u_n\|^p)_n$ is uniformly integrable and $u_n \overset{S}{\to} \tau$.

(a) According to Definition 1.85 and Theorem 1.84, $\{\|u_n\|^p : n \in \mathbb{N}\}$ is relatively weakly compact. So, with eventually extracting a subsequence, still noted $(u_n)_n$, we can suppose that $u_n^p \xrightarrow[L^1(\Omega)]{w} g \in L^1(\Omega)$.

(b) Let now $\varphi \in E^p$ and let $L = \lim_{\|x\|\to\infty} \frac{\varphi(x)}{1+\|x\|^p} \in \mathbb{R}$. There is $k > 0$ such that $|\frac{\varphi(x)}{1+\|x\|^p} - L| < 1$, for every $x \in \mathbb{R}^m$ with $\|x\| > k$. Since φ is continuous, $\sup_{\|x\|\le k} |\frac{\varphi(x)}{1+\|x\|^p}| = K \in \mathbb{R}_+$ and therefore

$$\left|\frac{\varphi(x)}{1 + \|x\|^p}\right| \le \max\{K, |L| + 1\} = M, \quad \text{for every} \quad x \in \mathbb{R}^m.$$

Then, for every $n \in \mathbb{N}$,

$$|\varphi(u_n)| = \left|\frac{\varphi(u_n)}{1 + \|u_n\|^p}\right| \cdot (1 + \|u_n\|^p) \le M \cdot (1 + \|u_n\|^p)$$

and, as $(\|u_n\|^p)_n$ is uniformly integrable, $(\varphi(u_n))_n$ is too uniformly integrable.

Let $\Psi : \Omega \times \mathbb{R}^m \to \mathbb{R}$, defined by $\Psi(t, x) = \varphi(x)$. Then Ψ is a Carathéodory integrand bounded from below for which $(\Psi(\cdot, u_n))_n = (\varphi(u_n))_n$ is uniformly integrable. According to (2) of Theorem 3.128, $u_\Psi = \int_{\mathbb{R}^m} \varphi(x) d\tau_\cdot(x) \in L^1(\Omega)$ and

$$\Psi(\cdot, u_n) = \varphi(u_n) \xrightarrow[L^1(\Omega)]{w} u_\Psi = u_\varphi.$$

(2)$\Longrightarrow$(3): Let $(u_n)_n \subseteq L^p(\Omega, \mathbb{R}^m)$ be a sequence which satisfies (a) and (b) of (2). According to 1.84, $(\|u_n\|^p)_n$ is uniformly integrable.

For every $f \in C_0(\mathbb{R}^m)$, $f \in E^p$ and so, from (b), $f(u_n) \xrightarrow[L^1(\Omega)]{w} u_f$. Therefore, according to Theorem 1.57, for every $A \in \mathcal{A}$, $\int_\Omega \chi_A \cdot f(u_n) dt \to \int_\Omega \chi_A \cdot u_f dt$. We can extend this to $\mathcal{E}(\mathcal{A})$—the set of all simple functions on $(\Omega, \mathcal{A}, \mu)$. Therefore,

$$\int_\Omega h \cdot f(u_n) dt \to \int_\Omega h \cdot u_f dt, \quad \text{for every} \quad h \in \mathcal{E}(\mathcal{A}).$$

Let now $h \in L^1(\Omega)$, arbitrary; for every $\varepsilon > 0$ there is $h_1 \in \mathcal{E}(\mathcal{A})$ such that $\|h - h_1\|_1 < \varepsilon$.

We remark that f is a bounded function on $\mathbb{R}^m$; then $(f(u_n))_n \subseteq L^\infty(\Omega)$ and $u_f \in L^\infty(\Omega)$. Moreover, $\|f(u_n) - u_f\|_\infty \leq 2\|f\|_\infty$, for every $n \in \mathbb{N}$. Then

$$\begin{aligned}
\left| \int_\Omega h f(u_n) dt - \int_\Omega h u_f dt \right| &\leq \left| \int_\Omega (h - h_1)(f(u_n) - u_f) dt \right| \\
+ \left| \int_\Omega h_1 (f(u_n) - u_f) dt \right| &\leq \|h - h_1\|_1 \|f(u_n) - u_f\|_\infty \\
&\quad + \left| \int_\Omega h_1 (f(u_n) - u_f) dt \right|.
\end{aligned}$$

Since the last term converges to 0, we obtain that

$$\int_\Omega h \cdot f(u_n) dt \xrightarrow[n \to +\infty]{} \int_\Omega h \cdot u_f dt, \quad \text{for every} \quad h \in L^1(\Omega).$$

Therefore, $f(u_n) \xrightarrow[L^\infty(\Omega)]{w*} u_f$.

(3)$\Longrightarrow$(1): Let $(u_n)_n \subseteq L^p(\Omega, \mathbb{R}^m)$ be a sequence p-uniformly integrable and such that

$$f(u_n) \xrightarrow[L^\infty(\Omega)]{w*} u_f, \quad \text{for every} \quad f \in C_c(\mathbb{R}^m) \subseteq C_0(\mathbb{R}^m).$$

Since, for every $A \in \mathcal{A}$, $\chi_A \in L^1(\Omega)$,

$$\int_A f(u_n) dt \to \int_A u_f dt = \int_\Omega \int_{\mathbb{R}^m} f(x) d\tau_t(x).$$

According to (7) of Proposition 3.22, this means that $u_n \overset{\mathcal{S}}{\to} \tau$. □

Remark 3.133.

(i) The equivalent form (2) for p-Young measures belongs to D. Kinderlehrer and P. Pedregal and has been given in [102].

(ii) Many papers use as definition the equivalent form (3); see for example the Definition 8.3 in [85] and definition 2.1 in [100]. We remark that, in these definitions, the condition (a) are missing; but without it, (3) does not imply (1). Indeed, the sequence $(u_n)_n \subseteq L^1([0,1]), u_n = n \cdot \chi_{[0,\frac{1}{n}]}$, is not uniformly integrable but it satisfies the condition (b) from (3).

In the case where a Young measure τ is generated by the gradients of a sequence $(u_n)_n \subseteq W^{1,p}(\Omega, \mathbb{R}^m)$, we say that τ is a gradient Young measure. This notion has been introduced by D. Kinderlehrer and P. Pedregal in [102]. This one is presented here in the equivalent form (1) (see the previous theorem).

Definition 3.134. Let $1 \leq p < +\infty$, $\Omega \subseteq \mathbb{R}^d$ be a bounded open set and let $W^{1,p}(\Omega, \mathbb{R}^m)$ be the Sobolev space (the definition of which we already presented in 3.98).

A Young measure $\tau \in \mathcal{Y}(\mathbb{R}^{md})$ is a *p-gradient Young measure* if there exists a sequence $(u_n)_n \subseteq W^{1,p}(\Omega, \mathbb{R}^m)$ such that:

$$\begin{cases} \text{(a)} \quad (\|\nabla u_n\|^p)_{n\in\mathbb{N}} \quad \text{is uniformly integrable in } L^1(\Omega, \mathbb{R}^{md}) \text{ and} \\ \text{(b)} \quad \nabla u_n \xrightarrow[\mathcal{Y}(\mathbb{R}^{md})]{\mathcal{S}} \tau. \end{cases}$$

In this definition, $\|\cdot\|$ notes the usual norm in $\mathbb{R}^{md}$.

Obviously, τ is generated by $(\nabla u_n)_n$.

Remark 3.135.

(i) Adapting Remark 3.131 to the case of p-gradient Young measures, we obtain: $\nabla u_n \xrightarrow[L^1(\Omega,\mathbb{R}^{md})]{w} \mathrm{bar}\tau_. \in L^1(\Omega, \mathbb{R}^{md})$.

(ii) Theorem 3.132 allows us to obtain easily characterizations of type (2) and (3) for p-gradient Young measures, replacing the sequence $(u_n)_n$ with the sequence $(\nabla u_n)_n$.

Proposition 3.136. *Let $\tau \in \mathcal{Y}(\mathbb{R}^{md})$ be a p-gradient Young measure and let $\mathrm{bar}\tau_. : \Omega \to \mathbb{R}^{md}$ be its barycenter:* $\mathrm{bar}\tau_t = \int_\Omega y\, d\tau_t(y)$. *Then*

$$\int_{\Omega\times\mathbb{R}^{md}} \|y\|^p d\tau(t,y) < +\infty \quad \textit{and so} \quad \mathrm{bar}\tau_. \in L^p(\Omega, \mathbb{R}^{md}).$$

Proof. According to (i) of previous remark, $\text{bar}\tau_{.} \in L^1(\Omega, \mathbb{R}^{md})$.

Let $(u_n)_n \subseteq W^{1,p}(\Omega, \mathbb{R}^m)$ such that $(\|\nabla u_n\|^p)_{n\in\mathbb{N}}$ is uniformly integrable in $L^1(\Omega, \mathbb{R}^{md})$ and $\nabla u_n \xrightarrow[\mathcal{Y}(\mathbb{R}^{md})]{\mathcal{S}} \tau$. The mapping $\Psi : \Omega \times \mathbb{R}^{md} \to \mathbb{R}$, defined by $\Psi(t, y) = \|y\|^p$, is a Carathéodory integrand and $\{\Psi(\cdot, \nabla u_n(\cdot)) : n \in \mathbb{N}\} = \{\|\nabla u_n\|^p : n \in \mathbb{N}\}$ is uniformly integrable. According to Corollary 3.36,

$$\begin{aligned}\int_{\Omega\times\mathbb{R}^{md}} \|y\|^p d\tau(t, y) &= \int_\Omega \int_{\mathbb{R}^{md}} \|y\|^p d\tau_t(y) dt \\ &= \lim_n \int_\Omega \|\nabla u_n(t)\|^p dt < +\infty\end{aligned}$$

(every uniformly integrable subset of $L^1(\Omega)$ is bounded).

Then the mapping $t \mapsto \int_{\mathbb{R}^{md}} \|y\|^p d\tau_t(y)$ is finite almost every where and integrable.

Now we can apply the Jensen's inequality to obtain:

$$\|\text{bar}\tau_t\|^p_{\mathbb{R}^{md}} = \left\| \int_{\mathbb{R}^{md}} y\, d\tau_t(y) \right\|^p \leq \int_{\mathbb{R}^{md}} \|y\|^p d\tau_t(y).$$

Therefore, $\text{bar}\tau_{.} \in L^p(\Omega, \mathbb{R}^{md})$. □

Proposition 3.137. *Let $\Omega \subseteq \mathbb{R}^d$ be a bounded open set with a Lipschitz boundary (see Definition* 12.10 *from* [50]*) and let $1 \leq p < +\infty$.*

Let $\tau \in \mathcal{Y}(\mathbb{R}^{md})$ be a p-gradient Young measure generated by $(\nabla u_n)_n \subseteq W_0^{1,p}(\Omega, \mathbb{R}^m)$ (or $(u_n - u_0)_n \subseteq W_0^{1,p}(\Omega, \mathbb{R}^m)$, where $u_0 \in W^{1,p}(\Omega, \mathbb{R}^m)$).

There exist $u \in W^{1,p}(\Omega, R^m)$ and a subsequence of $(u_n)_n$, still noted $(u_n)_n$, such that

(1) $u_n \xrightarrow[L^p(\Omega,\mathbb{R}^m)]{\|\cdot\|_p} u$.

(2) $\nabla u_n \xrightarrow[L^p(\Omega,\mathbb{R}^{md})]{w} \nabla u = \text{bar}\tau_{.}$.

(3) $u_n \xrightarrow[W^{1,p}(\Omega,\mathbb{R}^m)]{w} u$.

Proof. (1) Since $(\|\nabla u_n\|^p)_n$ is uniformly integrable in $L^1(\Omega, \mathbb{R}^{md})$, $(\nabla u_n)_n$ is bounded in $L^p(\Omega, \mathbb{R}^{md})$. Appealing to Poincaré inequality (see, for example, Corollary IX.19 in [36] or Theorem 1.47 in [49]), $(u_n)_n$ is bounded in $W^{1,p}(\Omega, \mathbb{R}^m)$ and therefore, from Rellich–Kondrachov theorem (Theorem 12.12 in [50]), up to a subsequence, $(u_n)_n$ converges strongly to $u \in L^p(\Omega, \mathbb{R}^m)$.

(2) Let now $u_n = (u_n^1, \dots, u_n^m)$ and $u = (u^1, \dots, u^m)$. For every $\varphi \in C_c^\infty(\Omega, \mathbb{R})$, every $i = 1, \dots, m$ and every $j = 1, \dots, d$,

$$\int_\Omega \frac{\partial u_n^i}{\partial t_j} \cdot \varphi dt = -\int_\Omega u_n^i \cdot \frac{\partial \varphi}{\partial t_j} dt. \tag{1}$$

From (i) of Remark 3.135, $\nabla u_n \xrightarrow[L^1(\Omega,\mathbb{R}^{md})]{w} \text{bar}\tau_. = (v_j^i)_{j=1,\dots,d}^{i=1,\dots,m} \in L^1(\Omega, \mathbb{R}^{md})$ and, since $\varphi \in C_c^\infty(\Omega, \mathbb{R}) \subseteq L^\infty(\Omega, \mathbb{R})$,

$$\int_\Omega \frac{\partial u_n^i}{\partial t_j} \cdot \varphi dt \to \int_\Omega v_j^i \cdot \varphi dt. \tag{2}$$

Since $u_n \xrightarrow[L^p]{\|\cdot\|_p} u$,

$$\int_\Omega u_n^i \cdot \frac{\partial \varphi}{\partial t_j} dt \to \int_\Omega u^i \cdot \frac{\partial \varphi}{\partial t_j} dt. \tag{3}$$

Using (2) and (3) in (1), we obtain, for every $\varphi \in C_c^\infty(\Omega, \mathbb{R})$:

$$\int_\Omega u^i \cdot \frac{\partial \varphi}{\partial t_j} dt = -\int_\Omega v_j^i \cdot \varphi dt, \quad \text{for every} \quad i = 1, \dots, m, j = 1, \dots, d.$$

Therefore $\nabla u = \text{bar}\tau_. \in L^p(\Omega, \mathbb{R}^{md})$ (see the previous proposition).

Then $u \in W^{1,p}(\Omega, \mathbb{R}^m)$. Moreover,

$$\int_\Omega \frac{\partial u_n^i}{\partial t_j} \cdot \varphi dt \to \int_\Omega \frac{\partial u^i}{\partial t_j} \cdot \varphi dt, \quad \text{for every} \quad \varphi \in C_c^\infty(\Omega, \mathbb{R}).$$

Since $C_c^\infty(\Omega, \mathbb{R})$ is dense in $L^q(\Omega, \mathbb{R})$ ($\frac{1}{p} + \frac{1}{q} = 1$), $\nabla u_n \xrightarrow[L^p(\Omega,\mathbb{R}^{md})]{w} \nabla u$.

(3) is a consequence of (1) and (2). □

Once we will introduce quasiconvexity, we will can present the Kinderlehrer–Pedregal's characterization of p-gradient Young measures.

3.10.2 Quasiconvex Functions

Quasiconvex functions were introduced in 1952 by Ch. B. Morrey, Jr. (see [121] and [122]) in order to characterize lower semicontinuity of functional

$$I(u) = \int_\Omega f(t, u(t), \nabla u(t)) dt,$$

commonly occurring in problems of variational calculus. Morrey's results were later extended by Acerbi and Fusco ([1], 1984) and Marcellini ([118], 1985).

In Chapter 5 of [50], we can find an excellent comparative study of various extensions of convexity.

Definition 3.138. Let $f : \mathbb{R}^{md} \to \mathbb{R}$ be a continuous function; f is said to be *quasiconvex* if, for every $y \in \mathbb{R}^{md}$, for every bounded open set $\Omega \subseteq \mathbb{R}^d$ and for every $\varphi \in C_c^\infty(\Omega, \mathbb{R}^m)$,

$$f(y) \leq \frac{1}{\mu(\Omega)} \cdot \int_\Omega f(y + \nabla\varphi(t))dt.$$

In the previous definition,

$$\nabla\varphi = \begin{pmatrix} \frac{\partial\varphi_1}{\partial t_1} \cdots \frac{\partial\varphi_1}{\partial t_d} \\ \cdots \\ \frac{\partial\varphi_m}{\partial t_1} \cdots \frac{\partial\varphi_m}{\partial t_d} \end{pmatrix}$$

is the gradient of function $\varphi = (\varphi_1, \ldots, \varphi_m) : \Omega \to \mathbb{R}^m$.

In the next proposition, we show that the definition of quasiconvexity does not depend on the choice of Ω.

Let $p^1, \ldots, p^d \in \mathbb{R}^d$ such that $< p^i, p^j >= \delta_{ij}$, for every $i, j \in \{1, \ldots, n\}$ and let $Q = Q(p^1, \ldots, p^d) = \{t \in \mathbb{R}^d : 0 << t, p^i >< 1$, for every $i = 1, \ldots, n\}$ be the open hypercube of side 1 which is based on versors $p^1, \ldots, p^d$.

Proposition 3.139. *The continuous function $f : \mathbb{R}^{md} \to \mathbb{R}$ is quasiconvex if and only if, for every $y \in \mathbb{R}^{md}$ and every $\varphi \in C_c^\infty(Q, \mathbb{R}^m)$,*

$$f(y) \leq \int_Q f(y + \nabla\varphi(t))dt.$$

Proof. The necessity is obvious. Let us prove the sufficiency; let $y \in \mathbb{R}^{md}$, let Ω be a bounded open subset of $\mathbb{R}^d$ and let $\varphi \in C_c^\infty(\Omega, \mathbb{R}^m)$. Let $t^0 \in \mathbb{R}^d$ and $r > 0$ such that $S(t^0, r) \supseteq \Omega$ and let $s^0 = -r \cdot \sum_{j=1}^d p^j \in \mathbb{R}^d$; then

$$\Omega \subseteq t^0 + s^0 + 2r \cdot Q \equiv Q_1.$$

Indeed, if $t \in \Omega \subseteq S(t^0, r)$, $\|t - t^0\| < r$ and so, for every $i \in \{1, \dots, n\}$,

$$\left\langle \frac{1}{2r} \cdot (t - t^0 - s^0), p^i \right\rangle = \frac{1}{2r} \cdot \left\langle t - t^0, p^i \right\rangle - \frac{1}{2r} \cdot \left\langle s^0, p^i \right\rangle = \frac{1}{2r} \cdot \left\langle t - t^0, p^i \right\rangle + \frac{1}{2}.$$

$$\left| \left\langle t - t^0, p^i \right\rangle \right| \leq \|t - t^0\| < r \Rightarrow -r < \left\langle t - t^0, p^i \right\rangle < r$$

$$\Rightarrow 0 = -r \cdot \frac{1}{2r} + \frac{1}{2} < \frac{1}{2r} \cdot \left\langle x - t^0, p^i \right\rangle + \frac{1}{2} < r \cdot \frac{1}{2r} + \frac{1}{2} = 1.$$

It follows that $\frac{1}{2r} \cdot (t - t^0 - s^0) \in Q$, from where $t \in t^0 + s^0 + 2r \cdot Q \equiv Q_1$.

We extend φ to Q_1, giving value 0 on $Q_1 \setminus \Omega$; obviously, $\varphi \in C_c^\infty(Q_1, \mathbb{R}^m)$.

$$\begin{aligned} \int_\Omega f(y + \nabla\varphi(t))dt &= \int_{Q_1} f(y + \nabla\varphi(t))dt - \int_{Q_1 \setminus \Omega} f(y)dt \\ &= \int_{Q_1} f(y + \nabla\varphi(t))dt - f(y) \cdot \mu(Q_1 \setminus \Omega). \end{aligned}$$

We make the change of variable $t = t^0 + s^0 + 2r \cdot s$; then

$$t \in Q_1 \iff s = \frac{1}{2r} \cdot (t - t^0 - s^0) \in Q$$

and so

$$\begin{aligned} \int_\Omega f(y + \nabla\varphi(t))dt = (2r)^d \cdot \int_Q & f(y + (\nabla\varphi)(t^0 + s^0 + 2rs))ds \\ & - f(y) \cdot \mu(Q_1 \setminus \Omega). \end{aligned}$$

Let $\psi(t) = \frac{1}{2r} \cdot \varphi(t^0 + s^0 + 2rs)$; $\mathrm{supp}\psi = -\frac{1}{2r}(t^0 + s^0) + \frac{1}{2r} \cdot \mathrm{supp}\varphi \subseteq Q$ $\nabla\psi(s) = (\nabla\varphi)(t^0 + s^0 + 2rs)$ and, therefore,

$$\begin{aligned} \int_\Omega f(y + \nabla\varphi(t))dt &= (2r)^d \cdot \int_Q f(y + \nabla\psi(s))ds - f(y) \cdot \mu(Q_1 \setminus \Omega) \\ &\geq \mu(Q_1) \cdot f(y) - \mu(Q_1 \setminus \Omega) \cdot f(y) = \mu(\Omega) \cdot f(y). \end{aligned}$$

□

Remark 3.140.

(i) The cube Q in the preceding proposition, may be replaced by a cube translated at any point of $\mathbb{R}^d$. So, if Q is a cube of side 1 in $\mathbb{R}^d$, then the continuous function $f : \mathbb{R}^{md} \to \mathbb{R}$ is quasiconvex if and only if, for every $y \in \mathbb{R}^{md}$ and every $\varphi \in C_c^\infty(Q, \mathbb{R}^m)$,

$$f(y) \leq \int_Q f(y + \nabla\varphi(t))dt.$$

(ii) If, in the previous proposition, $p^1, \dots, p^d$ are the versors $e^1, \dots, e^d$ of the coordinate system of $\mathbb{R}^d$, then $Q = (0,1)^d$.

(iii) Let $p^1 = \{p_1^1, \dots, p_d^1\}, \dots, p^d = \{p_1^d, \dots, p_d^d\}$ be d versors pairwise orthogonal in $\mathbb{R}^d$ and let $Q(p^1, \dots, p^d)$ be the hypercube which is based on them; we denote by

$$P = \begin{pmatrix} p_1^1 \cdots p_d^1 \\ \cdots \\ p_1^d \cdots p_d^d \end{pmatrix}$$

and by $T : \mathbb{R}^d \to \mathbb{R}^d$ the mapping defined by $T(t) = P \cdot t$, for every $t \in \mathbb{R}^d$. T is a linear change of basis in $\mathbb{R}^d$; $\forall i = 1, \dots, d, T(p^i) = e^i$ şi $T(Q(p^1, \dots, p^d)) = (0,1)^d = Q$.

For every $y \in \mathbb{R}^{md}$ and every $\varphi \in C_c^\infty(\Omega, \mathbb{R}^m), \varphi \circ T \in C_c^\infty(Q(p^1, \dots, p^d), \mathbb{R}^d)$ (T is a homeomorphism and so $\operatorname{supp}(\varphi \circ T) = T^{-1}(\operatorname{supp}\varphi)$ is a compact set in $Q(p^1, \dots, p^d)$); then $y \cdot P \in \mathbb{R}^{md}$ and therefore

$$\begin{aligned} f(y \cdot P) &\le \int_{Q(p^1,\dots,p^d)} f(y \cdot P + \nabla(\varphi \circ T)(t))dt \\ &= \int_Q f(y \cdot P + \nabla\varphi(s) \cdot P)ds = \int_Q f[(y + \nabla\varphi(s)) \cdot P]ds. \end{aligned}$$

If f is quasiconvex, then $f(\cdot T)$ is quasiconvex, also.

We can now prove that any convex function is quasiconvex.

Proposition 3.141. *Every convex continuous function $f : \mathbb{R}^{md} \to \mathbb{R}$ is quasiconvex.*

Proof. Let $Q = (0,1)^d$, let $y \in \mathbb{R}^{md}$ and let $\varphi \in C_c^\infty(Q, \mathbb{R}^m)$; applying Jensen's inequality, we get

$$f\left(\int_Q (y + \nabla\varphi(t))dt\right) \le \int_Q f(y + \nabla\varphi(t))dt.$$

On the other hand

$$\int_Q (y + \nabla\varphi(t))dt = y + \left(\int_Q \frac{\partial \varphi_j}{\partial t_i}(t)dt\right)_{\substack{1 \le i \le d \\ 1 \le j \le m}}$$

$$= y + \left(\int_{(0,1)^{d-1}} \left(\int_0^1 \frac{\partial \varphi_j}{\partial t_i}(t)dt_i\right) dt_1 \dots dt_{i-1} dt_{i+1} \dots dt_d\right)_{\substack{1 \le i \le d \\ 1 \le j \le m}} = y.$$

According to Proposition 3.139, f is quasiconvex. □

The converse of Proposition 3.141 is not true.

Example 3.142. Let $f : \mathbb{R}^{2\times 2} \to \mathbb{R}$ be the function defined by $f(x) = \det(x)$; i.e., if $x = \left(\begin{smallmatrix} x_1 & x_2 \\ x_3 & x_4 \end{smallmatrix}\right)$, then $f(x) = x_1x_4 - x_2x_3$.

$$|f(y) - f(z)| \le (\|y\| + \|z\|) \cdot \|y - z\|, \quad \text{for every} \quad y, z \in \mathbb{R}^{2\times 2}.$$

Therefore, f is continuous.

Let $Q = (0,1)^2 = (0,1) \times (0,1)$ and let $\varphi = (\varphi_1, \varphi_2) \in C_c^\infty(Q, \mathbb{R}^2)$;

$$\nabla\varphi(t) = \begin{pmatrix} \frac{\partial\varphi_1}{\partial t_1}(t) & \frac{\partial\varphi_1}{\partial t_2}(t) \\ \frac{\partial\varphi_2}{\partial t_1}(t) & \frac{\partial\varphi_2}{\partial t_2}(t) \end{pmatrix};$$

then, for every $y = \left(\begin{smallmatrix} y_1 & y_2 \\ y_3 & y_4 \end{smallmatrix}\right) \in \mathbb{R}^{2\times 2}$,

$$\begin{aligned} f(y + \nabla\varphi(t)) &= \left(y_1 + \frac{\partial\varphi_1}{\partial t_1}(t)\right) \cdot \left(y_4 + \frac{\partial\varphi_2}{\partial t_2}(t)\right) \\ &\quad - \left(y_2 + \frac{\partial\varphi_1}{\partial t_2}(t)\right) \cdot \left(y_3 + \frac{\partial\varphi_2}{\partial t_1}(t)\right) \\ &= f(y) + y_4\frac{\partial\varphi_1}{\partial t_1}(t) + y_1\frac{\partial\varphi_2}{\partial t_2}(t) - y_3\frac{\partial\varphi_1}{\partial t_2}(t) - y_2\frac{\partial\varphi_2}{\partial t_1}(t) \\ &\quad + \frac{\partial\varphi_1}{\partial t_1}(t) \cdot \frac{\partial\varphi_2}{\partial t_2}(t) - \frac{\partial\varphi_1}{\partial t_2}(t) \cdot \frac{\partial\varphi_2}{\partial t_1}(t). \end{aligned}$$

$$\int_Q \frac{\partial\varphi_1}{\partial t_1}(t)dt = \int_Q \frac{\partial\varphi_2}{\partial t_2}(t)dt = \int_Q \frac{\partial\varphi_1}{\partial t_2}(t)dt = \int_Q \frac{\partial\varphi_2}{\partial t_1}(t)dt = 0$$

and

$$\begin{aligned} &\int_Q \left(\frac{\partial\varphi_1}{\partial t_1}(t) \cdot \frac{\partial\varphi_2}{\partial t_2}(t) - \frac{\partial\varphi_1}{\partial t_2}(t) \cdot \frac{\partial\varphi_2}{\partial t_1}(t)\right) dt \\ &= -\int_Q \varphi_1(t) \cdot \frac{\partial^2\varphi_2}{\partial t_1\partial t_2}(t)dt + \int_Q \varphi_1(t) \cdot \frac{\partial^2\varphi_2}{\partial t_1\partial t_2}(t)dt = 0. \end{aligned}$$

It follows that

$$\int_Q f(y + \nabla\varphi(t))dt = f(y)$$

and, using Proposition 3.139, f is quasiconvex.

On the other hand, the function f is not convex; indeed, if $y = \left(\begin{smallmatrix} 1 & 0 \\ 0 & 0 \end{smallmatrix}\right), z = \left(\begin{smallmatrix} 0 & 0 \\ 0 & 1 \end{smallmatrix}\right) \in \mathbb{R}^{2\times 2}$, then $f(\frac{1}{2}y + \frac{1}{2}z) = \frac{1}{4} > 0 = \frac{1}{2}f(y) + \frac{1}{2}f(z)$.

In the following, we broad the class of admissible functions φ in the definition of quasiconvexity.

Theorem 3.143. *A continuous function $f : \mathbb{R}^{md} \to \mathbb{R}$ is quasiconvex if and only if, for every $y \in \mathbb{R}^{md}$, every bounded open subset $\Omega \subseteq \mathbb{R}^d$ and every $\varphi \in W_0^{1,\infty}(\Omega, \mathbb{R}^m)$,*

$$f(y) \leq \frac{1}{\mu(\Omega)} \cdot \int_\Omega f(y + \nabla\varphi(t))dt.$$

Proof. The sufficiency is a consequence of the inclusion $C_c^\infty(\Omega, \mathbb{R}^m) \subseteq W_0^{1,\infty}(\Omega, \mathbb{R}^m)$. Let us prove the necessity.

We recall the standard procedure of approximation.

Let $\rho \in C_c^\infty(\mathbb{R}^d)$, $\rho \geq 0$, $\rho(-t) = \rho(t)$, $\text{supp}\rho \subseteq B_1$, $\int_{\mathbb{R}^d} \rho(t)dt = 1$ and, for every $\varepsilon > 0$, $\rho_\varepsilon(t) = \frac{1}{\varepsilon^d} \cdot \rho(\frac{t}{\varepsilon})$. We extend every $u \in L^1(\Omega)$ to $\mathbb{R}^d$ giving the value 0 on the set $\mathbb{R}^d \setminus \Omega$ and we note the extension again with u.

Let $u_\varepsilon = u * \rho_\varepsilon$, i.e. $u_\varepsilon(t) = \int_{\mathbb{R}^d} u(s) \cdot \rho_\varepsilon(t-s)ds = \int_{\mathbb{R}^d} u(t - \varepsilon s) \cdot \rho(s)ds$. Then $u_\varepsilon \in C^\infty(\Omega)$, $u_\varepsilon \xrightarrow{L^1} u$ and $\nabla_i u_\varepsilon = \frac{1}{\varepsilon} \cdot \int_{\mathbb{R}^d} u(t - \varepsilon s)\nabla_i \rho(s)ds$.

For every $u \in W^{1,1}(\Omega, \mathbb{R}^m)$, $\nabla_i u_\varepsilon = u * \nabla_i \rho_\varepsilon = \nabla_i u * \rho_\varepsilon$.

For every $\varphi \in W_0^{1,\infty}(\Omega, \mathbb{R}^m)$, $\nabla\varphi \in L^\infty(\Omega, \mathbb{R}^{md})$ and so $(\nabla\varphi)_\varepsilon \xrightarrow{L^1} \nabla\varphi$; we recall that $\nabla\varphi = (\nabla_i \varphi^j)_{\substack{1\leq i\leq d \\ 1\leq j\leq m}}$.

It follows that $\varphi_\varepsilon \in C^\infty(\mathbb{R}^d, \mathbb{R}^m)$ and $\text{supp}\varphi_\varepsilon \subseteq \text{supp}\varphi + B_\varepsilon$ is compact. $\text{supp}\varphi = K \subseteq \Omega$; there exists $\varepsilon_0 > 0$ such that $K^\varepsilon = \text{supp}\varphi + B_\varepsilon \subseteq \Omega$, for every $\varepsilon \leq \varepsilon_0$ and then $\varphi_\varepsilon \in C_c^\infty(\Omega, \mathbb{R}^m)$.

From hypothesis, for every $y \in \mathbb{R}^{md}$ and every $\varepsilon \leq \varepsilon_0$,

$$f(y) \leq \frac{1}{\mu(\Omega)} \cdot \int_\Omega f(y + \nabla\varphi_\varepsilon(t))dt. \tag{*}$$

$\nabla\varphi_\varepsilon = (\nabla\varphi)_\varepsilon$; indeed, for every $1 \leq i \leq d, 1 \leq j \leq m$,

$$\begin{aligned}\left(\nabla_i \varphi^j\right)_\varepsilon(t) &= \int_{\mathbb{R}^d} \left(\nabla_i \varphi^j\right)(s) \cdot \rho_\varepsilon(t-s)ds = -\int_{\mathbb{R}^d} \varphi^j(s) \cdot \nabla_i(\rho_\varepsilon(t-s))ds \\ &= \int_{\mathbb{R}^d} \varphi^j(s) \cdot (\nabla_i \rho_\varepsilon)(t-s)ds = \nabla_i(\varphi_\varepsilon^i)(t).\end{aligned}$$

Since $\nabla\varphi_\varepsilon = (\nabla\varphi)_\varepsilon \xrightarrow[\varepsilon\to 0]{L^1} \nabla\varphi$, there exists a subsequence of $(\nabla\varphi_\varepsilon)_{\varepsilon>0}$ almost everywhere convergent to $\nabla\varphi$. On the other hand, $\nabla\varphi_\varepsilon$ is continuous with compact support in Ω; it follows that there is $M > 0$ such that $\|\nabla\varphi_\varepsilon\|_{\mathbb{R}^{md}} \leq M$. Since f is continuous on $\mathbb{R}^{md}$, f is bounded on $B(y, M)$ and then we can apply the Lebesgue's theorem

$$\int_\Omega f(y + \nabla\varphi_\varepsilon(t))dt \to \int_\Omega f(y + \nabla\varphi(t))dt.$$

According to $(*)$, it follows that

$$f(y) \leq \frac{1}{\mu(\Omega)} \cdot \int_{\Omega} f(y + \nabla\varphi(t))dt. \qquad \square$$

Remark 3.144. We cannot relax condition $\varphi \in W_0^{1,\infty}(\Omega, \mathbb{R}^m)$ of the previous theorem to condition $\varphi \in W^{1,\infty}(\Omega, \mathbb{R}^m)$. Indeed, as we remarked in Example 3.142, the function $f : \mathbb{R}^{2\times 2} \to \mathbb{R}$, $f(x) = \det(x)$ is quasiconvex. Let $Q = (0,1) \times (0,1)$ and $\varphi : Q \to \mathbb{R}^2$ defined by $\varphi(t_1, t_2) = (t_1 \cdot t_2, 0)$, for every $(t_1, t_2) \in Q$. Then $\varphi \in L^{\infty}(Q, \mathbb{R}^2)$ and $\nabla\varphi(t_1, t_2) = \left(\begin{smallmatrix} t_2 & t_1 \\ 0 & 0 \end{smallmatrix}\right)$. Therefore $\nabla\varphi \in L^{\infty}(Q, \mathbb{R}^{2\times 2})$, and so $\varphi \in W^{1,\infty}(Q, \mathbb{R}^2)$.

However, if $y = \left(\begin{smallmatrix} 0 & 0 \\ 2 & 0 \end{smallmatrix}\right) \in \mathbb{R}^{2\times 2}$, then

$$0 = f(y) > \int_Q f(y + \nabla\varphi(t))dt = -\int_0^1 2t_1 dt_1 = -1.$$

But we can extend the characterization theorem to the class of periodic functions of $W^{1,\infty}(\mathbb{R}^d, \mathbb{R}^m)$.

Definition 3.145. Let $Q = (0,1)^d$; the function $\varphi : \mathbb{R}^d \to \mathbb{R}^m$ is *Q - periodic* if, for every $t = (t_1, \dots, t_d) \in \mathbb{R}^d$ and every $i = 1, \dots, d$, the function $g_i : \mathbb{R} \to \mathbb{R}$, defined by $g_i(s) = \varphi(t_1, \dots, t_{i-1}, s, t_{i+1}, \dots, t_d)$, is periodic with period 1.

Theorem 3.146. *Let $f : \mathbb{R}^{md} \to \mathbb{R}$ be a continuous function; f is quasiconvex if and only if, for every $y \in \mathbb{R}^{md}$ and every Q - periodic function $\varphi \in W^{1,\infty}(\mathbb{R}^d, \mathbb{R}^m)$,*

$$f(y) \leq \int_Q f(y + \nabla\varphi(t))dt.$$

Proof. Sufficiency follows from the fact that any function $\varphi \in C_c^{\infty}(Q, \mathbb{R}^m)$ can be extended to a Q - periodic function of $W^{1,\infty}(\mathbb{R}^d, \mathbb{R}^m)$.

Let us prove the necessity. We suppose that $f : \mathbb{R}^{md} \to \mathbb{R}$ is quasiconvex, $y \in \mathbb{R}^{md}$ and $\varphi \in W^{1,\infty}(\mathbb{R}^d, \mathbb{R}^m)$ is a Q-periodic function; for every $k \in \mathbb{N}^*$, we define $\varphi_k : \mathbb{R}^d \to \mathbb{R}^m$ by $\varphi_k(t) = \frac{1}{k} \cdot \varphi(kt)$, $\forall t \in \mathbb{R}^d$. Then $\varphi_k \in W^{1\infty}(\mathbb{R}^d, \mathbb{R}^m)$, $\|\varphi_k\|_{L^\infty} = \frac{1}{k} \cdot \|\varphi\|_{L^\infty}$ and $\|\nabla\varphi_k\|_{L^\infty} = \|\nabla\varphi\|_{L^\infty}$.

For every $p \in \mathbb{N}^*$, let $Q_p = \{t \in Q : d(t, \partial Q) > \frac{1}{p}\}$; $\mu(Q \setminus Q_p) \to 0$.

Let $\psi_p \in C_c^{\infty}(Q, \mathbb{R})$ such that $0 \leq \psi_p \leq 1$, $\psi_p|_{Q_p} = 1$ and let $M_p = \|\nabla\psi_p\|_{L^\infty} < +\infty$. There exists a subsequence $(\varphi_{k_p})_p$ of $(\varphi_k)_k$ such that $\|\varphi_{k_p}\|_{L^\infty} \leq \frac{1}{M_p+1}$, for every $p \in \mathbb{N}$. We choose this subsequence such that

$$\liminf_k \int_Q f(y + \nabla\varphi_k(t))dt = \lim_p \int_Q f(y + \nabla\varphi_{k_p}(t))dt.$$

We define $\theta_p = \psi_p \cdot \varphi_{k_p} \in W_0^{1,\infty}(Q, \mathbb{R}^m)$. Since f is quasiconvex,

$$\begin{aligned}
f(y) &\leq \int_Q f(y+\nabla\theta_p(t))dt = \int_{Q_p} f(y+\nabla\theta_p(t))dt + \int_{Q\setminus Q_p} f(y+\nabla\theta_p(t))dt \\
&= \int_{Q_p} f(y + \nabla\varphi_{k_p}(t))dt \\
&\quad + \int_{Q\setminus Q_p} f(y + \psi_p(t)\cdot\nabla\varphi_{k_p}(t) + \varphi_{k_p}(t)\otimes\nabla\psi_p(t))dt \\
&= \int_Q f(y + \nabla\varphi_{k_p}(t))dt + \varepsilon_p,
\end{aligned}$$

where

$$\varepsilon_p = \int_{Q\setminus Q_p} \left[f(y+\psi_p(t)\cdot\nabla\varphi_{k_p}(t)+\varphi_{k_p}(t)\otimes\nabla\psi_p(t)) - f(y + \nabla\varphi_{k_p}(t))\right] dt.$$

$$\|\psi_p(t)\cdot\nabla\varphi_{k_p}(t)\|_{\mathbb{R}^{md}} \leq \|\nabla\varphi_{k_p}(t)\|_{\mathbb{R}^{md}} \dot{\leq} \|\nabla\varphi_{k_p}\|_{L^\infty} \leq \|\nabla\varphi\|_{L^\infty} < +\infty.$$

$$\begin{aligned}
\|\varphi_{k_p}(t)\otimes\nabla\psi_p(t)\|_{\mathbb{R}^{md}} &\leq \|\varphi_{k_p}(t)\|_{\mathbb{R}^m}\cdot\|\nabla\psi_p(t)\|_{\mathbb{R}^d} \dot{\leq} \|\varphi_{k_p}\|_{L^\infty}\cdot\|\nabla\psi_p\|_{L^\infty} \\
&\leq \frac{M_p}{M_p+1} < 1.
\end{aligned}$$

From the above inequalities, it follows that, when t belongs to $Q \setminus Q_p$, $y + \psi_p(t)\cdot\nabla\varphi_{k_p}(t) + \varphi_{k_p}(t)\otimes\nabla\psi_p(t)$ belongs to the bounded set $y + B(\underline{0}, \|\nabla\varphi\|_{L^\infty}) + B(\underline{0}, 1) \subseteq \mathbb{R}^{md}$; also, $y + \nabla\varphi_{k_p}(t) \in y + B(\underline{0}, \|\nabla\varphi\|_{L^\infty})$. Since f is continuous on $\mathbb{R}^{md}$, f preserves the bounded sets. Then let

$$M = \sup_{x\in Q\setminus Q_p} |f(y + \psi_p(t)\cdot\nabla\varphi_{k_p}(t) + \varphi_{k_p}(t)\otimes\nabla\psi_p(t)| \in \mathbb{R}_+.$$

Therefore, $|\varepsilon_p| \leq 2M\cdot\mu(Q \setminus Q_p) \to 0$ and so

$$\begin{aligned}
f(y) &\leq \lim_p \Big[\int_Q f(y + \nabla\varphi_{k_p}(t))dt + \varepsilon_p\Big] \qquad (1) \\
&= \lim_p \int_Q f(y + \nabla\varphi_{k_p}(t))dt + \lim_p \varepsilon_p \\
&= \liminf_k \int_Q f(y + \nabla\varphi_k(t))dt.
\end{aligned}$$

Now

$$\int_Q f(y+\nabla\varphi_k(t))dt = \int_Q f(y+(\nabla\varphi)(kt))dt \overset{(t=\frac{1}{k}s)}{=} \frac{1}{k^d}\cdot\int_{kQ} f(y+\nabla\varphi(s))ds.$$

But $kQ \doteq \cup_{j=1}^{k^d} (t^j + Q)$, where t^j is the bottom left corner of each sub-cube of side one obtained by dividing the cube kQ (we use the almost everywhere equality, $\doteq$, because we neglect the faces of these sub-cubes).

Since $\nabla\varphi$ is Q-periodic,

$$\int_Q f(y + \nabla\varphi_k(t))dt = \frac{1}{k^d} \cdot k^d \cdot \int_Q f(y + \nabla\varphi(t))dt$$

and so

$$\liminf_k \int_Q f(y + \nabla\varphi_k(t))dt = \int_Q f(y + \nabla\varphi(t))dt. \tag{2}$$

By (1) and (2), it follows

$$f(y) \leq \int_Q f(y + \nabla\varphi(t))dt. \qquad \square$$

Remark 3.147. Using (iii) of Remark 3.140, we can replace, in the previous theorem, the cube $Q = (0,1)^d$ by a cube $Q(p^1, \ldots, p^d)$ based on the pairwise orthogonal versors $p^1, \ldots, p^d \in \mathbb{R}^d$.

The following two lemmas are useful for characterizing the quasiconvex functions through the Lipschitz functions, zero on the boundary.

Lemma 3.148. *Let $\Omega \subseteq \mathbb{R}^d$ be a bounded open set; for every $t \in \bar{\Omega}$ and every $s \in \mathbb{R}^d \setminus \bar{\Omega}$, $d(t,s) \geq d(t, \partial\Omega)$.*

Proof. We assume, by absurd, that there exist $t \in \bar{\Omega}$, $s \in \mathbb{R}^d \setminus \bar{\Omega}$ such that $d(t,s) < d(t,\partial\Omega)$; then $t \in \Omega$ because, if we suppose that $t \in \bar{\Omega} \setminus \Omega = \partial\Omega$, then $d(t,s) < d(t,\partial\Omega) = 0$.

Let $\alpha_0 = \sup\{\alpha \in [0,1] : (1-\alpha)t + \alpha s \in \Omega\}$ and $t^0 = (1-\alpha_0)t + \alpha_0 s$; there exists $\alpha_n \uparrow \alpha_0$ such that $(1-\alpha_n)t + \alpha_n s \in \Omega$, form where $t^0 \in \bar{\Omega}$. If we assume that $t^0 \in \Omega$, then there exists $r > 0$ such that the closed ball $B(t^0, r) \subseteq \Omega$. In this situation, let $\alpha_1 = \alpha_0 + \frac{r}{\|t-s\|} > \alpha_0$ and $t^1 = (1-\alpha_1)t + \alpha_1 s = t^0 + \frac{r}{\|t-s\|} \cdot (s-t)$; then $\|t^1 - t^0\| = \frac{r}{\|t-s\|} \cdot \|s-t\| = r$ and so $t^1 \in B(t^0, r) \subseteq \Omega$, which is absurd because $\alpha_1 > \alpha_0$.

It follows that $t^0 \in \bar{\Omega} \setminus \Omega = \partial\Omega$ and then $d(t,s) = \|t-s\| < d(t,\partial\Omega) \leq d(t,t^0) = \alpha_0 \cdot \|t-s\|$, from where $\alpha_0 > 1$!

Thus the previous assumption is false. $\square$

Lemma 3.149. *Let $\varphi \in \mathrm{Lip}(\bar{\Omega}, \mathbb{R}^m)$ such that $\varphi|_{\partial\Omega} = 0$ and let $\Omega_1 \subseteq \mathbb{R}^d$ be a bounded open set with $\bar{\Omega} \subseteq \Omega_1$; we extend φ to Ω_1, giving value 0 on $\Omega_1 \setminus \bar{\Omega}$.*

Then $\varphi \in \operatorname{Lip}(\Omega_1, \mathbb{R}^d)$ *and*

$$L(\varphi, \Omega_1) = L(\varphi, \bar{\Omega}) = \sup\left\{\frac{\|\varphi(t) - \varphi(s)\|}{\|t - s\|} : t, s \in \bar{\Omega}, t \neq s\right\}.$$

Proof. Let $t, s \in \Omega_1$ arbitrary;

(1) If $t, s \in \bar{\Omega}$, then $\|\varphi(t) - \varphi(s)\| \leq L(\varphi, \bar{\Omega}) \cdot \|t - s\|$.

(2) If $t \in \bar{\Omega}, s \in \Omega_1 \setminus \bar{\Omega}$, then for every $r \in \partial\Omega$,

$$\|\varphi(t) - \varphi(s)\| = \|\varphi(t)\| = \|\varphi(t) - \varphi(r)\| \leq L(\varphi, \bar{\Omega}) \cdot \|t - r\|,$$

from where, using Lemma 3.148,

$$\|\varphi(t) - \varphi(s)\| \leq L(\varphi, \bar{\Omega}) \cdot d(t, \partial\Omega) \leq L(\varphi, \bar{\Omega}) \cdot \|t - s\|.$$

(3) Finally, if $t, s \in \Omega_1 \setminus \bar{\Omega}$, $\|\varphi(t) - \varphi(s)\| = 0 \leq L(\varphi, \bar{\Omega}) \cdot \|t - s\|$.

In any case $\|\varphi(t) - \varphi(s)\| \leq L(\varphi, \bar{\Omega}) \cdot \|t - s\|$; hence φ is a Lipschitz function and $L(\varphi, \Omega_1) = L(\varphi, \bar{\Omega})$. □

The following characterization of quasiconvexity is used as definition in [121] (see Definition 2.2).

Theorem 3.150. *A continuous function* $f : \mathbb{R}^{md} \to \mathbb{R}$ *is quasiconvex if and only if, for every* $y \in \mathbb{R}^{md}$*, every bounded open subset* $\Omega \subseteq \mathbb{R}^d$ *and every* $\varphi \in \operatorname{Lip}(\bar{\Omega}, \mathbb{R}^m)$ *with* $\varphi|_{\partial\Omega} = 0$*,*

$$f(y) \leq \frac{1}{\mu(\bar{\Omega})} \cdot \int_{\bar{\Omega}} f(y + \nabla\varphi(t))dt.$$

Proof. Necessity. We suppose that f is quasiconvex and let $y \in \mathbb{R}^{md}$, let a bounded open subset $\Omega \subseteq \mathbb{R}^d$ and let $\varphi \in \operatorname{Lip}(\bar{\Omega}, \mathbb{R}^m)$ with $\varphi|_{\partial\Omega} = 0$.

We consider an arbitrary bounded open set $\Omega_1 \subseteq \mathbb{R}^d$ with $\bar{\Omega} \subseteq \Omega_1$ and we extend φ on Ω_1, by giving value 0 on $\Omega_1 \setminus \bar{\Omega}$. Lemma 3.149 assures us that $\varphi \in \operatorname{Lip}(\Omega_1, \mathbb{R}^m)$; then φ is uniformly continuous on Ω_1 and then it is bounded (φ is bounded on $\bar{\Omega}$!). It follows that $\varphi \in W^{1,\infty}(\Omega, \mathbb{R}^m)$ and as $\operatorname{supp}\varphi$ is a subset of compact set $\bar{\Omega}$, $\varphi \in W_0^{1,\infty}(\Omega, \mathbb{R}^m)$.

According to Theorem 3.143,

$$\begin{aligned} f(y) &\leq \frac{1}{\mu(\Omega_1)} \cdot \int_{\Omega_1} f(y + \nabla\varphi(t))dt \\ &= \frac{1}{\mu(\Omega_1)} \cdot f(y) \cdot \mu(\Omega_1 \setminus \bar{\Omega}) + \frac{1}{\mu(\Omega_1)} \cdot \int_{\bar{\Omega}} f(y + \nabla\varphi(t))dt \end{aligned}$$

from where

$$\frac{\mu(\bar{\Omega})}{\mu(\Omega_1)} \cdot f(y) \leq \frac{1}{\mu(\Omega_1)} \cdot \int_{\bar{\Omega}} f(y + \nabla\varphi(t))dt.$$

Sufficiency. Let $y \in \mathbb{R}^{md}, \Omega \subseteq \mathbb{R}^d$ be a bounded open set and let $\varphi = (\varphi_1, \ldots, \varphi_m) \in W_0^{1,\infty}(\Omega, \mathbb{R}^m)$; we consider a bounded open convex set $\Omega_1 \subseteq \mathbb{R}^d$ such that $\Omega \subseteq \Omega_1$ and we extend φ and $\nabla\varphi$, giving to the both functions value 0 on $\Omega_1 \setminus \Omega$; then $\varphi \in W_0^{1,\infty}(\Omega_1, \mathbb{R}^m)$. Indeed, because $\operatorname{supp}\varphi = K \subseteq \Omega$ is compact, there exists an open set G such that $K \subseteq G \subseteq \bar{G} \subseteq \Omega$, $\bar{G}$ is compact and there exists $\psi_0 \in C_c^\infty(\Omega_1, \mathbb{R}), 0 \leq \psi_0 \leq 1$ such that $\psi_0|_{\bar{G}} = 1$ and $\psi_0|_{\mathbb{R}^d \setminus \Omega} = 0$. Then $\operatorname{supp}\psi_0 \subseteq \bar{\Omega}, \nabla\psi_0|_K = 0$ and $\nabla\varphi|_{\Omega \setminus G} = 0$.

Therefore, for every $\psi \in C_c^\infty(\Omega_1, \mathbb{R})$ and $i \in \{1, \ldots, d\}, j \in \{1, \ldots, m\}$,

$$\begin{cases} \int_{\Omega_1} \varphi_j(t) \cdot \nabla_i \psi(t) dt = \int_K \varphi_j(t) \cdot \nabla_i \psi(t) dt \\ = \int_K \varphi_j(t) \cdot \psi_0(t) \cdot \nabla_i \psi(t) dt = \int_K \varphi_j(t) \cdot \nabla_i (\psi \cdot \psi_0)(t) dt \\ - \int_K \varphi_j(t) \cdot \psi(t) \cdot \nabla_i \psi_0(t) dt = \int_\Omega \varphi_j(t) \cdot \nabla_i (\psi \cdot \psi_0)(t) dt. \end{cases} \tag{1}$$

Since $\psi \cdot \psi_0 \in C_c^\infty(\Omega, \mathbb{R})$,

$$\begin{cases} \int_\Omega \varphi_j(t) \cdot \nabla_i (\psi \cdot \psi_0)(t) dt = -\int_\Omega \nabla_i \varphi_j(t) \cdot \psi(t) \cdot \psi_0(t) dt \\ = -\int_G \nabla_i \varphi_j(t) \cdot \psi(t) \cdot \psi_0(t) dt = -\int_G \nabla_i \varphi_j(t) \cdot \psi(t) dt \\ = -\int_{\Omega_1} \nabla_i \varphi_j(t) \cdot \psi(t) dt. \end{cases} \tag{2}$$

By (1) and (2), it follows that, for every $i \in \{1, \ldots, d\}$, every $j \in \{1, \ldots, m\}$ and every $\psi \in C_c^\infty(\Omega_1, \mathbb{R})$,

$$\int_{\Omega_1} \varphi_j(t) \cdot \nabla_i \psi(t) dt = -\int_{\Omega_1} \nabla_i \varphi_j(t) \cdot \psi(t) dt$$

(we have extended $\nabla\varphi$ by $\nabla_i \varphi_j(t) = 0$, for every $t \in \Omega_1 \setminus \Omega$).

It follows that $\nabla_i \varphi_j$ is the i-weak derivative of φ_j on Ω_1; hence $\nabla\varphi$ is the gradient of φ.

It is obvious that $\varphi, \nabla\varphi \in L^\infty(\Omega_1, \mathbb{R}^m)$ and so $\varphi \in W^{1,\infty}(\Omega_1, \mathbb{R}^m)$.

Since Ω_1 is convex, $\varphi \in \operatorname{Lip}(\Omega_1, \mathbb{R}^m)$, hence $\varphi \in \operatorname{Lip}(\bar{\Omega}, \mathbb{R}^m)$ and $\varphi|_{\partial\Omega} = 0$.

In fact, we proved that $\varphi \in W_0^{1,\infty}(\Omega, \mathbb{R}^m) \Longrightarrow \varphi \in \operatorname{Lip}(\Omega_1, \mathbb{R}^m)$.

From the hypothesis, it follows that

$$\begin{aligned} f(y) &\leq \frac{1}{\mu(\bar{\Omega})} \cdot \int_{\bar{\Omega}} f(y + \nabla\varphi(t)) dt \\ &= \frac{1}{\mu(\bar{\Omega})} \cdot \int_\Omega f(y + \nabla\varphi(t)) dt + \frac{1}{\mu(\bar{\Omega})} \cdot f(y) \cdot \mu(\bar{\Omega} \setminus \Omega), \end{aligned}$$

from where

$$f(y) \leq \frac{1}{\mu(\Omega)} \cdot \int_\Omega f(y + \nabla\varphi(t)) dt.$$

According to Theorem 3.143, f is quasiconvex. □

In the following, we introduce a notion more general than quasiconvexity, notion which has the advantage, however, to speak in a language closer to that of convexity.

Definition 3.151. The function $f : \mathbb{R}^{md} \to \mathbb{R}$ is said *rank one convex* if the function $g : \mathbb{R} \to \mathbb{R}$, defined by $g(t) = f(y + t(a \otimes b))$, is convex, for every $y \in \mathbb{R}^{md}$, every $a \in \mathbb{R}^m$ and every $b \in \mathbb{R}^d$, where $a \otimes b$ is the matrix

$$a \otimes b = \begin{pmatrix} a_1 \\ \vdots \\ a_m \end{pmatrix} \cdot (b_1 \cdots b_d) = \begin{pmatrix} a_1 b_1 \cdots a_1 b_d \\ \cdots \\ a_m b_1 \cdots a_m b_d \end{pmatrix}_{m \times d} = a^T \cdot b.$$

We remark that $\operatorname{rank}(a \otimes b) \leq 1$.

If $f \in C^2(\mathbb{R}^{md})$, then f is rank one convex if and only if, for every $y \in \mathbb{R}^{md}$, every $a \in \mathbb{R}^m$ and every $b \in \mathbb{R}^d$,

$$(a \otimes b)^T \cdot d^2 f(y)(a \otimes b) \geq 0.$$

Proposition 3.152. *The function* $f : \mathbb{R}^{md} \to \mathbb{R}$ *is rank one convex if and only if, for every* $y, z \in \mathbb{R}^{md}$ *with* $\operatorname{rank}(y - z) \leq 1$,

$$f((1-\lambda)y + \lambda z) \leq (1-\lambda) f(y) + \lambda f(z), \quad \textit{for every} \quad \lambda \in [0,1].$$

Proof. We remark that if $y, z \in \mathbb{R}^{md}$, then $\operatorname{rank}(y - z) \leq 1$ if and only if there exist $a \in \mathbb{R}^m$ and $b \in \mathbb{R}^d$ such that $y - z = a \otimes b$.

Necessity. We suppose that the function f is rank one convex and let $y, z \in \mathbb{R}^{md}$ with $\operatorname{rank}(y - z) \leq 1$; let $a \in \mathbb{R}^m, b \in \mathbb{R}^d$ such that $y - z = a \otimes b$. Then the function $g : \mathbb{R} \to \mathbb{R}, g(t) = f(z + t(a \otimes b))$, is convex and so, for every $\lambda \in [0,1]$,

$$g((1-\lambda) \cdot 1 + \lambda \cdot 0) \leq (1-\lambda) \cdot g(1) + \lambda \cdot g(0)$$

which rewrites

$$f(z + (1-\lambda) \cdot (a \otimes b)) \leq (1-\lambda) \cdot f(z + a \otimes b) + \lambda \cdot f(z)$$

or

$$f((1-\lambda)y + \lambda z) \leq (1-\lambda) \cdot f(y) + \lambda \cdot f(z).$$

Sufficiency. For every $y \in \mathbb{R}^{md}$, every $a \in \mathbb{R}^m$ and every $b \in \mathbb{R}^d$, we define $g : \mathbb{R} \to \mathbb{R}$ by $g(t) = f(y + t(a \otimes b))$. Let us show that g is convex.

For every $t, s \in \mathbb{R}$ and every $\lambda \in [0,1]$, let $z = y + t(a \otimes b)$, $w = y + s(a \otimes b)$; then $z - w = (t-s) \cdot (a \otimes b) = [(t-s)a] \otimes b$ and so

$$f((1-\lambda)z + \lambda w) \leq (1-\lambda) \cdot f(z) + \lambda \cdot f(w),$$

from where

$$\begin{aligned} g((1-\lambda)z + \lambda w) &= f(y + (1-\lambda)t(a \otimes b) + \lambda s(a \otimes b)) \\ &= f((1-\lambda)z + \lambda w) \\ &\leq (1-\lambda) \cdot f(y + t(a \otimes b)) + \lambda f(y + s(a \otimes b)) \\ &= (1-\lambda) \cdot g(t) + \lambda \cdot g(s). \end{aligned}$$

Therefore the function g is convex and so f is rank one convex. □

Now, we analyze the relations between the three types of convexity: the classical convexity, quasiconvexity and rank one convexity(see also theorem of [50]).

Theorem 3.153. *Let $f : \mathbb{R}^{md} \to \mathbb{R}$ be a continuous function and let the following assertions:*

(i) *f is convex;*

(ii) *f is quasiconvex;*

(iii) *f is rank one convex.*

Then (i) $\Longrightarrow$ (ii) $\Longrightarrow$ (iii).

If $m = 1$ or $n = 1$ then the three statements are equivalent.

Proof. The implication (i)$\Longrightarrow$(ii) was demonstrated in Proposition 3.141.

(ii)$\Longrightarrow$(iii). We suppose that $f : \mathbb{R}^{md} \to \mathbb{R}$ is quasiconvex.

Let $y, z \in \mathbb{R}^{md}$, with $\operatorname{rank}(y - z) \leq 1$; according to Proposition 3.152, it must to show that

$$f((1-\lambda)y + \lambda z) \leq (1-\lambda) f(y) + \lambda f(z), \quad \text{for every} \quad \lambda \in [0,1].$$

Let $a \in \mathbb{R}^m$ and $b \in \mathbb{R}^d$ such that $y - z = a \otimes b$.

Since $a \otimes b = (\|b\| \cdot a) \otimes (\frac{1}{\|b\|} \cdot b)$, we can suppose that $\|b\| = 1$.

Choose an orthonormal basis $\{p^1, \ldots, p^d\}$ on $\mathbb{R}^d$ with $p^1 = b$ and let Q be the hypercube based on this basis.

If $\chi : \mathbb{R} \to \mathbb{R}$ is a periodic function with period 1 such that $\chi_{[0,1/2)} = 1$ and $\chi_{[1/2,1]} = -1$, then we define $\varphi : \mathbb{R}^d \to \mathbb{R}^m$ by

$$\varphi(t) = \frac{1}{2} \cdot \left(\int_0^{\langle t,b \rangle} \chi(r) dr \right) \cdot a.$$

For every $t, s \in \mathbb{R}^d$,

$$\begin{aligned}\|\varphi(t)-\varphi(s)\|_{\mathbb{R}^m} &= \frac{1}{2}\cdot\left|\int_0^{\langle t,b\rangle}\chi(r)dr-\int_0^{\langle s,b\rangle}\chi(r)dr\right|\cdot\|a\| \\ &= \frac{1}{2}\cdot\left|\int_{\langle s,b\rangle}^{\langle t,b\rangle}\chi(r)dr\right|\cdot\|a\| \\ &\le \frac{1}{2}\cdot|\langle t-s,b\rangle|\cdot\|a\| \le \frac{1}{2}\cdot\|a\|\cdot\|t-s\|_{\mathbb{R}^d}.\end{aligned}$$

It follows that φ is a bounded Lipschitz function on $\mathbb{R}^d$.

Then $\varphi \in W^{1,\infty}(\mathbb{R}^d, \mathbb{R}^m)$ and $\nabla\varphi(t) = \frac{1}{2}\cdot\chi(<t,b>)\cdot(a\otimes b)$.

Let us show that φ is Q-periodic.

Let $t \in \mathbb{R}^d, t = \alpha_1 p^1 + \ldots + \alpha_d p^d$; then $\varphi(t) = \frac{1}{2}\cdot(\int_0^{\alpha_1}\chi(r)dr)\cdot a$.

If $s = (\alpha_1+1)p^1 + \alpha_2 p^2 + \ldots + \alpha_d p^d$, then $\varphi(s) = \frac{1}{2}\cdot(\int_0^{\alpha_1+1}\chi(r)dr)\cdot a = \varphi(t)$; if $s = t + p^i$ with $1 < i \le d$, then $\varphi(s) = \varphi(t)$ because $< p^i, b >= 0$. Therefore φ is Q-periodic. According to Theorem 3.146 and to Remark 3.147,

$$f\left(\frac{y+z}{2}\right) \le \int_Q f\left(\frac{y+z}{2}+\nabla\varphi(t)\right)dt.$$

Let $Q_1 = \{t : 0 << t,b >< \frac{1}{2}\}$ and $Q_2 = \{t : \frac{1}{2} \le< t,b >< 1\}$; then $Q = Q_1 \cup Q_2$ and $\nabla\varphi|_{Q_1} = \frac{1}{2}(a\otimes b), \nabla\varphi|_{Q_2} = -\frac{1}{2}(a\otimes b)$. Therefore

$$\begin{aligned}f\left(\frac{y+z}{2}\right) &\le \int_{Q_1} f\left(\frac{y+z}{2}+\frac{a\otimes b}{2}\right)dt + \int_{Q_2} f\left(\frac{y+z}{2}-\frac{a\otimes b}{2}\right)dt \\ &= f(y)\mu(Q_1) + f(z)\mu(Q_2) = \frac{f(y)+f(z)}{2}.\end{aligned}$$

As f is continuous, it follows that

$$f((1-\lambda)y+\lambda z) \le (1-\lambda)f(y)+\lambda f(z), \quad \forall\lambda\in[0,1].$$

Therefore, f rank one convex.

Let now $m = 1$; then $a\otimes b = (ab_1\cdots ab_d)_{1\times d} = a\cdot b$. In this case, for every $y, z \in \mathbb{R}^{1\times d} = \mathbb{R}^d$, $y - z \in \mathbb{R}^d$ and we can find $a \in \mathbb{R}$ and $b \in \mathbb{R}^d$ such that $y - z = a\otimes b$ so that $\operatorname{rank}(y-z) \le 1$. According to Proposition 3.152, rank one convexity is equivalent in this case with convexity.

In the case $d = 1$,

$$a\otimes b = \begin{pmatrix} a_1 b \\ \vdots \\ a_m b\end{pmatrix}_{m\times 1} = b\cdot a$$

and again, for every $y, z \in \mathbb{R}^{m\times 1} = \mathbb{R}^m$, we can find $a \in \mathbb{R}^m$ and $b \in \mathbb{R}$ such that $y - z = a \otimes b$. □

Remark 3.154. In Example 3.142, we present a quasiconvex function which is not convex. In 1952, Ch. Morrey formulated the conjecture under which rank one convexity does not imply quasiconvexity if $m, d \geq 2$.

Šverák demonstrated in 1992 (see [156]) that Morrey's statement is true if $m \geq 3$ and $d \geq 2$. The case where $m = 2, d \geq 2$ is still open. A detailed discussion on this issue can be found in Chapter 5 of [50].

At the end of this section, we present the Kinderlehrer–Pedregal characterization of p-gradient Young measures (see [102]).

Theorem 3.155. *Let $\tau \in \mathcal{Y}(\mathbb{R}^{md})$ and let $1 \leq p < +\infty$; τ is a p-gradient Young measure if and only if the following conditions are satisfied:*

(1) *There is $u \in W^{1,p}(\Omega, \mathbb{R}^m)$ such that*

$$\nabla u(t) = \mathrm{bar}\tau_t \equiv \int_{\mathbb{R}^{md}} y\, d\tau_t(y), \quad \text{for almost every} \quad t \in \Omega.$$

(2) *Let X^p be the set of all bounded from below, continuous and quasiconvex functions $f : \mathbb{R}^{md} \to \mathbb{R}$, which satisfy $|f(y)| \leq c(1 + \|y\|^p)$, for every $y \in \mathbb{R}^{md}$. Then*

$$f(\nabla u(t)) \leq \int_{\mathbb{R}^{md}} f(y)d\tau_t(y), \text{ for every } f \in X^p \text{ and almost every } t \in \Omega.$$

(3) $\int_\Omega \int_{\mathbb{R}^{md}} \|y\|^p d\tau_t(y)dt < +\infty, \quad$ *almost for every* $\quad t \in \Omega.$

3.10.3 Lower Semicontinuity

We recall some notions and results about Sobolev spaces (for more complete information, see [36, 49, 50]).

Let $\Omega \subseteq \mathbb{R}^d$ be a bounded open set.

(1) The case $1 \leq p < +\infty$.
$W^{1,p}(\Omega, \mathbb{R}^m) = \{u : \Omega \to \mathbb{R}^m | u \in L^p(\Omega, \mathbb{R}^m), \nabla u \in L^p(\Omega, \mathbb{R}^{md})\}$.

The mapping $u \mapsto \|u\|_{W^{1,p}} = (\|u\|_p^p + \|\nabla u\|_p^p)^{\frac{1}{p}}$ is a norm on $W^{1,p}(\Omega, \mathbb{R}^m)$ and $(W^{1,p}(\Omega, \mathbb{R}^m), \|\cdot\|_{W^{1,p}})$ is a separable Banach space.

$$u_n \xrightarrow[W^{1,p}]{w} u \Longleftrightarrow \begin{cases} \text{a)} & u_n \xrightarrow[L^p]{w} u \\ \text{b)} & \nabla u_n \xrightarrow[L^p]{w} \nabla u \end{cases} ; \quad u_n \xrightarrow[W^{1,p}]{w} u \Longrightarrow u_n \xrightarrow[L^p]{\|\cdot\|_p} u.$$

If $1 < p < +\infty$, then this space is reflexive.

$$W_0^{1,p}(\Omega, \mathbb{R}^m) = \overline{C_c^\infty(\Omega, \mathbb{R}^m)}^{W^{1,p}}.$$

(2) The case $p = +\infty$.
$W^{1,\infty}(\Omega, \mathbb{R}^m) = \{u : \Omega \to \mathbb{R}^m | u \in L^\infty(\Omega, \mathbb{R}^m), \nabla u \in L^\infty(\Omega, \mathbb{R}^{md})\}$.

The mapping $u \mapsto \|u\|_{W^{1,\infty}} = \max\{\|u\|_{L^\infty}, \|\nabla u\|_{L^\infty}\}$ is a norm on $W^{1,\infty}$ and $(W^{1,\infty}(\Omega, \mathbb{R}^m), \|\cdot\|_{W^{1,\infty}})$ is a Banach space.

$$u_n \xrightarrow[W^{1,\infty}]{w*} u \iff \begin{cases} \text{a)} \quad u_n \xrightarrow[L^\infty]{w*} u \\ \text{b)} \quad \nabla u_n \xrightarrow[L^\infty]{w*} \nabla u \end{cases} ; \quad u_n \xrightarrow[W^{1,\infty}]{w*} u \implies u_n \xrightarrow[L^\infty]{\|\cdot\|_\infty} u.$$

$$W_0^{1,\infty}(\Omega, \mathbb{R}^m) = W^{1,\infty}(\Omega, \mathbb{R}^m) \cap W_0^{1,1}(\Omega, \mathbb{R}^m).$$

Remark 3.156.

(i) If $1 < p \leq +\infty$ and $(u_n)_n \subseteq W^{1,p}(\Omega, \mathbb{R}^m)$ is a bounded sequence, then there exist $u \in W^{1,p}(\Omega, \mathbb{R}^m)$ and a subsequence $(u_{k_n})_n$ of $(u_n)_n$ such that

$$u_{k_n} \xrightarrow[W^{1,p}]{w} u \;\; (u_{k_n} \xrightarrow[W^{1,\infty}]{w*} u, \text{ if } p = +\infty).$$

(ii) $W^{1,\infty}(\Omega, \mathbb{R}^m) \subseteq W^{1,p}(\Omega, \mathbb{R}^m)$, for every $1 \leq p < +\infty$.

For every $(u_n)_n \subseteq W^{1,\infty}(\Omega, \mathbb{R}^m)$ and every $u \in W^{1,\infty}(\Omega, \mathbb{R}^m)$,

$$u_n \xrightarrow[W^{1,\infty}]{w*} u \implies u_n \xrightarrow[W^{1,p}]{w} u, \quad \text{for every} \quad 1 \leq p < +\infty.$$

Definition 3.157. Let $f : \Omega \times \mathbb{R}^m \times \mathbb{R}^{md} \to \mathbb{R}$; f is a *Carathéodory function* if

(1) $t \mapsto f(t, x, y)$ is a measurable mapping, for every $(x, y) \in \mathbb{R}^m \times \mathbb{R}^{md}$.

(2) $(x, y) \mapsto f(t, x, y)$ is continuous on $\mathbb{R}^m \times \mathbb{R}^{md}$, for almost every $t \in \Omega$.

Definition 3.158. Let $f : \Omega \times \mathbb{R}^m \times \mathbb{R}^{md} \to \mathbb{R}$ be a Carathéodory function and let $I : W^{1,p}(\Omega, \mathbb{R}^m) \to \bar{\mathbb{R}}$ defined by $I(u) = \int_\Omega f(t, u(t), \nabla u(t))dt$ (we suppose that the integral exists).

Let $1 \leq p < +\infty$; I is *sequentially weakly lower semicontinuous* (wlsc) on $W^{1,p}(\Omega, \mathbb{R}^m)$ if, for every sequence $(u_n)_n \subseteq W^{1,p}(\Omega, \mathbb{R}^m)$ and every $u \in W^{1,p}(\Omega, \mathbb{R}^m)$ with $u_n \xrightarrow[W^{1,p}]{w} u$, $I(u) \leq \liminf_n I(u_n)$.

Let $p = +\infty$; I is *sequentially weak* lower semicontinuous* (w*lsc) on $W^{1,\infty}(\Omega, \mathbb{R}^m)$ if, for every sequence $(u_n)_n \subseteq W^{1,\infty}(\Omega, \mathbb{R}^m)$ and every $u \in W^{1,\infty}(\Omega, \mathbb{R}^m)$ with $u_n \xrightarrow[W^{1,\infty}]{w*} u$, $I(u) \leq \liminf_n I(u_n)$.

In the following, we usually omit the word "sequentially" in the above definitions.

Remark 3.159. According to (ii) of Remark 3.156, if $I : W^{1,p}(\Omega, \mathbb{R}^m) \to \bar{\mathbb{R}}$ is wlsc, then $I|_{W^{1,\infty}(\Omega,\mathbb{R}^m)}$ is w*lsc.

In the following, we present the relationships between the wlsc of I and the quasiconvexity of f; briefly, I is wlsc if and only if $f(t, x, \cdot)$ is quasiconvex.

We begin with a particular case (see Theorem 3.13 of [50]).

Theorem 3.160. *Let $1 \leq p \leq +\infty$, let $f : \mathbb{R}^{md} \to \mathbb{R}$ be a continuous function, and let $I : W^{1,p}(\Omega, \mathbb{R}^m) \to \bar{\mathbb{R}}$ be defined by $I(u) = \int_\Omega f(\nabla u(t))dt$.*

*If I is wlsc in $W^{1,p}(\Omega, \mathbb{R}^m)$ (w*lsc in $W^{1,\infty}(\Omega, \mathbb{R}^m)$), then f is a quasiconvex function.*

Proof. (1) Firstly, we suppose that $p = +\infty$. Let $Q \subseteq \Omega$ be a hypercube with the edge length a, let

$$y = \begin{pmatrix} y_{11} & \cdots & y_{1d} \\ & \cdots & \\ y_{m1} & \cdots & y_{md} \end{pmatrix}_{m \times d} \in \mathbb{R}^{md}$$

and let $\varphi \in W^{1,\infty}(\Omega, \mathbb{R}^m)$ be a Q-periodic function. We define $\varphi_n : \mathbb{R}^d \to \mathbb{R}^m$ letting $\varphi_n(t) = \frac{1}{n} \cdot \varphi(nt)$. Then $(\varphi_n)_{n \in \mathbb{R}^*}$ is a bounded sequence in $W^{1,\infty}(\Omega, \mathbb{R}^m)$. According to Remark 3.156 (i), there exist $\varphi_0 \in W^{1,\infty}(\Omega, \mathbb{R}^m)$ and a subsequence $(\varphi_{k_n})_n$ of $(\varphi_n)_n$ such that $\varphi_{k_n} \xrightarrow[W^{1,\infty}]{w*} \varphi_0$. Since $\varphi_n \to \underline{0}$, almost everywhere on$\mathbb{R}^d$ and $\varphi_{k_n} \xrightarrow[L^\infty]{\|\cdot\|_\infty} \varphi_0$, it follows that $\varphi_0 = \underline{0}$. Therefore, $\varphi_{k_n} \xrightarrow[W^{1,\infty}]{w*} \underline{0}$.

We remark that, every $t = \begin{pmatrix} t_1 \\ \vdots \\ t_d \end{pmatrix}_{d \times 1}$, $y \cdot t \in \mathbb{R}^m$.

Let now $u_n : \Omega \to \mathbb{R}^m$, defined by

$$u_n(t) = \begin{cases} y \cdot t, & t \in \Omega \setminus Q, \\ \varphi_n(t) + y \cdot t, & t \in Q. \end{cases}$$

Then $(u_n)_n \subseteq W^{1,\infty}(\Omega, \mathbb{R}^m)$ and

$$\nabla u_n(t) = \begin{cases} y, & t \in \Omega \setminus Q, \\ \nabla\varphi(nt) + y, & t \in Q. \end{cases}$$

Obviously, $u_{k_n} \xrightarrow[W^{1,\infty}]{w*} u$, where $u(t) = y \cdot t$.

Since I is w*lsc in $W^{1,\infty}(\Omega, \mathbb{R}^m)$,

$$I(u) \leq \liminf_n I(u_{k_n}). \tag{*}$$

Now, $I(u) = \int_\Omega f(y)dt = f(y) \cdot \mu(\Omega)$ and, for every $n \in \mathbb{N}^*$,

$$\begin{aligned} I(u_{k_n}) = \int_\Omega f(\nabla u_{k_n}(t))dt &= \int_{\Omega\setminus Q} f(\nabla u_{k_n}(t))dt + \int_Q f(\nabla u_{k_n}(t))dt \\ &= f(y) \cdot \mu(\Omega \setminus Q) + \int_Q f(y + \nabla\varphi(k_n t))dt \\ &= f(y) \cdot \mu(\Omega \setminus Q) + \frac{1}{k_n^d} \int_{k_n \cdot Q} f(y + \nabla\varphi(s))ds \\ &= f(y) \cdot \mu(\Omega \setminus Q) + \int_Q f(y + \nabla\varphi(s))ds. \end{aligned}$$

Therefore, by $(*)$,

$$f(y) \cdot \mu(\Omega) \leq f(y) \cdot \mu(\Omega \setminus Q) + \int_Q f(y + \nabla\varphi(s))ds,$$

from where

$$f(y) \cdot \mu(Q) \leq \int_Q f(y + \nabla\varphi(s))ds.$$

Now, if we suppose that the edge length a is so that the hypercube $Q_1 = (0,1)^d$ is an union of l^d hypercubes of edge length a, then we obtain

$$f(y) \leq \int_{Q_1} f(y + \nabla\varphi(t))dt.$$

By Theorem 3.146, f is quasiconvex.

(2) Let now $1 \leq p < +\infty$; if I is wlsc in $W^{1,p}(\Omega, \mathbb{R}^m)$, then, according to Remark 3.159, $I|_{W^{1,\infty}(\Omega,\mathbb{R}^m)}$ is w*lsc and so f is quasiconvex. □

For the general case, we need to define the following growth condition.

Definition 3.161 (see Definition 8.10 of [50]). Let $1 < p < +\infty$ and let $f : \Omega \times \mathbb{R}^m \times \mathbb{R}^{md} \to \mathbb{R}$ be a Carathéodory function. We say that f satisfies the condition (C_p) if, for almost every $t \in \Omega$ and every $(x, y) \in \mathbb{R}^m \times \mathbb{R}^{md}$,

$$-\alpha(\|x\| + \|y\|^q) - \beta(t) \le f(t, x, y) \le g(t, x)(1 + \|y\|^p), \qquad (C_p)$$

where $\alpha, \beta, g \ge 0, \beta \in L^1(\Omega)$, $1 \le q < p$ and g is a Carathéodory function.
In the case $p = 1$,

$$|f(t, x, y)| \le \alpha(1 + \|y\|), \qquad (C_1)$$

where $\alpha > 0$.

Theorem 3.162 (Morrey–Acerbi–Fusco). *Let Ω be a bounded open subset of $\mathbb{R}^d$ with a Lipschitz boundary, let $1 \le p < +\infty$ and let $f : \Omega \times \mathbb{R}^m \times \mathbb{R}^{md} \to \mathbb{R}$ be a Carathéodory function which satisfies the condition (C_p).*

The functional $I : W^{1,p}(\Omega, \mathbb{R}^m) \to \bar{\mathbb{R}}$ defined by $I(u) = \int_\Omega f(t, u(t), \nabla u(t))dt$, is sequentially weakly lower semicontinuous in $W^{1,p}(\Omega, \mathbb{R}^m)$ if and only if $f(t, x, \cdot)$ is quasiconvex, for almost every $t \in \Omega$ and every $x \in \mathbb{R}^m$.

We omit the laborious demonstration of this theorem; it can be found in [50] (Lemma 3.18 and Theorem 8.11).

Remark 3.163.

(i) This theorem is due to Morrey (see [121] and [122]). The variant for Carathéodory functions is due to Acerbi–Fusco (see [1] where it is proved for $f \ge 0$).

(ii) In [100], A. Kałamajska has proved the sufficiency of this theorem in the case $f \ge 0$ using a Jensen-type inequality for p-gradient Young measures.

3.11 Relaxed Solutions in Variational Calculus

Young measures were introduced as a need to provide generalized solutions to problems which do not admit classical solutions. With an illustrative purpose and without any intention of exhaustivity, we seek a classification of the bibliography, represented by works that use Young measures as a tool, in a few areas. Please note that these fields have ambiguous boundaries, some of the cited works could being able to belong to many of them.

(1) *Nonlinear differential equations and differential inclusions*:
Ch. Castaing, P. Raynaud de Fitte and A. Salvadori [47], S. Demoulini [55], H.-P. Gittel [90], P. Pedregal [135], G.-Q. Song and J. Xiao [155], S. Taheri, Q. Tang and K. Zhang [162], J. Yin and C. Wang [174], P. Zhang and Y. Zheng [180] and K. Zhang [182].

(2) *Nonlinear homogenization problems and* Γ*-convergence*:
L. Ambrosio, H. Frid and J. Silva [3], L. Ambrosio and H. Frid [4], M. Barchiesi [25], A. V. Demyanov [53], G. Michaille and M. Valadier [120], D. Piau [136], P. Pedregal [133], P. Pedregal and H. Serrano [134], D. Piau [136], H. Serrano [153], L. Tartar [165] and A. Visintin [172].

(3) *Continuum mechanics, nonlinear elasticity, microstructures theory, magnetic phenomena in fero-magnetic materials and hysteresis*:
A. L. Bessoud [27], C. Carstensen and A. Prohl [40], C. Carstensen and M. O. Rieger [41], G. Dal Maso, A. DeSimone, M. G. Mora and M. Morini [51], L. Freddi and R. Paroni [87, 88], M. Hillairet [94], P. Kamiński [101], Ph. G. LeFloch and M. Westdickenberg [111], C. De Lellis and L. Székelyhidi Jr. [112], C. Melcher [119], A. Münch, P. Pedregal and F. Periago [123, 124], S. Owczarek [127], P. Pedregal [132], M. O. Rieger and J. Zimmer [140, 142], M. O. Rieger [141], T. Roubiček and M. Kružik [146, 147] and F. Theil [166].

(4) *Economic equilibrium theory and games theory*:
E. J. Balder [17, 20, 21], A. M. Ramos and T. Roubiček [138] and A. Rustichini and P. Siconolfi [148].

However, we should note that the original reason for the emergence of Young measures has been to give relaxed solutions to variational problems. Many problems, formulated in the language of optimization theory, lack classical optimal solutions; the main reason is the rapid oscillations of its minimizing gradient sequences. This phenomenon requires the extension of the notion of solution of such problems. L. C. Young introduced in 1937 so called generalized curves (see [177]) and, afterwards, the generalized surfaces ([178]). The technique of Young measures was used then to obtain relaxed solutions by J. Warga ([173]), E. J. Balder ([9]), M. Valadier ([169]) and P. Pedregal ([128, 130]). We can also cite the works of C. Carstensen and T. Roubiček ([39]), C. Castaing, A. Jofre and A. Salvadori ([46]), M. Chipot ([48]), J. F. Edmond ([64]), J. J. Egozcue, R. Meziat and P. Pedregal ([67]) and J. Mach ([116]).

In this last section, we present applications of Young measures in this direction: the conditions for existence of relaxed solutions to problems of variational calculus.

The standard problem of the calculus of variations is to find minimizers of functional

$$I(u) = \int_{\Omega} f(t, u(t), \nabla u(t))dt,$$

where Ω is a bounded open subset of $\mathbb{R}^d$, the admissible mappings $u : \Omega \to \mathbb{R}^m$ belong to a set of competitors $H \subseteq W^{1,p}(\Omega, \mathbb{R}^m), 1 \leq p \leq +\infty$ and $f : \Omega \times \mathbb{R}^m \times \mathbb{R}^{md} \to \mathbb{R}$ is a Carathéodory integrand, i.e.:

a) $(x, y) \in \mathbb{R}^m \times \mathbb{R}^{md} \mapsto f(t, x, y)$ is continuous, for almost every $t \in \Omega$.

b) $t \in \Omega \mapsto f(t, x, y)$ is measurable, for every $(x, y) \in \mathbb{R}^m \times \mathbb{R}^{md}$.

In short, the direct method in variational calculus is the following:

Let a minimizing sequence $(u_n)_n \subseteq H$ for the problem:

$$\inf\{I(u) : u \in H\} = \mathbf{m}. \tag{P}$$

This means that $I(u_n) \to \mathbf{m}$.

Using some compactness results (to be specified), we can obtain a subsequence (still noted $(u_n)_n$) weakly convergent to u_0 in $W^{1,p}(\Omega, \mathbb{R}^m)$ (weak* convergent in the case $p = +\infty$).

If I is sequentially weakly lower semicontinuous (weak* l.s.c. in the case $p = +\infty$), then

$$I(u_0) \leq \liminf_{n \to +\infty} I(u_n) = \mathbf{m} \leq I(u_0).$$

Therefore, u_0 is a minimizer for the problem (P).

This method is fruitful when a convexity condition in the gradient variable is assumed on the integrand f (I is weakly lower semicontinuous if and only if $f(t, x, \cdot)$ is quasiconvex). For more detailed information on this approach, we indicate the books [49, 50, 85].

Unfortunately, many problems formulated in the language of optimization theory lack classical optimal solutions. So, non-convex optimization problems may not possess a classical minimizer because the minimizing sequences have typically rapid oscillations. This behavior requires a relaxation of classical solutions for such problems; often we can obtain a such relaxation by means of Young measures (see [45, 85, 128, 145, 163, 173, 176]).

We will extend the class of competitors to $\bar{H}$ and then we will extend the functional I to $\bar{I} : \bar{H} \to \mathbb{R}$; therefore we arrive at the relaxed problem:

$$\inf\{\bar{I}(\bar{u}) : \bar{u} \in \bar{H}\} = \bar{\mathbf{m}}. \tag{$\bar{P}$}$$

We will give some conditions that the problem $(\bar{P})$ to have a minimizer. If, moreover, F is quasiconvex in the gradient variable, then we obtain the known solutions for the problem (P).

Let $1 \le p < +\infty$, $\Omega \subseteq \mathbb{R}^d$ be a bounded open set with a Lipschitz boundary and let $f : \Omega \times \mathbb{R}^m \times \mathbb{R}^{md} \to \mathbb{R}$ be a Carathéodory integrand for which there exist $a, b > 0$ such that, for every $x \in \mathbb{R}^m$, every $y \in \mathbb{R}^{md}$ and almost every $t \in \Omega$:

$$a \cdot \|y\|^p - 1 \le f(t, x, y) \le b \cdot (1 + \|y\|^p). \tag{C}$$

We remark that, in this case, the condition (C_p) from the above section is satisfied, for every $1 \le p < +\infty$.

The classical problem of variational calculus is:

$$\inf \left\{ I(u) = \int_\Omega f(t, u(t), \nabla u(t)) dt : u \in W_0^{1,p}(\Omega, \mathbb{R}^m) \right\} = \mathbf{m}. \tag{P}$$

We remark that, for every $u \in W_0^{1,p}(\Omega, \mathbb{R}^m)$,

$$|f(t, u(t), \nabla u(t))| \le c \cdot (1 + \|u(t)\|^p), \quad \text{for almost every} \quad t \in \Omega,$$

where $c = \max\{a, b, 1\}$. Therefore $f(\cdot, u(\cdot), \nabla u(\cdot)) \in L^1(\Omega)$.

Moreover, for every $u \in W_0^{1,p}(\Omega, \mathbb{R}^m)$ and for almost every $t \in \Omega$,

$$\begin{aligned} -1 \le f(t, u(t), \nabla u(t)) \quad &\text{hence} \quad -\mu(\Omega) \le I(u) \quad \text{and} \\ f(t, \underline{0}, \nabla \underline{0}) \le b \quad &\text{and so} \quad I(\underline{0}) \le b \cdot \mu(\Omega). \end{aligned}$$

Then $-\infty < \mathbf{m} < +\infty$.

The following example shows that, generally, the problem (P) has no minimizer.

Example 3.164 (Bolza example). Let $d = m = 1$ and let $f : (0,1) \times \mathbb{R} \times \mathbb{R} \to \mathbb{R}$, defined by $f(t, x, y) = (y^2 - 1)^2 + \min\{x^2, 1\}$, for every $(t, x, y) \in (0,1) \times \mathbb{R} \times \mathbb{R}$.

Then f satisfies the condition (C):

$$\frac{1}{2} \cdot y^4 - 1 \le f(t, x, y) \le 3(1 + y^4), \quad \forall t \in (0,1), \quad \forall x, y \in \mathbb{R}.$$

If we consider $u_0 = \underline{0} \in W^{1,4}((0,1), \mathbb{R})$, then the problem becomes:

$$\inf \left\{ I(u) = \int_0^1 \left[((u'(t))^2 - 1)^2 + \min\{u^2(t), 1\} \right] dt : u \in W^{1,4} \right\} = \mathbf{m}.$$

Let $(u_n)_n \subseteq W_0^{1,4}((0,1),\mathbb{R})$, defined by

$$u_n(t) = \sum_{k=0}^{2^{n-1}-1} \left[\left(t - \frac{2k}{2^n}\right) \cdot \chi_{[\frac{2k}{2^n}, \frac{2k+1}{2^n})}(t) + \left(\frac{2k+2}{2^n} - t\right) \cdot \chi_{[\frac{2k+1}{2^n}, \frac{2k+2}{2^n})}(t) \right].$$

Obviously, $0 \le u_n \le \frac{1}{2^n}$, for every $n \in \mathbb{N}^*$. Then $(u_n)_n$ is uniformly convergent to $\underline{0}$ on $(0,1)$ and $u_n'(t) = \operatorname{sign}[\sin(2^n \pi t)]$, for a.e. $t \in (0,1)$ ($(u_n')_n$ is the Rademacher sequence). Therefore $u_n' = \pm 1$, for a.e. $t \in (0,1)$ and so $I(u_n) \to 0$; then $\mathbf{m} = 0$. The sequence $(u_n)_n$ is a minimizing one, but there is no function $u \in W^{1,4}((0,1),\mathbb{R})$ for which $I(u) = 0$; hence the problem (P) has no minimizer.

We note that the integrand f is not convex in the gradient variable and, therefore, it is not quasiconvex (see Theorem 3.153). According to Theorem 3.162, I is not wlsc on $W^{1,4}((0,1),\mathbb{R})$; we can remark that $u_n \xrightarrow[W^{1,4}]{w} \underline{0}$ but $I(\underline{0}) = 1 > 0 = \liminf_n I(u_n)$.

To obtain a satisfactory minimum, we will enlarge the set of competitors $W_0^{1,p}(\Omega,\mathbb{R}^m)$ to a set $\bar{H}$ and we will extend the functional I on $\bar{H}$.

The following results can be easily extended if we replace $W_0^{1,p}(\Omega,\mathbb{R}^m)$ with the broader class of competitors $u_0 + W_0^{1,p}(\Omega,\mathbb{R}^m)$, where $u_0 \in W^{1,p}(\Omega,\mathbb{R}^m)$ is a fixed function.

Definition 3.165. Let $\bar{H} \subseteq W_0^{1,p}(\Omega,\mathbb{R}^m) \times \mathcal{Y}(\mathbb{R}^{md})$ be the subset of all (u,τ) for which there exists $(u_n)_n \subseteq W_0^{1,p}(\Omega,\mathbb{R}^m)$ such that

(a) $(u_n, \nabla u_n) \xrightarrow[\mathbb{R}^m \times \mathbb{R}^{md}]{\mathcal{S}} \tau^u \otimes \tau$,

(b) $(\|\nabla u_n\|^p)_n$ is uniformly integrable in $L^1(\Omega)$.

We say that $(u_n)_n$ generates (u,τ).

Proposition 3.166. $(u_n, \nabla u_n) \xrightarrow[\mathbb{R}^m \times \mathbb{R}^{md}]{\mathcal{S}} \tau^u \otimes \tau$ *if and only if*

$$u_n \xrightarrow[\mathcal{M}(\mathbb{R}^m)]{\mu} u \quad \textit{and} \quad \nabla u_n \xrightarrow[\mathbb{R}^{md}]{\mathcal{S}} \tau.$$

Proof. First, we recall that τ^u is the Young measure associated to u (see Definition 3.15) and $\tau^u \otimes \tau$ is the fiber product of τ^u and τ (see Definition 3.86).

Therefore, $(u_n, \nabla u_n) \xrightarrow[\mathbb{R}^m\times\mathbb{R}^{md}]{\mathcal{S}} \tau^u \otimes \tau$ if and only if, for every $A \in \mathcal{A}$ and every $f \in C_b(\mathbb{R}^m \times \mathbb{R}^{md})$,

$$\int_A f(u_n(t), \nabla u_n(t))dt \longrightarrow \int_A \int_{\mathbb{R}^m\times\mathbb{R}^{md}} f(x, y)d(\tau_t^u \otimes \tau_t)(x, y)dt$$
$$= \int_A \int_{\mathbb{R}^{md}} f(u(t), y)d\tau_t(y)dt.$$

Let us suppose that $(u_n, \nabla u_n) \xrightarrow[\mathbb{R}^m\times\mathbb{R}^{md}]{\mathcal{S}} \tau^u \otimes \tau$; let $A \in \mathcal{A}$ and $f \in C_b(\mathbb{R}^m)$. We define $f_1 : \mathbb{R}^m \times \mathbb{R}^{md} \to \mathbb{R}$ letting $f_1(x, y) = f(x)$, for every $(x, y) \in \mathbb{R}^m \times \mathbb{R}^{md}$; then $f_1 \in C_b(\mathbb{R}^m \times \mathbb{R}^{md})$ and so

$$\int_A f(u_n(t))dt \longrightarrow \int_A f(u(t))dt.$$

Therefore, $\tau^{u_n} \xrightarrow[\mathbb{R}^m]{\mathcal{S}} \tau^u$ and, according to Proposition 3.40, $u_n \xrightarrow[\mathcal{M}(\mathbb{R}^m)]{\mu} u$.

On the other hand, if $A \in \mathcal{A}$, $f \in C_b(\mathbb{R}^{md})$ and if we note $f_2 : \mathbb{R}^m \times \mathbb{R}^{md} \to \mathbb{R}$ letting $f_2(x, y) = f(y)$, then $f_2 \in C_b(\mathbb{R}^m \times \mathbb{R}^{md})$ and so

$$\int_A f(\nabla u_n)dt \longrightarrow \int_A \int_{\mathbb{R}^{md}} f(y)d\tau_t(y)dt.$$

Therefore, $\tau^{\nabla u_n} \xrightarrow[\mathbb{R}^{md}]{\mathcal{S}} \tau$.

The sufficiency is a direct consequence of fiber product lemma (see Theorem 3.87). □

Remark 3.167. According to the previous proposition, $(u, \tau) \in \bar{H}$ if and only if there exists $(u_n)_n \subseteq W_0^{1,p}(\Omega, \mathbb{R}^m)$ such that:

(1) $u_n \xrightarrow[\mathcal{M}(\mathbb{R}^m)]{\mu} u$.

(2) $\nabla u_n \xrightarrow[\mathbb{R}^{md}]{\mathcal{S}} \tau$.

(3) $(\|\nabla u_n\|^p)_n$ is uniformly integrable in $L^1(\Omega)$.

We remark that τ is a p-gradient Young measure (see Definition 3.134) generated by $(\nabla u_n)_n$ and $\nabla u = \text{bar}\tau_{.}$. According to (i) of Remark 3.135, $\nabla u_n \xrightarrow[L^1(\Omega,\mathbb{R}^{md})]{w} \text{bar}\tau_{.}$.

Using Proposition 3.137, we can obtain:

Proposition 3.168. *Let* $(u, \tau) \in \bar{H}$*; there exists* $(u_n)_n \subseteq W_0^{1,p}(\Omega, \mathbb{R}^m)$ *such that:*

(1) $u_n \xrightarrow[L^p]{\|\cdot\|_p} u.$

(2) $\nabla u_n \xrightarrow[L^p]{w} \nabla u = \operatorname{bar}\tau_{\cdot}.$

(3) $u_n \xrightarrow[W^{1,p}]{w} u.$

Remark 3.169. If $(u, \tau) \in \bar{H}$, then $\tau^u \otimes \tau$ is a p-Young measure on $\mathbb{R}^m \times \mathbb{R}^{md}$.

Indeed, let $(u_n)_n \subseteq W_0^{1,p}(\Omega, \mathbb{R}^m)$ such that $(u_n, \nabla u_n) \xrightarrow[\mathbb{R}^m \times \mathbb{R}^{md}]{\mathcal{S}} \tau^u \otimes \tau$ and $(\|\nabla u_n\|^p)_n$ is uniformly integrable in $L^1(\Omega)$. According to the previous proposition, $u_n \xrightarrow[L^p]{\|\cdot\|_p} u$; then $(\|u_n\|^p)_n$ is uniformly integrable in $L^1(\Omega)$. Since

$$\|(u_n, \nabla u_n)\|^p_{\mathbb{R}^m \times \mathbb{R}^{md}} \leq 2^p \cdot (\|u_n\|^p_{\mathbb{R}^m} + \|\nabla u_n\|^p_{\mathbb{R}^{md}}), \quad \text{for every} \quad n \in \mathbb{N},$$

$(\|(u_n, \nabla u_n)\|^p_{\mathbb{R}^m \times \mathbb{R}^{md}})_n$ is uniformly integrable in $L^1(\Omega)$ and then $\tau^u \otimes \tau$ is a p-Young measure.

Proposition 3.170.

(1) *For every* $u \in W_0^{1,p}(\Omega, \mathbb{R}^m)$, $(u, \tau^{\nabla u}) \in \bar{H}$*; the mapping* $u \mapsto (u, \tau^{\nabla u})$ *is an injection of* $W_0^{1,p}(\Omega, \mathbb{R}^m)$ *in* $\bar{H}$*. Therefore,* $W_0^{1,p}(\Omega, \mathbb{R}^m) \hookrightarrow \bar{H}$.

(2) *Let* $v \in \mathcal{M}(\mathbb{R}^{md})$*; then*

$$(u, \tau^v) \in \bar{H} \iff v \in L^p(\Omega, \mathbb{R}^{md}) \quad \textit{and} \quad v = \nabla u.$$

Proof. (1) For every $u \in W_0^{1,p}(\Omega, \mathbb{R}^m)$, let $(u_n)_n \subseteq W_0^{1,p}(\Omega, \mathbb{R}^m)$ be defined by $u_n = u$, for every $n \in \mathbb{N}$. Then $u_n \xrightarrow[\mathcal{M}(\mathbb{R}^m)]{\mu} u$ and $\nabla u_n \xrightarrow[\mathcal{M}(\mathbb{R}^{md})]{\mu} \nabla u$, so that $\tau^{\nabla u_n} \overset{\mathcal{S}}{\to} \tau^{\nabla u}$. Obviously, $(\|u_n\|^p)_n$ is uniformly integrable and so $(u, \tau^{\nabla u}) \in \bar{H}$.

(2) Suppose that $(u, \tau^v) \in \bar{H}$; there exists $(u_n)_n \subseteq W_0^{1,p}(\Omega, \mathbb{R}^m)$ such that $u_n \xrightarrow[L^p]{\|\cdot\|_p} u$, $\nabla u_n \xrightarrow[\mathbb{R}^{md}]{\mathcal{S}} \tau^v$ and $(\|\nabla u_n\|^p)_n$ is uniformly integrable. Using Propo-

sition 3.40, $\nabla u_n \xrightarrow[\mathcal{M}(\mathbb{R}^{md})]{\mu} v$. Now we can apply the Vitali's theorem (see Theorem 3.118) to obtain that $v \in L^p(\Omega, \mathbb{R}^{md})$ and $\nabla u_n \xrightarrow[L^p]{\|\cdot\|_p} v$.

For every $\varphi \in C_c^\infty(\Omega, \mathbb{R})$ and every $n \in \mathbb{N}$,

$$\int_\Omega u_n^T \cdot \nabla\varphi dt = -\int_\Omega \nabla u_n \cdot \varphi dt \tag{*}$$

(for the definition of $u_n^T \cdot \varphi$, see 3.151).

Since $u_n \xrightarrow[L^p]{\|\cdot\|_p} u$, $\int_\Omega u_n^T \cdot \nabla\varphi dt \to \int_\Omega u^T \cdot \nabla\varphi dt$. On the other hand,

$$\left|\int_\Omega (v - \nabla u_n) \cdot \varphi dt\right| \le \|v - \nabla u_n\|_p \cdot \|\varphi\|_q \quad \left(\frac{1}{p} + \frac{1}{q} = 1\right)$$

and, therefore, $\int_\Omega \nabla u_n \cdot \varphi dt \to \int_\Omega v \cdot \varphi dt$. Passing to limit in $(*)$, we get

$$\int_\Omega u^T \cdot \nabla\varphi dt = -\int_\Omega v \cdot \varphi dt, \quad \text{for every} \quad \varphi \in C_c^\infty(\Omega, \mathbb{R}).$$

Therefore $v = \nabla u$.

Conversely, if $v \in L^p(\Omega, \mathbb{R}^{md})$ and $v = \nabla u$, then, from 1), we obtain $(u, \tau^v) \in \bar{H}$. □

Remark 3.171. $(u, \tau), (u, \sigma) \in \bar{H} \not\Rightarrow \tau = \sigma$.

Indeed, let $(u_n)_n \subseteq W_0^{1,4}((0,1), \mathbb{R}) \subseteq W^{1,1}((0,1), \mathbb{R})$ be the sequence defined in Example 3.164. $(u_n)_n$ is uniformly convergent to $\underline{0}$ on $(0,1)$ and $(\nabla u_n)_n$ is the Rademacher' sequence. In the Example 3.44 we have shown that $\nabla u_n \overset{\mathcal{S}}{\to} \tau = \mu \otimes \frac{1}{2} \cdot (\delta_1 + \delta_{-1})$. Obviously, $(|\nabla u_n|)_n$ is uniformly integrable. Therefore $(\underline{0}, \tau) \in \bar{H}$ and $\mathrm{bar}\tau_. = \underline{0} = \nabla\underline{0}$.

By (1) of the previous proposition, $(\underline{0}, \tau^{\nabla\underline{0}}) = (\underline{0}, \tau^{\underline{0}}) \in \bar{H}$. But the disintegration of $\tau^{\underline{0}}$ is $\delta_0 \ne \frac{1}{2} \cdot (\delta_1 + \delta_{-1})$ which is the disintegration of τ.

We consider now the following relaxation of problem (P):

$$\inf\left\{\bar{I}(u, \tau) = \int_{\Omega \times \mathbb{R}^{md}} f(t, u(t), y) d\tau(t, y) : (u, \tau) \in \bar{H}\right\} = \bar{\mathrm{m}}, \tag{$\bar{P}$}$$

where f is a Carathéodory integrand satisfying, for every $x \in \mathbb{R}^m$, every $y \in \mathbb{R}^{md}$ and almost every $t \in \Omega$, the condition:

$$a \cdot \|y\|^p - 1 \le f(t, x, y) \le b \cdot (1 + \|y\|^p), \tag{C}$$

with $a, b > 0$.

For every $u \in W_0^{1,p}(\Omega, \mathbb{R}^m)$, $(u, \tau^{\nabla u}) \in \bar{H}$ and $I(u) = \bar{I}(u, \tau^{\nabla u})$; therefore $\bar{I}$ extends I from $W_0^{1,p}(\Omega, \mathbb{R}^m)$ to $\bar{H}$ and then $\bar{\mathrm{m}} \leq \mathrm{m} < +\infty$.

Next we present some existence results for the relaxed problem $(\bar{P})$; in the particular case where f is quasiconvex in the gradient variable, we obtain the existence results for the classical problem (P). For similar results, see Pedregal's monograph [128] (Theorem 4.4 and Corollary 4.5).

Proposition 3.172. $\mathrm{m} = \bar{\mathrm{m}}$.

Proof. We have remarked that $\bar{\mathrm{m}} \leq \mathrm{m}$.

For every $(u, \tau) \in \bar{H}$, there exists a sequence $(u_n)_n \subseteq W_0^{1,p}(\Omega, \mathbb{R}^m)$ such that $(u_n, \nabla u_n) \xrightarrow[\mathbb{R}^m \times \mathbb{R}^{md}]{\mathcal{S}} \tau^u \otimes \tau$ and $(\|\nabla u_n\|^p)_n$ is uniformly integrable in $L^1(\Omega)$. According to condition (C), $(f(., u_n(.), \nabla u_n(.)))_n$ is uniformly integrable and then, by Corollary 3.36,

$$\begin{aligned}\bar{I}(u, \tau) &= \int_\Omega \int_{\mathbb{R}^{md}} f(t, u(t), y) d\tau_t(y) dt \\ &= \int_\Omega \int_{\mathbb{R}^m \times \mathbb{R}^{md}} f(t, x, y) d(\tau_t^u \otimes \tau_t)(x, y) dt \\ &= \lim_n \int_\Omega f(t, u_n(t), \nabla u_n(t)) dt \geq \mathrm{m}.\end{aligned}$$

Therefore, $\bar{\mathrm{m}} \geq \mathrm{m}$ and then $\mathrm{m} = \bar{\mathrm{m}}$. □

Theorem 3.173. *Let $f : \Omega \times \mathbb{R}^m \times \mathbb{R}^{md} \to \mathbb{R}$ be a Carathéodory integrand satisfying the condition (C); there exists $(u, \tau) \in \bar{H}$ such that $\bar{I}(u, \tau) = \bar{\mathrm{m}}$.*

If, moreover, almost for every $t \in \Omega$ and for every $x \in \mathbb{R}^m$, $f(t, x, \cdot)$ is quasiconvex, then (P) admits a minimizer.

Proof. Let $(u_n)_n \subseteq W_0^{1,p}(\Omega, \mathbb{R}^m)$ be a minimizing sequence for problem (P); i.e. $I(u_n) \to \mathrm{m} = \bar{\mathrm{m}}$. According to Lemma 4.3 from [128] (p. 66), we can always suppose that $(\|\nabla u_n\|^p)_n$ is uniformly integrable. By condition (C),

$$a \cdot \|\nabla u_n\|^p \leq f(t, u_n(t), \nabla u_n(t)) + 1, \quad \text{for every} \quad n \in \mathbb{N},$$

from where

$$a \cdot \int_\Omega \|\nabla u_n\|^p dt \leq I(u_n) + \mu(\Omega), \quad \text{for every} \quad n \in \mathbb{N}.$$

Therefore $(\nabla u_n)_n$ is a bounded sequence in $L^p(\Omega, \mathbb{R}^{md})$. According to the Poincaré inequality (see Theorem 1.47 in [49], $(u_n)_n$ is bounded in $W_0^{1,p}$ and, by Rellich–Kondrachov theorem (Theorem 12.12 of [50]), up to a subsequence, $(u_n)_n$ is strongly convergent to a function $u \in L^p(\Omega, \mathbb{R}^m)$.

Let us show that $u \in W_0^{1,p}(\Omega, \mathbb{R}^m)$.

(a) Let first suppose that $p > 1$; then $W^{1,p}(\Omega, \mathbb{R}^m)$ is reflexive and, since $(u_n)_n$ is bounded in $W^{1,p}(\Omega, \mathbb{R}^m)$, up to a subsequence, $(u_n)_n$ is weakly convergent to a function $u_0 \in W_0^{1,p}(\Omega, \mathbb{R}^m)$; then $u = u_0 \in W_0^{1,p}(\Omega, \mathbb{R}^m)$.

(b] Let $p = 1$. For every $\varphi \in C_c^\infty(\Omega, \mathbb{R})$, every $i \in \{1, \ldots, m\}$ and every $j \in \{1, \ldots, d\}$,

$$\int_\Omega u_n^i \cdot \frac{\partial \varphi}{\partial t_j} dt = -\int_\Omega \frac{\partial u_n^i}{\partial t_j} \cdot \varphi dt, \quad \text{for every} \quad n \in \mathbb{N}, \tag{*}$$

where $u_n = (u_n^1, \ldots, u_n^m)$ and $u = (u^1, \ldots, u^m)$.

$(\nabla u_n)_n$ is uniformly integrable in $L^1(\Omega, \mathbb{R}^{md})$ and then, up to a subsequence, it is weakly convergent to a function $v = (v_j^i)_{1 \le i \le m, 1 \le j \le d} \in L^1(\Omega, \mathbb{R}^{md})$. Since $u_n \xrightarrow[L^1]{w} u$ and $\nabla u_n \xrightarrow[L^1]{w} v$, for every $i \in \{1, \ldots, m\}$ and every $j \in \{1, \ldots, d\}$,

$$\lim_n \int_\Omega u_n^i \cdot \frac{\partial \varphi}{\partial t_j} dt = \int_\Omega u^i \cdot \frac{\partial \varphi}{\partial t_j} dt \quad \text{and} \quad \lim_n \int_\Omega \frac{\partial u_n^i}{\partial t_j} \cdot \varphi dt = \int_\Omega v_j^i \cdot \varphi dt.$$

Passing to limit in $(*)$, we obtain

$$\int_\Omega u^i \cdot \frac{\partial \varphi}{\partial t_j} dt = -\int_\Omega v_j^i \cdot \varphi dt, \quad \text{for every} \quad \varphi \in C_c^\infty(\Omega, \mathbb{R}).$$

Then $\frac{\partial u^i}{\partial t_j} = v_j^i \in L^1(\Omega)$, for every $i \in \{1, \ldots, m\}$ and every $j \in \{1, \ldots, d\}$. Therefore, $u \in W^{1,1}(\Omega, \mathbb{R}^m)$. Since $u_n \xrightarrow[L^1]{\|\cdot\|_1} u, u|_{\partial\Omega} = 0$ and then $u \in W_0^{1,1}(\Omega, \mathbb{R}^m)$.

Therefore, for every $1 \le p < +\infty, u \in W_0^{1,p}(\Omega, \mathbb{R}^m)$ and $u_n \xrightarrow[L^p]{\|\cdot\|_p} u$.

$(\nabla u_n)_n$ being bounded in $L^p(\Omega, \mathbb{R}^{md})$ it is bounded in $L^1(\Omega, \mathbb{R}^{md})$.

According to Proposition 3.56, it is tight and then we can apply Prohorov's theorem (see Theorem 3.64). Therefore, $(\nabla u_n)_n$ is $\mathcal{S}$-convergent, up to a subsequence, to a Young measure $\tau \in \mathcal{Y}(\mathbb{R}^{md})$.

Then $(u_n, \nabla u_n)_n \xrightarrow{\mathcal{S}} (u, \tau)$ and so $(u, \tau) \in \bar{H}$.

On the other hand, using again the condition (C), we obtain, for every $n \in \mathbb{N}$ and for almost every $t \in \Omega$,

$$|f(t, u_n(t), \nabla u_n(t))| \le c \cdot (1 + \|\nabla u_n(t)\|^p), \quad \text{where} \quad c = \max\{a, b, 1\}.$$

Therefore, $(f(.,u_n(.),\nabla u_n(.))_n$ is uniformly integrable in $L^1(\Omega)$ and then, Corollary 3.36 allow us to write:

$$\begin{aligned}\bar{I}(u,\tau) &= \int_\Omega \int_{\mathbb{R}^{md}} f(t,u(t),y)d\tau_t(y)dt \\ &= \lim_n \int_\Omega f(t,u_n(t),\nabla u_n(t))dt = \lim_n I(u_n) = \bar{\mathbf{m}}.\end{aligned}$$

Now, let us suppose that, almost for every $t \in \Omega$ and for every $x \in \mathbb{R}^m$, $f(t,x,\cdot)$
is quasiconvex. At begin of this subsection, we remarked already that (C) implies (C_p) and, therefore the conditions of Morrey–Acerbi–Fusco theorem are accomplished (see Theorem 3.162). Then I is sequentially weakly lower semicontinuous in $W^{1,p}(\Omega,\mathbb{R}^m)$.

With above notations, we can decide that $u_n \xrightarrow[W^{1,p}]{w} u$ (see also Proposition 3.168). Therefore,

$$I(u) \leq \liminf_n I(u_n) = \lim_n I(u_n) = \mathbf{m}$$

and so $I(u) = \mathbf{m}$. □

Remark 3.174. In the case of Bolza problem, the relaxed problem is:

$$\inf\left\{\bar{I}(u,\tau) = \int_{(0,1)\times\mathbb{R}} \left[(y^2-1)^2 + \min\{u^2(t),1\}\right] d\tau(t,y) : (u,\tau) \in \bar{H}\right\}.$$

According to Theorem 3.173, $(\bar{P})$ admits as minimizer.

The sequence $(u_n)_n$ from the Example 3.164 is uniformly convergent to the constant function $u_0 = \underline{0}$ and the sequence of gradients (the Rademacher sequence) is stably convergent to Young measure τ^0 whose disintegration is constant $\tau_t^0 = \frac{1}{2}\cdot(\delta_1 + \delta_{-1})$, for every $t \in (0,1)$. Therefore $\mathbf{m} = \bar{I}(u_0,\tau^0)$.

We note that we have no solution for problem (P).

Bibliography

[1] E. Acerbi and N. Fusco, *Semicontinuity problems in the calculus of variations*, Arch. Rational Mech. Anal. **86** (1984), pp. 125–145.

[2] J. J. Alibert and G. Bouchitté, *Non-uniform integrability and generalized Young measures*, J. Convex Anal. **4** (1997), pp. 129–147.

[3] L. Ambrosio, H. Frid and J. Silva, *Multiscale Young measures in homogenization of continuous stationary processes in compact spaces and applications*, Journal of Functional Analysis **256** (2009), pp. 1962–1997.

[4] L. Ambrosio and H. Frid, *Multiscale Young measures in almost periodic homogenization and applications*, Arch. Rational Mech. Anal. **192** (2009), pp. 37–85.

[5] Z. Artstein, *A note on Fatou's lemma in several dimensions*, J. Math. Econom. **6** (1979), pp. 277–282.

[6] Z. Artstein, *Relaxed multifunctions and Young multimeasures*, Set-Valued Anal. **6** (1998), pp. 237–255.

[7] J. P. Aubin, *Un théorème de compacité*, C. R. Acad. Sci. Paris, Sér. A, **256** (1963), pp. 5042–5044.

[8] J. Babadjian, M. Baía and P. M. Santos, *Characterization of two-scale gradient Young measures and application to homogenization*, Appl. Math. Optim. **57** (2008), pp. 69–97.

[9] E. J. Balder, *A general approach to lower semicontinuity and lower closure in optimal control theory*, SIAM J. Control Optim. **22** (1984), pp. 570–598.

[10] E. J. Balder, *A general denseness result for relaxed control theory*, Bull. Austral. Math. Soc. **30** (1984), pp. 463–475.

[11] E. J. Balder, *On weak convergence implying strong convergence in L_1-spaces*, Bull. Austral. Math. Soc. **33** (1986), pp. 363–368.

[12] E. J. Balder, *Generalized equilibrium results for games with incomplete information*, Math. Oper. Res. **13** (1988), pp. 265–276.

[13] E. J. Balder, *On Prohorov's theorem for transition probabilities*, Sém. Anal. Convexe **19** (1989), 9.1-9.11.

[14] E. J. Balder, *On equivalence of strong and weak convergence in L^1–spaces under extreme point conditions*, Israel J. Math. **75** (1991), pp. 21–47.

[15] E. J. Balder, *Lectures on Young Measures*, Cahiers de Mathématiques, CEREMADE, Université Paris-Dauphine, Paris, 1995.

[16] E. J. Balder, *Consequences of denseness of Dirac Young measures*, J. Math. Anal. Appl. **207** (1997), pp. 536–540.

[17] E. J. Balder, *On the existence of Cournot-Nash equilibria in continuum games*, Journal of Mathematical Economics **32** (1999), pp. 207–223.

[18] E. J. Balder, *Lectures on Young measure theory and its applications in economics*, Proceedings Grado School on Measure Theory and Real Analysis, Rendiconti dell'Istituto di Matematico dell'Università di Trieste **31** Suppl. 1 (2000), pp. 1–69.

[19] E. J. Balder, *On ws-convergence of product measures*, Math. Oper. Res. **26** (2001), pp. 494–518.

[20] E. J. Balder, *An equilibrium existence result for games with incomplete information and indeterminate outcomes*, Journal of Mathematical Economics **40** (2004), pp. 297–320.

[21] E. J. Balder, *Comments on purification in continuum games*, Int. J. Game Theory **37** (2008), pp. 73–92.

[22] J. M. Ball, *A version of the fundamental theorem for Young measures*, in PDE's and Continuum Models of Phase Transitions, M. Rascle, D. Serre and M. Slemrod (eds.), Lecture Notes in Phys. **344**, Springer, Berlin, 1989.

[23] J. M. Ball and F. Murat, *Remarks on Chacon's biting lemma*, Proc. Amer. Math. Soc. **107** (1989), pp. 655–663.

[24] J. M. Ball, B. Kirchheim and J. Kristensen, *Regularity of quasiconvex envelopes*, Calc. Var. Partial Dif. **11** (2000), pp. 333–359.

[25] M. Barchiesi, *Loss of polyconvexity by homogenization: a new example*, Calc. Var. Partial Dif. **30** (2007), pp. 215–230.

[26] H. Berliocchi and J. M. Lasry, *Intégrandes normales et mesures paramétrées en calcul des variations*, Bull. Soc. Math. France **101** (1973), pp. 129–184.

[27] A. L. Bessoud, *Multi-materials with strong interface: Young measures formulation*, J. Math. Pures Appl. **93** (2010), pp. 518–535.

[28] P. Billingsley, *Convergence of Probability Measures*, John Wiley& Sons, Inc., New-Yor. London. Sydney. Toronto, 1968.

[29] G. Birkhoff, *Lattice Theory*, Amer. Math. Soc. Colloq. Publ., Vol. XXV, 2nd edition; Amer. Math. Soc., New–York, 1948.

[30] V. I. Bogachev, *Measure Theory*, vol. 1-2, Springer-Verlag, Berlin. Heidelberg, 2007.

[31] N. Bourbaki, *Topologie Générale. Chap. 9: Utilisation des nombres réels en topologie générale*, Livre III, Hermann, Paris, 1958.

[32] N. Bourbaki, *Intégration. Chap. IX: Intégration sur les espaces topologiques séparés*, Livre VI, Hermann, Paris, 1969.

[33] J. Bourgain and H. P. Rosenthal, *Martingales valued in certain subspaces of L^1*, Israel J. Math. **37** (1980), pp. 54–75.

[34] O. Boussaid and P. Pedregal, *Quasiconvexification of sets in optimal design*, Calc. Var. Partial Dif. **34** (2009), pp. 139–152.

[35] A. Braides, I. Fonseca and G. Leoni, *$\mathcal{A}$-quasiconvexity: relaxation and homogenization*, ESAIM: COCV **5** (2000), pp. 539–577.

[36] H. Brézis, *Analyse Fonctionnelle - Théorie et Applications*, Collection Mathématiques Appliquées pour la Maîtrise, Masson, Paris, 1983.

[37] J. K. Brooks and R. V. Chacon, *Continuity and compactness of measures*, Adv. Math. **37** (1980), pp. 16–26.

[38] J. K. Brooks and M. Rao, *Compactness and convergence of measures*, Bull. Polish Acad. Sci. Math. **38** (1990), pp. 229–234.

[39] C. Carstensen and T. Roubiček, *Numerical approximation of Young measures in non-convex variational problems*, Numer. Math. **84** (2000), pp. 395–415.

[40] C. Carstensen and A. Prohl, *Numerical analysis of relaxed micromagnetics by penalised finite elements*, Numer. Math. **90** (2001), pp. 65–99.

[41] C. Carstensen and M. O. Rieger, *Young-measure approximations for elastodynamics with non-monotone stress-strain relations*, ESAIM: Mathematical Modelling and Numerical Analysis **38** (2004), pp. 397–418.

[42] Ch. Castaing and M. Valadier, *Convex Analysis and Measurable Multifunctions*, Lecture Notes in Math. **580**, Springer-Verlag, Berlin, 1977.

[43] Ch. Castaing and M. Valadier, *Weak convergence using Young measures*, Université Montpellier II, Laboratoire d'Analyse Convexe, Pré publication 1998/08.

[44] Ch. Castaing and P. Raynaud de Fitte, *On the fiber product of Young measures with application to a control problem with measures*, Adv. Math. Econ. **6** (2004), pp. 1–38.

[45] Ch. Castaing, P. Raynaud de Fitte and M. Valadier, *Young Measures on Topological Spaces. With applications in control theory and probability theory*, Kluwer Academic Publ. Dordrecht.Boston.London, 2004.

[46] C. Castaing, A. Jofre and A. Salvadori, *Control problems governed by functional evolution inclusions with Young measures*, J. Nonlinear Convex Anal. **5** (2004), pp. 131–152.

[47] Ch. Castaing, P. Raynaud de Fitte and A. Salvadori, *Some variational convergence results for a class of evolution inclusions of second order using Young measures*, Adv. Math. Econ. **7** (2005), pp. 1–32.

[48] M. Chipot, *The use of Young measures for constructing minimizing sequences in the calculus of variations*, J. Appl. Anal. **1** (1995), pp. 71–81.

[49] B. Dacorogna, *Introduction to the Calculus of Variations*, Imperial College Press, London, 2004.

[50] B. Dacorogna, *Direct Methods in the Calculus of Variations*, Applied Math. Sci. vol. 78, Second Edition, Springer, New York, 2008.

[51] G. Dal Maso, A. DeSimone, M. G. Mora and M. Morini, *A vanishing viscosity approach to quasistatic evolution in plasticity with softening*, Arch. Rational Mech. Anal. **189** (2008), pp. 469–544.

[52] C. Delacherie and P.-A. Meyer, *Probabilités et Potentiel*, Chapitres I à IV, Hermann, Paris, 1975.

[53] A. V. Demyanov, *Quasiconvex pair: necessary and sufficient condition for the weak lower semicontinuity of some integral functionals*. Journal of Mathematical Sciences **124** (2004), pp. 4941–4957.

[54] A. V. Demyanov, *Lower semicontinuity of some functionals under PDE constraints: an $\mathcal{A}$-quasiconvex pair*, Journal of Mathematical Sciences **136** (2006), pp. 3706–3717.

[55] S. Demoulini, *Young measure solutions for nonlinear evolutionary systems of mixed type*, Ann. Inst. Henri Poincaré **14** (1997), pp. 143–162.

[56] J. Diestel and J. J. Uhl, *Vector Measures*, Mathematical Surveys and Monographs **15**, AMS, Providence, Rhode Island, 1977.

[57] J. Diestel, *Sequences and Series in Banach Spaces*, Springer-Verlag, New-York. Berlin. Heidelberg.Tokyo, 1984.

[58] J. Diestel, *Uniform Integrability: An Introduction*, Rend. Istit. Mat. Univ. Trieste, **XXIII** (1991), pp. 41–80.

[59] J. Dieudonné, *Un exemple d'espace normal non susceptible d'une structure uniforme d'espace complet*, C. R. Acad. Sci. Paris, Sér. A, **209** (1939), pp. 145–147.

[60] R. M. Dudley, *Convergence of Baire measures*, Studia Mathematica **XXVII** (1966), pp. 251–268.

[61] R. M. Dudley, *Real Analysis and Probability*, Cambridge Studies in Advanced Mathematics **74**, Cambridge University Press, 2004.

[62] N. Dunford and J. T. Schwartz, *Linear Operators, Part I*, Interscience, New-York, 1958.

[63] G. A. Edgar, *Disintegration of measures and the vector-valued Radon-Nikodym theorem*, Duke Math. J. **42** (1975), pp. 447–450.

[64] J. F. Edmond, *Relaxation of a Bolza problem governed by a time-delay sweeping process*, Set-Valued Anal. **16** (2008), pp. 563–580.

[65] R. E. Edwards, *Functional Analysis. Theory and Applications*, Holt, Rinehart and Winston, Inc., New-York.Chicago. San Francisco. Toronto. London, 1965.

[66] A. A. Egorov, *Quasiconvex functions and null lagrangians in the stability problems of classes of mappings*, Siberian Mathematical Journal **49** (2008), pp. 637–649.

[67] J. J. Egozcue, R. Meziat and P. Pedregal, *From a nonlinear, nonconvex variational problem to a linear, convex formulation*, Appl. Math. Optim. **47** (2003), pp. 27–44.

[68] R. Engelking, *General Topology*, Heldermann Verlag, Berlin, 1989.

[69] L. C. Evans, *Weak Convergence Methods for Nonlinear Partial Differential Equations*, A. M. S., Providence, Rhode Island, CBMS **74** (1990).

[70] L. C. Evans and R. F. Gariepy, *Measure Theory and Fine Properties of Functions*, CRC Press, USA, 1992.

[71] D. Faraco and X. Zhong, *Quasiconvex functions and hessian equations*, Arch. Rational Mech. Anal. **168** (2003), 245–252.

[72] K. Fey, *Young measures generated by sequences in Morrey spaces*, Calc. Var. Partial Dif. **39** (2010), pp. 183–201.

[73] L. C. Florescu, *Pseudouniform structures and weak compactness on L^1*, Boll. Un. Mat. Ital. **3-A** (1989), pp. 87–94.

[74] L. C. Florescu, *Probabilistic convergence structures*, Aeq. Math. **38** (1989), pp. 123–145.

[75] L. C. Florescu, *w^2–Pseudouniformities and biting lemma*, Boll. Un. Mat. Ital. **3-A** (1993), pp. 87–96.

[76] L. C. Florescu, *Weak compactness results on $L^1(X)$*, An.Ştiinţ.Univ. Al. I. Cuza Iaşi Mat.(N. S.) **XLV** (1999), pp. 75–86.

[77] L. C. Florescu and C. Godet–Thobie, *Quelques propriétés des mesures de Young*, An. Ştiinţ. Univ. Al. I.Cuza Iaşi Mat. **XLVI** (2000), pp. 393–412.

[78] L. C. Florescu, *Compacity in narrow limit tower spaces*, Int. J. Math. Math. Sci. **26** (2001), pp. 359–370.

[79] L. C. Florescu, *Extensions of Fatou's inequality*, Math. Ineq. Appl. **5(3)** (2002), pp. 525–534.

[80] L. C. Florescu and C. Godet–Thobie, *A version of biting lemma for unbounded sequences in L^1 with applications*, Mathematical Analysis and Applications, AIP Conference Proceedings, New York, **835** (2006), pp. 58-73.

[81] L. C. Florescu, *Finite-tight sets*, Central European Journal of Mathematics **5(4)** (2007), pp. 619-638.

[82] L. C. Florescu, *Existence results in relaxed variational calculus*, An. Univ. Vest Timişoara, **XLV** (2007), pp. 231–244.

[83] I. Fonseca and S. Müller, *$\mathcal{A}$-quasiconvexity, lower semicontinuity and Young measures*, SIAM J. Math. Anal. **30** (1999), pp. 1355–1390.

[84] I. Fonseca, G. Leoni and R. Paroni, *On lower semicontinuity in BH^p and 2-quasiconvexification*, Calc. Var. Partial Dif. **17** (2003), pp. 283–309.

[85] I. Fonseca and G. Leoni, *Modern Methods in the Calculus of Variations: L^p-Spaces*, Springer Verlag, New York, 2007.

[86] M. Fréchet, *Sur les ensembles compacts de fonctions mesurables*, Fund. Math. **IX** (1927), pp. 25–32.

[87] L. Freddi and R. Paroni, *A 3D-1D Young measure theory of an elastic string*, Asymptotic Analysis **39** (2004), pp. 61–89.

[88] L. Freddi and R. Paroni, *The energy density of martensitic thin films via dimension reduction*, Interfaces and Free Boundaries **6** (2004), pp. 439–459.

[89] V. F. Gaposhkin, *Convergence and limit theorems for sequences of random variables*, Theory Probab. Appl. **17-3** (1972), pp. 379–400 (translated from Russian).

[90] H.-P. Gittel, *Young measure solutions of some nonlinear mixed type equations*, Mathematical Methods in the Applied Sciences **33** (15) (2010), pp. 1882–1894.

[91] C. Godet-Thobie, *Multimesures et multimesures de transition*, Thèse de Doctorat d'Etat es Sciences Mathematiques, L'Université des Sciences et Techniques du Languedoc, Montpellier, 1975.

[92] S. Gutman, *Compact perturbations of m–accretive operators in general Banach spaces*, SIAM J. Math. Anal. **13** (1982), pp. 789–800.

[93] P. R. Halmos, *Measure Theory*, Springer-Verlag, New-York. Heidelberg. Berlin, 1974.

[94] M. Hillairet, *Propagation of density-oscillations in solutions to the barotropic compressible Navier-Stokes system*, J. Math. Fluid Mech. **9** (2007),pp. 343–376.

[95] J. Hoffmann-Jørgensen, *Convergence in law of random elements and random sets*, High dimensional probability (Oberwolfach, 1996), Progress in Probability no. **43** (1998), pp. 151–189.

[96] N. Hungerbühler, *A refinement of Ball's theorem on Young measures*, New York J. Math. **3** (1997), pp. 48–53.

[97] A. Ionescu-Tulcea and C. Ionescu-Tulcea, *Topics in the Theory of Lifting*, Springer-Verlag, Berlin.Hedelberg.New York, 1969.

[98] J. Jacod and J. Mémin, *Sur un type de convergence intermédiaire entre la convergence en loi et la convergence en probabilité*, Séminaire de Probabilités **XV** (J. Azéma and M. Yor, eds.) Lecture Notes in Math. **986**, Springer-Verlag, Berlin, 1983, pp. 529–546.

[99] A. Jawhar, *Mesures de transition et applications*, Sém. Analyse Convexe **14** (1984), 13.1–13.62.

[100] A. Kałamajska, *On lower semicontinuity of multiple integrals*, Colloq. Math. **74** (1997), pp. 71–78.

[101] P. Kamiński, *Nonlinear quasistatic problems of gradient type in inelastic deformations theory*, J. Math. Anal. Appl. **357** (2009), pp. 284–299.

[102] D. Kinderlehrer and P. Pedregal, *Characterizations of Young measures generated by gradients*, Arch. Rat. Mech. Anal. **115** (1991), pp. 329-365.

[103] D. Kinderlehrer and P. Pedregal, *Gradient Young measures generated by sequences in Sobolev spaces*, The Journal of Geometric Analysis **4** (1994), pp. 59–90.

[104] H.-A. Klei, *Convergences faible en mesure et au sens de Cesaro dans $L^1(\mathbb{R})$*, C. R. Acad. Sci. Paris, Sér. A, **315** (1992), pp. 9–12.

[105] H.-A. Klei, *Convergence and extraction of bounded sequences in $L^1(\mathbb{R})$*, J. Math. Anal. Appl. **198** (1996), pp. 636–642.

[106] H.-A. Klei, *Measure convergent sequences in Lebesgue spaces and Fatou's lemma*, Bull. Austral. Math. Soc. **54** (1996), pp. 197–202.

[107] H.-A. Klei, *Fatou's identity and Lebesgue's convergence theorem*, Proc. Amer. Math. Soc. **127** (1999), pp. 2297–2302.

[108] J. Kristensen, *Lower semicontinuity in spaces of weakly differentiable functions*, Math. Ann. **313** (1999), pp. 653–710.

[109] J. Kristensen and F. Rindler, *Characterization of generalized gradient Young measures generated by sequences in $W^{1,1}$ and BV*, Arch. Rational Mech. Anal. **197** (2010), pp. 539–598.

[110] M. Kružík – *DiPerna–Majda measures and uniform integrability*, Comment. Math. Univ. Carolin. **39** (1998), pp. 511–523.

[111] Ph. G. LeFloch and M. Westdickenberg, *Finite energy solutions to the isentropic Euler equations with geometric effects*, J. Math. Pures Appl. **88** (2007), pp. 389–429.

[112] C. De Lellis and L. Székelyhidi Jr., *Simple proof of two-well rigidity*, C. R. Acad. Sci. Paris, Ser. I **343** (2006), pp. 367–370.

[113] Z. Li, *Computation of the lower semicontinuous envelope of integral functionals and non-homogeneous microstructures*, Nonlinear Analysis: Theory, Methods & Applications **68** (2008), pp. 2058–2071.

[114] J. L. Lions, *Équations Différentielles Opérationnelles et Problèmes aux Limites*, Die Grundlehren der mathematischen Wissenschaften **111**, Springer-Verlag, Berlin–Göttingen–Heidelberg, 1961.

[115] D. Q. Luu, *A short proof of biting lemma*, Sém. Anal. Convexe **19** (1989), 1.1–1.13.

[116] J. Mach, *On optimality conditions of relaxed non-convex variational problems*, J. Math. Anal. Appl. **298** (2004), pp. 157–170.

[117] F. Maestre and P. Pedregal, *Quasiconvexification in 3-D for a variational reformulation of an optimal design problem in conductivity*, Nonlinear Analysis **64** (2006), pp. 1962–1976.

[118] P. Marcellini, *Approximation of quasiconvex functions and lower semicontinuity of multiple integrals*, Manuscripta Math. **51** (1985), pp. 1–28.

[119] C. Melcher, *A dual approach to regularity in thin film micromagnetics*, Calc. Var. Partial Dif. **29** (2007), pp. 85–98.

[120] G. Michaille and M. Valadier, *Young measures generated by a class of integrands: a narrow epicontinuity and applications to homogenization*, J. Math. Pures Appl. **81** (2002), pp. 1277–1312.

[121] Ch. B. Morrey Jr., *Quasi-convexity and lower semicontinuity of multiple integrals*, Pacific J. Math. **2** (1952), pp. 25–53.

[122] Ch. B. Morrey Jr., *Multiple Integrals in the Calculus of Variations*, Springer, Berlin, 1966.

[123] A. Münch, P. Pedregal and F. Periago, *A variational approach to a shape design problem for the wave equation*, C. R. Acad. Sci. Paris, Ser. I **343** (2006), pp. 371–376.

[124] A. Münch, P. Pedregal and F. Periago, *Relaxation of an optimal design problem for the heat equation*, J. Math. Pures Appl. **89** (2008), pp. 225–247.

[125] J. Muñoz, *Characterisation of the weak lower semicontinuity for a type of nonlocal integral functional: the n-dimensional scalar case*, J. Math. Anal. Appl. **360** (2009), pp. 495–502.

[126] C. Olech, *The characterization of the weak* closure of certain sets of integrable functions*, SIAM J. Control Optim. **12** (1974), pp. 311–318.

[127] S. Owczarek, *Convergence of a monotonisation procedure for a non-monotone quasi-static model in poroplasticity*, J. Math. Anal. Appl. **364** (2010), pp. 599–608.

[128] P. Pedregal, *Parametrized Measures and Variational Principles*, Birkhäuser Verlag, Basel.Boston.Berlin, 1997.

[129] P. Pedregal, *Equilibrium conditions for Young measures*, SIAM J. Control Optim. **36** (1998), pp. 797–813.

[130] P. Pedregal, *Optimization, relaxation and Young measures*, Bull. Amer. Math. Soc. (N. S.) **36** (1999), pp. 27–58.

[131] P. Pedregal, *Constrained quasiconvexity and structural optimization*, Arch. Rational Mech. Anal. **154** (2000), pp. 325–342.

[132] P. Pedregal, *Relaxation in magnetostriction*, Calc. Var. Partial Dif. **10** (2000), pp. 1–19.

[133] P. Pedregal, *From micro to macroenergy through Young measures*, Meccanica **40** (2005), pp. 329–338.

[134] P. Pedregal and H. Serrano, *Γ-convergence of quadratic functionals with oscillating linear perturbations*, Nonlinear Analysis: Theory, Methods & Applications **70** (2009), pp. 4178–4189.

[135] P. Pedregal, *On a generalization of compact operators and its application to the existence of critical points without convexity*, Arch. Rational Mech. Anal. **197** (2010), pp. 965–983.

[136] D. Piau, *Large deviations and Young measures for a poissonian model of biphased material*, The Annals of Applied Probability **9** (1999), pp. 706–718.

[137] L. Piccinini and M. Valadier, *Uniform integrability and Young measures*, J. Math. Anal. Appl. **195** (1995), pp. 428–439.

[138] A. M. Ramos and T. Roubiček, *Nash equilibria in noncooperative predator-prey games*, Appl. Math. Optim. **56** (2007), pp. 211–241.

[139] P. Raynaud de Fitte, *Compactness criteria for stable topology*, Bull. Polish Acad. Sci. Math. **51** (2003), pp. 343–363.

[140] M. O. Rieger and J. Zimmer, *Global existence for nonconvex thermoelasticity*, Adv. Math. Sci. Appl. **15** (2005), pp. 559–569.

[141] M. O. Rieger, *A model for hysteresis in mechanics using local minimizers of Young measures*, Progress in Nonlinear Differential Equations and Their Applications **63** (2005), pp. 403–414.

[142] M. O. Rieger and J. Zimmer, *Young measure flow as a model for damage*, Z. Angew. Math. Phys. **60** (2009), pp. 1–32.

[143] H. P. Rosenthal, *Sous-espaces de L^1*, Lectures Univ. Paris **VI** (1979).

[144] R. Rossi and G. Savaré, *Tightness, integral equicontinuity and compactness for evolution problems in Banach spaces*, Ann. Sc. Norm. Super. Pisa, Cl. Sci. **(5) 2** (2003), pp. 395–431.

[145] T. Roubiček, *Relaxation in Optimization Theory and Variational Calculus*, Walter de Gruyter, Berlin.New York, 1997.

[146] T. Roubiček and M. Kružik, *Microstructure evolution model in micromagnetics*, Z. Angew. Math. Phys. **55** (2004), pp. 159–182.

[147] T. Roubiček and M. Kružik, *Mesoscopic model for ferromagnets with isotropic hardening*, Z. Angew. Math. Phys. **56** (2005), pp. 107–135.

[148] A. Rustichini and P. Siconolfi, *Growth in economies with non convexities: sunspots and lottery equilibria*, Economic Theory **24** (2004), pp. 701–726.

[149] M. Saadoune and M. Valadier, *Extraction of a "good" subsequence from a bounded sequence of integrable functions*, J. Convex Anal. **2** (1995), pp. 345–357.

[150] J. Saint-Pierre, *Une remarque sur les espaces sousliniens réguliers*, C. R. Acad. Sci. Paris, Sér. A, **282** (1976), pp. 1425-1427.

[151] J. Schwartz, *A note on the space L_p^**, Proc.Amer. Math. Soc. **2** (1951), pp. 270–275.

[152] L. Schwartz, *Radon Measures on Arbitrary Topological Spaces and Cylindrical Measures*, Oxford University Press, 1973.

[153] H. Serrano, *On Γ-convergence in divergence-free fields through Young measures*, J. Math. Anal. Appl. **359** (2009), pp. 311–321.

[154] M. Słaby *Strong convergence of vector–valued pramarts and subpramarts*, Probab. Math. Statist. **5** (1985), pp. 187–196.

[155] G.-Q. Song and J. Xiao, *Existence of global weak solutions to a special system of Euler equation with a source (II): General case*, J. Math. Anal. Appl. **352** (2009), pp. 943–953.

[156] V. Šverák, *Rank-one convexity does not imply quasiconvexity*, Proc. Roy. Soc. Edinburgh, Sect. A, **120** (1992), pp. 185–189.

[157] V. Šverák, *New examples of quasiconvex functions*, Arch. Rational Mech. Anal. **119** (1992), pp. 293–300.

[158] M. A. Sychev, *Young measures as measurable functions and their applications to variational problems*, Journal of Mathematical Sciences **132** (2006), pp. 359–370.

[159] M. A. Sychev, *Theorems on lower semicontinuity and relaxation for integrands with fast growth*, Siberian Mathematical Journal **46** (2005), pp. 540–554.

[160] L. Székelyhidi, Jr., *On quasiconvex hulls in symmetric* 2×2 *matrices*, Ann. I. H. Poincaré **23** (2006), pp. 865–876.

[161] A. Taheri, *Local minimizers and quasiconvexity - the impact of topology*, Arch. Rational Mech. Anal. **176** (2005), pp. 363–414.

[162] S. Taheri, Q. Tang and K. Zhang, *Young measure solutions and instability of the one-dimensional Perona-Malik equation*, J. Math. Anal. Appl. **308** (2005), pp. 467–490.

[163] L. Tartar, *Compensated Compactness and Applications to Partial Differential Equations*, Research Notes in Math., Nonlinear Analysis and Mechanics: Heriot-Watt Symposium, **4**, Pitman Press, San Francisco, 1979.

[164] L. Tartar, *Beyond Young measures*, Meccanica **30** (1995), pp. 505–526.

[165] L. Tartar, *The General Theory of Homogenization*, Springer-Verlag Berlin Heidelberg 2009.

[166] F. Theil, *Young-measure solutions for a viscoelastically damped wave equation with nonmonotone stress-strain relation*, Arch. Rational Mech. Anal. **144** (1998), pp. 47–78.

[167] M. Valadier, *Désintégration d'une mesure sur un produit*, C. R. Acad. Sci. Paris, Sér. A, **276** (1973), pp. 33–35.

[168] M. Valadier, *Young Measures*, Methods of nonconvex analysis (Varenna, 1989). Lecture Notes in Mathematics, **1446**, Springer, Berlin (1990), pp. 152–188.

[169] M. Valadier, *A Course on Young Measures*, Rend. Istit. Mat. Univ. Trieste, **XXVI** (1994), Supplemento, pp. 349–394.

[170] D. Vasiliu and B. Yan, *(PS)-weak lower semicontinuity in one dimension: A necessary and sufficient condition*, Nonlinear Analysis **66** (2007), pp. 582–590.

[171] A. Visintin, *Strong convergence results related to strict convexity*, Comm. Partial Differential Equations **9** (1984), pp. 439–466.

[172] A. Visintin, *Scale-integration and scale-disintegration in nonlinear homogenization*, Calc. Var. Partial Dif. **36** (2009), pp. 565–590.

[173] J. Warga, *Optimal Control of Differential and Functional Equations*, Academic Press, New York. London, 1972.

[174] J. Yin and C. Wang, *Young measure solutions of a class of forward-backward diffusion equations*, J. Math. Anal. Appl. **279** (2003), pp. 659–683.

[175] K. Yosida and E. Hewitt, *Finitely additive measures*, Trans. Amer. Math. Soc. **72** (1952), pp. 46–66.

[176] L. C. Young, *Lectures on the Calculus of Variations and Optimal Control Theory*, Saunders, Philadelphia.London. Toronto, 1969.

[177] L. C. Young, *Generalized curves and the existence of an attained absolute minimum in the calculus of variations*, C. R. Soc. Sci. Lettres de Varsovie, Cl. III , **30** (1937), pp. 212–234.

[178] L. C. Young, *Generalized surfaces in the calculus of variations, I-II*, Ann. of Math., **43** (1942), pp. 84–103; 530–544.

[179] K. Zhang, *Biting lemma for Jacobians and their applications*, Ann. Inst. H. Poincaré Anal. Non Linéaire **7** (1990), pp. 345–365.

[180] P. Zhang and Y. Zheng, *Existence and uniqueness of solutions of an asymptotic equation arising from a variational wave equation with general data*, Arch. Rational Mech. Anal. **155** (2000), pp. 49–83.

[181] K. Zhang, *On separation of gradient Young measures*, Calc. Var. Partial Dif. **17** (2003), pp. 85–103.

[182] K. Zhang, *On existence of weak solutions for one-dimensional forward-backward diffusion equations*, J. Differential Equations **220** (2006), pp. 322–353.

Index

About the Authors

Prof. Dr. Liviu C. Florescu teaches measure theory, harmonic analysis on topological groups, compactness in measure spaces and Young measures, general topology, and functional analysis at the "Alexandru Ioan Cuza" University of Iaşi (Romania). In published works, he studied problems related to probabilistic topological structures, compactness in measure spaces, Young measures and applications to relaxed variational calculus.

Prof. Dr. Christiane Godet-Thobie is member of the Math Research Laboratory UMR 6205 (CNRS) of the Université de Bretagne Occidentale, Brest (France). She was member of the Scientific Committee of the international conference "Mesures de Young et Contrôle Stochastique" and organizer of the part "Measures de Young" (Brest, 2002). Her research interests are Young measures, set-valued analysis, fixed point theory, and differential inclusions.

Bei Fragen zur Produktsicherheit wenden Sie sich bitte an:
If you have any questions regarding product safety,
please contact:

Walter de Gruyter GmbH
Genthiner Straße 13
10785 Berlin
productsafety@degruyterbrill.com